# Lecture Notes in Computer Scien

T0076525

*Commenced Publication in 1973*
Founding and Former Series Editors:
Gerhard Goos, Juris Hartmanis, and Jan van Leeuwen

Chenyang Lu   Toshimitsu Masuzawa
Mohamed Mosbah (Eds.)

# Principles of Distributed Systems

14th International Conference, OPODIS 2010
Tozeur, Tunisia, December 14-17, 2010
Proceedings

 Springer

Volume Editors

Chenyang Lu
Washington University, Department of Computer Science and Engineering
Campus Box 1045, One Brookings Drive, St. Louis, MO 63130, USA
E-mail: lu@cse.wustl.edu

Toshimitsu Masuzawa
Osaka University, Graduate School of Information Science and Technology
1-5 Yamadaoka, Suita 5650871, Japan
E-mail: masuzawa@ist.osaka-u.ac.jp

Mohamed Mosbah
University of Bordeaux, LaBRI
351 cours de la Libération, 33405 Talence, France
E-mail: mosbah@labri.fr

Library of Congress Control Number: 2010940206

CR Subject Classification (1998): C.2.4, C.2, F.2, D.2, I.2.11, G.2.2

LNCS Sublibrary: SL 1 – Theoretical Computer Science and General Issues

ISSN        0302-9743
ISBN-10     3-642-17652-6 Springer Berlin Heidelberg New York
ISBN-13     978-3-642-17652-4 Springer Berlin Heidelberg New York

springer.com

© Springer-Verlag Berlin Heidelberg 2010
Printed in Germany

Typesetting: Camera-ready by author, data conversion by Scientific Publishing Services, Chennai, India
Printed on acid-free paper        06/3180

# Preface

The 14th International Conference on Principles of Distributed Systems (OPODIS 2010) took place during December 14–17, 2010 in Tozeur, Tunisia. It continued a tradition of successful conferences; Chantilly (1997), Amiens (1998), Hanoi (1999), Paris (2000), Mexico (2001), Reims (2002), La Martinique (2003), Grenoble (2004), Pisa (2005), Bordeaux (2006), Guadeloupe (2007), Luxor (2008) and Nîmes (2009).

The OPODIS conference constitutes an open forum for the exchange of state-of-the-art knowledge on distributed computing and systems among researchers from around the world. Following the tradition of the previous events, the program was composed of high-quality contributed papers. The program call for papers looked for original and significant research contributions to the theory, specification, design and implementation of distributed systems, including:

- Communication and synchronization protocols
- Distributed algorithms, multiprocessor algorithms
- Distributed cooperative computing
- Embedded systems
- Fault-tolerance, reliability, availability
- Grid and cluster computing
- Location- and context-aware systems
- Mobile agents and autonomous robots
- Mobile computing and networks
- Peer-to-peer systems, overlay networks
- Complexity and lower bounds
- Performance analysis of distributed systems
- Real-time systems
- Security issues in distributed computing and systems
- Sensor networks: theory and practice
- Specification and verification of distributed systems
- Testing and experimentation with distributed systems

In response to this call for papers, 122 papers were submitted. Each paper was reviewed by at least three reviewers, and judged according to scientific and presentation quality, originality and relevance to the conference topics. The Program Committee selected 32 papers for regular presentations at the conference and 4 papers as brief announcements.

It is impossible to organize a successful program without the help of many individuals. We would like to express our appreciation to the authors of the submitted papers, the Program Committee members and the external referees. We would also like to thank the OPODIS Steering Committee members, in particular Hacene Fouchal, who supervised and supported the continuation of this event. We owe special thanks to the Organizing Committee for the hard work they did locally in Tozeur.

December 2010                                              Chenyang Lu
                                                    Toshimitsu Masuzawa
                                                     Mohamed Mosbah

# Organization

## General Chair

Mohamed Mosbah             Institut Polytechnique de Bordeaux,
University of Bordeaux, France

## Program Co-chairs

Toshimitsu Masuzawa      University of Osaka, Japan
Chenyang Lu              Washington University, USA

## Program Committee

| | |
|---|---|
| Marcos Aguilera | Microsoft, USA |
| Anish Arora | Ohio State University, USA |
| Ted Baker | Florida State University, USA |
| Doina Bein | Pennsylvania State University, USA |
| Thibault Bernard | University of Reims, France |
| Zinaida Benenson | University of Mannheim, Germany |
| Wei Chen | Microsoft Research Asia, China |
| Xavier Defago | JAIST, Japan |
| Carole Delporte | University of Paris 7, France |
| Khalil Drira | LAAS-CNRS, France |
| Pascal Felber | University of Neuchatel, Switzerland |
| Antonio Fernández | University of Juan Carlos, Spain |
| Satoshi Fujita | Hiroshima University, Japan |
| Chris Gill | Washington University, USA |
| Zonghua Gu | Zhejiang University, China |
| Ted Herman | University of Iowa, USA |
| Rachid Guerraoui | EPFL, Switzerland |
| Mohamed Jemni | University of Tunis, Tunisia |
| Mohamed Jmaiel | University of Sfax, Tunisia |
| Hirotsugu Kakugawa | Osaka University, Japan |
| Boris Koldehofe | University of Stuttgart, Germany |
| Xenofon Koutsoukos | Vanderbilt University, USA |
| Shay Kutten | Technion, Israel |
| Franck Petit | University of Pierre et Marie Curie, France |
| Giuseppe Prencipe | University of Pisa, Italy |
| Michel Raynal | IRISA, France |

Alex Shvartsman               University of Connecticut, USA
Sebastien Tixeuil             University of Pierre et Marie Curie, France
Eduardo Tovar                 ISEP-IPP, Portugal
Xiaorui Wang                  University of Tennessee, USA
Koichi Wada                   Nagoya Institute of Technology, Japan

## Publicity Chair

Ying Lu                       University of Nebraska, USA
Fukuhito Ooshita              Osaka University, Japan

## Organizing Chair

Mohamed Jmaiel                University of Sfax, Tunisia

## Organizing Committee

Soufien Chtourou              University of Sfax, Tunisia
Ahmed Hadj Kacem              University of Sfax, Tunisia
Mohamed Hadj Kacem            University of Sfax, Tunisia
Moez Krichen                  University of Sfax, Tunisia

## Steering Committee

Tarek Abdelzaher              University of Illinois at Urbana-Champaign,
                                 USA
Alain Bui                     University of Versailles S.Q., France
Marc Bui                      EPHE, France
Hacène Fouchal                University of Reims C.A., France
Roberto Gomez                 ITESM-CEM, Mexico
Michel Raynal                 IRISA, France
Nicola Santoro                Carleton University, Canada
Sebastien Tixeuil             University of Pierre et Marie Curie, France
Philippas Tsigas              Chalmers University of Technology, Sweden

# Table of Contents

## Brief Announcements II

## Fault-Tolerance

## Distributed Programming

# Real-Time

# Shared Memory

# Concurrency

# Pattern Formation through Optimum Matching by Oblivious CORDA Robots

Nao Fujinaga, Hirotaka Ono, Shuji Kijima, and Masafumi Yamashita

Department of Informatics, Graduate School of Information Science and Electrical
Engineering, Kyushu University, Fukuoka 819-0395, Japan
{fuji,ono,shuji,mak}@tcslab.csce.kyushu-u.ac.jp

**Abstract.** A (new) geometric pattern formation problem by a set of oblivious, anonymous, asynchronous (i.e., CORDA) robots is investigated in this paper. The conventional pattern formation problem assumes that the target pattern is given as a set of the positions by their coordinates in the global coordinate system, under the assumption that the robots are not aware of it. In the pattern formation problem we discuss in this paper, the points comprising the pattern are assumed to be "visible" to all robots, like landmarks. However, the robots still cannot obtain their positions in the global coordinate system. This paper shows that this pattern formation problem is solvable by oblivious asynchronous robots through the optimum matching between the robots and the pattern's points.

Our study is partly motivated by the state-of-arts of the conventional pattern formation problem by oblivious asynchronous robots; description and correctness proof of a formation algorithm is usually complicated and ambiguous, because of the oblivious and asynchronous natures of the robots. A modular method is thus looked for to describe and prove algorithm in a clearer and more concrete way. Our pattern formation problem and the formation algorithm based on the optimum matching are used as a primitive building block in the modular method.

## 1 Introduction

This paper considers a system $R$ of anonymous mobile robots, each represented by a point in 2D Euclidean space. A robot, given an algorithm, repeats a "Look–Compute–Move" cycle, to observe the other robots' positions (in Look phase), to compute the next position by using the algorithm (in Compute phase), and to move toward the next position (in Move phase). The robots are anonymous in the sense that they do not have identifiers (and are not identified just by their looks neither), and are controlled by the same algorithm. A basic and crucial assumption on the system is that they are not aware of the global coordinate system, and all the actions by robots are via their local coordinate systems, which may be inconsistent each other.

The problem of forming a given pattern $F$ by a set of robots $R$ is called the (geometric) pattern formation problem and has been studied extensively. The difficulty of solving the problem depends mainly on two factors; the degree

C. Lu, T. Masuzawa, and M. Mosbah (Eds.): OPODIS 2010, LNCS 6490, pp. 1–15, 2010.
© Springer-Verlag Berlin Heidelberg 2010

of asynchrony and the amount of memory. As for the asynchrony, three kinds of robots have been discussed. A (fully) synchronous robot synchronously executes a Look–Compute–Move cycle; all robots simultaneously start and finish the Look, Compute and Move phases in each iteration, and they always reach their next positions (computed in their Compute phases) in their Move phases. We call this robot model FSYNCH. An asynchronous robot, on the other hand, asynchronously executes a Look–Compute–Move cycle. Moreover, a Move phase may finish when a robot is still on the way to its next position. We call this robot model ASYNCH (or CORDA). Finally, a semi-synchronous robot is the same as an asynchronous robot, except that a Look and a Move phases of two robots never overlap, or informally, no robots observe other robots moving. We call this robot model SSYNCH (or SYM). As for the amount of memory, we assume that a robot has sufficient memory in general. However, an oblivious robot does not have a memory, so that any algorithm to compute the next position must be a function of the current robots' positions.

Main known results about formation are summarized as follows. Assume that $|R| = |F|$. $F$ may be decomposed into $k/|F|$ $k$-sets such that these $k$-sets form co-centered $k$-gons. $\rho(F)$ is defined as the maximum of such $k$. A configuration $C$ of $R$ contains for each robot its position and local coordinate system. Let $F_C$ be the set of the robots' positions in $C$.

1. For anonymous FSYNCH robots (and hence for anonymous SSYNCH and ASYNCH robots), a pattern $F$ is formable from an initial configuration $I$, only if the symmetricity $\sigma(I)$ of $I$ divides $\rho(F)$ [14, 16][1]

2. There is a pattern formation algorithm for (non-oblivious) anonymous ASYNCH robots (and hence for anonymous SSYNCH and FSYNCH robots) that forms pattern $F$ from any initial configuration $I$ such that $\sigma(I)$ divides $\rho(F)$ [11].[2]

3. There is a pattern formation algorithm for oblivious anonymous FSYNCH robots that forms a pattern $F$ from any initial configuration $I$ such that $\sigma(I)$ divides $\rho(F)$ [16]. Oblivious anonymous FSYNCH robots thus have exactly the same formation power as the non-oblivious anonymous FSYNCH robots.

4. Except the gathering problem for two robots (i.e., $F$ is a singleton with multiplicity 2), there is a pattern formation algorithm for oblivious anonymous SSYNCH robots that forms a pattern $F$ from any initial configuration $I$ such that $\sigma(I)$ divides $\rho(F)$ [16]. Oblivious anonymous SSYNCH robots thus have almost the same formation power as the non-oblivious anonymous FSYNCH robots.

5. Any pattern $F$ is formable by oblivious anonymous ASYNCH robots, if $\rho(F_I) = 1$. [5] It is however still open to decide whether or not oblivious anonymous ASYNCH robots have (almost) the same formation power as the non-oblivious anonymous FSYNCH robots.

---

[1] See [14, 16] for the definition of symmetricity $\sigma$.

[2] The algorithm in [11] solves the agreement problem on their initial positions $F_I$, but it is easy to modify so that all robots agree on $I$. From $I$, each robot can determine whether or not $F$ is formable, compute its final position, and move there.

The formation problem for oblivious ASYNCH robots when $\rho(F_I) > 2$ is thus a promising target in this area. We however hesitate to dash to the target, since oblivious formation algorithms in the current form look to be very complicated, and both their descriptions and their correctness proofs are hard to understand as in [16]. The authors of [16] introduced many tricks to overcome the difficulties appeared in different phases in the algorithm, but they might fail to clearly state their meanings to readers.

In this paper, we propose a building block to construct an easy-understanding algorithm, motivated by the following observation: The formation algorithm in [16] first agrees on a coordinate system defined by the positions of three robots. Let $R'$ be the set of the robots supporting the coordinate system. Then all robots can put pattern $F$ in their working space in terms of the coordinate system, and all robots in $R \setminus R'$ move their corresponding positions in $F$. Note that the robots in $R'$ stop motionless to maintain the coordinate system unchanged, while the robots in $R \setminus R'$ are moving. When the robots in $R \setminus R'$ reach their final positions, the robots in $R'$ agree on a coordinate system defined by the positions of the robots in $R \setminus R'$, and move to their final positions in $F$.

The algorithm in [16] is thus constructed from two sub-formation tasks, one for the robots in $R \setminus R'$, and the other for the robots in $R'$. If we could treat these tasks more elegantly, we might be able to make the whole algorithm clearer. Emphasizing that a common coordinate system exists in each task, we formulate our formation problem: Given a geometric pattern $F$, we discuss the pattern formation problem for oblivious ASYNCH robots. The crucial difference between our problem and the conventional problem is that in our problem, the points in $F$ are assumed to be "visible" to all robots in their working space like landmarks, or equivalently, every robot observes their positions, as well as the positions of the robots, in its local coordinate system, whenever it executes the Look phase.

Apart from the above motivation, this formation problem has many applications. Some of them are when the points of $F$ are anonymous home bases of the robots, and when they are tasks assigned to the robots.

This paper proposes a formation algorithm CWM-Form and proves its correctness. CWM-Form makes use of the optimum matching $M$ between $R$ and $F$, and shows that asynchronous behaviors of the robots obeying CWM-Form maintain $M$ as the optimum matching.

*Related work.* In SSYNCH, since point gathering problem by two robots is not solvable in general and solvable if robots agree on the north and have common handedness, as intermediate model, unreliable compass model was considered [8, 9, 13]. Izumi et al. showed in [9] that 1) the gathering problem by two SSYNCH robots can be solved if and only if the maximum deviation angle of the compasses is less than $\pi/2$ for static compass model, and $\pi/4$ for dynamic compass model, 2) the gathering problem by two ASYNCH robots can be solved if and only if the maximum deviation angle of the compasses is less than $\pi/2$ for static compass model.

Other inaccurate sensor/control model was discussed in [4, 15]. They measured errors by a pair of the maximum angle and distance errors, and showed necessary and sufficient conditions of the pair for robots to have a convergence algorithm.

The model in which robots have limited visibility was considered in [7, 13]. Flocchini et al. showed in [7] that gathering problem can be solved by robots with limited visibility if they agree on the north and have common handedness. Souissi et al. in [13], extended the result to the eventually consistent compass model.

For other related works that focused on fault-tolerant formation/convergence for anonymous robots, see [1–3].

*Organization.* The rest of paper is organized as follows: In section 2.1 we present formal definition of our model of robots. In section 2.2 we give the outlook of our algorithm CWM-Form and present the main theorem (Theorem 1). In section 2.3 we define terminology that we use in this paper. For formal definition and discussion of "clockwise matching" which plays key role in our algorithm, whole section 3 is devoted. In section 4 we give the proof of Theorem 1.

## 2    Preliminary

### 2.1    Model

In this section we introduce our model of robots. For $x, y \in \mathbb{R}^2$, $d(x, y)$ denotes the Euclidean distance between coordinates $x$ and $y$, and $\overline{xy}$ denotes the line segment $\{x + t(y - x) : t \in [0, 1]\}$ and $\overrightarrow{xy}$ denotes the half line $\{x + t(y - x) : t \in [0, \infty)\}$. We call a set of finite number of points in $\mathbb{R}^2$ a pattern. Let $\mathcal{L}_n$ denotes all patterns of $n$ points. Let $\mathbb{T}$ be a set of all transformations consisting of rotation, translation and uniform scaling. An element of $\mathbb{T}$ is called a *local coordinate system*. Let $R$ denotes a set of robots.

In our model, every robot observes other robots' positions $A$ and goal pattern $B$ via its own (local) coordinate system (Look), and computes bijective maps (i.e., perfect bipartite matchings) between $A$ and $B$ minimizing total Euclidean distance and decides its own target point of $B$ by an *algorithm* $\psi$ (Compute), and move to the direction of the target with $\epsilon$ (or larger) length if there is no other robot(s) along the way (Move). Note that robots cannot communicate each other, and use distinct local coordinate systems where each robot only knows its own system, hence does not know others' ones. Thus if there uniquely exists the minimum perfect matching $M = \{(a_i, b_i) \in A \times B : i = 1, \ldots, n\}$ between $A$ and $B$, minimizing $\sum_{(a_i, b_i) \in M} d(a_i, b_i)$, a formation is easily done. In this paper, we are concerned with symmetrical patterns $A$ and $B$, for which there are some minimum perfect matchings between $A$ and $B$ (See e.g. Fig. 1 appearing later).

An algorithm $\psi$ is described by a function, which takes two patterns $A, B \in \mathcal{L}_n$ as an input (where $A$ is positions of robots and $B$ is the points of target pattern), and returns a bijective map $\psi(A, B) : A \to \mathbb{R}^2$ which indicates that the robot at $a \in A$ calculates the destination $\psi(A, B)(a) \in \mathbb{R}^2$ if activated. Additionally, we require an algorithm $\psi$ to satisfy

$$Z(\psi(A, B)(a)) = \psi(Z(A), Z(B))(Z(a)) , \tag{1}$$

for all $Z \in \mathbb{T}$, $A, B \in \mathcal{L}_n$ and $a \in A$. We call this property *local coordinate system independency*. In this paper, since elements in $\mathbb{T}$ consist of rotations, translations and uniform scalings, this suggests that the robots do not have common knowledge about the north (directions of $y$-axis of local coordinate systems may not agree) and the unit length. However, since we do not allow mirror transformation, the robots have common handedness, i.e., bases of their local coordinate system have same orientation.

For an algorithm $\psi$ and patterns $A, B \in \mathcal{L}_n$, let $\mathcal{P}(\psi, A, B)$ denotes a set of all executions which we consider, i.e., $\mathcal{P}(\psi, A, B)$ is a set of all executions $P : R \times \mathbb{N} \to \mathbb{R}^2$ where $P(r, t)$ denotes the position of robot $r$ at time $t$. Expansively, let $P(R, t)$ denotes the positions of all robots in $R$ at time $t$. Note that $P(r, t)$ includes uncertainty even when an algorithm $\psi$ and patterns $A$ and $B$ are fixed since robots in our model are (oblivious and) asynchronous.

In our model, all robots in $R$ at time 0 are located at $A$, i.e., $P(R, 0) = A$. We assume that there exists a minimal moving distance $\epsilon$ which is a positive real number. For each robot $r \in R$, there exists monotonically increasing infinite sequence $t_1^{(r)}, t_2^{(r)}, \cdots \in \mathbb{N}$ where $t_i^{(r)}$ express the starting time of the $i^{\text{th}}$ cycle of $r$. The position of $r$ does not change till the beginning of the first cycle, i.e., $P(r, t) = P(r, 0)$ for all $t \leq t_1^{(r)}$. For the $i^{\text{th}}$ cycle, let $a_{r,i}$ denotes the position of $r$ at the beginning of the $i^{\text{th}}$ cycle, and $b_{r,i}$ denotes the destination $r$ calculated in the $i^{\text{th}}$ cycle, i.e., $a_{r,i} = P(r, t_i^{(r)})$ and $b_{r,i} = \psi(P(R, t_i^{(r)}), B)(a_{r,i})$. Robot $r$ moves along the line $\overline{a_{r,i}b_{r,i}}$, i.e., $P(r, t) \in \overline{a_{r,i}b_{r,i}}$ for all $t \in [t_i^{(r)}, t_{i+1}^{(r)}]$, and $d(a_{r,i}, P(r, t)) \leq d(a_{r,i}, P(r, t'))$ for all $t, t' \in [t_i^{(r)}, t_{i+1}^{(r)}]$ such that $t < t'$. The robot $r$ arrives at the destination $b_{r,i}$ or moves away from $a_{r,i}$ at least $\epsilon$ at the end of the $i^{\text{th}}$ cycle, i.e., $P(r, t_{i+1}^{(r)}) = b_{r,i}$ or $\epsilon < d(a_{r,i}, P(r, t_{i+1}^{(r)}))$ .

Pattern formation is defined as follow.

**Definition 1.** *An algorithm $\psi$ forms a pattern $B$ from a pattern $A$ if for all possible executions $P \in \mathcal{P}(\psi, A, B)$, there exists a time $t^* \in \mathbb{N}$ such that for all $t > t^*$, $P(R, t) = B$.*

## 2.2 Algorithm — CWM-Form

Now we describe an outline of our algorithm CWM-Form.

Given a pair of patterns $A$ and $B$, let $M$ be a bijective map from $A$ to $B$ which we call *clockwise matching* (shortly CWM) uniquely defined for $A$ and $B$. We show an example of CWM in Fig. 1 where black dots represent positions of the robots and white dots represent the points of the target pattern. We will give a formal definition of CWM in Section 3. With $M$, CWM-Form is defined by

$$\text{CWM-Form}(A, B)(a) = \begin{cases} a & \text{if } \exists a' \in A \text{ s.t. } a' \in \overline{aM(a)} \\ M(a) & \text{otherwise} \end{cases} \tag{2}$$

That is, the robot at $a \in A$ calculates its matched destination $M(a) \in B$, and if there is other robot(s) along the way, it does not move, otherwise, it moves to $M(a)$.

**Theorem 1.** *Given any patterns $A, B \in \mathcal{L}_n$, CWM-Form forms $B$ from $A$.*

We show Theorem 1 in the rest of this paper. Since our robots are oblivious and asynchronous, the point of Theorem 1 is that every robot can reach its own point of $B$ assigned by CWM during formation steps. In Section 3, we define a *clockwise matching* which is local coordinate system independent and show that it presents a kind of "canonical matching" in our formation step.

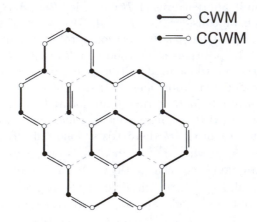

**Fig. 1.** An example of CWM

### 2.3 Terminology

To describe "clockwise matching" we define our terminology (c.f., [6, 10]). Throughout this paper, we consider bipartite graph $G = (V, E)$ with vertex set $V = A \cup B$ and edge set $E \subseteq A \times B$. We draw vertices of $A$ with black and vertices of $B$ with white in Figures 1-7. The vertex set of a graph $G$ is referred to as $V(G)$ and its edge set as $E(G)$. $K_2$ denotes complete graph with 2 vertices. A graph $G$ is *elementary* if the union of all perfect matchings of $G$ is a connected subgraph of $G$. For a bipartite graph $G$ and its perfect matchings $M$ and $M'$, the cycle $C$ of $G$ is an *alternating cycle of $M$ and $M'$* if the edges of $C$ appear alternately in $M$ and $M'$. For a plane graph, the boundary of exterior face is called the *periphery*. A bijective map from $A$ to $B$ (i.e., matching) is denoted by a set of pairs of vertices of $A$ and $B$; e.g., if $M$ is a bijection defined by $M(a_i) = b_i$ for $i = 1, 2, \ldots, n$ then we may also consider $M = \{(a_i, b_i) : i = 1, 2, \ldots, n\}$. Since we only consider graphs which are union of perfect matchings of $A$ and $B$, and do not allow isolated vertices, we sometimes specify a graph only by its edge set. For $S \subseteq A \times B$, we define $A(S)$ to be a set $\{a : (a, b) \in S\}$ and $B(S)$ to be a set $\{b : (a, b) \in S\}$. For a pair of patterns $A$ and $B$ such that $|A| = |B|$, $\mathcal{U}(A, B)$ denotes a set of all perfect matchings of $A$ and $B$ (bijection from $A$ to $B$). For a matching $M \in \mathcal{U}(A, B)$ we define cost of the matching by

$$d(M) = \sum_{(a,b) \in M} d(a, b) . \tag{3}$$

Let $\mathcal{M}(A, B)$ denotes a set of all matchings $M \in \mathcal{U}(A, B)$ which minimize $d(M)$ and for all distinct $(a, b), (a', b') \in M$, $a, a', b', b$ does not reside on a line in the order. Note that as an element of $\mathcal{M}(A, B)$, we do not allow matching whose edge includes its another edge as in Fig. 2 (a), while allowing parallel edges as in Fig. 2 (b). See that $\mathcal{M}(A, B) \neq \emptyset$ (except $A = B = \emptyset$), but not necessarily $|\mathcal{M}(A, B)| = 1$. We call a matching $M \in \mathcal{U}(A, B)$ to be optimum if $M \in \mathcal{M}(A, B)$. $G(A, B)$ denotes a bipartite graph with vertex set $V = (A, B)$ and edge set $E = \bigcup_{M \in \mathcal{M}(A,B)} M$. See Fig. 3 for an example of $G(A, B)$. For a Jordan curve $C$ on Euclidean plane, tour $T$ of $C$ is a continuous injective map $T : [0, 1) \to \mathbb{R}^2$ which satisfies $T([0, 1)) = C$. We call $T$ to be CCW tour of $C$, if $T$ is tour of $C$ and $\int_0^1 \frac{d\theta_{T,p}}{dt} dt = 2\pi$ where $(x(t), y(t)) = T(t)$ and $p = (p_x, p_y)$ is some point of inner face of $C$ and $\theta_{T,p}(t) = \arctan \frac{y(t) - p_y}{x(t) - p_x}$. Expansively, for a cycle $C$ of a plane graph, a sequence of vertices $v_0 v_1 \ldots v_n v_0$ of $C$ is CCW tour if there exists CCW tour $T$ of $\hat{C} = V(C) \cup \bigcup E(C)$ and monotonically increasing sequence $t_0, t_1, \ldots, t_n \in [0, 1)$ such that $T(t_i) = v_i$, and CW tour otherwise.

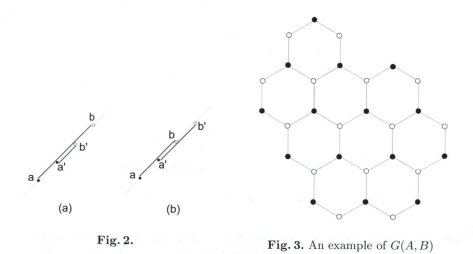

(a)　　　　　　　　(b)

**Fig. 2.**

**Fig. 3.** An example of $G(A, B)$

## 3 Clockwise Matching

In this section, we define the "clockwise-matching" $CWM(A, B)$ of pattern $A$ and $B$, which plays a key role in our algorithm. To begin with, we analyze a graph $G = G(A, B)$, in order to define CWM.

### 3.1 Graph $G(A, B)$

Though we want to consider $G$ to be a plane graph by replacing each edge $(a, b)$ with a line $\overline{ab}$, $G$ is not a plane graph in general unfortunately. First, we look at the relation between two edges of $G$. The following holds even if $G$ is not connected graph.

**Lemma 2.** *There are four cases[3] between two edges $e = (a, b)$ and $e' = (a', b')$ of $G$. See Fig. 4 for illustration.*

- *(separate)* $\overline{ab} \cap \overline{a'b'} = \emptyset$.
- *(fold)* $e$ *and* $e'$ *have exactly one common end vertex and* $a' \in \overline{ab}$.
- *(adjacent)* $e$ *and* $e'$ *have exactly one common end vertex and not (fold)*.
- *(parallel)* $a, a', b, b'$ *reside on one line in the order.*

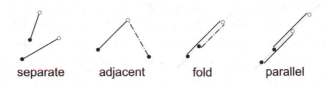

separate  adjacent  fold  parallel

**Fig. 4.** Possible relation of two edges of $G(A, B)$

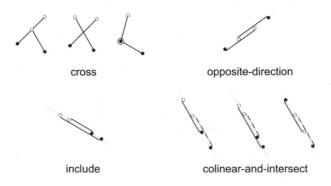

cross      opposite-direction

include    colinear-and-intersect

**Fig. 5.** Impossible relation of two edges of $G(A, B)$

*Proof.* We classify the other cases which do not satisfy the above relation as follow. For each case, we derive contradiction. See Fig. 5. Let $e = (a, b)$ and $e' = (a', b')$ be two edges that are not adjacent and let $M, M' \in \mathcal{M}(A, B)$ be matchings such that $e \in M$, $e' \in M'$. Since $e$ and $e'$ are not adjacent, $a \neq a'$ and $b \neq b'$.

- Case 1. $M = M'$.
    - (cross) Two lines $\overline{ab}$ and $\overline{a'b'}$ cross at a point.
    - (opposite-direction) $a, b', b, a'$ reside on a line in the order.
    - (include) $a, a', b', b$ reside on a line in the order.
- Case 2. $M \neq M'$.
    - (cross) Two lines $\overline{ab}$ and $\overline{a'b'}$ cross at a point.
    - (colinear-and-intersect) $a, b, a', b'$ are colinear and $\overline{ab} \cap \overline{a'b'} \neq \emptyset$.

---

[3] We do not consider the cases in which the color of each the vertices is converted to the other color, since those are symmetric.

Let $\tilde{e} = (a, b'), \tilde{e}' = (a', b)$.

- Case 1. $M = M'$. Let $W = M \setminus \{e, e'\} \cup \{\tilde{e}, \tilde{e}'\}$. Obviously, $W \in \mathcal{U}(A, B)$.
  - (cross) Let $p$ be the crossing point. By the triangle inequality, $d(a, b') + d(a', b) < d(a, p) + d(p, b) + d(a', p) + d(p, b') = d(a, b) + d(a', b')$. Thus $d(W) = d(M \setminus \{e, e'\} \cup \{\tilde{e}, \tilde{e}'\}) = d(M) - \{d(a, b) + d(a', b')\} + \{d(a, b') + d(a', b)\} < d(M)$. This contradict with $M \in \mathcal{M}(A, B)$.
  - (opposite-direction) Same as the case (cross).
  - (include) This contradict with $M \in \mathcal{M}(A, B)$.
- Case 2. $M \neq M'$.
  - (cross) Let $C = (M \oplus M') \setminus \{e, e'\} \cup \{\tilde{e}, \tilde{e}'\}$. Then, each connected component of $C$ forms an alternating cycle and $d(C) < d(M \oplus M')$ by the same argument as the case (cross) of Case 1. Thus, there exists $W, W' \in \mathcal{U}(A, B)$ such that $d(W \oplus W') < d(M \oplus M')$ and $M \cap M = W \cap W'$. This is because, as edges of $W$ and $W'$, you can take edges of the alternating cycle alternately for each connected component of $C$ and for the rest of the $W$ and $W'$, you can take edges both from $M \cap M'$. Thus, $d(W) + d(W') = d(W \oplus W') + 2d(W \cap W') < d(M \oplus M') + 2d(M \cap M') = d(M) + d(M')$. Furthermore, since $M$ and $M'$ are optimum matchings, $d(M) = d(M')$. Therefore, $d(W) < d(M)$ or $d(W') < d(M)$. This contradict with $M \in \mathcal{M}(A, B)$.
  - (colinear-and-intersect) In this case, though each connected component of $M \oplus M'$ must form alternating cycle, this is impossible because two edges in the same matching cannot have relation (cross) or (opposite-direction) as we proved in Case 1. □

We consider the connected component of $G$. When $G$ is connected, let's see, what kind of graph we can draw on the plane. To begin with edge $e_1$, by Lemma 2, as the next edge $e_2$, we can draw edge which have either (adjacent) or (fold) relation with $e_1$. If we choose (fold), the next edge $e_3$ must be the (parallel) relation with $e_1$. Then, for the next edge $e_4$, you can choose (adjacent) or (fold). However in case you choose (fold) you have to be careful not to draw the line too long and include $e_2$, and so on, and of course any of two edges could never cross each other.

With that observation, let us define the plane graph representation $D(G)$ of $G$ as follow. We call an alternating path $a_1 b_1 \ldots a_m b_m$ of $G$ which satisfy $a_{i+1} \in \overline{a_i b_i}$ and $b_i \in \overline{a_{i+1} b_{i+1}}$ for all $i = 1 \ldots m - 1$, a *folded-path*. Any edge is a folded-path with length 1. A *maximal folded-path* is a folded-path, which by extending the path with one more vertex, no longer holds above condition. $D(G)$ is a plane graph which is produced by replacing each maximal folded-path $aPb$ of $G$ with a line $\overline{ab}$. See that those two lines never intersect with each other except for end points and for any perfect matching of $D(G)$, there is a corresponding perfect matching of $G$ since each $aPb$ is an alternating path without branch from inner vertices.

From the definition, Lemma 2 and the above arguments, we obtain the following Corollaries.

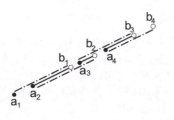

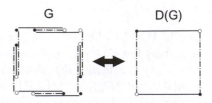

**Fig. 6.** An example of folded-path

**Fig. 7.** An example of $G$ and $D(G)$

**Corollary 3.** *$G$ is bipartite elementary.*

**Corollary 4.** *$D(G)$ is plane bipartite elementary.*

We also remark the followings.

**Theorem 5.** *[10] Elementary bipartite graph is 2-connected.*

**Theorem 6.** *[6] Any face of 2-connected plane graph with more than 4 vertices is bounded by a cycle of the graph.*

The geometrical property defined on plane graph $D(G)$ is naturally lifted to the original graph $G$. i.e., if the tour $x_0 x_1 \ldots x_{m-1} x_0$ of the cycle $C$ of $D(G)$ is CW, as there is unique maximal folded-path $P_i$ of $G$ that connect $x_i$ and $x_{i+1 \bmod m}$, we define the tour $x_0 P_0 x_1 P_1 \ldots x_{m-1} P_{m-1} x_0$ of $G$ to be CW and CCW otherwise. The periphery of $G$ is defined in the same manner. By the above theorems, we can say that the periphery of any connected component of $G$ is either $K_2$ or an alternating cycle.

### 3.2  Definition of CWM

In order to define $\mathrm{CWM}(A, B)$, we first define functions ToCycles and CWMC which are building blocks of CWM.

Given a pair of patterns $A$ and $B$ which satisfies $|A| = |B|$, let $G_1, G_2, \ldots, G_N$ be connected components of $G(A, B)$, and let $C_i$ be the periphery of $G_i$, and let $A_i = A(G_i \setminus C_i)$ and $B_i = B(G_i \setminus C_i)$. ToCycles$(A, B)$ is a subgraph of $G(A, B)$ each connected component of which is a cycle or $K_2$, defined by

$$\mathrm{ToCycles}(A, B) = \begin{cases} \emptyset & \text{if } A = B = \emptyset \\ \bigcup_{i=1}^{N} \{ C_i \cup \mathrm{ToCycles}(A_i, B_i) \} & \text{otherwise} \end{cases} \quad (4)$$

Note that ToCycles$(A, B)$ consists of even cycles and $K_2$s. In Fig. 1 ToCycles$(A, B)$ is union of CWM and CCWM.

Given a graph $C$ which is a cycle or $K_2$, $\mathrm{CWMC}'(C)$ is a matching of $C$ defined by

$$\mathrm{CWMC}'(C) = \begin{cases} E(C) & \text{if } C = K_2 \\ \{(a_1, b_1), (a_2, b_2), \ldots, (a_m, b_m)\} & \text{otherwise} \end{cases}, \quad (5)$$

where sequence $a_1 b_1 \ldots a_m b_m a_1$ is CW tour of $C$ and $a_i \in A$.

Extending the definition of CWMC', we define a matching CWMC($G$) of a graph $G$ consisting of even cycles and $K_2$s by

$$\text{CWMC}(G) = \bigcup_{i=1}^{N} \text{CWMC}'(G_i) \,, \tag{6}$$

where $G_1, G_2, \ldots, G_N$ is connected components of $G$.

Finally, for a pair of patterns $A$ and $B$ which satisfies $|A| = |B|$, let $G = \text{ToCycles}(A, B)$. CWM($A, B$) is a perfect matching of $A$ and $B$ defined by

$$\text{CWM}(A, B) = \text{CWMC}(G) \,. \tag{7}$$

## 3.3   Properties of CWM

Here, we show the properties of CWM($A, B$) (Corollary 10 and Lemma 11) which we will use to prove Theorem 1.

**Lemma 7.** *Any perfect matching of $G(A, B)$ is optimum.*

*Proof.* Let $|\mathcal{M}(A, B)| = k$ and cost of each matching of $\mathcal{M}(A, B)$ to be $w$. Here, we consider a multi graph $\hat{G}(A, B)$ consisting of (disjoint) union of $k$ perfect matchings, hence $\hat{G}(A, B)$ is $k$-regular. Suppose that there exists perfect matching $M$ of $G(A, B)$ that is not optimum. Since $M$ is not optimum, $w < d(M)$. The sum of weight of all edges of $G$ is $kw$. Consider $G \setminus M$. The sum of weight of all edges of $G \setminus M$ is less than $(k-1)w$. However, since $G \setminus M$ is $(k-1)$-regular graph, by Hall's marriage theorem, $G \setminus M$ can be decomposed into $k-1$ perfect matchings. Thus at least one of the $k-1$ perfect matchings has the cost less than $w$. This contradict with optimality of $\mathcal{M}(A, B)$.      □

**Theorem 8.** *[17] The periphery of a plain elementary bipartite graph is an alternating cycle of two perfect matchings.*

**Lemma 9.** *The periphery of $G(A, B)$ is an alternating cycle of two optimum matchings.*

*Proof.* Consider $G = G(A, B)$. By Corollary 4 and Theorem 8, the periphery of $D(G)$ is an alternating cycle of two perfect matchings. Thus by its corresponding two perfect matchings of $G$, the periphery of $G$ is an alternating cycle of the two perfect matchings. Furthermore by Lemma 7, these two matchings of $G$ are optimum.      □

By Lemma 9 and definition of CWM($A, B$), we obtain the following.

**Corollary 10.** CWM($A, B$) *is an optimum matching.*

We define binary relation ($\preceq$) on $\mathcal{M}(A, B)$ as follow.

**Definition 2.** $M \preceq M' \overset{\text{def}}{\Longleftrightarrow} \text{CWMC}(M \cup M') = M$.

Intuitively, $M \preceq M'$ means $M$ is "closer to clockwise" than $M'$ and the following Lemma states that CWM($A, B$) is "exactly clockwise" matching of $\mathcal{M}(A, B)$.

**Lemma 11.** *For all $M \in \mathcal{M}(A, B)$, $\mathrm{CWM}(A, B) \preceq M$.*

*Proof.* We prove $\mathrm{CWMC}(\mathrm{CWM}(A, B) \cup M) = \mathrm{CWM}(A, B)$. Since it is obvious that CWMC takes the edges of $\mathrm{CWM}(A, B) \cap M$. We only need to focus on the edges of $\mathrm{CWM}(A, B) \oplus M$ each connected component of which is an alternating cycle. Let $C$ be such a cycle. $C$ is an alternating cycle of $\mathrm{CWM}(A, B)$ and $M$. Of the edges of $C$, either edges of $C \cap \mathrm{CWM}(A, B)$ or $C \cap M$ will be chosen as edges of $\mathrm{CWMC}(\mathrm{CWM}(A, B) \cup M)$. By the definition of the algorithm CWM, some cycle $C'$ in $\mathrm{ToCycles}(A, B)$ has common edges with $C$ while inner face of $C$ is included in that of $C'$. This is because, if not, it contradicts with the definition of ToCycles, which takes the periphery of the connected component of the graph one by one. This suggests that edges of $C \cap \mathrm{CWM}(A, B)$ will be chosen for the cycle $C$ since any point of inner face of $C$ is also a point of inner face of $C'$ and clockwise tour of these cycles never contradict with each other. The same argument holds for other alternating cycles of $\mathrm{CWM}(A, B) \oplus M$. Therefore, $\mathrm{CWMC}(\mathrm{CWM}(A, B) \cup M) = \mathrm{CWM}(A, B)$. □

## 4    Proof of Theorem 1

In this section, we prove Theorem 1, by using the properties of CWM we presented in the previous section.

**Lemma 12.** *Let $A, B, X$ be patterns which satisfy following,*

1. $\{(a_1, b_1), (a_2, b_2), \dots, (a_n, b_n)\} \in \mathcal{M}(A, B)$.
2. $X = \{x_1, x_2, \dots, x_n\}$ where $x_i \in \overline{a_i b_i}$.
3. *if there exists $i' \neq i$ such that $a_{i'} \in \overline{a_i b_i}$ then $x_i = a_i$.*

*then $\{(x_1, b_1), (x_2, b_2), \dots, (x_n, b_n)\} \in \mathcal{M}(X, B)$.*

Lemma 12 states that an optimum matching continues to be optimum even after some robots moved according to the matching.

*Proof.* Let $M = \{(a_1, b_1), \dots, (a_n, b_n)\}$, $M' = \{(x_1, b_1), \dots, (x_n, b_n)\}$. First, consider the case in which only one point has moved. Without loss of generality, let $a_j \neq x_j$ and $a_i = x_i$ for all $i \neq j$. i.e., $a_j$ has moved to $x_j$ along the line $\overline{a_j b_j}$ and the rest of the points has not moved. Since $a_j \neq x_j$, we can assume there is no element of $A$ on $\overline{x_j b_j}$. (That is, for all distinct $(a, b), (a', b') \in M'$, we can assume that $a, a', b', b$ does not reside on a line in the order.) Suppose $M' \notin \mathcal{M}(X, B)$. This implies that there exists $W' \in \mathcal{U}(X, B)$ such that $d(W') < d(M')$. Let $\{(a_1, b'_1), \dots, (x_j, b'_j), \dots, (a_n, b'_n)\}$ be such $W'$ and $W$ be $\{(a_1, b'_1), (a_2, b'_2), \dots, (a_n, b'_n)\}$. Since $W \in \mathcal{U}(A, B)$, $d(M) \leq d(W)$. Let $\delta = d(a_j, x_j)$, $a = d(x_j, b'_j)$, $d(a_j, b'_j) = b$. By the triangle inequality, $-\delta \leq a - b$. Therefore, $d(M') = d(M) - \delta \leq d(W) + a - b = d(W) - d(x_j, b'_j) + d(a_j, b'_j) = d(W')$. This contradict with $d(W') < d(M')$.

In case that arbitrary number of points have moved, inductively applying the same argument for the one point case, we obtain the claim. □

**Lemma 13.** *Let* $A, B, X$ *be patterns which satisfy following,*

1. $\mathrm{CWM}(A, B) = \{(a_1, b_1), (a_2, b_2), \ldots, (a_n, b_n)\}$.
2. $X = \{x_1, x_2, \ldots, x_n\}$ *where* $x_i \in \overline{a_i b_i}$.
3. *if there exists* $i'(\neq i)$ *such that* $a_{i'} \in \overline{a_i b_i}$ *then* $x_i = a_i$.

*then* $\mathrm{CWM}(X, B) = \{(x_1, b_1), (x_2, b_2), \ldots, (x_n, b_n)\}$.

Lemma 13 states that CWM of the robots and the pattern never change while the execution. Thus justifies Theorem 1.

*Proof.* Let $M = \{(a_1, b_1), \ldots, (a_n, b_n)\}$, $M' = \{(x_1, b_1), \ldots, (x_n, b_n)\}$. Suppose $W' = \{(x_1, b'_1), \ldots, (x_n, b'_n)\} = \mathrm{CWM}(X, B)$ and $W = \{(a_1, b'_1), \ldots, (a_n, b'_n)\}$ and $M' \neq W'$. First consider the case in which only one point has moved. Without loss of generality, let $a_j \neq x_j$ and $a_i = x_i$ for all $i \neq j$. Since $a_j \neq x_j$, we can assume there is no element of $A$ on $\overline{x_j b_j}$. Now we are considering the following four matchings.

$$M = \{(a_1, b_1), (a_2, b_2), \ldots, (a_j, b_j), \ldots, (a_n, b_n)\},$$
$$W = \{(a_1, b'_1), (a_2, b'_2), \ldots, (a_j, b'_j), \ldots, (a_n, b'_n)\},$$
$$M' = \{(a_1, b_1), (a_2, b_2), \ldots, (x_j, b_j), \ldots, (a_n, b_n)\},$$
$$W' = \{(a_1, b'_1), (a_2, b'_2), \ldots, (x_j, b'_j), \ldots, (a_n, b'_n)\}.$$

By Lemma 12 and Lemma 10, $M', W' \in \mathcal{M}(X, B)$. Therefore, by Lemma 11, $\mathrm{CWM}(X, B) = W' \prec M'$, also since $W, M \in \mathcal{M}(A, B)$, $\mathrm{CWM}(A, B) = M \prec W$.

- Case $b_j = b'_j$. In this case $W' \oplus M' = W \oplus M$ and for all $e \in W \oplus M$, $e \in W$ if and only if $e \in W'$ and $e \in M$ if and only if $e \in M'$. Thus $W \prec M$. This contradict with $\mathrm{CWM}(A, B) = M \prec W$.
- Case $b_j \neq b'_j$. In this case, $b'_j$ is on the half line $\overrightarrow{x_j b_j}$ since if not, it contradict with $M', W' \in \mathcal{M}(X, B)$. Thus $b_j a_j b'_j$ and $b_j x_j b'_j$ is a folded-path and the rest of the alternating cycle is the same as in Case $b_j = b'_j$. Therefore $W \prec M$. This contradict with $\mathrm{CWM}(A, B) = M \prec W$.

In case that arbitrary number of points have moved, inductively applying the same argument for the one point case, we obtain the claim. $\square$

## 5 Conclusion

In this paper, we considered autonomous mobile robots and constructed the algorithm CWM-Form which forms any given pattern under assumption of CORDA, anonymous and oblivious autonomous mobile robots. We assumed that they do not have common knowledge about the north and the unit length, but have common handedness. Besides, we assumed that they can observe target pattern according to their view of the world. By CWM-Form, robots always calculate matching of the robots and the pattern which minimize the sum of distances.

CWM-Form precedes CW matching to CCW. We showed such simple algorithm can form any given pattern even if robots are anonymous, oblivious and asynchronous.

Although we have not considered in this paper, in the conventional problem, robots only know *shape* of target pattern and cannot see the pattern itself. Whether the idea of pattern formation by matching we considered in this paper can be applied to these models and is effective approach in pattern formation in general, remains to be investigated.

# References

1. Agmon, N., Peleg, D.: Fault-tolerant gathering algorithms for autonomous mobile robots. SIAM Journal on Computing 36, 56–82 (2006)
2. Bouzid, Z., Potop-Butucaru, M.G., Tixéuil, S.: Optimal Byzantine resilient convergence in asynchronous robot networks. In: Guerraoui, R., Petit, F. (eds.) SSS 2009. LNCS, vol. 5873, pp. 165–179. Springer, Heidelberg (2009)
3. Bouzid, Z., Potop-Butucaru, M.G., Tixéuil, S.: Byzantine-resilient convergence in oblivious robot networks: The price of asynchrony. In: Garg, V., Wattenhofer, R., Kothapalli, K. (eds.) ICDCN 2009. LNCS, vol. 5408, pp. 275–280. Springer, Heidelberg (2009)
4. Cohen, R., Peleg, D.: Convergence of autonomous mobile robots with inaccurate sensors. SIAM Journal on Computing 38, 276–302 (2008)
5. Dieudonné, Y., Petit, F., Villain, V.: Leader election problem versus pattern formation problem. In: Lynch, N.A., Shvartsman, A.A. (eds.) Distributed Computing. LNCS, vol. 6343, pp. 267–281. Springer, Heidelberg (2010)
6. Diestel, R.: Graph Theory, 2nd edn. Springer, Heidelberg (2000)
7. Flocchini, P., Prencipe, G., Santoro, N., Widmayer, P.: Gathering of asynchronous mobile robots with limited visibility. Theoretical Computer Science 337, 147–168 (2005)
8. Izumi, T., Samia, S., Katayama, Y., Inuzuka, N., Défago, X., Wada, K., Yamashita, M.: The Gathering Problem for Two Oblivious Robots with Unreliable Compasses (to appear)
9. Katayama, Y., Tomida, Y., Imazu, H., Inuzuka, N., Wada, K.: Dynamic compass models and gathering algorithms for autonomous mobile robots. In: Prencipe, G., Zaks, S. (eds.) SIROCCO 2007. LNCS, vol. 4474, pp. 274–288. Springer, Heidelberg (2007)
10. Lovasz, L., Plummer, M.: Matching Theory. AMS Chelsea Publishing, Providence (2009)
11. Nagamochi, H., Yamashita, M., Ibaraki, T.: Distributed algorithms for cooperative controlling of anonymous mobile robots. Technical Reports of IEICE, COMP95-24, pp. 31–40 (1995) (in Japanese)
12. Prencipe, G.: Distributed coordination of a set of autonomous mobile robots. PhD Thesis, Universita di Pisa (2002)
13. Souissi, S., Défago, X., Yamashita, M.: Using eventually consistent compasses to gather memory-less mobile robots with limited visibility. ACM Trans. Autonomous and Adaptive Systems 4 (2009)

14. Suzuki, I., Yamashita, M.: Distributed anonymous mobile robots: Formation of geometric patterns. SIAM Journal on Computing 28, 1347–1363 (1999)
15. Yamamoto, K., Izumi, T., Katayama, Y., Inuzuka, N., Wada, K.: Convergence of mobile robots with uniformly-inaccurate sensors. In: Kutten, S., Žerovnik, J. (eds.) SIROCCO 2009. LNCS, vol. 5869, pp. 320–333. Springer, Heidelberg (2009)
16. Yamashita, M., Suzuki, I.: Characterizing geometric patterns formable by oblivious anonymous mobile robots. Theoretical Computer Science 411, 2433–2453 (2010)
17. Zhang, H., Zhang, F.: Plane elementary bipartite graphs. Discrete Applied Mathematics 105, 291–311 (2000)

# RoboCast: Asynchronous Communication in Robot Networks

Zohir Bouzid[1,*], Shlomi Dolev[2,**],
Maria Potop-Butucaru[1], and Sébastien Tixeuil[1]

[1] Université Pierre et Marie Curie - Paris 6, France
[2] Ben Gurion University of the Negev, Israel

**Abstract.** This paper introduces the *RoboCast* communication abstraction. The RoboCast allows a swarm of non oblivious, anonymous robots that are only endowed with visibility sensors and do not share a common coordinate system, to asynchronously exchange information. We propose a generic framework that covers a large class of asynchronous communication algorithms and show how our framework can be used to implement fundamental building blocks in robot networks such as gathering or stigmergy. In more details, we propose a RoboCast algorithm that allows robots to broadcast their local coordinate systems to each others. Our algorithm is further refined with a local collision avoidance scheme. Then, using the RoboCast primitive, we propose algorithms for deterministic asynchronous gathering and binary information exchange.

## 1 Introduction

Existing studies in robots networks focus on characterizing the computational power of these systems when robots are endowed with visibility sensors and communicate using *only* their movements without relying on any sort of agreement on a global coordinate system. Most of these studies [1,5,4] assume oblivious robots (*i.e.* robots have no persistent memory of their past actions), so the "memory" of the network is implicit and generally deduced from the current positions of the robots. Two computation models are commonly used in robot networks: ATOM [9] and CORDA [7]. In both models robots perform in Look-Compute-Move cycles. The main difference is that these cycles are executed in a fully asynchronous manner in the CORDA model while each phase of the Look-Compute-Move cycle is executed in a lock step fashion in the ATOM model. These computation models have already proved their limitations. That is, the

* Supported by DIGITEO project PACTOLE and the ANR projects R-DISCOVER and SHAMAN.
** Part of the research was done during a supported visit of Shlomi Dolev at LIP6 Universit Pierre et Marie Curie - Paris 6. Partially supported by Rita Altura trust chair in computer sciences, ICT Programme of the European Union under contract number FP7-215270 (FRONTS), and US Air Force European Office of Aerospace Research and Development, grant number FA8655-09-1-3016.

C. Lu, T. Masuzawa, and M. Mosbah (Eds.): OPODIS 2010, LNCS 6490, pp. 16–31, 2010.

deterministic implementations of many fundamental abstractions such as gathering or leader election are proved impossible in these settings without additional assumptions ([8]). The purpose of this paper is to study how the addition of *bounded* memory to each individual robot can increase the computational power of an *asynchronous* swarm of robots. We focus on an *all-to-all* communication primitive, called RoboCast, which is a basic building block for the design of any distributed system. A positive answer to this problem is the open gate for solving fundamental problems for robot networks such as gathering, scattering, election or exploration.

In robot networks, using motion to transmit information is not new [9,10,6]. In [9], Suzuki and Yamashita present an algorithm for broadcasting the local coordinate system of each robot (and thus build a common coordinate system) under the ATOM model. The algorithm heavily relies on the phase atomicity in each Look-Compute-Move cycle. In particular, a robot $a$ that observes another robot $b$ in four distinct positions has the certitude that $b$ has in turn already seen $a$ in at least two different positions. The situation becomes more intricate in the asynchronous CORDA model. Indeed, the number of different positions observed for a given robot is not an indicator on the number of complete cycles executed by that robot since cycles are completely uncorrelated. By contrast, our implementation of RoboCast is designed for the more general CORDA model and uses a novel strategy: the focus moves from observing robots in different positions to observing robots moving in different *directions*. That is, each robot changes its direction of movement when a particular stage of the algorithm is completed; this change allows the other robots to infer information about the observed robot.

Another non trivial issue that needs to be taken care of without explicit communication is *collisions avoidance*, since colliding robots could be confused due to indistinguishability. Moreover, robots may physically collide during their Move phase. One of the techniques commonly used to avoid collisions consists in computing a Voronoi diagram [2] and allowing robots to move *only* inside their Voronoi cells [5]. Since the Voronoi cells do not overlap with one another, robots are guaranteed to not collide. This simple technique works well in the ATOM model but heavily relies on the computation of the same Voronoi diagram by the robots that are activated concurrently, and thus does not extend to the CORDA model where different Voronoi diagrams may be computed by different robots, inducing possible collisions. Our approach defines a collision-free zone of movement that is compatible with the CORDA model constraints.

Applications of our RoboCast communication primitive include fundamental services in robot networks such as gathering and stigmergy. Deterministic gathering of two stateless robots has already been proved impossible when robots have no common orientation [9]. In [9], the authors also propose non-oblivious solutions for deterministic gathering in the ATOM model. Our RoboCast permits to extend this result to the CORDA model, using bounded memory and a limited number of movements. Recently, in [6], the authors extend the work of [9] to efficiently implement stigmergy in robot networks in the ATOM model.

Stigmergy is the ability for robots to exchange binary information that is encoded in the way they move. This scheme is particularly appealing for secure communication in robot networks, since *e.g.* jamming has no impact on robot communication capability. The RoboCast primitive allows to extend this mechanism to the CORDA model, with a collision-free stigmergy scheme.

***Our contribution.*** We formally specify a robot network communication primitive, called RoboCast, and propose implementation variants for this primitive, that permit anonymous robots not agreeing on a common coordinate system, to exchange various information (*e.g.* their local coordinate axes, unity of measure, *rendez-vous* points, or binary information) using only motion in a two dimensional space. Contrary to previous solutions, our protocols all perform in the *fully asynchronous* CORDA model, use *constant* memory and a *bounded* number of movements. Then, we use the RoboCast primitive to efficiently solve some fundamental open problems in robot networks. We present a fully asynchronous deterministic gathering and a fully asynchronous stimergic communication scheme. Our algorithms differ from previous works by several key features: they are totally asynchronous (in particular they do not rely on the atomicity of cycles executed by robots), they make no assumption on a common chirality or knowledge of the initial positions of robots, and finally, each algorithm uses only a bounded number of movements. Also, for the first time in these settings, our protocols use CORDA-compliant collision avoidance schemes.

***Roadmap.*** The paper is made up of six sections. Section 2 describes the computing model and presents the formal specification of the RoboCast problem. Section 3 presents our protocol and its complexity. The algorithm is enhanced in Section 4 with a collision-avoidance scheme. Using the Robocast primitive, Section 5 proposes algorithms for deterministic asynchronous gathering and binary information exchange. Finally, Section 6 provides concluding remarks. Due to space limitations, some proofs are given in a companion technical report [3].

## 2   Model

We consider a network that consists of a finite set of $n$ robots arbitrarily deployed in a two dimensional space, with no two robots located at the same position. Robots are devices with sensing, computing and moving capabilities. They can observe (sense) the positions of other robots in the space and based on these observations, they perform some local computations that can drive them to other locations.

In the context of this paper, the robots are *anonymous*, in the sense that they can not be distinguished using their appearance and they do not have any kind of identifiers that can be used during the computation. In addition, there is no direct mean of communication between them. Hence, the only way for robots to acquire information is by observing their positions. Robots have *unlimited visibility*, *i.e.* they are able to sense the entire set of robots. We assume that robots are *non-oblivious*, *i.e.* they can remember observations, computations and motions performed in previous steps. Each robot is endowed with a local coordinate system and a local unit measure which may be different from those of other robots. This local coordinate system is assumed to be fixed during a run unless it is explicitly modified by the corresponding robot as a result of a computation. We say in this case that robots *remember* their own coordinate systems. This is a common assumption when studying non-oblivious robot networks [9,6].

A *protocol* is a collection of $n$ *programs*, one operating on each robot. The program of a robot consists in executing *Look-Compute-Move cycles* infinitely many times. That is, the robot first observes its environment (Look phase). An observation returns a snapshot of the positions of all robots within the visibility range. In our case, this observation returns a snapshot of the positions of *all* robots. The observed positions are *relative* to the observing robot, that is, they use the coordinate system of the observing robot. Based on its observation, a robot then decides — according to its program — to move or to stay idle (Compute phase). When a robot decides a move, it moves to its destination during the Move phase.

The local state of a robot is defined by the content of its memory and its position. A configuration of the system is the union of the local states of all the robots in the system. An *execution* $e = (c_0, \ldots, c_t, \ldots)$ of the system is an infinite sequence of configurations, where $c_0$ is the initial configuration of the system, and every transition $c_i \to c_{i+1}$ is associated to the execution of a non empty subset of *actions*. The granularity (or atomicity) of those actions is model-dependent and is defined in the sequel of this section.

A *scheduler* is a predicate on computations, that is, a scheduler defines a set of *admissible* computations, such that every computation in this set satisfies the scheduler predicate. A *scheduler* can be seen as an entity that is external to the system and selects robots for execution. As more power is given to the scheduler for robot scheduling, more different executions are possible and more difficult it becomes to design robot algorithms. In the remainder of the paper, we consider that the scheduler is fair and *fully asynchronous*, that is, in any infinite execution, every robot is activated infinitely often, but there is no bound on the ratio between the most activated robot and the least activated one. In each cycle, the scheduler determines the distance to which each robot can move in this cycle, that is, it can stop a robot before it reaches its computed destination. However, a robot $r_i$ is guaranteed to be able to move a distance of at least $\delta_i$ towards its destination before it can be stopped by the scheduler.

We now review the main differences between the ATOM [9] and CORDA [7] models. In the ATOM model, whenever a robot is activated by the scheduler, it performs a *full* computation cycle. Thus, the execution of the system can be viewed as an infinite sequence of rounds. In a round one or more robots are activated by the scheduler and perform a computation cycle. The *fully-synchronous ATOM* model refers to the fact that the scheduler activates all robots in each round, while the regular *ATOM* model enables the scheduler to activate only a subset of the robots. In the CORDA model, robots may be interrupted by the scheduler after performing only a portion of a computation cycle. In particular, phases (Look, Compute, Move) of different robots may be interleaved. For example, a robot $a$ may perform a Look phase, then a robot $b$ performs a Look-Compute-Move complete cycle, then $a$ computes and moves based on its previous observation (that does not correspond to the current configuration anymore). As a result, the set of executions that are possible in the CORDA model are a strict superset of those that are possible in the ATOM model. So, an impossibility result that holds in the ATOM model also holds in the CORDA model, while an algorithm that performs in the CORDA model is also correct in the ATOM model. Note that the converse is not necessarily true.

**The RoboCast Problem.** The RoboCast communication abstraction provides a set of robots located at arbitrary positions in a two-dimensional space the possibility to broadcast their local information to each other. The RoboCast abstraction offers robots two communication primitives: *RoboCast(M)* sends *Message* M to all other robots, and *Deliver(M)* delivers *Message* M to the local robot. The message may consists in the local coordinate system, the robot chirality, the unit of measure, or any binary coded information.

Consider a run at which each robot $r_i$ in the system invokes $RoboCast(m_i)$ at some time $t_i$ for some message $m_i$. Let $t$ be equal to $max\{t_1, \ldots, t_n\}$. Any protocol solving the RoboCast Problem has to satisfy the following two properties:

*Validity*: For each message $m_i$, there exists a time $t'_i > t$ after which every robot in the system has performed $Deliver(m_i)$.

*Termination*: There exists a time $t_T \geq max\{t'_1, \ldots, t'_n\}$ after which no robot performs a movement that causally depends on the invocations of $RoboCast(m_i)$.

## 3   Local Coordinate System RoboCast

In this section we present algorithms for robocasting the local coordinate system. For ease of presentation we first propose an algorithm for two-robots then the general version for systems with $n$ robots.

The local coordinate system is defined by two axes (abscissa and ordinate), their positive directions and the unity of measure. In order to robocast this information we use a modular approach. That is, robots invoke first the robocast primitive (*LineRbcast1* hereafter) to broadcast a line representing their abscissa. Then, using a parametrized module (*LineRbcast2*), they robocast three successive lines encoding respectively their ordinate, unit of measure and the positive direction of axes. This invocation chain is motivated by the dependence between the transmitted lines. When a node broadcasts a line, without any additional knowledge, two different points have to be sent in order to uniquely identify the line at the destination. However, in the case of a coordinate system, only for the first transmitted axis nodes need to identify the two points. The transmission of the subsequent axes needs the knowledge of a unique additional point.

## 3.1    Line RoboCast

In robot networks the broadcast of axes is not a new issue. Starting with their seminal paper [9], Suzuki and Yamashita presented an algorithm for broadcasting the axes via motion that works in the ATOM model. Their algorithm heavily relies on the atomicity of cycles and the observation focus on the different positions of the other robots during their Move phase.

This type of observation is totally useless in asynchronous CORDA model. In this model, when a robot $r$ moves towards its destination, another robot $r'$ can be activated $k > 1$ times with $k$ arbitrarily large, and thus observe $r$ in $k$ different positions without having any clue on the number of complete cycles executed by $r$. In other words, the number of different positions observed for a given robot is not an indicator on the number of complete executed cycles since in CORDA cycles are completely uncorrelated.

Our solution uses a novel strategy. That is, the focus moves from observing robots in different positions to observing their change of direction: each robot changes its direction of movement when a particular stage of the algorithm is completed; this change allows the other robots to infer information about the observed robot.

**Line RoboCast Detailed Description.** Let $r_0$ and $r_1$ be the two robots in the system. In the sequel, when we refer to one of these robots without specifying which, we denote it by $r_i$ and its peer by $r_{1-i}$. In this case, the operations on the indices of robots are performed modulo 2. For ease of presentation we assume that initially each robot $r_i$ translates and rotates its local coordinate system such that its x-axis and origin coincide with the line to be broacast and its current location respectively. We assume also that each robot is initially located in the origin of its local coordinate system.

At the end of the execution, each robot must have broadcast its own line and have received the line of its peer. A robot "receives" the line broadcast by its peer when it knows at least two distinct positions of this line. Thus, to send its line, each robot must move along it (following a scheme that will be specified later) until it is sure that it has been observed by the other robot.

The algorithm idea is simple: each robot broadcasts its line by moving along it in a certain direction (considered to be positive). Simultaneously, it observes the different positions occupied by its peer $r_{1-i}$. Once $r_i$ has observed $r_{1-i}$ in two distinct positions, it informs it that it has received its line by changing its direction of movement, that is, by moving along its line in the reverse direction (the negative direction if the first movement have been performed in the positive direction of the line). This change of direction is an acknowledgement for the reception of the peer line. A robot finishes the algorithm once it changed its direction and observed that the other robot also changed its direction. This means that both robots have sent their line and received the other's line.

The algorithm is described in detail as Algorithm 1. Its proof can be found in [3]. Each robot performs four stages referred in Algorithm 1 as states:

- state $S_1$: This is the initial state of the algorithm. At this state, the robot $r_i$ stores the position of its peer in the variable $pos_1$ and heads towards the position $(1.0)$ of its local coordinate system. That is, it moves along its line in the positive direction. Note that $r_i$ stays only one cycle in this state and then goes to state $S_2$.
- state $S_2$: A this point, $r_i$ knows only one point of its peer line (recorded in $pos_1$). To be able to compute the whole peer line, $r_i$ must observe $r_{1-i}$ in another (distinct) position of this line. Hence, each time it is activated, $r_i$ checks if $r_{1-i}$ is still located in $pos_1$ or if it has already changed its position. In the first case (line 2.$a$ of the code), it makes no movement by selecting its current position as its destination. Otherwise (line 2.$b$), it saves the new position of $r_{1-i}$ in $pos_2$ and delivers the line formed by $pos_1$ and $pos_2$. Then, it initiates a change of direction by moving towards the point $(-1.0)$ of its local coordinate system, and moves to state $S_3$.
- state $S_3$: at this point $r_i$ knows the line of its peer locally derived from $pos_1$ and $pos_2$. Before finishing the algorithm, $r_i$ must be sure that also $r_{1-i}$ knows its line. Therefore, it observes $r_{1-i}$ until it detects a change of direction (the condition of line 3.$a$). If this is not the case and if $r_i$ is still in the positive part of its x-axis, then it goes to the position $(-1, 0)$ of its local coordinate system (line 3.$b$). Otherwise (if $r_i$ is already in the negative part of its x-axis), it performs a null movement (line 3.$c$). When $r_i$ is in state $S_3$ one is sure, as we shall show later, that $r_{1-i}$ knows at least one position of $l_i$, say $p$. Recall that $l_i$ corresponds to the x-axis of $r_i$. It turns out that $p$ is located in the positive part of this axis. In moving towards the negative part of its x-axis, $r_i$ is sure that it will eventually be observed by $r_{1-i}$ in a position distinct from $p$ which allows $r_{1-i}$ to compute $l_i$.
- state $S_4$: At this stage, both $r_i$ and $r_{1-i}$ received the line sent by each others. That is, $r_i$ has already changed its own direction of movement, and observed that $r_{1-i}$ also changed its direction. But nothing guarantees that at this step $r_{1-i}$ knows that $r_i$ changed its direction of movement. If $r_i$ stops now, $r_{1-i}$ may remain stuck forever (in state $S_3$). To announce the end of the algorithm to its peer, $r_i$ heads towards a position located outside $l_i$, That is, it will move on a line $nextl_i$ (distinct from $l_i$) which is given as parameter to the algorithm. During the move from $l_i$ to $nextl_i$, $r_i$ should avoid points outside these lines. To this end, $r_i$ must first pass through $myIntersect$ - which is the intersection of $l_i$ and $nextl_i$ - before moving to a point located in $nextl_i$ but not on $l_i$ (refer to lines 3.$a$.2, 3.$a$.3 and 4.$a$ of the code).

Note that the robocast of a line is usually followed by the robocast of other information (e.g. other lines that encode the local coordinate system). To helps this process the end of the robocast of $l_i$ should mark the beginning of the next line, $nextl_i$, robocast. Therefore, once $r_i$ reaches $myIntersect$, $r_i$ rotates its local coordinate system such that its x-axis matches now with $nextl_i$, and then it moves toward the point of $(1,0)$ of its (new) local coordinate system. When $r_{1-i}$ observes $r_i$ in a position that is not on $l_i$, it learns that $r_i$ knows that $r_{1-i}$ learned $l_{1-i}$, and so it can go to state $S_4$ (lines 3.$a$.$*$) and finish the algorithm.

---

**Algorithm 1.** Line RoboCast **LineRbcast1** for two robots: Algorithm for robot $r_i$

---

**Variables:**
*state*: initially $S_1$
$pos_1, pos_2$: initially $\bot$
*destination, myIntersect*: initially $\bot$

**Actions:**
1. **State** $[S_1]$: %*Robot $r_i$ starts the algorithm*%

        a. $pos_1 \leftarrow observe(1-i)$
        b. $destination \leftarrow (1,0)_i$
        c. $state \leftarrow S_2$
        d. Move to destination

2. **State** $[S_2]$: %*$r_i$ knows one position of $l_{1-i}$*%

        a. **if** $(pos_1 = observe(1-i))$ **then** $destination \leftarrow observe(i)$
        b. **else**
            1. $pos_2 \leftarrow observe(1-i)$
            2. $l_{1-i} \leftarrow line(pos_1, pos_2)$
            3. Deliver $(l_{1-i})$
            4. $destination \leftarrow (-1,0)_i$
            5. $state \leftarrow S_3$ **endif**
        c. Move to destination

3. **State** $[S_3]$: %*$r_i$ knows the line robocast by robot $r_{1-i}$*%

        a. **if** $(pos_2$ is not inside the line segment $[pos_1, observe(1-i)])$ **then**
            1. $state \leftarrow S_4$
            2. $myIntersect \leftarrow intersection(l_i, nextl_i)$
            3. $destination \leftarrow myIntersect$
        b. **else if** $(observe(i) \geq (0,0)_i)$ **then** $destination \leftarrow (0,-1)_i$
        c. **else** $destination \leftarrow observe(i)$ **endif endif**
        d. Move to destination

4. **State** $[S_4]$: %*$r_i$ knows that robot $r_{1-i}$ knows its line $l_i$*%

        a. **if** $(observe(i) \neq myIntersect)$ **then** $destination \leftarrow myIntersect$
        b. **else**
            1. $r_i$ rotates its coordinate system such that its x-axis and the origin match with $nextl_i$ and $myIntersect$ respectively.
            2. $destination \leftarrow (1,0)_i$; return **endif**
        c. Move to destination

---

## 3.2   Line RoboCast: A Composable Version

Line RoboCast primitive is usually used as a building block for achieving more complex tasks. For example, the RoboCast of the local coordinate system requires the transmission of four successive lines representing respectively the abscissa, the ordinate, the value of the unit measure and a forth line to determine the positive direction of axes. In stigmergic communication a robot has to transmit at least a line for each binary information it wants to send. In all these examples, the transmitted lines are dependent one of each other and therefore their successive transmission can be accelerated by directly exploiting this dependence. Indeed, the knowledge of a unique point (instead of two) is sufficient for the receiver to infer the sent line. In the following we propose modifications of the Line RoboCast primitive in order to exploit contextual information that are encoded in a set of predicates that will be detailed in the sequel.

In the case of the local coordinate system, the additional information the transmission can exploit is the fact that the abscissa is perpendicular to the ordinate. Once the abscissa is transmitted, it suffices for a robot to simply send a single position of its ordinate, say $pos1$. The other robots can then calculate the ordinate by finding the line that passes through $pos1$ and which is perpendicular to the previously received abscissa. In the modified version of the Line RoboCast algorithm the predicate *isPerpendicular* encodes this condition.

For the case of stigmergy, a robot transmits a binary information by robo-casting a line whose angle to the abscissa encodes this information. The lines transmitted successively by a single robot are not perpendicular to each others. However, all these lines pass through the origin of the coordinate system of the sending robot. In this case, it suffices to transmit only one position located on this line as long as it is distinct from the origin. We say in this case that the line satisfies the predicate *passThrOrigin*.

A second change we propose relates to the asynchrony of the algorithm. In fact, even if robots execute in unison, they are not guaranteed to finish the execution of *LineRbcast*1 at the same time (by reaching $S_4$). A robot $r_i$ can begin transmitting its $k$-th line $l_i$ when its peer $r_{1-i}$ is still located in its $(k-1)$-th line $ancientl_{1-i}$ that $r_i$ has already received. $r_i$ should ignore the positions transmitted by $r_{1-i}$ until it leaves $ancientl_{1-i}$ for a new line. It follows that to make the module composable, the old line that the peer has already received from its peer should be supplied as an argument ($ancientl_{1-i}$) to the function. Thus, it will not consider the positions occupied by $r_{1-i}$ until the latter leaves $ancientl_{1-i}$.

In the following, we present the code of the new Line RoboCast function that we denote by *LineRbcast*2. Its description and its formal proof are omitted since they follow the same lines as those of *LineRbcast*1.

### 3.3   RoboCast of the Local Coordinate System

To robocast their two axes (abscissa and ordinate), robots call LineRbcast1 to robocast the abscissa, then LineRbcast2 to robocast the ordinate. The parameter $\neq myOrdinate$ of *LineRbcast*2 stands for the next line to be robocast and it can be set to any line different from *myOrdinate*. The next line to robocast (*unitLine*) is a line whose angle with the x-axis encodes the unit of measure. This angle will be determined during the execution *LineRbcast*2.

1. $peerAbscissa \leftarrow LineRbcast1(myAbscissa, myOrdinate)$
2. $peerOrdinate \leftarrow$
   $LineRbcast2(myOrdinate, \neq myOrdinate, peerAbscissa, isPerpendicular)$

After executing the above code, each robot knows the two axes of its peer coordinate system but not their positive directions neither their unit of measure. To robocast the unit of measure we use a technique similar to that used by [9]. The idea is simple: each robot measures the distance $d_i$ between its origin and the peer's origin in terms of its local coordinate system. To announce the value of $d_i$ to its peer, each robot robocast via LineRbcast2 a line, *unitLine*, which passes through its origin and whose angle with its abscissa is equal to $f(d_i)$ where for $x > 0, f(x) = (1/2x) \times 90°$ is a monotonically increasing function with range $(0°, 90°)$. The receiving robot $r_{1-i}$ can then infer $d_i$ from $f(d_i)$ and compute the unit measure of $r_i$ which is equal to $d_{1-i}/d_i$. The choice of $(0°, 90°)$ as a range for $f(x)$ (instead of $(0°, 360°)$) is motivated by the fact that the positive

---

**Algorithm 2.** Line RoboCast **LineRbcast2 for two robots: Algorithm for robot** $r_i$

---

**Inputs:**
$l_i$ : the line to robocast
$nextl_i$: the next line to robocast after $l_i$
$precedentl_{1-i}$: the line robocast precedently by $r_{1-i}$
$predicate$: a predicate on the output $l_{1-i}$, for example $isPerpendicular$ and $passThrOrigin$.

**Outputs:**
$l_{1-i}$ : the line robocast by $r_{1-i}$

**Variables:**
$state$: initially $S_1$
$pos_1$: initially $\perp$
$destination, myIntersect, peerIntersect$: initially $\perp$

**Actions:**
1. **State** $[S_2]$: *%$r_i$ starts robocasting its line $l_i$ %*

       a. **if** $(observe(1 - i) \in precedentl_{1-i})$ **then** $destination \leftarrow observe(i)$
       b. **else**
           1. $pos3 \leftarrow observe(1 - i)$
           2. $l_{1-i} \leftarrow$ the line that passes through $pos3$ and satisfies $predicate$.
           3. Deliver $(l_{1-i})$
           4. $peerIntersect \leftarrow$ intersection between $l_{1-i}$ and $precedentl_{1-i}$
           5. $destination \leftarrow (0, -1)_i$
           6. $state \leftarrow S_3$ **endif**
       c. Move to destination

2. **State** $[S_3]$: *%$r_i$ knows the line robocast by robot $r_{1-i}$ %*

       a. **if** $(pos3$ is not inside the line segment $[peerIntersect, observe(1 - i)])$ **then**
           1. $state \leftarrow S_4$
           2. $myIntersect \leftarrow intersection(l_i, nextl_i)$
           3. $destination \leftarrow myIntersect$
       b. **else if** $(observe(i) \geq (0, 0)_i)$ **then** $destination \leftarrow (0, -1)_i$
       c. **else** $destination \leftarrow observe(i)$ **endif endif**
       d. Move to destination

3. **State** $[S_4]$: similar to state $S_4$ of the $lineRbcast1$ function.

---

directions of the two axes are not yet known to the robots. It is thus impossible to distinguish between an angle $\alpha$ with $\alpha \in (0°, 90°)$ and the angles $\Pi - \alpha$, $-\alpha$, and $\Pi + \alpha$. To overcome the ambiguity and to make $f(x)$ injective, we restrict the range to $(0°, 90°)$. In contrast, Suzuki and Yamashita [9] use a function $f'(x)$ slightly different from ours: $(1/2x) \times 360°$. That is, its range is equal to $(0°, 360°)$. This is because in ATOM, robots can robocast at the same time the two axis and their positive directions, for example by restricting the movement of robots to only the positive part of their axes. Since the positive directions of the two axes are known, $unitLine$ can be an oriented line whose angle $f'(x)$ can take any value in $(0°, 360°)$ without any possible ambiguity.

***Positive directions of axes.*** Once the two axes are known, determining their positive directions amounts to selecting the upper right quarter of the coordinate system that is positive for both $x$ and $y$. Since the line used to robocast the unit of distance passes through two quarters (the upper right and the lower left), it remains to choose among these two travelled quarters which one corresponds to the upper right one. To do this, each robot robocast just after the line encoding the unit distance another line which is perpendicular to it such that their intersection lays inside the upper right quarter.

***Generalization to $n$ robots.*** The generalization of the solution to the case of $n > 2$ robots has to use an additional mechanism to allow robots to "recognize" other robots and distinguish them from each others despite anonymity.

Let us consider the case of three robots $r_1, r_2, r_3$. When $r_1$ looks the second time, $r_2$ and $r_3$ could have moved (or be moving), each according to its local coordinate system and unit measure. At this point, even with memory of past observations, $r_1$ may be not able to distinguish between $r_2$ and $r_3$ in their new positions given the fact that robots are anonymous. Moreover, $r_2$ and $r_3$ could even switch places and appear not to have moved. Hence, the implementation of the primitive *observe*($i$) is not trivial. For this, we use the collision avoidance techniques presented in the next section to instruct each robot to move only in the vicinity of its initial position. This way, other robots are able to recognize it by using its past positions. The technical details of this mechanism are given at the end of the next section.

Apart from this, the generalization of the protocol with $n$ robots is trivial. We present its detailed description in [3].

### 3.4   Motion Complexity Analysis

Now we show that the total number of robot moves in the coordinate system RoboCast is upper bounded. For the sake of presentation, we assume for now that the scheduler does not interrupt robots execution before they reach their planned destination. Each robot is initially located at the origin of its local coordinate system. To robocast each axis, a robot must visit two distinct positions: one located in the positive part of this axis and the other one located in its negative part. For example, to robocast its $x$-axis, a robot has first to move from its origin to the position $(1.0)_i$, then from $(1.0)_i$ to the $(-1, 0)_i$. Then, before initiating a robocast for the other axis, the robot must first return back to its origin. Hence, at most 3 movements are needed to robocast each axis. This implies that to robocast the whole local coordinate system, at most 12 movements have to be performed by a particular robot.

In the general CORDA model, the scheduler is allowed to stop robots before they reach their destination, as long as a minimal distance of $\delta_i$ has been traversed. In this case, the number of necessary movements is equal to at most $8 * (1 + 1/\delta_i)$. This worst case is obtained when a robot is *not* stopped by the scheduler when moving from its origin towards another position (thus letting it go the farthest possible), but stopped whenever possible when returning back from this (far) position to the origin.

This contrasts with [9] and [6] where the number of positions visited by each robot to robocast a line is unbounded (but finite). This is due to the fact that in both approaches, robots are required to make a non null movement whenever activated until they know that their line has been received. Managing an arbitrary large number of movements in a restricted space to prevent collisions yields severe requirements in [6]: either robots are allowed to perform infinitely small movements (and such movements can be seen by other robots with infinite precision), or the scheduler is restricted in its choices for activating robots (no robot can be activated more than $k$ times, for a given $k$, between any two activations of another robot) and yields to a setting that is not fully asynchronous. Our solution does not require any such hypothesis.

# 4   Collision-Free RoboCast

In this section we enhance the algorithms proposed in Section 3 with the collision-free feature. In this section we propose novel techniques for collision avoidance that cope with the system asynchrony.

Our solution is based on the same principle of locality as the Voronoi Diagram based schemes. However, acceptable moves for a robot use a different geometric area. This area is defined for each robot $r_i$ as a local *zone of movement* and is denoted by $ZoM_i$. We require that each robot $r_i$ moves only inside $ZoM_i$. The intersection of different $ZoM_i$ must remain empty at all times to ensure collision avoidance. We now present three possible definitions for the zone of movement: $ZoM_i^1$, $ZoM_i^2$ and $ZoM_i^3$. All three ensure collision avoidance in CORDA, but only the third one can be computed in a model where robots do not know the initial position of their peers.

Let $P(t) = \{p_1(t), p_2(t) \ldots, p_n(t)\}$ be the configuration of the network at time $t$, such that $p_i(t)$ denotes the position of robot $r_i$ at time $t$ expressed in a global coordinate system. This global coordinate system is unknown to individual robots and is only used to ease the presentation and the proofs. Note that $P(t_0)$ describes the initial configuration of the network.

**Definition 1.** *(**Voronoi Diagram**) [2] The Voronoi diagram of a set of points $P = \{p_1, p_2, \ldots, p_n\}$ is a subdivision of the plane into $n$ cells, one for each point in $P$. The cells have the property that a point $q$ belongs to the Voronoi cell of point $p_i$ iff for any other point $p_j \in P$, $dist(q, p_i) < dist(q, p_j)$ where $dist(p, q)$ is the Euclidean distance between $p$ and $q$. In particular, the strict inequality means that points located on the boundary of the Voronoi diagram do not belong to any Voronoi cell.*

**Definition 2.** *($ZoM_i^1$) Let $DV(t_0)$ be the Voronoi diagram of the initial configuration $P(t_0)$. For each robot $r_i$, the zone of movement of $r_i$ at time $t$, $ZoM_i^1(t)$, is the Voronoi cell of point $p_i(t_0)$ in $DV(t_0)$.*

**Definition 3.** *($ZoM_i^2$) For each robot $r_i$, define the distance $d_i = min\{dist(p_i(t_0), p_j(t_0))$ with $r_j \neq r_i\}$. The zone of movement of $r_i$ at time $t$, $ZoM_i^2(t)$, is the circle centered in $p_i(0)$ and whose diameter is equal to $d_i/2$. A point $q$ belongs to $ZoM_i^2(t)$ iff $dist(q, p_i(t_0)) < d_i/2$.*

**Definition 4.** *($ZoM_i^3$) For each robot $r_i$, define the distance $d_i(t) = min\{dist(p_i(t_0), p_j(t))$ with $r_j \neq r_i\}$ at time $t$. The zone of $r_i$ at time $t$, $ZoM_i^3(t)$, is the circle centered in $p_i(t_0)$ and whose diameter is equal to $d_i(t)/3$. A point $q$ belongs to $ZoM_i^3(t)$ iff $dist(q, p_i(t_0)) < d_i(t)/3$.*

Note that $ZoM^1$ and $ZoM^2$ are defined using information about the initial configuration $P(t_0)$, and thus cannot be used with the hypotheses of Algorithm 2. In contrast, robot $r_i$ only needs to know its *own* initial position and the *current* positions of other robots to compute $ZoM_i^3$. As there is no need for $r_i$ to know the *initial* positions of other robots, $ZoM_i^3$ can be used with Algorithm 2. It

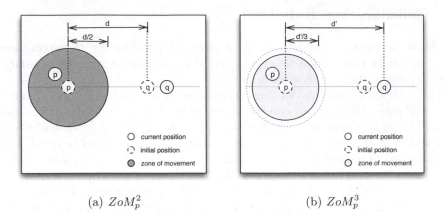

(a) $ZoM_p^2$                                 (b) $ZoM_p^3$

**Fig. 1.** Example zones of movement: The network is formed of two robots: $p$ and $q$. $d$ is the distance between the initial positions of $p$ and $q$ (dashed circles), $d'$ is the distance between the initial position of $p$ and the current position of $q$. The diameter of $ZoM_p^2$ (blue) is $d/2$ and that of $ZoM_p^3$ (yellow) is $d'/3$.

remains to prove that $ZoM_i^3$ guarantees collision avoidance. We first prove that $ZoM_i^1$ does, which is almost trivial because its definition does not depend on time. Then, it suffices to prove that $ZoM_i^3 \subseteq ZoM_i^2 \subseteq ZoM_i^1$. Besides helping us in the proof, $ZoM_i^2$ can be interesting in its own as a cheap collision avoidance scheme in the ATOM model, as computing a cycle of radius half the distance to the nearest neighbor is much easier that computing a full blown Voronoi diagram.

**Lemma 4.1.** *If $\forall t$, for each robot $r_i$, the destination point computed by $r_i$ at $t$ remains inside $ZoM_i^1(t)$, then collisions are avoided.*

*Proof.* By definition of Voronoi diagram, different Voronoi cells do not overlap. Moreover, for a given $i$, $ZoM_i^1$ is static and does not change over time. Hence, $\forall i, j \in \Pi, \forall t, t', ZoM_i^1(t) \cap ZoM_j^1(t') = \emptyset$.

Clearly, $ZoM_i^2 \subseteq ZoM_i^1$ which means that $ZoM_i^2$ ensures also collision avoidance.

**Lemma 4.2.** *If $\forall t$, for each robot $r_i$, the destination point computed by $r_i$ at $t$ always remains inside $ZoM_i^2(t)$, then collisions are avoided.*

The proof of the above lemma follows directly from the fact that $\forall t ZoM_i^2(t) \subseteq ZoM_i^1(t)$ and Lemma 4.1.

**Lemma 4.3.** $\forall t, ZoM_i^3(t) \subseteq ZoM_i^2(t)$.

***Ensuring Collision-freedom in Line Robocast Algorithms.*** To make LineRbcast1 and LineRbcast2 collision-free, it is expected that any destination computed by a robot $r_i$ at $t$ be located within its $ZoM_i^3(t)$. The computation of destinations is modified as follows: Let $dest_i(t)$ be the destination computed by a robot $r_i$ at time $t$. Based on $dest_i(t)$, $r_i$ computes a new destination $dest_i'(t)$

that ensures collision avoidance. $dest'_i(t)$ can be set to any point located in $[p_i(t_0), dest_i(t)] \cap ZoM_i^3(t)$. For example, we can take $dest'_i(t)$ to be equal to the point located in the line segment $[p_i(t_0), dest_i(t)]$ and distant from $p_i(t_0)$ by a distance of $d_i(t)/2$ with $d_i(t)$ computed as explained in Definition 4.

This modification of the destination computation method does not impact algorithms correctness since it does not depend on the exact value of computed destinations, but on the relationship between the successive positions occupied by each robot. The algorithms remain correct as long as robots keep the capability to freely change their direction of movement and to move in both the positive and the negative part of each such direction. This capability is not altered by the collision avoidance scheme since the origin of the coordinate system of each robot - corresponding to its original position - is strictly included in its zone of movement, be it defined by $ZoM^1$, $ZoM^2$ or $ZoM^3$.

***Generalisation of the Protocols to $n$ Robots.*** As explained at the end of Section 3, the generalisation of our algorithms to the case of $n$ robots has to deal with the issue of distinguishing robots from each others despite their anonymity. The solution we use is to instruct each robot to move in the close neighbourhood of its original position. Thus, other robots can recognize it by comparing its current position with past ones. For this solution to work, it is necessary that each robot always remains the closest one to all the positions it has previously occupied. Formally speaking, we define the zone of movement $ZoM^4$ in a similar way as $ZoM^3$ except that the diameter is this time equal to $d_i(t)/6$ (vs. $d_i(t)/3$). We now show that $ZoM^4$ provides the required properties. Let $r_i$ and $r_j$ be an arbitrary pair of robots and Let $d_{ij}$ denotes the distance between their initial positions. It can easily shown, using the same arguments as the proof of Lemma 4.3, that:

1. Neither of the two robots moves away from its initial position by a distance greater than $d_{ij}/4$. This implies that each robot remains always at a distance strictly smaller than $d_{ij}/2$ from all the positions it has previously held.
2. The distance between $r_i$ (resp. $r_j$) and all the positions held by $r_j$ ($r_i$) is strictly greater than $d_{ij}/2$.

Hence, $r_i$ can never be closer than $r_j$ to a position that was occupied by $r_j$, and vice versa. This implies that it is always possible to recognize a robot by associating it with the position which is closest to it in some previously observed configuration.

# 5   RoboCast Applications

## 5.1   Asynchronous Deterministic 2-Gathering

Given a set of $n$ robots with arbitrary initial locations and no agreement on a global coordinate system, $n$-Gathering requires that all robots eventually reach the same unknown beforehand location. $n$-Gathering was already solved when

$n > 2$ in both ATOM [9] and CORDA [4] oblivious models. The problem is impossible to solve for $n = 2$ even in ATOM, except if robots are endowed with persistent memory [9]. In this section we present an algorithm that uses our RoboCast primitive to solve 2-Gathering in the non-oblivious CORDA model.

A first "naive" solution is for each robot to robocast its abscissa and ordinate axes and to meet the other robot at the midpoint $m$ of their initial positions. RoboCasting the two axes is done using our Line RoboCast function described above in conjunction with the $ZoM^3$–based collision avoidance scheme.

A second possible solution is to refine Algorithm $\psi_{f-point(2)}$ of [9,10] by using our Line RoboCast function to "send" lines instead of the one used by the authors. The idea of this algorithm is that each robot which is activated for the first time translates and rotates its coordinate system such that the other robot is on its positive $y$-axis, and then it robocasts its (new) $x$-axis to the other robot using our Line Robocast function. In [9], the authors give a method that allows each robot to compute the initial position of one's peer by comparing their two robocast $x$-axes defined above. Then each robot moves toward the midpoint of their initial positions. Our Line RoboCast routine combined with the above idea achieves gathering in asynchronous systems within a bounded (*vs.* finite in [9]) number of movements of robots and using only two (*vs.* four) variables in their persistent memory.

**Theorem 5.1.** *There is an algorithm for solving deterministic gathering for two robots in non-oblivious asynchronous networks (CORDA).*

## 5.2    Asynchronous Stigmergy

Stigmergy [6] is the ability of a group of robots that communicate only through vision to exchange binary information. Stigmergy comes to encode bits in the movements of robots. Solving this problem becomes trivial when using our Robo-Cast primitive. First, robots exchange their local coordinate system as explained in Section 3. Then, each robot that has a binary packet to transmit robocasts a line to its peers whose angle with respect to its abscissa encodes the binary information. Theoretically, as the precision of visual sensors is assumed to be infinite, robots are able to observe the exact angle of this transmitted line, hence the size of exchanged messages may be infinite also. However, in a more realistic environment in which sensor accuracy and calculations have a margin of error, it is wiser to discretize the measuring space. For this, we divide the space around the robot in several sectors such that all the points located in the same sector encode the same binary information (to tolerate errors of coding). For instance, to send binary packets of 8 bits, each sector should have an angle equal to $u = 360°/2^8$. Hence, when a robot moves through a line whose angle with respect to the abscissa is equal to $\alpha$, the corresponding binary information is equal to $\lfloor \alpha/n \rfloor$. Thus, our solution works in asynchronous networks, uses a bounded number of movements and also allows robots to send binary packets and not only single bits as in [6].

# 6   Conclusion and Perspectives

We presented a new communication primitive for robot networks, that can be used in fully asynchronous CORDA networks. Our scheme has the additional properties of being motion, memory, and computation efficient. We would like to raise some open questions:

1. The solution we presented for collision avoidance in CORDA can be used for protocols where robots remain in their initial vicinity during the whole protocol execution. A collision-avoidance scheme that could be used with all classes of protocol is a challenging issue.
2. Our protocol assumes that a constant number of positions is stored by each robot. Investigating the minimal number of stored positions for solving a particular problem would lead to interesting new insights about the computing power that can be gained by adding memory to robots.

# References

1. Ando, H., Oasa, Y., Suzuki, I., Yamashita, M.: Distributed memoryless point convergence algorithm for mobile robots with limited visibility. IEEE Transactions on Robotics and Automation 15(5), 818–828 (1999)
2. Aurenhammer, F.: Voronoi diagrams a survey of a fundamental geometric data structure. ACM Computing Surveys (CSUR) 23(3), 405 (1991)
3. Bouzid, Z., Dolev, S., Potop-Butucaru, M., Tixeuil, S.: RoboCast: Asynchronous Communication in Robot Networks. Technical report (2010)
4. Cieliebak, M., Flocchini, P., Prencipe, G., Santoro, N.: Solving the robots gathering problem. In: Automata, Languages and Programming, pp. 192–192 (2003)
5. Défago, X., Konagaya, A.: Circle formation for oblivious anonymous mobile robots with no common sense of orientation. In: POMC, pp. 97–104 (2002)
6. Dieudonné, Y., Dolev, S., Petit, F., Segal, M.: Deaf, dumb, and chatting asynchronous robots. In: OPODIS, pp. 71–85 (2009)
7. Flocchini, P., Prencipe, G., Santoro, N., Widmayer, P.: Hard tasks for weak robots: The role of common knowledge in pattern formation by autonomous mobile robots. In: Aggarwal, A.K., Pandu Rangan, C. (eds.) ISAAC 1999. LNCS, vol. 1741, pp. 93–102. Springer, Heidelberg (1999)
8. Prencipe, G.: On the feasibility of gathering by autonomous mobile robots. In: Pelc, A., Raynal, M. (eds.) SIROCCO 2005. LNCS, vol. 3499, pp. 246–261. Springer, Heidelberg (2005)
9. Suzuki, I., Yamashita, M.: Distributed anonymous mobile robots: Formation of geometric patterns. SIAM J. of Computing 28(4), 1347–1363 (1999)
10. Suzuki, I., Yamashita, M.: Erratum: Distributed anonymous mobile robots: Formation of geometric patterns. SIAM J. of Computing 36(1), 279–280 (2006)

# Biased Selection for Building Small-World Networks*

Andrés Sevilla[1], Alberto Mozo[2], M. Araceli Lorenzo[2], Jose Luis López-Presa[3],
Pilar Manzano[1], and Antonio Fernández Anta[4]

[1] Dpto Informática Aplicada, U. Politécnica de Madrid, Madrid, Spain
asevilla@eui.upm.es, pmanzano@eui.upm.es
[2] Dpto Arquitectura y Tecnología de Computadores, U. Politécnica de Madrid, Madrid, Spain
amozo@eui.upm.es, alorenzo@eui.upm.es
[3] DIATEL, U. Politécnica de Madrid, Madrid, Spain
jllopez@diatel.upm.es
[4] Institute IMDEA Networks, Leganés, Spain
antonio.fernandez@imdea.org

**Abstract.** Small-world networks are currently present in many distributed applications and can be built augmenting a base network with long-range links using a probability distribution. Currently available distributed algorithms to select these long-range neighbors are designed ad hoc for specific probability distributions. In this paper we propose a new algorithm called Biased Selection (BS) that, using a uniform sampling service (that could be implemented with, for instance, a gossip-based protocol), allows to select long-range neighbors with *any* arbitrary distribution in a distributed way. This algorithm is of iterative nature and has a parameter $r$ that gives its number of iterations. We prove that the obtained sampling distribution converges to the desired distribution as $r$ grows. Additionally, we obtain analytical bounds on the maximum relative error for a given value of this parameter $r$. Although the BS algorithm is proposed in this paper as a tool to sample nodes in a network, it can be used in any context in which sampling with an arbitrary distribution is required, and only uniform sampling is available.

The BS algorithm has been used to choose long-range neighbors in complete and incomplete tori, in order to build Kleinberg's small-world networks. We observe that using a very small number of iterations (1) BS has similar error as a simulation of the Kleinberg's harmonic distribution and (2) the average number of hops with greedy routing is no larger with BS than in a Kleinberg network. Furthermore, we have observed that before converging to the performance of a Kleinberg network, the average number of hops with BS is significantly smaller (up to 14% smaller in a $1000 \times 1000$ network).

## 1 Introduction

Overlay networks are currently present in many distributed global applications and services. Overlay networks based on a small-world topology are an efficient and flexible alternative to structured overlays. Small-world networks can be built augmenting a base

---

* This research was supported in part by Comunidad de Madrid grant S2009TIC-1692 and Spanish MICINN grant TIN2008–06735-C02-01.

C. Lu, T. Masuzawa, and M. Mosbah (Eds.): OPODIS 2010, LNCS 6490, pp. 32–47, 2010.

network with long-range links, where the long-range neighbors are chosen using a probability distribution [10,7]. In order to obtain small-world networks by augmentation, a few distributed protocols have been proposed [5]. Among them, the simplest are epidemic protocols based on gossiping. However, these protocols only implement specific probability distributions to select long-range neighbors, i.e., each protocol is designed in an ad-hoc way for a given distribution. Gossip-based protocols have been designed for the uniform distribution and an approximation of Kleinberg's harmonic distribution [9,3,2]. In this paper we propose a local algorithm that, using a uniform sampling service (that could be implemented with, for instance, a gossip-based protocol), allows to select long-range neighbors with *any* arbitrary distribution. This algorithm is of iterative nature and we have found experimentally that in a small number of iterations converges to the desired distribution.

## 1.1 Related Work

Small world networks have been introduced in an attempt to explain the properties of social networks, and in particular the surprisingly small diameter and short routing in these networks [11]. One line of work on small-world networks has to do with synthesizing networks that have these properties. This can be done by starting from a base network (representing acquaintances geographically close) and adding long-range links (representing distant acquaintances). This process is called *augmentation* of the base network. Watts and Strogatz [14] considered a network augmentation in which the long-range neighbors are chosen uniformly at random. However, Kleinberg [10] has shown that a polylogarithmic greedy routing is achieved only if the long-range neighbors are chosen with specific distributions. A network built this way is commonly called a Kleinberg network. This seminal result has led to a large amount of subsequent work on construction of small-world networks [1,5,6,7]. Most of the algorithms that have been proposed are centralized. As far as we know, the first distributed algorithm to build a small world network is due to Duchon et al. [5]. Bonnet et al. [3] have proposed two gossip-based protocols to select long-range links, which are modified versions of Cyclon [13]. One selects the long-range neighbors with uniform probability, while the other selects them with an approximation to the Kleinberg distribution. This latter protocol has been improved in [2]. There have been other gossip-based protocols that sample the network nodes with uniform probability [4,9,13].

## 1.2 Contributions

In this paper we provide an algorithm, called *Biased Selection* (BS), that implements a sampling service in a set $S$ with any probability distribution. The probability distribution is proportional to probabilities (represented as weights) assigned to the element in $S$. The algorithm BS is very simple and completely local. It only needs access to a uniform sampling service (that could be implemented with a gossip-based protocol) and to the weight assigned to each element returned by this service. This algorithm has a parameter $r$ that determines the number of times the uniform sampling service is used (number of rounds) before returning a sample. (In fact, the times the uniform sampling service is used is exactly $r + 1$.) We prove that the obtained sampling distribution converges to the desired distribution as $r$ grows. Additionally, we obtain analytical bounds on the maximum relative error for a given value of this parameter $r$. Although the BS algorithm is proposed in this paper as a tool to sample nodes in a network, it can be used in any context in which sampling with an arbitrary distribution is required, and only uniform sampling is available.

To evaluate by simulation the performance of the algorithm, we use it to choose long range neighbors in a torus, in order to build small-world networks similar to Kleinberg's [10]. In this network, every node $i$ in the torus chooses another node $j$ as its long-range neighbor with a probability proportional to $1/d(i,j)^2$, where $d(i,j)$ is the Euclidean distance[1] from $i$ to $j$. For simplicity, we will call this the *Kleinberg distribution*. The obvious way to choose the long range neighbor of a node $i$ implies to know all the nodes, the distance from $i$ to each of them, and to compute the associated probabilities, which requires $\Omega(n)$ operations in a network of $n$ nodes. To do this for all nodes requires $\Omega(n^2)$ operations. On the other hand, BS does not need to know all the nodes, and only requires the distance from $i$ to the nodes returned by the uniform sampling service. If BS uses $r$ rounds, the overall number of operations needed for each node $i$ is $O(r)$, and $O(rn)$ operations for the whole network. We have observed experimentally that the value of $r$ required is much smaller than $n$.

Comparing the samples obtained simulating BS and samples from a simulation of the Kleinberg distribution, we observe that in a very small number of rounds (10 in a $100 \times 100$ torus), both simulations have matching average relative error and very similar maximum relative error. Then, we build networks by adding to each node in the torus a long range neighbor. We evaluate the performance of greedy routing in networks in which the long range neighbor is chosen with the BS algorithm (BS-network) when different number of rounds is used. We compare these results with networks that use the Kleinberg distribution (K-network) and the uniform distribution (U-network). We observe that the average number of hops of greedy routes in the BS network converges to the values obtained in the K-network as the number of rounds increases. Furthermore, with only a few rounds, the average number of hops in the BS network is sensibly smaller than in the U-network (especially for large networks). Surprisingly, we have observed that before converging to the K-network performance, the average number of hops of the BS-network is significantly smaller (up to $14\%$ smaller in a $1000 \times 1000$ network). In fact, the best performance of BS-networks is achieved with a small number of rounds. The origin of this behavior is left for future study. Finally, we have done similar experiments adding long range neighbors to incomplete tori. These are obtained by deciding whether to remove each node with a fixed probability. The experiments have been done with two probability values, namely 0.8 and 0.3, obtaining dense and sparse networks. The results observed are consistent with those obtained in the complete torus.

### 1.3  Structure of the Rest of the Paper

In Section 2 we introduce concepts and notation that will be used in the rest of the paper, along with the description of the experimental environment that will be used. In Section 3 the BS algorithm is presented, its correctness is proven, and a bound on its convergence rate is derived. Finally, in Section 4 simulation and experimental results are presented.

---

[1] Observe that Kleinberg used, instead, Manhattan distance on a grid.

## 2   Definitions and Experimental Setup

### 2.1   Definitions

Although the BS algorithm is proposed in this paper as a tool to sample nodes in a network, it will be presented in a more general form, to emphasize the fact that it could possibly be used in other contexts. Let $S$ be a set of $n$ elements such that each element $i \in S$ has an associated weight $w(i) > 0$. The problem to solve is to sample the set $S$ with a probability distribution $p$ such that the probability of choosing $i$ is proportional to $w(i)$. Let us denote $\eta = \sum_{j \in S} w(j)$. Then, the sampling probability of $i \in S$ has to be $p(i) = w(i)/\eta$. The challenges of sampling $S$ are the following:

1. We assume that the whole set is not available.
2. The weight values can only be consulted for individual previously known elements.

These restrictions prevent, for instance, from even computing the value $\eta$. However, in order to be able to solve the problem, we assume the availabilty of a sampling primitive $USel_S$ that returns an element $i$ of $S$ chosen with uniform probability. Once an element $i \in S$ is obtained using $USel_S$, its weight $w(i)$ can also be obtained.

### 2.2   Experimental Setup

In the experiments conducted in this paper, we consider a 2-dimensional torus topology. A pair of integer values $(x, y)$ is used to locate each node into the 2-dimensional space. The former node coordinates range from 0 to $m - 1$, and so the number of nodes in the network is $m^2$. In this topology, the distance between two nodes located at positions $(x_1, y_1)$ and $(x_2, y_2)$ is the Euclidean distance in the torus, computed as:

$$d_e = \sqrt{\left(\min\left(|x_1 - x_2|, m - |x_1 - x_2|\right)\right)^2 + \left(\min\left(|y_1 - y_2|, m - |y_1 - y_2|\right)\right)^2}$$

To test the BS algorithm, we design two different types of experiments. The first experiment (Section 4.1) shows the average and maximum relative error values of the BS algorithm with respect to the Kleinberg probability distribution. These values are also compared with the relative error values obtained with a real simulation of the ideal Kleinberg distribution. The second group of experiments (Sections 4.2 and 4.3) compares BS with the Kleinberg simulator with respect to the average number of hops when these algorithms are used to greedily route packets in a network. This group of experiments is executed in two different scenarios. Firstly, a complete torus with $m^2$ nodes is used. In this torus, each node has four local neighbors and one long range neighbor. Secondly, we use an incomplete torus, where nodes are eliminated using a random uniform probability. In this network, the expected number of present nodes is $m^2 q$, being $q$ the probability of node presence. Note that $q$ is equal for every node, and the presence of each node is independent of the presence of other nodes. In this case, to allow for greedy routing, each node $i$ has links to seven neighbors. These include one long range neighbor, and six local neighbors, that are the closest node in each of the six $60°$ wedge in a circle centered on node $i$ [15].

Additionally, when using the BS algorithm, the number of rounds $r$ will determine the accuracy of the values obtained. Hence, experiments are executed several times. We start with $r = 0$ rounds (uniform distribution) and we gradually increase $r$ until the BS algorithm converges to the Kleinberg distribution. Each experiment is repeated 10 times with different seeds, and we present the average of these executions.

## 3   Biased Selection

### 3.1   The Biased Selection Algorithm

We present here the algorithm that can be used to sample the set $S$ as defined in Section 2 with the desired probability distribution. The algorithm is called *Biased Selection* and presented in Figure 1. The input of the algorithm is a value $r$ that specifies the number of rounds the algorithm must execute before returning the sample. As will be shown, the larger the number of rounds $r$, the closer the output of $BSel_S(r)$ gets to the desired probability distribution.

1   **function** $BSel_S(r)$
2   $\quad x \leftarrow USel_S$
3   $\quad$**for** $i \leftarrow 1$ **to** $r$ **do**
4   $\quad\quad y \leftarrow USel_S$
5   $\quad\quad$**set** $x \leftarrow y$ *with probability* $\frac{w(y)}{w(x)+w(y)}$
6   $\quad$**end for**
7   $\quad$**return** $x$

**Fig. 1.** Biased Selection Algorithm for set $S$

### 3.2   Correctness

We first show that, as $r$ goes to infinity, the probability distribution of the output values of $BSel_S(r)$ converges to the desired probability distribution $p$. Let $x_i$ be the value stored in variable $x$ after $i$ iterations of the for loop, being $x_0$ the value assigned to $x$ in Line 2. Let us consider the infinite run of $BSel_S(\infty)$. The infinite sequence of values $x_0, x_1, \ldots$ can be seen as a Markov chain[2] $M$ on the finite state space $S$. We will first show that $M$ has a unique stationary distribution $\pi$ such that $\pi = A\pi$, where $A$ is the transition matrix of $M$. Finally, we show that $\pi$ matches the probability distribution $p$.

The transition matrix $A = [a_{ij}]$ of the Markov chain $M$ can be obtained from the algorithm of Figure 1 in the following way. For each $i, j \in S$ and $i \neq j$,

$$a_{ij} = \Pr[x_{t+1} = j | x_t = i] = \frac{1}{n}\frac{w(j)}{w(j)+w(i)} = \frac{1}{n}\frac{p(j)}{p(j)+p(i)}.$$

Additionally, $a_{ii} = 1 - \sum_{j \neq i} a_{ij}$. Observe that for all $j \neq i$, $a_{ij} < 1/n$ (recall that $n = |S|$), and hence $a_{ii} > 0$. We show now that $M$ is *ergodic*. Let $a_{ij}^{(s)}$ be the probability

<hr />

[2] Some familiarity of the reader with Markov chains is assumed.

of reaching state $j$ in $s$ steps starting from state $i$. A Markov chain is ergodic if it is finite, *irreducible* ($\forall i, j \in S, \exists s : a_{ij}^{(s)} > 0$), and *aperiodic* ($\forall i, j \in S, \gcd\{s : a_{ij}^{(s)} > 0\} = 1$) [12]. Ergodicity implies that the stationary distribution is unique.

**Lemma 1.** *The Markov chain $M$ is ergodic, and hence has a unique stationary distribution $\pi$ that satisfies $\pi = \pi A$.*

*Proof.* The fact that the Markov chain $M$ is irreducible follows directly from the fact that $a_{ij} > 0$ for all $i, j \in S$. Additionally, since $a_{ii} > 0$ for all $i \in S$, $M$ is aperiodic. Hence, it is ergodic, and has a unique stationary distribution $\pi$ that satisfies $\pi = \pi A$ [12].

Let us now prove that the stationary distribution is in fact the desired distribution $p$.

**Theorem 1.** *The output of $BSel_S(r)$ converges to the probability distribution $p$ as $r$ tends to infinity.*

*Proof.* From the above lemma, the distribution of values $x_r$ output by the algorithm converge to the stationary distribution $\pi$ of Markov chain $M$ as $r$ goes to infinity. All that has to be proven is that $p = \pi$, i.e., that $p = pA$. Consider any $i \in S$, we need to prove that $p(i) = \sum_{j \in S} p(j) a_{ji}$. Replacing,

$$
\frac{w(i)}{\eta} = \sum_{j \in S} \frac{w(j)}{\eta} a_{ji}
$$

$$
= \sum_{j \neq i} \frac{w(j)}{\eta} \frac{1}{n} \frac{w(i)}{w(j) + w(i)} + \frac{w(i)}{\eta} \left(1 - \sum_{j \neq i} a_{ij}\right)
$$

$$
= \sum_{j \neq i} \frac{w(i)}{\eta} \frac{1}{n} \frac{w(j)}{w(j) + w(i)} + \frac{w(i)}{\eta} \left(1 - \sum_{j \neq i} a_{ij}\right)
$$

$$
= \frac{w(i)}{\eta} \left(\sum_{j \neq i} a_{ij} + 1 - \sum_{j \neq i} a_{ij}\right)
$$

$$
= \frac{w(i)}{\eta}
$$

Since this holds for all $i \in S$, the proof is complete.

## 3.3  Convergence Rate

We study now the number of rounds $r$ that are needed for the distribution of the output values of $BSel_S(r)$ to be *almost* the same as the probability distribution $p$. To measure the distance between both distributions, we will use the *relative pointwise distance* as defined in [12]. This parameter measures the largest relative error between the distributions, for all possible final and initial values. Observe that $a_{ij}^{(r)}$ is the probability that

$BSel_S(r)$ outputs $j$ if the initial value of $x$ is $i$, i.e. $a_{ij}^{(r)} = \Pr[x_r = j | x_0 = i]$. Then, the *maximum relative error* is defined as

$$\Delta(r) = \max_{i,j \in S} \frac{|a_{ij}^{(r)} - p(j)|}{p(j)}.$$

In order to bound $\Delta(r)$ we first prove that the Markov chain $M$ is *time-reversible*, which holds if $a_{ij}p(i) = a_{ji}p(j)$ [12].

**Lemma 2.** *The Markov chain $M$ is time-reversible.*

*Proof.* Replacing in $a_{ij}p(i) = a_{ji}p(j)$, we get

$$a_{ij}p(i) = \frac{1}{n}\frac{w(j)}{w(j)+w(i)}\frac{w(i)}{\eta} = \frac{1}{n}\frac{w(i)}{w(j)+w(i)}\frac{w(j)}{\eta} = a_{ji}p(j)$$

Lemmas 1 and 2 are useful to bound $\Delta(r)$ because of the following result, derived from Proposition 3.1 in [12].

**Lemma 3 ([12]).** *Let $A$ be the transition matrix of an ergodic time-reversible Markov chain, $p$ its stationary distribution, and $1 = \lambda_0 > \lambda_1 \geq \lambda_2 \geq \cdots \geq \lambda_{n-1}$ its (real) eigenvalues. Then, for all $r \geq 1$ the maximum relative error satisfies*

$$\Delta(r) \leq \frac{\lambda^r}{p_{\min}},$$

*where $\lambda = \max_{k \geq 1} |\lambda_k|$ and $p_{\min} = \min_{i \in S} p(i)$.*

Clearly, $\lambda = \max(\lambda_1, |\lambda_{n-1}|)$, where $\lambda_1 < |\lambda_{n-1}|$ only if $\lambda_{n-1} < 0$. Let us define $a_{\min} = \min_{i \in S} a_{ii}$. From the Gershgorin Circle Theorem [8], we have that $\lambda_{n-1} \geq 2a_{\min} - 1$. Then, $\lambda \leq \max\{\lambda_1, 1 - 2a_{\min}\}$. To bound $\lambda_1$ we use the conductance of $M$.

**Definition 1.** *Consider the Markov chain $M$. For any set $B \subset S$, denote $C(B) = \sum_{i \in B} p(i)$ and $F(B) = \sum_{i \in B, j \notin B} a_{ij}p(i)$. The* conductance *of $M$ is defined as*

$$\Phi = \min_{\emptyset \subset B \subset S : C(B) \leq 1/2} \frac{F(B)}{C(B)}.$$

Lemma 3.3 in [12] shows that $\lambda_1 \leq 1 - \frac{\Phi^2}{2}$. Then, we can bound $\lambda_1$ as follows.

**Lemma 4.** *The eigenvalue $\lambda_1$ of the ergodic time-reversible Markov chain $M$ satisfies*

$$\lambda_1 \leq 1 - \frac{1}{2}\left(\max\{(1 - 1/n)p_{\min}, 1/(4np_{\max})\}\right)^2,$$

*where $p_{\min} = \min_{i \in S} p(i)$ and $p_{\max} = \max_{i \in S} p(i)$.*

*Proof.* Consider any set $B$ such that $\emptyset \subset B \subset S$ and $C(B) \leq 1/2$. Let us denote $\Phi(B) = \frac{F(B)}{C(B)}$. We will obtain two lower bounds for $\Phi(B)$.

First, since $C(B) \leq 1/2$, then $\Phi(B) \geq 2F(B)$. In order to bound $F(B)$, we observe that $\frac{p(i)p(j)}{p(j)+p(i)}$ decreases with the values of $p(i)$ and $p(j)$, which implies that $\frac{p(i)p(j)}{p(j)+p(i)} \geq \frac{p_{\min}^2}{2p_{\min}} = \frac{p_{\min}}{2}$. On the other hand, $|B| \cdot |S \setminus B| \geq n-1$. Then,

$$F(B) = \sum_{i \in B, j \notin B} a_{ij}p(i) = \frac{1}{n} \sum_{i \in B, j \notin B} \frac{p(i)p(j)}{p(j)+p(i)} \geq \frac{1}{n} \sum_{i \in B, j \notin B} \frac{p_{\min}}{2} \geq \frac{n-1}{n} \frac{p_{\min}}{2}.$$

This implies that $\Phi(B) \geq (1-1/n)p_{\min}$. For the second bound, observe that

$$F(B) = \frac{1}{n} \sum_{i \in B, j \notin B} \frac{p(i)p(j)}{p(j)+p(i)} \geq \frac{1}{2np_{\max}} \sum_{i \in B, j \notin B} p(i)p(j) = \frac{\left(\sum_{i \in B} p(i)\right)\left(\sum_{j \notin B} p(j)\right)}{2np_{\max}}.$$

Then, since $C(B) = \sum_{i \in B} p(i)$,

$$\Phi(B) = \frac{F(B)}{C(B)} \geq \frac{\sum_{j \notin B} p(j)}{2np_{\max}} \geq \frac{1}{4np_{\max}},$$

where the second inequality follows from $\sum_{j \notin B} p(j) = 1 - C(B)$ and $C(B) \leq 1/2$.

Since both bounds hold for any $B$, we obtain that $\Phi \geq \max\{(1-1/n)p_{\min}, 1/(4np_{\max})\}$. Combining this bound with Lemma 3.3 in [12], the claim follows.

From the above results we can bound the relative pointwise distance as follows.

**Theorem 2.** *The maximum relative error of the Markov chain $M$ satisfies*

$$\Delta(r) \leq \left(1 - \min\{\frac{1}{2}\left(\max\{(1-1/n)p_{\min}, 1/(4np_{\max})\}\right)^2, 2a_{\min}\}\right)^r / p_{\min},$$

*where $p_{\min} = \min_{i \in S} p(i)$, $p_{\max} = \max_{i \in S} p(i)$, and $a_{\min} = \min_{i \in S} a_{ii}$.*

## 4   Experimental Results

### 4.1   Accuracy of Biased Selection

To be able to evaluate the goodness of BS, we compare its relative error with the one measured in a simulation of the Kleinberg distribution. We define the relative error $e_i$ for node $i$ in a collection $C$ of $s$ samples as $e_i = \frac{|fsim_i - fkl_i|}{fkl_i}$, where $fsim_i$ is the number of instances of $i$ in collection $C$, and $fkl_i = p(i) \cdot s$ is the expected number of instances of $i$ with the ideal Kleinberg distribution. The experiments have been done in a $100 \times 100$ torus. For each experiment, a collection of around 13.5 million samples has been used in order to guarantee that every node appears on average at least 100 times. Additionally, in the BS algorithm, experiments have been performed using an increasing number of rounds, trying to reach a behavior similar to that of the Kleinberg

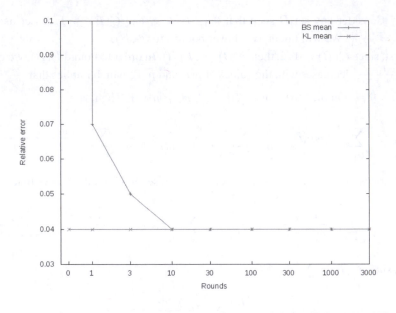

**Fig. 2.** Average relative error distribution of the Kleinberg simulator and BS ($100 \times 100$ torus)

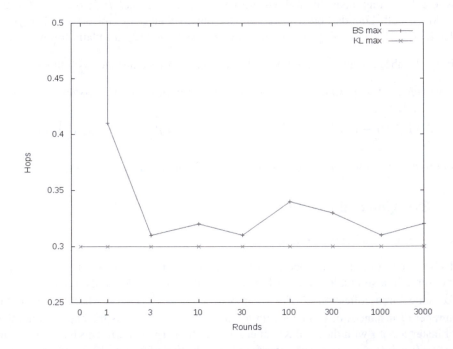

**Fig. 3.** Maximum relative error distribution of the Kleinberg simulator and BS ($100 \times 100$ torus)

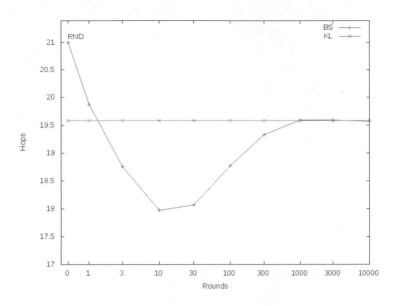

**Fig. 4.** Comparative of average routing hops ($100 \times 100$ torus)

simulator. In the 10,000-node torus used this happens for relatively small number of rounds ($r = 10$). We have not performed this experiment on larger networks (more than $100 \times 100$ nodes), due to limitations in the execution time needed to handle experiments above that size.

In Figures 2 and 3 we show, respectively, the average and maximum relative error values obtained using the BS algorithm against a Kleinberg simulator. Round numbers approximately follow an exponential sequence of $r = 0, 1, 3, 10, 30, 100, 300, 1000$, and 3000. It must be noted that for $r = 0$, our BS algorithm is equivalent to a uniform random distribution. We can observe that, for $r \geq 10$ the mean values of the relative errors in the BS algorithm match those obtained using the Kleinberg simulator.

## 4.2   Building Small-World Networks with Biased Selection

As previously commented, in this scenario we built a complete torus with $m^2$ nodes. In the experiments we perform 500,000 search operations, choosing source and destination uniformly at random for each search. We measure the average number of hops needed to reach the destination with greedy routing. Let $h_K$ be the average number of hops measured with the Kleinberg simulator, and $h_{BS}(r)$ the average number of hops measured when using BS with $r$ rounds. We consider that BS and Kleinberg have converged for $r$ rounds when, for all $r' \geq r$, it holds that $|h_{BS}(r') - h_K|/h_K < 0.05$ (they are off by less than 5%). First, we run the Kleinberg simulator, and then we run the BS algorithm, starting from $r = 0$ and gradually increasing the number of rounds until convergence is reached. We use round numbers that approximately follow an exponential sequence, $r = 0, 1, 3, 10, 30, 100, 300, 1000, 3000, 10,000$, and 30,000. We denote the smallest of these values of $r$ that satisfy convergence as $r_{conv}$. It must be noted that

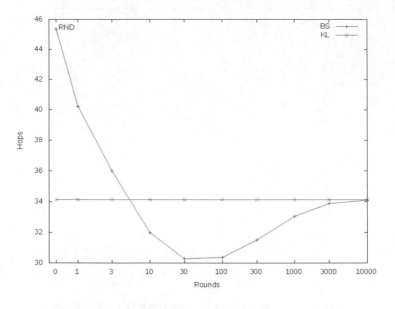

**Fig. 5.** Comparative of average routing hops ($300 \times 300$ torus)

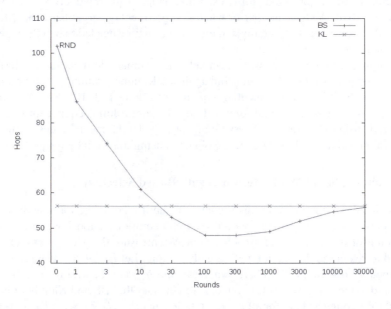

**Fig. 6.** Comparative of average routing hops ($1000 \times 1000$ torus)

using 0 rounds in BS is equivalent to using a uniform distribution to choose the long range neighbor. The experiment was run using three different torus sizes, $100 \times 100$ (Figure 4), $300 \times 300$ (Figure 5) and $1000 \times 1000$ (Figure 6).

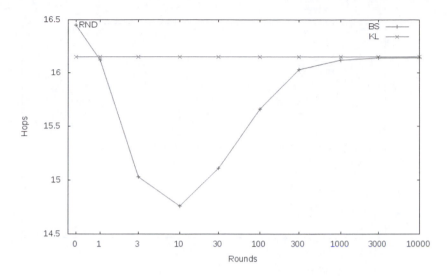

**Fig. 7.** Comparative of average routing hops ($100 \times 100$ Incomplete torus $q = 0.8$)

From these experiments, we can conclude that:

- As analytical results showed in section 3.3, the execution of the BS algorithm converges to the Kleinberg simulator results when using a sufficient number of rounds. The approximate number of rounds needed is 100 in the $100 \times 100$ torus, 1,000 in the $300 \times 300$ torus, and 10,000 in the $1000 \times 1000$ torus. Note that the number of rounds needed grows when the torus size increases.
- Using a uniform distribution to choose the long range link (equivalent to BS with 0 rounds) produces worse results than the Kleinberg and BS simulators with $r \geq 1$. The difference increases as the network size grows.
- With a relatively small number of rounds, BS outperforms the Kleinberg simulator. This singular behavior appears in all the experiment executions. In a $100 \times 100$ torus, with 10 rounds, BS results are 8% better than the Kleinberg results and 10% better than the uniform distribution. In a $300 \times 300$ torus, with 30 rounds, we get improvements of 12% and 25% when comparing to the Kleinberg simulation and the uniform distribution, respectively. Finally, in a $1000 \times 1000$ torus, with 100 rounds, improvements are of 14% and 45% when comparing to the Kleinberg simulation and the uniform distribution, respectively. It can be seen that the number of rounds required by BS to obtain the minimum average number of hops grows as the network size increases. We denote this number of rounds as $r_{min}$.

In Table 1, we present $r_{min}$ and $r_{conv}$ as a function of the size of the network.

### 4.3   Building Incomplete Small-World Networks with Biased Selection

In this scenario we built an incomplete torus using two different probability values (0.8 and 0.3) to determine node presence in $100 \times 100$ and $300 \times 300$ topologies. As

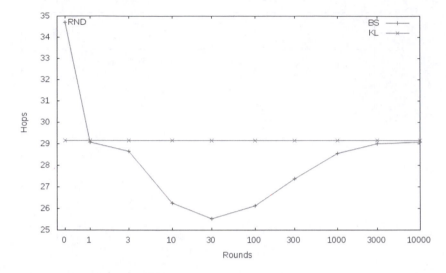

**Fig. 8.** Comparative of average routing hops ($300 \times 300$ Incomplete torus $q = 0.8$)

before, in this experiment, we perform 500,000 search operations, using alternatively the Kleinberg and the BS simulators. The number of rounds in the BS algorithm started with 0 and continued until reaching convergence with the Kleinberg simulator.

Figure 7 ($100 \times 100$ torus) and Figure 8 ($300 \times 300$ torus) compare the performance of the BS algorithm with that of the Kleinberg simulator in a topology with a probability value $q = 0.8$ of node presence in the network. Additionally, Figure 9 ($100 \times 100$ torus) and Figure 10 ($300 \times 300$ torus) compare the former simulators using a topology with a probability value $q = 0.3$ for node presence in the network. Finally, in Table 1, we show $r_{min}$ and $r_{conv}$ for each network considered.

From these experiments, we can conclude that:

– The results are similar to those obtained in the previous section when using a complete torus. The values of $r_{min}$ and $r_{conv}$ for networks with roughly the same number of nodes are similar. These results are especially interesting because an incomplete torus with $q = 0.3$ is almost a random network, so it seems that the results obtained are not associated only to a torus topology. In an incomplete network with $q = 0.8$ the obtained improvements are: (a) in the $100 \times 100$ torus, with 10 rounds, BS results are 8% better than the Kleinberg results and 11% better than the uniform distribution; and (b) in the $300 \times 300$ torus, with 30 rounds, BS results are 12% better than the Kleinberg results and 26% better than the uniform distribution. Using $q = 0.3$: (a) in the $100 \times 100$ torus, with 3 rounds, BS results are 8% better than the Kleinberg results and 4% better than the uniform distribution; and (b) in the $300 \times 300$ torus, with 10 rounds, BS results are 12% better than the Kleinberg results and 15% better than the uniform distribution.

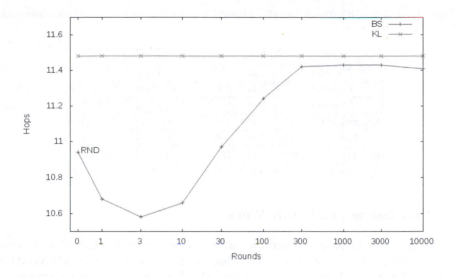

**Fig. 9.** Comparative of average routing hops ($100 \times 100$ Incomplete torus $q = 0.3$)

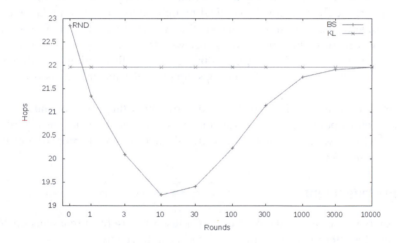

**Fig. 10.** Comparative of average routing hops ($300 \times 300$ Incomplete torus $q = 0.3$)

– In these experiments, the average number of routing hops and the number of rounds needed to converge seem to be somewhat smaller than the values obtained in the complete torus experiments. We presume that this result may be due to two factors: (a) the number of local neighbors is greater than in the previous experiment (6 neighbors versus 4); and (b) each routing hop in this network generates a larger advance than in a complete network.

**Table 1.** Number of BS rounds for minimal number of routing hops $r_{min}$ and for convergence $r_{conv}$ in the different experiments

| Topology | Torus size | Nodes (avg) | $r_{min}$ | $r_{conv}$ |
|---|---|---|---|---|
| Complete | $100 \times 100$ | $10,000$ | 10 | 100 |
| Complete | $300 \times 300$ | $90,000$ | 30 | 1,000 |
| Complete | $1000 \times 1000$ | $1,000,000$ | 300 | 10,000 |
| Incomplete ($q = 0.8$) | $100 \times 100$ | $8,000$ | 10 | 100 |
| Incomplete ($q = 0.8$) | $300 \times 300$ | $72,000$ | 30 | 1,000 |
| Incomplete ($q = 0.3$) | $100 \times 100$ | $3,000$ | 3 | 30 |
| Incomplete ($q = 0.3$) | $300 \times 300$ | $27,000$ | 10 | 300 |

## 5   Conclusions and Future Work

In this paper we proposed a simple, iterative and local algorithm (BS) that allows us to select long-range neighbors with any arbitrary distribution to build small-world networks. BS uses a uniform sampling service and only needs one parameter determining the number of rounds needed by the algorithm to converge to the desired distribution. In this work, we use the Kleinberg distribution as the target. We also proved the algorithm convergence and obtained analytical bounds on the maximum relative error for a given value of the algorithm parameter. We evaluated the algorithm by simulation in different scenarios, obtaining convergence with the Kleinberg simulator results. We also observed that, before converging to the Kleinberg distribution, the BS algorithm provided a smaller average number of hops, up to 14 % smaller in a $1000 \times 1000$ network.

Future work will provide an analytical description of this behavior and evaluate the algorithm with topologies different from the torus. We will also study the algorithm in a dynamic network scenario. Finally, we would like to compare the properties of BS with aggregation protocols.

## Acknowledgement

The authors would like to thank Juan A. Carrasco for useful discussions on Markov chains and for referring them to the Gershgorin Circle Theorem.

## References

1. Barrière, L., Fraigniaud, P., Kranakis, E., Krizanc, D.: Efficient routing in networks with long range contacts. In: Welch, J.L. (ed.) DISC 2001. LNCS, vol. 2180, pp. 270–284. Springer, Heidelberg (2001)
2. Bertier, M., Bonnet, F., Kermarrec, A.-M., Leroy, V., Peri, S., Raynal, M.: D2ht: The best of both worlds, integrating rps and dht. In: EDCC, pp. 135–144. IEEE Computer Society, Los Alamitos (2010)
3. Bonnet, F., Kermarrec, A.-M., Raynal, M.: Small-world networks: From theoretical bounds to practical systems. In: Tovar, E., Tsigas, P., Fouchal, H. (eds.) OPODIS 2007. LNCS, vol. 4878, pp. 372–385. Springer, Heidelberg (2007)

4. Bonnet, F., Tronel, F., Voulgaris, S.: Brief announcement: Performance analysis of cyclon, an inexpensive membership management for unstructured P2P overlays. In: Dolev, S. (ed.) DISC 2006. LNCS, vol. 4167, pp. 560–562. Springer, Heidelberg (2006)
5. Duchon, P., Hanusse, N., Lebhar, E., Schabanel, N.: Towards small world emergence. In: Gibbons, P.B., Vishkin, U. (eds.) SPAA, pp. 225–232. ACM, New York (2006)
6. Fraigniaud, P., Gavoille, C., Paul, C.: Eclecticism shrinks even small worlds. Distributed Computing 18(4), 279–291 (2006)
7. Fraigniaud, P., Giakkoupis, G.: On the searchability of small-world networks with arbitrary underlying structure. In: Schulman, L.J. (ed.) STOC, pp. 389–398. ACM, New York (2010)
8. Gerschgorin, S.: Über die abgrenzung der eigenwerte einer matrix. Izv. Akad. Nauk. USSR Otd. Fiz.-Mat. Nauk. 7, 749–754 (1931)
9. Jelasity, M., Voulgaris, S., Guerraoui, R., Kermarrec, A.-M., van Steen, M.: Gossip-based peer sampling. ACM Trans. Comput. Syst. 25(3) (2007)
10. Kleinberg, J.M.: Navigation in a small world. Nature 406(6798) (August 2000)
11. Milgram, S.: The small world problem. Psychology Today 2, 60–67 (1967)
12. Sinclair, A., Jerrum, M.: Approximate counting, uniform generation and rapidly mixing markov chains. Inf. Comput. 82(1), 93–133 (1989)
13. Voulgaris, S., Gavidia, D., van Steen, M.: Cyclon: Inexpensive membership management for unstructured P2P overlays. J. Network Syst. Manage. 13(2) (2005)
14. Watts, D.J., Strogatz, S.H.: Collective dynamics of 'small-world' networks. Nature 393(6684), 440–442 (1998)
15. Yao, A.C.C.: On constructing minimum spanning trees in k-dimensional space and related problems. SIAM Journal of Computing (1982)

# Application of Random Walks to Decentralized Recommender Systems*

Anne-Marie Kermarrec[1], Vincent Leroy[2],
Afshin Moin[1], and Christopher Thraves[1]

[1] INRIA Rennes - Bretagne Atlantique, Rennes, France
[2] INSA de Rennes, UEB, Rennes, France

**Abstract.** The need for efficient decentralized recommender systems
has been appreciated for some time, both for the intrinsic advantages of
decentralization and the necessity of integrating recommender systems
into P2P applications. On the other hand, the accuracy of recommender
systems is often hurt by data sparsity. In this paper, we compare different
decentralized user-based and item-based Collaborative Filtering (CF) algorithms with each other, and propose a new user-based random walk
approach customized for decentralized systems, specifically designed to
handle sparse data. We show how the application of random walks to
decentralized environments is different from the centralized version. We
examine the performance of our random walk approach in different settings by varying the sparsity, the similarity measure and the neighborhood size. In addition, we introduce the *popularizing* disadvantage of the
significance weighting term traditionally used to increase the precision of
similarity measures, and elaborate how it can affect the performance of
the random walk algorithm. The simulations on MovieLens 10,000,000
ratings dataset demonstrate that over a wide range of sparsity, our algorithm outperforms other decentralized CF schemes. Moreover, our results show decentralized user-based approaches perform better than their
item-based counterparts in P2P recommender applications.

## 1 Introduction

*Recommender systems* are crucial to the success of e-commerce websites like
Amazon, eBay or Netflix. Different theoretical [4,14,3] or empirical [15,6,7,13]
approaches have addressed recommender systems. *Collaborative Filtering (CF)*
is the most popular strategy in recommender systems. The reason behind this
popularity is that CF requires no information about the content of the items.
*Neighborhood Model* [7,18] is the most widly used model of CF due to some
of its advantages like better explainability. It is important for a recommender
system to be capable of explaining the reason behind a given recommendation.
Consequently, the users may increase the quality of future predictions by giving
feedback about received recommendations.

---

\* This work is supported by the ERC Starting Grant GOSSPLE number 204742.

C. Lu, T. Masuzawa, and M. Mosbah (Eds.): OPODIS 2010, LNCS 6490, pp. 48–63, 2010.

A neighborhood model consists of two phases: neighborhood formation and rating estimation. In the neighborhood formation phase a set of similar items (item-based approach) is formed for each item or alternatively a set of similar users (user-based approach) is formed for each user based on some similarity measure like Cosine similarity or Pearson correlation. Then, the neighborhood is input to a prediction function in the rating estimation phase to predict scores for items unseen by the client. Item-based approach has recently received more attention in the domain of centralized recommenders for its better scalability. More specifically, the number of users is usually larger and grows faster than the number of items. These schemes also benefit from better explainability because users have a better knowledge of items than of users.

Yet, recommender systems are confronted to a growing amount of data to process as the number of online users increases, and typically require expensive computational operations and significant storage to provide accurate results. While this combination of factors may saturate centralized systems, fully decentralized approaches provide an attractive alternative with multiple advantages. Firstly, the computation of the predictions can be distributed among all users, removing the need for a costly central server and enhancing scalability. Secondly, a decentralized recommender improves the *privacy* of the users for there is no central entity storing and owning the private information of the users. Several existing algorithms [5], which are out of the scope of this paper, can eventually be deployed in decentralized environments to communicate users' opinions in encrypted form without disclosing their identity. Finally, a distributed recommender service is a valuable feature for peer-to-peer (P2P) applications like BitTorrent and Gnutella as very popular media for users to share their content.

Beside scalability, *sparsity* is another well-known issue of recommender systems. Typically, each user only rates a small amount of items. Consequently, the number of ratings given by the users is very small in comparison with the total number of (user, item) pairs in the system. For example, the MovieLens 10,000,000 ratings dataset has a density of 1.31%. Therefore, the efficient use of the data at hand is an essential matter to recommender systems.

Despite the numerous advantages that decentralized recommenders offer, the majority of work on recommendation algorithms has been focused on centralized systems so far. These algorithms are then not directly applicable to distributed settings. *In this paper, we investigate decentralized neighborhood-based CF recommenders for P2P applications.* Each user can only leverage her own information and data provided by a small (wrt the size of the system) number of other peers[1]. We rely on epidemic algorithms as a decentralized method to form the neighborhood. CF is particularly suitable for the P2P context where no assumption can be made on the content of the items because of the incoherence of meta-data. The contributions of this paper are as follows:

First, decentralized user-based and item-based CF algorithms are implemented and compared in a P2P context using different similarity measures. We show that decentralized user-based approaches deliver better precision and less complexity

---

[1] The terms peer and user are interchangeable in this paper.

than decentralized item-based approaches. In fact, decentralized user-based approach does not suffer from drawbacks usually attributed to their centralized counterpart.

Second, we propose a new decentralized recommender system based on random walks. We explain how the decentralized nature of P2P complicates the application of random walks compared to centralized settings. In our algorithm, each peer is provided with a neighborhood composed of a small (wrt the size of the system) set of similar peers by means of an epidemic protocol. Then, the ratings for unknown items of the neighborhood is estimated by running a random walk on this neighborhood. Once the peers have formed their neighborhood, i.e. the epidemic protocol has converged, each peer is thoroughly independent from other peers in generating her recommendations. This algorithm has the best performance over previous decentralized CF algorithms when the data is sparse.

Third, the behavior of the random walk algorithm is discussed in detail in function of three parameters: sparsity, similarity measure, and neighborhood size. This latter strongly affects the precision and complexity of the algorithm in a P2P context. The optimal parameter values of the algorithm is empirically found for MovieLens 10,000,000 ratings dataset. Fortunately, our algorithm significantly improves the precision over a wide range of sparsity while keeping the execution time affordable for peers. At the end of the paper, we show how significance weighting can be a barrier against the success of random walk algorithms.

The rest of this paper is organized as follows. In Section 2 we provide the preliminaries necessary for understanding our approach. Related work is summarized in Section 3. Decentralization of CF algorithms and our user-based random walk recommender system are described in Sections 4 and 5 respectively. In Section 6, we represent the simulation results and compare the performance of different algorithms. The behavior of random walk is also analyzed in this section. Section 7 concludes the paper.

## 2   Preliminaries

Traditionally, recommender systems are modeled by a two-dimensional matrix denoted by $R$, with rows representing users and columns representing items. Each entry $r_{ui}$ of $R$ contains the rating of user $u$ for item $i$. We assume an $M$ user and $N$ item system, that is $u \in \{1, 2, ..., M\}$ and $i \in \{1, 2, ..., N\}$. Each row $R_{u*}$ is called the *rating vector of user $u$*, and each column $R_{*i}$ the *rating vector of item $i$*. The goal of the recommender system is to predict the missing entries of this matrix. In this section, we provide some necessary background on the CF approach. Moreover, epidemic protocols [10] are briefly discussed as the decentralization method we use to form a neighborhood of similar users.

### 2.1   Collaborative Filtering

User-based CF is presented in [7]. In this approach, a neighborhood of similar users is assigned to each user using some *similarity measure*. One popular coefficient is *Cosine similarity*. For users $u$ and $v$ it is defined as:

$$cos(u, v) = \frac{\sum_{i \in I_u \cap I_v} r_{ui} r_{vi}}{\sqrt{\sum_{i \in I_u \cap I_v} r_{ui}^2} \sqrt{\sum_{i \in I_u \cap I_v} r_{vi}^2}}$$

where $I_u$ and $I_u$ are the set of items rated by $u$ and $v$ respectively. A disadvantage of Cosine similarity is that it does not take into account the differences in users' rating behaviors. For example in a 5-star rating system, a user may rate from 3 to 5, but another one rates from 1 to 3 to reflect the same opinion on items.

The *Pearson correlation* lifts this drawback by considering the offset of each rating from the user's mean rating. It is defined as:

$$\rho_{uv} = \frac{\sum_{i \in I_u \cap I_v} (r_{ui} - \overline{r}_u)(r_{vi} - \overline{r}_v)}{\sqrt{\sum_{i \in I_u \cap I_v} (r_{ui} - \overline{r}_u)^2} \sqrt{\sum_{i \in I_u \cap I_v} (r_{vi} - \overline{r}_v)^2}}$$

where $\overline{r}_u$ is the mean rating of user $u$. Pearson correlation considers only the items rated by both users, but does not take into account the number of such items. As a result, one may choose a user in her neighborhood while having very few items in common.

To deal with this shortage, some authors opt for integrating a factor of *trust* to Pearson correlation known as *significance weighting* [7]. This is achieved by multiplying the Pearson correlation by a term reflecting the number of common items. In [7], this term is defined as $min(|I_v \cap I_u|/50, 1)$. Choosing 50 as the minimum number of ratings not to be attenuated is achieved empirically and must be updated with the growth of the dataset and evolution of user ratings. In this paper we use *log* as the term of significance weighting. Since the steep of logarithmic function decreases constantly, it is more discriminating for smaller numbers of common items. We call this *modified Pearson coefficient* and define it as:

$$corr(u, v) = \rho_{uv} \log(|I_v \cap I_u|). \tag{1}$$

Significance weighting has a *popularizing* disadvantage discussed in Section 6.

Once the neighborhood is formed, the rating estimation phase is accomplished following some prediction rule, usually a weighted sum aggregation function:

$$\hat{r}_{ui} = \overline{r}_u + \frac{\sum_{v \in N(u,i)} \omega_{uv}(r_{vi} - \overline{r}_v)}{\sum_{v \in N(u,i)} \omega_{uv}} \tag{2}$$

where $\hat{r}_{ui}$ is the estimated rating of user $u$ for item $i$, and $N(u, i)$ the set of users in the neighborhood of $u$ having rated $i$. Henceforth, we call $\omega_{uv}$ the *similarity weight* between users $u$ and $v$. In this paper, depending on the setting, it can be either of the similarity measures presented in this section, or is the output of the random walk algorithm.

Item-based CF [18] is quite similar to user-based CF. However, the rating vectors of items are used instead of the rating vectors of users to form a neighborhood of similar items for each item. More details and the relevant equations are provided in the technical report [12].

## 2.2   Epidemic Protocols

In centralized recommender systems, the entire rating matrix $R$ is known to the central recommender. Consequently, the recommender algorithm can search among all the users to assign a neighborhood to a client. This is not efficiently achievable in a decentralized system. Instead, we use epidemic protocols to create users' neighborhood.

In epidemic protocols (also known as *gossip protocols*), peers have access to a Random Peer Sampling service (RPS) [10] providing them with a continuously changing random subset of the peers of the network. When a peer joins the network, her view is initialized at random through the RPS. Each peer also maintains a *view* of the network. This view contains information about the $c$ peers that maximize a clustering function. In this paper, this clustering function reflects how much the peers exhibit a similar rating behavior. It can be either of the similarity measures presented in previous sections depending on the context. In order to converge to the ideal view, each peer runs a *clustering protocol* [20,9]. A peer periodically selects a gossip target from her view and exchanges her view information with her. Upon reception of new information, the peer compares the new candidates with her actual view, and a set of random peers suggested by RPS. Then, keeping only the $c$ most similar entries, she updates her view in order to improve its quality. While the clustering algorithm increases the risks of network partition, the RPS ensures connectivity with high probability. Gossip clustering protocols are known for converging quickly to high quality views. By regularly communicating with the peers in the view, gossip protocols also ensure their liveness and eliminate disconnected nodes. Gossip protocols are fully decentralized, can handle high churn rates, and do not require any specific protocol to recover from massive failures.

## 3   Related Work

In this section, we review the previous work on decentralized recommender systems and suggested solutions to sparsity. The research on decentralized recommender systems has remained modest although the need for them grows rapidly. Notable works on the context are as follows: Tribler [2], a decentralized search engine using BitTorrent protocol, is capable of recognizing the user's taste and give recommendations after a few search queries by the user. Each entry of the binary rating vector is 1 if the user has ever downloaded the corresponding item, and 0 otherwise. Tribler uses epidemic protocols to form the neighborhood, and Cosine function as similarity measure. The significance weighting term is defined as $min(1, |I_v| / 40)$, where $v$ is the corresponding neighbor. A non-normalized score is computed for each item through user-based CF approach, being consequently used to generate an ordered recommendation list.

PocketLens [16] is a decentralized recommender algorithm developed by GroupeLens research group. In [16], different architectures from centralized to fully decentralized are suggested for neighborhood formation. PocketLens uses the

Cosine similarity to estimate the neighborhood quality. Once a neighborhood of similar users is formed, an item-based algorithm is applied on the ratings existing in the neighborhood. The Cosine similarity is used as the similarity weight between items, and predictions are made using a normalized weighted sum.

All of these works use classic similarity measures to predict the ratings. The distinctive point of our work is to apply a *model* to introduce a decentralized model-based CF algorithm.

Several solutions have been suggested to alleviate the problem of sparsity. Some works exploit content information of items or *demographic* information [11] of user's profiles like age, gender or code area to improve the recommendations when the data is not dense enough. Such information is not easy to collect in P2P applications. Furthermore, providing demographic data endangers the users' privacy. *Default rating* [17] is another method for dealing with sparsity. This solution slightly improves the precision of the recommendations by assuming some default value for missing ratings. The disadvantage of this method is in creation of dense input data matrix, hugely increasing the complexity of computations. Hence, this is not a proper solution for P2P either, because the computational power of P2P processors is in general much less than central servers.

Hence, a lot of effort has been made to develop models to mine further the existing data in order to detect potential hidden links between items or users. In [8] trust-based and item-based approaches are combined by means of a random walk model. The algorithm is centralized and the trust is explicitly expressed by the users. The information about trust does not exist in the majority of datasets including MovieLens. Authors in [21] suggest a random walk model as a solution to sparsity in a centralized item-based CF approach. Their algorithm is to some extent similar to an item-based version of our random walk algorithm, but does not lend itself well to decentralized environments.

## 4    Decentralization of CF Algorithms

The main difficulty in decentralization of the user-based CF algorithm is the neighborhood formation phase. Contrary to the central recommender algorithms, each user of a P2P network can only access the data related to a limited number of other users. It is therefore critical to devise a protocol able to efficiently navigate through the P2P system and gather the most similar peers. Epidemic protocols described in Section 2.2 are very suitable for this task, and converge to a view of the most similar users in only a few cycles. Once the neighborhood is formed, the rating estimation is done locally at each user. While scalability is an issue in centralized user-based recommender systems, decentralized approaches do not suffer from this drawback as each user computes her own recommendations.

The decentralization of the item-based CF algorithm is more of a challenge because the algorithm needs the rating vector of the items to find the similarity between them. This vector can not be known by P2P users as they do not know the ratings of the majority of other peers. Consequently, similar to the user-based approach, each peer should find a neighborhood of similar users as a first step.

A *partial* rating vector is then constructed for each item based on the ratings available in the neighborhood, and the item-based CF algorithm is applied.

The complexity of CF algorithms is mostly due to the similarity computation between users or items. For decentralized user-based algorithms, a similarity *vector* between the central user and all peers in the neighborhood is calculated. The complexity of each similarity calculation depends on the number of common items between two users, which may go up to a considerable fraction of all items in the system. The complexity of the operation is then $O(SN)$ for each user, where $S$ is the neighborhood size, and $N$ the number of items in the whole P2P system. In decentralized item-based algorithms, the similarity between unknown items of the neighborhood and the items rated by the user is calculated to form a similarity *matrix*. Provided the neighborhood is big enough, it often contains some users having rated the majority of items. Then, user $u$ needs to compute $L(N - L)$ similarities where $L$ equals $|I_u|$. The complexity of each similarity calculation is proportional to the size of the item rating vectors , being up to the neighborhood size. Hence, the complexity of the decentralized item-based approach is at most $O(N^2S)$ for each peer, where the worst case happens when $L = N/2$. Therefore, it is seen that the decentralized user-based approach is much less complex than the decentralized item-based approach.

For the above reasons, user-based approaches seem to match better a P2P setting. In Section 6, it is empirically shown that decentralized user-based approaches also have better precision than decentralized item-based schemes.

## 5   Decentralized Prediction through Random Walk

In CF recommender algorithms, the similarity weight ($\omega_{uv}$ or $\omega_{ij}$) is usually the same as similarity measure. In our algorithm, this is computed through random walks. Random walk has been used to design decentralized search engines [19]. In the context of recommender systems, some centralized approaches [21] have used random walks to improve the precision of recommendations. In general, the recommendation problem is modeled by a weighted and directed graph where vertices represent the entity of interest. This entity is items in item-based recommenders or webpages in PageRank algorithm for example. The application of random walks to centralized recommenders is relatively obvious. Since the whole graph topology is known to the central algorithm, this latter can launch random walks from a vertex and output a similarity score for each of the other vertices. In other words, the random walk acts as a clustering mechanism on its own to form the neighborhood. In P2P however, this can not be done because the knowledge of each peer about the P2P network is limited to its neighborhood.

In our decentralized algorithm, each peer first locally executes a neighborhood formation phase through clustering gossip protocols as described in Section 2.2. Once the protocol has converged, each peer holds in its view the rating information of the $c$ closest peers according to the similarity measure used for clustering. Note that only peers that get a strictly positive similarity score are inserted in the view. When all the peers have selected their views, we define the *P2P network* (or the topology of it) as the network created by the peers connected via edges to the peers in their views. $c$ is typically small with respect to the size of the network for scalability reasons. Gathering information from only $c$ peers is not enough to achieve good precision and high coverage due to data sparsity. In order to obtain more data at a low network cost, each peer also uses information of the peers in the view of her neighbors. Therefore, we define the *neighborhood* of each user as the peers directly connected and the peers connected within a distance of two hops in the P2P network. Depending on the clustering function, the size of the neighborhood can be up to $c^2 + c$. We evaluate the size of the neighborhood on the MovieLens dataset in Section 6.2.

To compute a personalized score prediction for an item, a user $a$ leverages all the scores that users in her neighborhood have assigned to that item. Each contribution is weighted to reflect the similarity between $a$ and the corresponding user. The users in the neighborhood are modeled as Markov Chain graph vertices, and a random walk is applied on this graph. A Markov chain can be represented by a directed graph where vertices are the states of the chain and edges represent the transition probabilities from one state to another. In our case, the states symbolize the users in the neighborhood of a peer, let us say peer $a$. Since the vertices represent the users of the neighborhood, we call our algorithm *user-based random walk algorithm*. The neighborhood size will consequently be an important parameter of the algorithm, while in centralized algorithms it is always fixed to the size of the complete graph, i.e. the graph containing all users or items of the system. We will see in Section 6 that increasing the neighborhood size until some threshold raises the precision of recommendations while keeping the execution time in a reasonable level. Intuitively, the benefit of random walks is to consider the whole graph topology when estimating the similarity between users, while classic similarity meaures may only take advantage of the explicit intersection between the rating vectors of each two users.

Let $P^a$ be the transition probability matrix corresponding to the graph of user $a$'s neighborhood. Each element $p_{uv}^a$ of $P^a$ represents the probability that $u$ would ask $v$ for recommendations. This probability is defined as the normalized similarity of the tail peer to the head. Another parameter $\beta \in (0,1)$ is also added to the equation to consider the case where each peer jumps randomly to any other peer in the neighborhood during the random walk. Choosing very high values of $\beta$ leads to assignment of equal transition probability towards all users in the neighborhood regardless of their similarity. It means that the ratings of all users will have the same weight in predictions. The value $p_{uv}^a$ is computed using the following equation:

$$p_{uv}^a = (1 - \beta)\frac{s'_{uv}}{\sum_{z \in K(a)} s'_{uz}} + \frac{\beta}{m}, \quad s'_{uv} = \begin{cases} s_{uv} & \text{if } s_{uv} \geq 0 \text{ and } u \neq v \\ \gamma_u & \text{if } u = v \\ 0 & otherwise \end{cases} \quad (3)$$

$K(a)$ is the list of all users in the neighborhood of $a$, $s_{uv}$ is the similarity between two users $u$ and $v$. In the experiments presented in Section 6, $s_{uv}$ is either Pearson correlation or Modified Pearson correlation. $\gamma_u$ is the self loop parameter, modeling the case where a user answers the recommendation query before forwarding it to other users of the neighborhood. Since each user is logically more confident in her own opinion than that of any other user, we fixed $\gamma_u$ as twice the similarity measure between $u$ and the most similar user in her view.

The random walk starts from the users directly connected to the active user $a$, that is, peers having a one hop distance with the active peer in the network. The vector of initial probability distribution over the neighborhood is represented by $\boldsymbol{d}_a$. Each entry of $\boldsymbol{d}_a$ is defined as:

$$\boldsymbol{d}_a(v) = \frac{s''_{av}}{\sum_{z \in K(a)} s''_{az}}, \quad s''_{av} = \begin{cases} s_{av} & \text{if } v \in \text{clustering view of } a \\ 0 & \text{otherwise.} \end{cases} \tag{4}$$

Since each user computes her own predictions, we omit the index of $a$ for the sake of simplicity. We use a finite length random walk where each peer decides to continue the walk with probability $\alpha$. In Markov chains, the probability of being in a state at step $k$ depends only on its previous state. Therefore, the probability of being in state $u$ at step $k$ is:

$$\Pr(X_k = u) = \alpha \sum_{v=1}^{m} \Pr(X_{k-1} = v) p_{vu} = \alpha^k \sum_{v=1}^{m} d(v) P^k_{vu}$$

where $m$ is the size of the active peer's neighborhood, and $P^k$ the power $k$ of the transition probability matrix. This is equal to the inner product of the initial distribution vector by column $u$ of the $P^k$ matrix. The overall probability of being in state $u$ is then:

$$\Pr(X = u) = \sum_{k=1}^{\infty} \alpha^k \boldsymbol{d} \cdot \boldsymbol{P}^k_{*u}.$$

At last, the final probability distribution vector over the neighborhood is:

$$\hat{R} = \sum_{k=1}^{\infty} \alpha^k \boldsymbol{d} P^k = \boldsymbol{d} \alpha P (I - \alpha P)^{-1}. \tag{5}$$

We use Equation (5) to estimate the final distribution vector, and $\alpha$ is optimized empirically. Note that even in the real implementation of the algorithm, Equation (5) may still be used instead of launching real random walks. Once the final distribution vector is output by the random walk model, its entries are used as similarity weights $w_{uv}$ in Equation (2) in order to generate the recommendations.

The computation of transition similarity matrix and the matrix inversion of Equation (5) are the main sources of complexity of the algorithm. The similarity must be calculated between each two users. The complexity of matrix inversion is $O(S^3)$, where $S$ is the neighborhood size. Each similarity computation depends on the number of items in the neighborhood. If the set of items of the neighborhood gets close to the set of items in the whole system, the complexity of both operations becomes $O(S^2 N + S^3)$. With a correct selection of neighborhood size, the algorithm gives excellent performance with reasonable execution time.

In the same way, we can also imagine applying the same algorithm on the graph of items, then having an item-based random walk algorithm. The complexity of this algorithm will be $O(N^2S + N^3)$. Unfortunately, the execution time of item-based random walk algorithm is far from being affordable for the peers in real settings. The lack of efficiency of item-based random walk algorithm pushed us through suggesting the user-based random walk as a better approach for P2P applications. Furthermore, we will see in next section that item-based approaches have in general poor results in P2P systems.

# 6 Experiments and Results

In this section we compare our algorithm with other decentralized CF algorithms. Besides, the behavior of the random walk is analyzed.

## 6.1 Evaluation Methodology and Results

In P2P systems the users do not report any feedback to a central server. As a result, no trace of real P2P data is available. In our experiments we use the MovieLens 10,000,000 ratings dataset [1]. It consists of $10,000,054$ ratings on $10,681$ movies, rated by $71,567$ real users of the MovieLens website, where each user has rated at least 20 movies. A 5-star scale is used to ask for ratings. To the best of our knowledge, this is the second biggest dataset available after the Netflix dataset for research on recommender systems.

Since MovieLens is a central database, we adopt the following strategy to adjust it for our P2P experiments: For each user in the database, a peer object is instanced. This peer is attributed with the profile of the corresponding user in the database. This profile contains the list of films and corresponding ratings of the user. Consequently, each peer can access directly only her own ratings, and needs to rely on the epidemic protocol described in Section 2.2 to find and retrieve the profiles of similar peers. This strategy enables us to simulate a P2P network of MovieLens users, as if each of them had registered her ratings on her own computer instead of reporting them to the website.

We evaluate different recommender algorithms by cross validation. Namely, each MovieLens user profile is split into 20 regular random slices. 20 comes from the minimum number of ratings per user in the MovieLens dataset. Consequently, each profile slice contains at least one rating. A number of slices form the *training profile* input to the algorithm as the learning data. The predictions are made on the *test profile* composed of the remaining slices. Different levels of sparsity are modeled by changing the proportion of the test and training profiles.

We use Root Mean Squared Error (RMSE) to measure the *precision* of the recommendations. For user $u$ it is defined as $\sqrt{(\sum_{r_{ui} \in I_{T_u}} (\hat{r}_{ui} - r_{ui})^2)/|I_{T_u}|}$, where $|I_{T_u}|$ is the size of the test profile of $u$. Each peer computes its own RMSE, and the total RMSE of the system is defined as the mean of RMSEs.

*Coverage* is another important measure of usefulness for recommender systems. It shows the proportion of items for which the algorithm can predict a

**Table 1.** Short description of decentralized CF algorithms with their abbreviations

| | |
|---|---|
| P-RW | decentralized user-based random walk algorithm described in Section 5 |
| MP-U | decentralized version of the user-based algorithm in [7] with Modified Pearson correlation |
| P-U | decentralized version of the user-based algorithm in [7] with Pearson correlation |
| Tribler[2] | decentralized user-based approach with Cosine similarity and significance weighting |
| PocketLens [16] | decentralized item-based approach using Cosine similarity |
| MP-I | decentralized version of the item-based algorithm in [18] with Modified Pearson correlation |
| P-I | decentralized version of the item-based algorithm in [18] with Pearson correlation |

rating. Since the total number of items of a P2P network is not known to the users, we define the coverage for user $u$ as ($\left|\hat{I}_{T_u}\right| / |I_{T_u}|$), where $\hat{I}_{T_u}$ is the set of predictable items in the test profile of u. The total coverage of the system is then defined as the mean coverage of all peers.

The simulations are run for three view sizes: 10, 20 and 30. All results were obtained after 30 cycles of gossip, and the epidemic protocol had converged. $\beta$ was fixed to 0.15 in the random walk algorithm. $\alpha$ was optimized by trying values in $(0, 1)$ with a step of 0.1 in different levels of sparsity. In general, we observe that the optimal length of the random walk increases (larger $\alpha$) as the data becomes sparser. Even though the similarity between users is most often transitive, it happens in few cases that users in a two hop distance have negative similarity. We do not take such users into account when making predictions in user-based approaches, although they exist in the neighborhood. This problem never happens in the random walk algorithm because the similarity weights generated by the algorithm are non-negative probabilities. In the same way, only items with positive similarity are used for prediction in item-based methods.

We compare our algorithm with 6 decentralized recommender algorithms. The description of these algorithms and corresponding abbreviations are listed in Table 1. The best results, obtained with a view size of 30, is reported in Tables 2 and 3. The results for view sizes of 10 and 20 is found in the technical report [12]. The item scores computed by Tribler are not scaled. Hence, we generated a score for each item using Equation (2) to be able to compare it with other algorithms. In P-I and MP-I, both neighborhood formation and item-based prediction are done using the same type of similarity measure, and the predictions are made using the item-based version of Equation (2).

As seen in Table 2, P-RW algorithm outperforms all other decentralized algorithms when the sparsity is less than 70%. P-U and MP-U approaches significantly outperform all item-based approaches. Tribler shows the poorest performance among user-based approaches, but still improves over item-based approaches when sparsity is more than 5% and less than 25%. This shows that Pearson correlation is a better choice than Cosine similarity in user-based approaches. PocketLens shows the best performance among item-based approaches. Therefore, Cosine similarity seems to perform better in item-based approaches. Moreover, comparing MP-U and MP-I with P-U and P-I proves that significance weighting is efficient for both item-based and user-based approaches. As a general term, we can state that provided the right similarity measure is used,

**Table 2.** RMSE in different levels of sparsity, view = 30

| Training Profile | 5% | 10% | 15% | 20% | 25% | 30% | 40% | 50% | 70% | 90% |
|---|---|---|---|---|---|---|---|---|---|---|
| P-RW | 1.0719 | 1.0147 | 0.9869 | 0.9693 | 0.9575 | 0.9513 | 0.9423 | 0.9327 | 0.9196 | 0.8842 |
| MP-U | 1.1164 | 1.0481 | 1.0081 | 0.9841 | 0.9717 | 0.9662 | 0.9522 | 0.9408 | 0.9168 | 0.8752 |
| P-U | 1.1288 | 1.0594 | 1.0220 | 0.9980 | 0.9812 | 0.9725 | 0.9594 | 0.9477 | 0.9294 | 0.8903 |
| Tribler | 1.2301 | 1.0946 | 1.0439 | 1.0234 | 1.0166 | 1.0119 | 1.0050 | 0.9988 | 0.9892 | 0.9489 |
| PocketLens | 1.2036 | 1.1110 | 1.0595 | 1.0296 | 1.0119 | 0.9998 | 0.9833 | 0.9721 | 0.9553 | 0.9174 |
| MP-I | 1.2218 | 1.1410 | 1.0867 | 1.0493 | 1.0211 | 1.0011 | 0.9732 | 0.9559 | 0.9338 | 0.8985 |
| P-I | 1.2508 | 1.1601 | 1.0984 | 1.0524 | 1.0255 | 1.0062 | 0.9805 | 0.9656 | 0.9441 | 0.9038 |

**Table 3.** Coverage in different levels of sparsity, view = 30

| Training Profile | 5% | 10% | 15% | 20% | 25% | 30% | 40% | 50% | 70% | 90% |
|---|---|---|---|---|---|---|---|---|---|---|
| P-RW | 0.8429 | 0.9272 | 0.9370 | 0.9487 | 0.9492 | 0.9506 | 0.9560 | 0.9540 | 0.9511 | 0.9474 |
| MP-U | 0.8324 | 0.9642 | 0.9854 | 0.9917 | 0.9943 | 0.9956 | 0.9969 | 0.9979 | 0.9983 | 0.9986 |
| P-U | 0.6971 | 0.8657 | 0.892 | 0.9220 | 0.9264 | 0.9316 | 0.9415 | 0.9407 | 0.9394 | 0.9364 |
| Tribler | 0.7469 | 0.9669 | 0.9881 | 0.9933 | 0.9952 | 0.9966 | 0.9978 | 0.9984 | 0.9990 | 0.9993 |
| PocketLens | 0.7435 | 0.9023 | 0.9337 | 0.9453 | 0.9515 | 0.9549 | 0.9583 | 0.9598 | 0.9582 | 0.9558 |
| MP-I | 0.8265 | 0.9612 | 0.9853 | 0.9924 | 0.9951 | 0.9963 | 0.9974 | 0.9979 | 0.9985 | 0.9989 |
| P-I | 0.6872 | 0.8844 | 0.9192 | 0.9406 | 0.9434 | 0.9470 | 0.9543 | 0.9521 | 0.9488 | 0.9458 |

user-based approach is preferable to item-based approach in P2P recommenders. All methods have good coverage when the training profile is more than 5%. However, the coverage of the methods using significance weighting, that is MP-U, MP-I and Tribler, is slightly better than others. P-RW improves the coverage over P-U although they use the same neighborhood. This is because P-RW can also use the ratings of users with negative direct similarity.

In most recommender systems, the predicted scores are used to propose a recommendation list of top-N items to the user. The quality of this list strongly depends on the RMSE of the system. The achievable RMSE lies in a very restricted range in available datasets, but it is proven that only slight improvement in RMSE yields much more satisfactory recommendation lists [13]. Hence, the improvement of our algorithm over the best of previous algorithms is absolutely valuable specifically because we are very close to the limit of achievable RMSE.

The precision and coverage of all approaches increase with the size of the neighborhood. This is due to the fact that algorithms rely on more users for making predictions. We observed in simulations that increasing the view size over 30 does not yield any significant improvement. Note there is no advantage in choosing very large views. Not only does it exponentially increase the execution time, but also renders the recommendations less personalized. A view size about 30, allows for good precision and coverage while keeping the computation time quite affordable. This value may be different for datasets other than MovieLens.

## 6.2 Analysis of the Behavior of Random Walk

In this section we discover further the behavior of random walk in function of sparsity, neighborhood size and similarity measure.

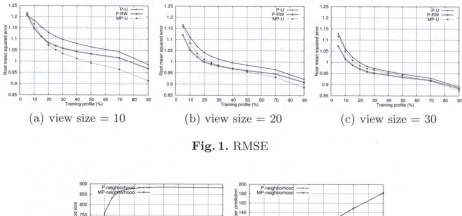

(a) view size = 10        (b) view size = 20        (c) view size = 30

**Fig. 1.** RMSE

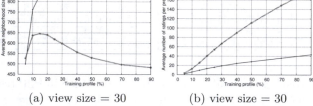

(a) view size = 30        (b) view size = 30

**Fig. 2.** MP-neighborhood vs. P-neighborhood

*Random Walk vs. Sparsity.* Random walk works well when the data is so sparse that classic similarity measures fail to detect meaningful relation between users. By increasing the training set proportion, classic similarity measures deliver better performance than the random walk algorithm. For the view size of 30, P-RW gives the best results until when the training set proportion is below 70%. However, when the training set proportion goes beyond 70%, the direct similarities become more reliable than random walk similarities.

*Random Walk vs. Neighborhood Size.* The precision of the three approaches with the best precision is plot in Figure 1. Before the training profile arrives at a threshold, P-RW delivers the best precision outperforming the MP-U algorithm as the second best approach. This threshold increases rapidly with incrementing the size of the neighborhood. It is 15% for a view size of 10, and goes up to 40% for a view size of 20. The threshold reaches 70% for the view size of 30, suggested as the best view size by our experiments. Furthermore, the amount of improvement of P-RW over other approaches increases with the neighborhood size. In fact, random walk reevaluates the similarity weight between users by mining longer paths in the neighborhood to find implicit transitive similarities. However, classic similarity measures can only capture direct similarity. The chance of detecting the transitive similarities is naturally higher for larger neighborhoods. It is why P-RW outperforms MP-U, but P-U has poorer precision than the latter, while both P-U and P-RW use the same type of neighborhood.

*Random Walk vs. Similarity Measure.* To see how significance weighting of the similarity measure can influence the quality of the neighborhood, we compared the average neighborhood size and the average number of ratings per prediction for two types of neighborhood formed either through Pearson correlation or Modified Pearson correlation (see Figure 2). It is seen that MP-neighborhood has more ratings per prediction than P-neighborhood while its size is smaller. It means that Modified Pearson correlation prefers over-active users having rated a large number of items. Note P-RW has better precision than MP-U although it uses less ratings, showing that P-RW *learns* faster than MP-U. With increasing the training profile, the P-neighborhood approaches its maximum size (about 900) very soon. Unlike P-neighborhood, the size of MP-neighborhood decreases continuously when the training profile goes beyond 15%. This indicates that the P2P network becomes more clustered because the views of directly-connected peers contain many common neighbors. In fact, since over-active users have more ratings in each profile slice the significance weighting term grows faster for them with incrementing the training profile. Consequently, their chance being put in the neighborhood becomes more than moderate users, and their indegree increases quickly. We call it the *popularizing effect* of significance weighting.

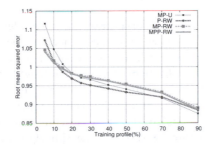

**Fig. 3.** Error v/s proportion of data set used

Although the popularizing effect leads to better coverage, it prevents the random walk algorithm from working in two ways: first, it decreases the ability of the algorithm in similarity estimation by decreasing the neighborhood size and omitting users with few ratings but *implicit* similarity to the central user. Second, over-active users act as a *sink* in the Markov Chain model during the random walk. Then, their state probability at the end of the random walk is higher than other peers. In other words, random walk intensifies the influence of over-active users in predictions with respect to the users with less ratings. This significantly decreases the quality of random walk predictions in the MP-neighborhood. The size of MP-neighborhood has a peak when the training profile is 15%. This shows that the sinking behavior of significance weighting starts at this point. For smaller training profiles, the P2P network is not still well clustered, and peers continue to add new users to their views.

To investigate the performance of random walk on an MP-neighborhood, we implemented two new variants of our algorithm. The first variant is MP-RW.

Being quite similar to P-RW, it uses Modified Pearson correlation instead of Pearson correlation for neighborhood formation and also as user similarity weight $s_{uv}$ in Markov chain model (see Equations (3) and (4)). The second one is MPP-RW where the neighborhood is formed through Modified Pearson correlation, while user similarity weight in Markov Chain model is assigned using Pearson correlation. The results are plot in Figure 3. The exact RMSE values can be found in the technical report [12].

When the training profile is more than 15%, MP-RW starts to show poorer results than P-RW. Its performance is even worse than MP-U when the training set is more than 20%. The reason hides behind the popularizing effect of significance weighting. The slightly better precision of MPP-RW than MP-RW is due to the fact that the transition probability of the edges pointing towards over-active users decreases when significance weighting is not used for user similarity assignment. Hence, the sink role of such users is partly alleviated. It is also observed that MPP-RW can outperform P-RW when the training profile is extremely sparse (below 15%). This is due to the fact that the sinking behavior of Modified Pearson correlation is not still severe in this range.

## 7   Conclusion

In this paper, we propose a user-based random walk algorithm to enhance the precision of previous decentralized CF recommender systems. We use epidemic protocols to assign each user with a neighborhood of similar peers. Each user locally runs the random walk algorithm on her neighborhood, and computes her recommendations. The algorithm is fully decentralized, and users are totally independent from each other in computing their own recommendations.

We implemented decentralized CF recommenders using different similarity measures and compared them with our algorithm. Our algorithm had the best precision over a wide range of sparsity. Decentralized user-based algorithms showed better precision and less complexity than their item-based counterparts. Moreover, Cosine similarity performed better in decentralized item-based algorithms, while Pearson correlation worked better for decentralized user-based algorithms.

Simulating a P2P network using the MovieLens 10,000,000 ratings dataset, we empirically showed how sparsity, neighborhood size, and similarity measure are determining parameters of the random walk algorithm. This algorithm delivers better precision when the data gets sparser. It works better for larger neighborhood sizes. The view size of 30 was given as a good trade-off between precision and execution time for MovieLens dataset. In the end, the behavior of the random walk was studied for two types of neighborhood formed either through Pearson correlation or Modified Pearson correlation. We showed how popularizing effect related to significance weighting term of Modified Pearson correlation is a barrier against the performance of random walk.

**Acknowledgements.** We are very grateful to GroupLens research group for providing MovieLens datasets.

# References

1. MovieLens Datasets (2010), http://www.grouplens.org/node/73#attachments
2. Tribler (2010), http://www.tribler.org
3. Azar, Y., Fiat, A., Karlin, A.R., Mcsherry, F., Saia, J.: Spectral analysis of data. In: ACM Symposium on Theory of Computing, pp. 619–626 (2001)
4. Biau, G., Cadre, B., Rouviere, L.: A stochastic model for collaborative recommendation. The Annals of Statistics (2009)
5. Canny, J., Sorkin, S.: Practical large-scale distributed key generation. In: Cachin, C., Camenisch, J.L. (eds.) EUROCRYPT 2004. LNCS, vol. 3027, pp. 138–152. Springer, Heidelberg (2004)
6. Deshpande, M., Karypis, G.: Item-based top-n recommendation algorithms. ACM Trans. Inf. Syst. 22(1), 143–177 (2004)
7. Herlocker, J.L., Konstan, J.A., Borchers, A., Riedl, J.: An algorithmic framework for performing collaborative filtering. In: ACM SIGIR, pp. 230–237 (1999)
8. Jamali, M., Ester, M.: Trustwalker: a random walk model for combining trust-based and item-based recommendation. In: ACM SIGKDD, pp. 397–406 (2009)
9. Jelasity, M., Montresor, A., Babaoglu, O.: T-man: Gossip-based fast overlay topology construction. IJCNC (2009)
10. Jelasity, M., Voulgaris, S., Guerraoui, R., Kermarrec, A.-M., van Steen, M.: Gossip-based peer sampling. ACM Trans. Comput. Syst. 25(3), 8 (2007)
11. Pazzani, M.J.: A framework for collaborative, content-based and demographic filtering. Artificial Intelligence Review, 393–408 (1999)
12. Kermarrec, A.-M., Leroy, V., Moin, A., Thraves, C.: Addressing sparsity in decentralized recommender systems through random walks. Technical report, INRIA (2010)
13. Koren, Y.: Factorization meets the neighborhood: a multifaceted collaborative filtering model. In: Proc. of the 14th ACM SIGKDD, pp. 426–434 (2008)
14. Kumar, R., Raghavan, P., Rajagopalan, S., Tomkins, A.: Recommendation systems: A probabilistic analysis. In: Proc. IEEE Symp. on Foundations of Computer Science (1998)
15. Linden, G., Smith, B., York, J.: Amazon.com recommendations: item-to-item collaborative filtering. IEEE Internet Computing, 76–80 (2003)
16. Miller, B.N., Konstan, J.A., Riedl, J.: Pocketlens: Toward a personal recommender system. ACM Trans. Inf. Syst. 22(3), 437–476 (2004)
17. Breeze, J.S., Heckerman, D., Kadie, C.: Empirical analysis of predictive algorithms for collaborative filtering. In: Uncertainty in Artificial Intelligence (1998)
18. Sarwar, B., Karypis, G., Konstan, J., Reidl, J.: Item-based collaborative filtering recommendation algorithms. In: World Wide Web, pp. 285–295 (2001)
19. Tsoumakos, D., Roussopoulos, N.: Adaptive probabilistic search for peer-to-peer networks. In: P2P, pp. 102–109 (2003)
20. Voulgaris, S., Steen, M.V.: Epidemic-style management of semantic overlays for content-based searching. In: Cunha, J.C., Medeiros, P.D. (eds.) Euro-Par 2005. LNCS, vol. 3648, pp. 1143–1152. Springer, Heidelberg (2005)
21. Yildirim, H., Krishnamoorthy, M.S.: A random walk method for alleviating the sparsity problem in collaborative filtering. In: Proc. of the ACM Conf. on Recommender Systems, pp. 131–138 (2008)

# Uniform and Ergodic Sampling in Unstructured Peer-to-Peer Systems with Malicious Nodes

Emmanuelle Anceaume[1], Yann Busnel[2], and Sébastien Gambs[3]

[1] IRISA / CNRS Rennes, France
emmanuelle.anceaume@irisa.fr
[2] LINA / Université de Nantes, France
Yann.Busnel@univ-nantes.fr
[3] IRISA / Université de Rennes 1 - INRIA, France
sebastien.gambs@irisa.fr

**Abstract.** We consider the problem of uniform sampling in large scale open systems. Uniform sampling is a fundamental primitive that guarantees that any individual in a population has the same probability to be selected as sample. An important issue that seriously hampers the feasibility of uniform sampling in open and large scale systems is the unavoidable presence of malicious nodes. In this paper we show that restricting the number of requests that malicious nodes can issue and allowing for a full knowledge of the composition of the system is a necessary and sufficient condition to guarantee uniform and ergodic sampling. In a nutshell, a uniform and ergodic sampling guarantees that any node in the system is equally likely to appear as a sample at any non malicious node in the system and that infinitely often any node has a non-null probability to appear as a sample of honest nodes.

**Keywords:** Uniform sampling, unstructured peer-to-peer systems, ergodicity, Byzantine adversary.

## 1 Introduction

We consider the problem of uniform sampling in large scale open systems with adversarial (Byzantine) nodes. Uniform sampling is a fundamental primitive guaranteeing that any individual in a population has the same probability to be selected as sample. This property is of utmost importance in systems in which the population is continuously evolving and where it is impossible to capture the full complexity of the network through global snapshots. By collecting random subsets of information over the network, one can infer at almost no cost some global characteristic of the whole population (such as its size, its topological organization, its resources, ...). Therefore uniform sampling finds its root in many problems such as data collection, dissemination, load balancing, and data-caching [1–4].

Providing unbiased (*i.e.*, uniform) sampling in these open systems is a challenging issue. First, this primitive must cope with the continuous change of the

C. Lu, T. Masuzawa, and M. Mosbah (Eds.): OPODIS 2010, LNCS 6490, pp. 64–78, 2010.

network structure caused by nodes departures and arrivals. Nevertheless, it has been shown through simulations [1, 5] and analytic studies [6–8] that simply maintaining a partial and small local view of node identifiers (ids) is sufficient to provide near uniform sampling. This can be achieved through gossip-based algorithms [1, 9, 10] or through random walks [5, 11–13]. Gossip-based algorithms mainly consist, for each node $v$ in the system, in periodically selecting some other node $w$ in $v$'s local view and exchanging information. Information can either be pushed to other nodes or pulled from other nodes. Over time, information spreads over the system in an epidemic fashion allowing each node to continuously update its local view with fresh node ids. On the other hand, a random walk on a network (which can be represented as a graph) is a sequential process, starting from an initial node $v$, which consists in visiting a node in $v$'s neighborhood according to some randomized order. In its simpler form, the next node is chosen uniformly at random among the neighbors, while more sophisticated choices are implemented to cope with the bias introduced towards high degree nodes (for instance, through the Metropolis-Hastings algorithm [14]).

An important issue that seriously hampers the feasibility of uniform sampling in open and large scale systems is the unavoidable presence of malicious nodes. Malicious (or Byzantine) nodes typically try to manipulate the system by exhibiting undesirable behaviors [15]. In our context, they try to subvert the system by launching targeting attacks against nodes in the aim of biasing uniformity by isolating honest nodes within the system. This is quickly achieved by poisoning local views of honest nodes with malicious node ids. For instance in unstructured graphs, a number of push operations logarithmic in the size of local views is sufficient to fully eclipse honest nodes from the local view of a node [16], while in structured graphs, a linear number of join operations is required [17]. Recent works have been proposed to detect and exclude these adversarial behaviors [18–20] by observing that malicious nodes try to get an in-degree much higher than honest nodes in order to isolate them. Extensive simulations [18] have shown that this approach is only highly effective for a very small number of malicious nodes (*i.e.*, in $\mathcal{O}(\log|\mathcal{S}|)$ where $|\mathcal{S}|$ is the size of the network $\mathcal{S}$), otherwise detection mechanisms may boil down to false positive detection (*i.e.*, detection of honest nodes).

On the other hand, when the system is harmed by a large number of malicious peers (*i.e.*, a linear proportion of the nodes of the system), which is definitely a realistic assumption in peer-to-peer systems [15, 21], additional mechanisms are required to prevent targeted attacks from succeeding. Specifically, in structured peer-to-peer systems, analytical studies have shown that applying the "induced churn" principle allows to defend the system against adversarial behaviors, either through competitive induced churn strategies [22], or through global induced churn [23]. Briefly, this principle states that, by forcing nodes to periodically change their position in the graph, malicious peers cannot predict the evolution of the state of the system after a given sequence of join and leave operations. By taking advantage of the properties of structured graphs, the authors of both papers have shown that, with high probability, any node is equally likely to appear in the local view of each other honest node in a number of rounds polynomial in the size of the system. Unfortunately, in unstructured peer-to-peer systems, nodes cannot rely on the topological nature of structured graphs to reject new node ids that do not conform to the imposed distance function (contrary to structured networks [22, 23]). To circumvent this issue, Bortnikov *et al.* [16] rely on the properties of min-wise independent permutations, which are fed by the streams of gossiped node ids, to eventually converge towards uniform sampling on the node ids. More precisely, these authors have derived an upper bound on the expected time $T_s$ to converge towards unbiased (uniform) samples. However, by construction, this convergence is definitive in the sense that once a random sample has been locally observed it is kept as a local sample forever. As a consequence, beyond the time limit $T_s$, no other node ids received in the input stream can ever appear in the random sample. The property of a sampler to guarantee that each received node id infinitely often has a non-null probability to locally appear as a sample is called the *ergodic sampling* property (this property is formally defined later in the paper).

Intuitively, this lack of adaptivity seems to be the only defense against adversarial behavior when considering bounded resources (memory and bandwidth). This paper is devoted to the formal analysis of the conditions under which uniform and ergodic sampling is feasible or not. More precisely, *the main contribution of this paper* is to show necessary and sufficient conditions under which uniform and ergodic sampling is achievable in unstructured peer-to-peer systems potentially populated with a large proportion of Byzantine nodes. Specifically, let $\mathcal{S}$ represent the wide collection of nodes in the system, and $k < 1$ the proportion of malicious nodes in $\mathcal{S}$. Let $\delta_m$ be the number of (not necessarily unique) malicious node ids gossiped by malicious nodes during a time interval $T_s$, and $\Gamma$ denote the local memory of any honest node $u$ in $\mathcal{S}$. In this context, we prove the following assertions:

- If the number $\delta_m$ of (non-unique) malicious ids received at node $u$ during a given period of time $T_s$ is strictly greater than $T_s - |\mathcal{S}|(1 - k)$ then, neither uniform sampling nor ergodic sampling can be achieved;
- If $\delta_m \leq T_s - |\mathcal{S}|(1 - k)$ and the size of the memory $\Gamma$ is greater than or equal to $|\mathcal{S}|$ then, both uniform and ergodic sampling can be achieved;
- If $\delta_m \leq T_s - |\mathcal{S}|(1 - k)$, and $|\Gamma| < |\mathcal{S}|$ then, uniform and ergodic sampling cannot be achieved.

Briefly, these conditions show that if the system cannot provide the means to limit the number of messages an adversary can periodically send, then solving either uniform sampling or ergodic sampling is impossible. On the other hand, if this assumption holds and if all honest nodes in the system have access to a very large memory (in the size of the network) then, the problem becomes trivially solvable. Unfortunately, as will be shown, both conditions are necessary and sufficient to solve the uniform and ergodic sampling problem. Clearly, these strong conditions highlight the damage that adversarial behavior can cause in large-scale unstructured systems.

To the best of our knowledge, we are not aware of any previous work that has specified the conditions for which uniform and ergodic sampling is reachable in presence of adversarial behaviors.

The outline of this paper is the following. In the next section, we describe the model of the system and how it is vulnerable to malicious nodes. Afterwards in Section 3, we define uniform and ergodic sampling, while in Section 4, related work is presented. Finally, Section 5 identifies the two conditions for which uniform and ergodic sampling is achievable, before concluding in Section 6.

## 2    System Model

An overlay network is a logical network built on top of a physical network. We consider an overlay network $S$ populated with nodes labelled through a system wide identifier. We assume that a unique and permanent identifier is assigned to each node. In the following, nodes identifiers are abbreviated by node ids. Nodes communicate among each other along the edges of the overlay by using the communication primitives provided by the underlying network (e.g., IP network service). Nodes are free to join and leave the overlay at any time. The particular algorithms use by nodes to choose their neighbors and to route messages induce the resulting overlay topology. In particular, the topology of unstructured overlays conforms with that of random graphs (i.e. relationships among nodes are mostly set according to a random process).

### 2.1    Adversary

A fundamental issue faced by any practical open system is the inevitable presence of nodes that try to manipulate the system by exhibiting undesirable behaviors [15]. Such nodes are called malicious or Byzantine nodes. Malicious nodes can simply display behaviors such as simply dropping or re-routing messages towards other malicious nodes, or they can devise more complex strategies such as mounting eclipse attacks (also called routing-table poisoning [15, 24]) by having honest nodes redirecting outgoing links towards malicious ones. Moreover, they can magnify the impact of their attacks by colluding and coordinating their actions. In our work, we do not consider Sybil attacks [21], which mainly consist in flooding the system with numerous fake identifiers. We assume the existence of some external mechanism for solving this problem (for instance an off-line certification authority, cf. Section 2.2). We model malicious behaviors through a strong adversary that fully controls these malicious nodes. The adversary has the ability to inspect the whole overlay and strategizes on the time at which malicious nodes operations must be issued. We assume that the adversary cannot control more than a fraction $k < 1$ of malicious nodes in the overlay. A node which always follows the prescribed protocols is called *honest*. Note that honest nodes cannot *a priori* distinguish honest nodes from malicious ones, which would otherwise render the problem trivial.

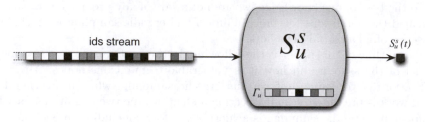

**Fig. 1.** Sampling component of node $u \in \mathcal{N}$

## 2.2   Security Mechanisms

We assume the availability of a signature scheme that enables to verify the validity of a signature on a message (i.e. the authenticity and integrity of this message with respect to a particular node). Recipients of a message ignore any message that is not signed properly. Nodes ids and keys (private and public) are acquired via a registration authority [24] and it is assumed that honest nodes never reveal their private keys to other nodes. We also assume the existence of private channels (obtained through cryptographics means) between each pair of nodes preventing an adversary from eavesdropping and unnoticeably tampering with the content of a message exchanged between two honest nodes through this channel. However of course, a malicious node has complete control over the messages it sends and receives.

## 3   Uniform and Ergodic Sampling

In this section, we describe the terminology and assumptions used in this paper and then define uniform and ergodic sampling.

### 3.1   Assumptions and Terminology

Similarly to Bortnikov *et al.* [16], we consider the following assumptions. There exists a time $T_0$ such that from time $T_0$ onwards, the churn of the system ceases. This assumption is necessary to make the notion of uniform sample meaningful. Thus from $T_0$ onwards, the population of the system $\mathcal{S}$ is composed of $|\mathcal{S}|$ nodes, such that at least $(1 - k)|\mathcal{S}|$ of them are honest and no more than $k|\mathcal{S}|$ of them are controlled by the adversary (for $k < 1$). The subset of honest nodes in the overlay is denoted by $\mathcal{N}$ and we assume that all the nodes in $\mathcal{N}$ are weakly connected from time $T_0$ onwards.

Each node $u \in \mathcal{N}$ has locally access to a *sampling component*[1] as presented in Figure 1. The sampling component implements a *strategy s* and has uniquely access to a data structure $\Gamma_u$, referred to as the *sampling memory*. The size

---

[1] Although malicious nodes have also access to a sampling component, we cannot impose any assumptions on how they feed it or use it as their behavior can be totally arbitrary.

of the sampling memory $\Gamma_u$ is bounded and is denoted by $|\Gamma_u|$. The sampling component $S_u^s$ is fed with (non unique) node ids that correspond to the node ids periodically received (either through gossip algorithms or through random walks). This stream of node ids may contain repetition of the same node id, which can be particularly frequent for malicious node ids, as discussed later. At each time $t$, the following three steps are executed: the first element of the stream, say node id $v$, is given as input to the sampler component. The sampling component $S_u^s$ reads $v$, and removes it from the stream. According to its strategy $s$, $S_u^s$ may store or not $v$ in $\Gamma_u$ (for example, the strategy $s$ may consist in storing $v$ if $\Gamma_u$ is not full or in substituting $v$ for a randomly chosen node id in $\Gamma_u$), and outputs at most one node id $v'$. The output at time $t$ is denoted $S_u^s(t)$. The produced node id $v'$ is chosen among the node ids in $\Gamma_u$ according to the strategy $s$ (for instance, strategy $s$ may choose the smallest node id in $\Gamma_u$ or the smallest node id under a given min-wise permutation [16]). Note that these three steps are atomically done. The maximum finite hitting time needed for the sampling component $S_u^s$ to reach a uniform sample is denoted by $T_s$. Clearly $T_s$ depends on the strategy $s$ implemented by the sampling component and also on the stream of node ids the sampling component has access to. Finally, $\delta_m$ represents the number of malicious node ids received (possibly multiple times) in the stream of node ids at node $u$ during the time interval $T_s$.

## 3.2   Sampling Properties

We consider the problem of achieving an unbiased (uniform) and ergodic sampling in large scale unstructured peer-to-peer systems subject to adversarial attacks. A strategy $s$ that solves this problem has to meet the following two properties: *i)* Uniformity, which states that any node in the overlay should have the same probability to appear in the sample of honest nodes in the overlay, and *ii)* Ergodicity, which states that any node should have a non-null probability to appear infinitely often in the sample of any honest nodes in the overlay. More formally, strategy $s$ should guarantee:

*Property 1 (Uniformity).* Let $\mathcal{N}$ be a weakly connected graph from time $T_0$ onwards, then for any time $t \geq T_s$, for any node $u \in \mathcal{S}$, and for any node $v \in \mathcal{N}$,

$$\mathbb{P}[u \in S_v^s(t)] = \frac{1}{|\mathcal{S}|}.$$

*Property 2 (Ergodicity).* Let $\mathcal{N}$ be a weakly connected graph from time $T_0$ onwards, then for any time $t \geq T_0$, for any node $u \in \mathcal{S}$, and for any node $v \in \mathcal{N}$,

$$\mathbb{P}\left[\{t'|t' > t \wedge u \in S_v^s(t')\} = \emptyset\right] = 0,$$

where $\emptyset$ represents the empty set. In the following, Properties 1 and 2 are respectively denoted $\mathcal{U}$ and $\mathcal{E}$.

*Remark 1.* Uniformity by itself does not imply ergodicity and conversely, ergodicity by itself does not imply uniformity. Indeed, Property 1 guarantees that any node (honest or not) has an equal probability to be sampled by any honest node in the system. Nonetheless, once convergence to a random sample locally holds, this property does not say that this sample must change over time to provide a fresh and random node id (this is definitively important for data-caching applications which continuously require fresh node id). Guaranteeing this dynamicity is formalized by Property 2 which states that each node has a non-null probability to be selected as a sample at any time, guaranteeing the access of new sample graphs.

## 4   Related Work

In the literature, different approaches have been proposed to deal with malicious behaviors, each one focusing on a particular adversarial strategy.

With respect to eclipse attacks, a very common technique, called *constrained routing table*, relies on the uniqueness and impossibility of forging nodes identifiers. It consists in selecting as neighbors only the nodes whose identifiers are closer to some particular points in the identifier space [24]. Such an approach has been successfully implemented into several overlays (*e.g.*, CAN, Chord, Pastry). More generally, to prevent messages from being misrouted or dropped, the seminal works of Castro *et al.* [24] and Sit and Morris [15] on distributed hash tables based overlays combine routing failure tests and redundant routing as a solution to ensure robust routing. Their approach has then been successfully implemented in different structured-based overlays (*e.g.*, [25–27]). In all these previous works, it is assumed that at any time, and anywhere in the overlay, the proportion of compromised nodes is bounded and known, allowing powerful building blocks such as Byzantine tolerant agreement protocols to be used among peers subsets [26, 27]. When such an assumption fails, additional mechanisms are needed. For instance, Awerbuch *et al.* [22] propose the *Cuckoo&flip* strategy, which consists in introducing local induced churn (*i.e.*, forcing a subset of nodes to leave the overlay) upon each join and leave operation. This strategy prevents malicious nodes from predicting what is going to be the state of the overlay after a given sequence of join and leave operations. Subsequently to this theoretical work, experiments have been conducted to verify the practical feasibility of global induced churn, which consists in having all the nodes of the overlay periodically leaving their positions. These experiments assume that the overlay is populated by no more than $k = 25\%$ of compromised nodes [28]. Authors of [23] have analyzed several adversarial strategies, and show that an adversary can very quickly subvert DHT-based overlays (DHT for Distributed Hash Tables) by simply never triggering leave operations. They also show that when all nodes (honest and malicious ones) are imposed a limited lifetime, the system eventually reaches a stationary regime where the ratio of corrupted clusters is bounded, independently from the initial amount of corruption in the system.

Jesi *et al.* [18] propose a random sampling algorithm that deals with malicious nodes. Their solution assumes that the ultimate goal of the malicious nodes is to mutate the random graph in a hub-based graph, hub for which malicious nodes gain the lead. Once this goal is reached, malicious nodes can very quickly and easily subvert the whole overlay by performing denial-of-service attacks. Conducting a hub attack mainly consists for malicious nodes in increasing their in-degree. Jesi *et al.* [18] propose to detect highly popular nodes by extending classic node sampling services with a module that identifies and blacklists nodes that have an in-degree much higher than the other nodes of the overlay. This approach, also adopted in several structured based overlays [19] through auditing mechanisms, or in sensor networks [20], is effective only if the number of malicious nodes is very small with respect to the size of the overlay, typically of $\mathcal{O}(\log |\mathcal{S}|)$.

Recently, Bortnikov *et al.* [16] have proposed a uniform sampling algorithm that tolerates up to a linear number of malicious nodes. Their sampling mechanism exploits the properties offered by min-wise permutations. Specifically, the sampling component is fed with the stream of node ids periodically gossiped by nodes, and outputs the node id whose image value under the randomly chosen permutation is the smallest value ever encountered. Thus eventually, by the property of min-wise permutation, the sampler converges towards a random sample. By limiting the number of requests malicious nodes can periodically issue, their solution requires a single node id to be stored in the local memory. Nevertheless, their solution does not satisfy the ergodicity property as convergence toward a random sample is permanent. It is worth noting that our results complement two previous results [6, 7], in which both papers propose an analysis of the class of uniform and ergodic sampling protocols. Each paper provides a complete analytical proof of a gossip-based protocol that reaches both $\mathcal{U}$ and $\mathcal{E}$. However, in contrast to the present work, adversarial behaviors were not considered.

Finally, taking a completely different approach from the previously mentioned papers, which are based on gossip algorithms or on distance function properties, the techniques presented in [29, 30] rely on social network topologies to guard against Sybil attacks. Both protocols take advantage of the fact that Sybil attacks try to alter the fast mixing property of social networks to defend against these attacks. However, in presence of malicious nodes with a high degree, performance of both protocols degrade drastically.

Note that the analysis presented in this paper is independent from the way the stream of node ids at each node $u$ has been generated. That is, it may result from the propagation of node ids through gossip-based algorithms (namely through push, pull or push-pull mechanisms initiated by $u$ and its neighbors), from the node ids received during random walks initiated at $u$, or even from the induced churn imposed in structured-based overlays.

## 5 Characterization of the Uniform and Ergodic Sampling Problem

We start our characterization by showing that the adversary can bias the input stream in such a way that neither uniform nor ergodic properties can be met. This is achieved by flooding the input stream with sufficiently many malicious node ids. Specifically, Lemma 1 states that for any strategy $s$, if the number

$\delta_m$ of non unique malicious node ids that appear in the input stream of node $u \in \mathcal{N}$ during $T_s$ time units exceeds a given threshold then it is impossible for any node in the overlay to equally likely appear as a sample of node $u$, and this holds forever. Let $(\mathcal{C}_1)$ be a condition on $\delta_m$ value

$$\delta_m \leq T_s - (1-k)|\mathcal{S}|. \tag{$\mathcal{C}_1$}$$

Condition $(\mathcal{C}_1)$ characterizes the fact that for any honest node $v \in \mathcal{N}$, during the time interval $T_s$, $v$ has a non-null probability to appear in the input stream. We have

**Lemma 1**

$$\neg(\mathcal{C}_1) \implies \neg\mathcal{U} \wedge \neg\mathcal{E}.$$

*Proof.* Let $v \in \mathcal{N}$. Suppose that Condition $(\mathcal{C}_1)$ does not hold, namely it exists an adversarial behavior such that

$$\delta_m > T_s - (1-k)|\mathcal{S}|.$$

In this case, the number of honest node ids in the input stream at $v$ (*i.e.*, $T_s - \delta_m$) is strictly lower than $(1-k)|\mathcal{S}|$, which means formally that

$$T_s - \delta_m < (1-k)|\mathcal{S}|.$$

By assumption (*cf.* Section 3.1) the overlay is populated by $(1-k)|\mathcal{S}|$ honest nodes. Thus, as the adversary manages to flood the input stream at $v$, there exists at least one node id $u \in \mathcal{S}$ that will never appear in the stream. Therefore, whatever the strategy $s$, $v$'s sampling component can never output $u$. Thus,

$$\forall t > T_0, \ \mathbb{P}[u \in S_v(t)] = 0, \tag{1}$$

which clearly violates Property $\mathcal{U}$.

Equation (1) can be rewritten as $\exists t > T_0, \exists u \in \mathcal{S}, \forall t' > t, \mathbb{P}[u \in S_v(t')] = 0$, which has for consequence that the set of instants $t'$ for which $u$ can be sampled by $v$ is empty. Formally,

$$\mathbb{P}[\{t'|t' > T_0 \wedge u \in S_v(t')\} = \emptyset] = 1,$$

which violates Property $\mathcal{E}$, and completes the proof of the lemma. □

We now assume that Condition $(\mathcal{C}_1)$ holds. The second lemma states that if the size of the sampling memory is large enough, then whatever the constrained adversarial behavior, the sampling component succeeds in exhibiting uniform and ergodic samples. This makes a sufficient condition to solve our problem. Specifically, let $(\mathcal{C}_2)$ be defined as follows

$$|\Gamma| < |\mathcal{S}|. \tag{$\mathcal{C}_2$}$$

Condition $(\mathcal{C}_2)$ characterizes the fact that nodes cannot maintain the full knowledge of the population overlay (essentially for scalability reasons). Then,

**Lemma 2**

$$(\mathcal{C}_1) \wedge \neg(\mathcal{C}_2) \Longrightarrow \mathcal{U} \wedge \mathcal{E}.$$

*Proof.* Proof of the lemma is straightforward. By Condition $(\mathcal{C}_1)$, any node $u \in \mathcal{S}$ has a non-null probability to appear in the input stream of any node $v \in \mathcal{N}$. By assumption of the lemma, $|\Gamma_v| \geq |\mathcal{S}|$. Consider the basic strategy $s$ of $v$'s sampling component that consists in storing into $\Gamma_v$, any new id read from the input stream. Then eventually, all the node ids will be present into $\Gamma_v$, and thus any node $u$ is equally likely to be chosen in $\Gamma_v$, which guarantees Property $\mathcal{U}$.

Moreover, $v$ has the possibility to return infinitely often any node id $u$ present in $\Gamma_v$. Thus for any time $t$, the set of instants $t'$, with $t' > t$, such that $u$ is chosen has a zero probability to be empty, which provides Property $\mathcal{E}$ and completes the proof. □

The following Lemma completes the characterization of the problem, specifically:

**Lemma 3**

$$(\mathcal{C}_1) \wedge (\mathcal{C}_2) \Longrightarrow \neg(\mathcal{U} \wedge \mathcal{E}).$$

*Proof.* Suppose that both Conditions $(\mathcal{C}_1)$ and $(\mathcal{C}_2)$ hold. Proving that $\neg(\mathcal{U} \wedge \mathcal{E})$ is equivalent to showing that $(\neg\mathcal{E} \vee \neg\mathcal{U})$ holds, and thus, that $(\mathcal{E} \Longrightarrow \neg\mathcal{U})$ holds. Suppose that $(\mathcal{C}_1) \wedge (\mathcal{C}_2) \wedge \mathcal{E}$ is met, we now show that $\mathcal{U}$ cannot hold.

Consider any node $v \in \mathcal{N}$ (the set of honest nodes) and let $\Gamma_v(t)$ denote the content of $v$'s sampling memory at the instant $t$. From Condition $(\mathcal{C}_2)$,

$$\forall t' \geq T_0, \exists u \in \mathcal{S}, \quad u \notin \Gamma_v(t'). \tag{2}$$

In particular, Equation (2) is true for $t' = T_s$. Let node $w \in \mathcal{S}$ be such that $w \notin \Gamma_v(T_s)$, then by assumption, Property $\mathcal{E}$ holds. Thus

$$\exists t > T_s, \quad w \notin \Gamma_v(T_s) \wedge w \in \Gamma_v(t). \tag{3}$$

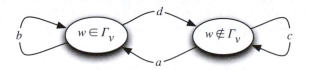

**Fig. 2.** Markov chain that represents the evolution of $w$'s presence in the sampling memory $\Gamma_v$ of node $v \in \mathcal{N}$

The presence of a node id in the local memory of the sampling component can be represented by a Markov chain. Figure 2 depicts the evolution of $w \in \Gamma_v$ as a function of the time. Labels $a, b, c$ and $d$ on the edges represent the probability of transitions from both states. We have $a + c = b + d = 1$. From Equation (3), we have $a > 0$ and thus, $c < 1$. We prove by contradiction that $d > 0$.

Suppose that $d = 0$, then $\forall t'' \geq t$, $w \in \Gamma_v(t'')$, the state $(w \in \Gamma_v)$ is absorbing. Suppose that the overlay contains only two nodes, $v$ and $w$. By assumption, at least one of the two nodes is honest ($k < 1$). Let us assume that $v$ is honest (the proof is similar for $w$). Then, by Condition $(\mathcal{C}_2)$, we have $|\Gamma_v| = 1$ (the case $|\Gamma_v| = 0$ trivially leads to impossibility). By assumption, we have $\forall t'' \geq t$, $w \in \Gamma_v(t'')$ and as $|\Gamma_v| = 1$, we also have $\forall t'' \geq t$, $\Gamma_v(t'') = \{w\}$. As a consequence, whatever the strategy $s$ implemented in $v$'s sampling component,

$$\forall t'' \geq t,\ \mathbb{P}[v \in S_v^s(t'')] = 0 \quad \Longrightarrow \quad \mathbb{P}[\{t''|t'' > t \wedge v \in S_v^s(t'')\} = \emptyset] > 0,$$

contradicting $\mathcal{E}$, and thus contradicting the assumption of the lemma. Thus $d > 0$ and, *a fortiori*, $b < 1$, and no state is absorbing.

Suppose now that $\mathcal{U}$ holds. We prove the lemma by contradiction. Consider again the case where the overlay is populated by only two nodes, $v$ and $w$. As above suppose that node $v$ is honest and that $|\Gamma_v| = 1$. The evolution of the sampling memory at node $v$ can be modeled by a Markov chain as represented in Figure 3. By assumption, $\mathcal{E}$ holds, thus infinitely often, and successively, both $v$ and $w$ appear in $\Gamma_v$. Moreover also by assumption, $\mathcal{U}$ holds, that is, $\forall t \geq T_s, \mathbb{P}[w \in S_v^s(t)] = \mathbb{P}[v \in S_v^s(t)] = \frac{1}{2}$. As a consequence, $w$ has the same probability as $v$ to be in $\Gamma_v$, whatever the number of times $w$ and $v$ appear in the stream before time $T_s$.

**Fig. 3.** Markov chain that represents the state of the local memory $\Gamma_v$ of $v$

Suppose now that node $w$ is malicious. By Condition $(\mathcal{C}_1)$, node id $w$ can appear in $v$'s stream no more than $T_s - 1$ times during any sliding window of $T_s$ time units. As $|\Gamma_v| = 1$, a single node id can be stored, and beyond this node id, no other additional information can be stored. We show that whatever the strategies $s$ implemented by $v$'s sampling component, they all lead to a contradiction.

**Blind replacement.** At any time $t$, the sampling component reads the first node id in the stream, and stores it in $\Gamma_v$ in place of the previous one. By construction, any strategy has to select its output among the elements stored in $\Gamma_v$, thus the output of the sampling component follows the same probability distribution as the one observed in the stream. As the adversary can flood the stream with up to $T_s - 1$ malicious node ids, this means that property $\mathcal{U}$ cannot be met.

**No replacement.** Similarly to the blind replacement strategy, node ids are read from the stream, and stored in $\Gamma_v$ up to time $t$, where $t$ is the first time at which a specific node id is read. From time $t$ onwards, this specific node id is kept in $\Gamma_v$, independently from the node ids read from the stream after $t$, leading to an absorbing state of the Markov chain. For instance, this specific node id can be the smallest image value under a random min-wise independent function, such as the min-wise permutation [16]. Clearly, this strategy violates property $\mathcal{E}$.

**Probabilistic replacement.** This strategy consists in substituting the current node id in $\Gamma_v$ with the next one read from the stream according to a given probability law. To guarantee that $\forall t, \mathbb{P}[w \in S_v^s(t)] = \mathbb{P}[v \in S_v^s(t)] = \frac{1}{2}$, then either both $v$ and $w$ have an equal probability to appear in the stream or the sampling component must be able to remember the node ids it has seen in the past to guarantee that, at any time $t$, each node id has the same probability to be chosen as sample. The former case does not hold as by assumption, the adversary can flood the stream with up to $T_s - 1$ malicious ids. Moreover, the latter case is impossible as by assumption $|\Gamma_v = 1|$, and thus a single information can be stored which prevents to store more than a single piece of information (*e.g.*, it is impossible to store both a node id and a counter), therefore property $\mathcal{U}$ cannot hold.

Thus $(\mathcal{C}_1) \wedge (\mathcal{C}_2) \implies \neg(\mathcal{U} \wedge \mathcal{E})$, which concludes the proof of the lemma. $\square$

The last lemma reformulates the necessary condition of the problem characterization by combining Lemmata 1 and 3.

**Lemma 4**
$$\mathcal{U} \wedge \mathcal{E} \implies (\mathcal{C}_1) \wedge \neg(\mathcal{C}_2).$$

*Proof.* The contrapositive form of writing Lemma 3 is $\mathcal{U} \wedge \mathcal{E} \implies \neg((\mathcal{C}_1) \wedge (\mathcal{C}_2))$, and thus, by distributivity,

$$\mathcal{U} \wedge \mathcal{E} \implies \neg(\mathcal{C}_1) \vee \neg(\mathcal{C}_2). \tag{4}$$

On the other hand, the contraposition of Lemma 1 leads to $\mathcal{U} \vee \mathcal{E} \implies (\mathcal{C}_1)$. As $(\mathcal{U} \wedge \mathcal{E} \Rightarrow \mathcal{U} \vee \mathcal{E})$, we have

$$\mathcal{U} \wedge \mathcal{E} \implies (\mathcal{C}_1). \tag{5}$$

By combining Equations 4 and 5, the following holds

$$\mathcal{U} \wedge \mathcal{E} \implies (\mathcal{C}_1) \wedge (\neg(\mathcal{C}_1) \vee \neg(\mathcal{C}_2)).$$

Thus,
$$\mathcal{U} \wedge \mathcal{E} \implies ((\mathcal{C}_1) \wedge \neg(\mathcal{C}_1)) \vee ((\mathcal{C}_1) \wedge \neg(\mathcal{C}_2)).$$

Due to the principle of contradiction, $(\mathcal{C}_1) \wedge \neg(\mathcal{C}_1)$ cannot hold, leading to

$$\mathcal{U} \wedge \mathcal{E} \implies (\mathcal{C}_1) \wedge \neg(\mathcal{C}_2),$$

which completes the proof. $\square$

The Uniform and Ergodic Sampling Problem defined in Sections 2 and 3 is completely characterized by the following theorem:

**Theorem 1.** $(\mathcal{C}_1) \wedge \neg(\mathcal{C}_2)$ *is a necessary and sufficient condition for Uniform and Ergodic Sampling Problem to hold.*

*Proof.* This result follows directly from the statements of Lemma 2 and 4. $\square$

# 6  Conclusion

In this paper, we have investigated the sampling problem of large-scale unstructured peer-to-peer systems in adversarial environments. We have first shown that, if the system cannot provide the means to limit resources of an adversary, then solving either uniform sampling or ergodic sampling is impossible. We have then demonstrated that, if this assumption holds and if all honest nodes in the system have access to a very large memory (in the size of the system) then, the problem becomes trivially solvable but not yet realistic. Unfortunately, we have shown that both conditions are necessary and sufficient ingredients to solve the uniform and ergodic sampling problem in potentially adversarial environments. Clearly, these strong conditions highlight the damage that adversarial behavior can cause in large-scale unstructured systems.

As future work, first we intend to study to which extent the adversary model needs to be weaken to achieve uniform and ergodic sampling in a setting where the nodes themselves have limited ressources (for instance in terms of memory). Second, we plan to investigate an approximate version of the sampling primitive to achieve near uniform and/or near ergodic sampling despite the presence of a strong adversary. Both studies should have a positive impact for applications exhibiting different requirements in terms of resources (*i.e.* memory, computational power and communication complexity) and for settings in which probabilistic guarantees on samples are sufficient.

# Acknowledgements

We are very grateful to the anonymous reviewers for their constructive comments that have help us to improve the quality of this paper.

# References

1. Jelasity, M., Voulgaris, S., Guerraoui, R., Kermarrec, A.M., van Steen, M.: Gossip-based Peer Sampling. ACM Transaction on Computer System 25(3) (2007)
2. Bertier, M., Busnel, Y., Kermarrec, A.M.: On Gossip and Populations. In: Kutten, S., Žerovnik, J. (eds.) SIROCCO 2009. LNCS, vol. 5869, pp. 72–86. Springer, Heidelberg (2009)
3. Karger, D.R., Ruhl, M.: Simple Efficient Load Balancing Algorithms for Peer-to-Peer. In: Voelker, G.M., Shenker, S. (eds.) IPTPS 2004. LNCS, vol. 3279, pp. 131–140. Springer, Heidelberg (2004)
4. Lv, Q., Cao, P., Cohen, E., Li, K., Shenker, S.: Search and Replication in Unstructured Peer-to-Peer Networks. In: Proceedings of the International Conference on Supercomputing (ICS), pp. 84–95 (2002)
5. Massoulié, L., Merrer, E.L., Kermarrec, A.M., Ganesh, A.: Peer Counting and Sampling in Overlay Networks: Random Walk Methods. In: Proceedings of the 25th Annual Symposium on Principles of Distributed Computing (PODC), pp. 123–132. ACM Press, New York (2006)

6. Busnel, Y., Beraldi, R., Baldoni, R.: A Formal Characterization of Uniform Peer Sampling Based on View Shuffling. In: Proceedings of the International Conference on Parallel and Distributed Computing, Applications and Technologies (PDCAT), pp. 360–365. IEEE Computer Society, Los Alamitos (2009)
7. Gurevich, M., Keidar, I.: Correcness of Gossip-Based Membership under Message Loss. In: Proceedings of the 28th Annual Symposium on Principles of Distributed Computing (PODC), Calgary, AL, Canada. ACM Press, New York (2009)
8. Bakhshi, R., Gavidia, D., Fokkink, W., van Steen, M.: An Analytical Model of Information Dissemination for a Gossip-based Protocol. Computer Networks 53(13), 2288–2303 (2009)
9. Karp, R., Schindelhauer, C., Shenker, S., Vocking, B.: Randomized Rumor Spreading. In: the 41st Annual Symposium on Foundations of Computer Science (FOCS), p. 565. IEEE Computer Society, Los Alamitos (2000)
10. Voulgaris, S., Gavidia, D., van Steen, M.: CYCLON: Inexpensive Membership Management for Unstructured P2P Overlays. Journal of Network System Management 13(2), 197–217 (2005)
11. Stutzbach, D., Rejaie, R., Duffield, N., Sen, S., Willinger, W.: On Unbiased Sampling for Unstructured Peer-to-Peer Networks. IEEE/ACM Transactions on Networking 17(02), 377–390 (2009)
12. Bollobás, B.: Random Graphs, 2nd edn. Cambridge University Press, Cambridge (2001)
13. Zhong, M., Shen, K., Seiferas, J.: Non-uniform Random Membership Management in Peer-to-Peer Networks. In: Proceedings of the 24th Annual Joint Conference of the Computer and Communications Societies (INFOCOM). IEEE Press, Los Alamitos (2005)
14. Awan, A., Ferreira, R.A., Jagannathan, S., Grama, A.: Distributed Uniform Sampling in Unstructured Peer-to-Peer Networks. In: Proceedings of the 39th Annual Hawaii International Conference on System Sciences, HICSS (2006)
15. Sit, E., Morris, R.: Security Considerations for Peer-to-Peer Distributed Hash Tables. In: Druschel, P., Kaashoek, M.F., Rowstron, A. (eds.) IPTPS 2002. LNCS, vol. 2429, pp. 261–269. Springer, Heidelberg (2002)
16. Bortnikov, E., Gurevich, M., Keidar, I., Kliot, G., Shraer, A.: Brahms: Byzantine Resilient Random Membership Sampling. Computer Networks 53, 2340–2359 (2009); A former version appeared in the 27th ACM Symposium on Principles of Distributed Computing (PODC) (2008)
17. Awerbuch, B., Scheideler, C.: Group Spreading: A Protocol for Provably Secure Distributed Name Service. In: Díaz, J., Karhumäki, J., Lepistö, A., Sannella, D. (eds.) ICALP 2004. LNCS, vol. 3142, pp. 183–195. Springer, Heidelberg (2004)
18. Jesi, G.P., Montresor, A., van Steen, M.: Secure Peer Sampling. Computer Networks (2010) (to appear)
19. Singh, A., Ngan, T.W., Druschel, P., Wallach, D.S.: Eclipse Attacks on Overlay Networks: Threats and Defenses. In: Proceedings of the 25th IEEE International Conference on Computer Communications, INFOCOM (2006)
20. Liu, D., Ning, P., Du, W.: Detecting Malicious Beacon Nodes for Secure Location Discovery in Wireless Sensor Networks. In: Proceedings of the 25th IEEE International Conference on Distributed Computing Systems, ICDCS (2005)
21. Douceur, J., Donath, J.S.: The Sybil Attack. In: Druschel, P., Kaashoek, M.F., Rowstron, A. (eds.) IPTPS 2002. LNCS, vol. 2429, pp. 251–260. Springer, Heidelberg (2002)

22. Awerbuch, B., Scheideler, C.: Towards a Scalable and Robust Overlay Network. In: Proceedings of the 6th International Workshop on Peer-to-Peer Systems, IPTPS (2007)
23. Anceaume, E., Brasileiro, F.V., Ludinard, R., Sericola, B., Tronel, F.: Analytical Study of Adversarial Strategies in Cluster-based Overlays. In: Proceedings of the International Conference on Parallel and Distributed Computing, Applications and Technologies (PDCAT), pp. 293–298 (2009)
24. Castro, M., Druschel, P., Ganesh, A., Rowstron, A., Wallach, D.S.: Secure Routing for Structured Peer-to-peer Overlay Networks. In: Proceedings of the 5th Symposium on Operating Systems Design and Implementation (OSDI), pp. 299–314. ACM, New York (2002)
25. Hildrum, K., Kubiatowicz, J.: Asymptotically Efficient Approaches to Fault-tolerance in Peer-to-Peer Networks. In: Fich, F.E. (ed.) DISC 2003. LNCS, vol. 2848, pp. 321–336. Springer, Heidelberg (2003)
26. Fiat, A., Saia, J., Young, M.: Making Chord Robust to Byzantine Attacks. In: Brodal, G.S., Leonardi, S. (eds.) ESA 2005. LNCS, vol. 3669, pp. 803–814. Springer, Heidelberg (2005)
27. Anceaume, E., Brasileiro, F., Ludinard, R., Ravoaja, A.: PeerCube: an Hypercube-based P2P Overlay Robust against Collusion and Churn. In: Proceedings of the IEEE International Conference on Self-Adaptive and Self-Organizing Systems, pp. 15–24 (2008)
28. Condie, T., Kacholia, V., Sank, S., Hellerstein, J.M., Maniatis, P.: Induced Churn as Shelter from Routing-Table Poisoning. In: Proceedings of the International Network and Distributed System Security Symposium, NDSS (2006)
29. Yu, H., Kaminsky, M., Gibbons, P.B., Flaxman, A.: SybilGuard: Defending against Sybil Attacks via Social Networks. In: Proceedings of the ACM Conference on Applications, Technologies, Architectures, and Protocols for Computer Communications (SIGCOMM), pp. 267–278 (2006)
30. Yu, H., Gibbons, P.B., Kaminsky, M., Xiao, F.: SybilLimit: A Near-Optimal Social Network Defense against Sybil Attacks. In: Proceedings of the IEEE Symposium on Security and Privacy (SP), pp. 3–17 (2008)

# Self-stabilizing
# (k,r)-Clustering in Wireless Ad-hoc
# Networks with Multiple Paths*

Andreas Larsson and Philippas Tsigas

Chalmers University of Technology and Göteborg University
{larandr,tsigas}@chalmers.se

**Abstract.** Wireless Ad-hoc networks are distributed systems that often reside in error-prone environments. Self-stabilization lets the system recover autonomously from an arbitrary state, making the system recover from errors and temporarily broken assumptions. Clustering nodes within ad-hoc networks can help in many ways like forming backbones, facilitating routing, improving scaling, aggregating information and saving power. A (k,r)-clustering assigns cluster heads so that exists $k$ cluster heads within $r$ communication hops for all nodes in the network while trying to minimize the total number of cluster heads. We present the first self-stabilizing distributed (k,r)-clustering algorithm. The algorithm uses synchronous communication rounds and uses multiple paths to different cluster heads for providing improved security, availability and fault tolerance. From any starting configuration the algorithm quickly assigns enough cluster heads and stabilizes towards a local minimum using a randomized scheme.

**Keywords:** Clustering, Self-Stabilization, Ad-hoc Networks, (k,r)-dominating sets.

## 1 Introduction

An algorithm for clustering nodes together in an ad-hoc network serves an important role. It can be used for back bone formation, routing, data aggregation, improve scaling and energy saving by taking turns. Clustering is a well studied problem. Due to space constraints we point to the survey of the area with regard to wireless ad-hoc networks by Chen et al. in [2] for references to the area in general. We will focus on self-stabilization, redundancy and some security aspects. One way of clustering nodes in a network is for nodes to associate themselves with one or more cluster heads. In the (k,r)-clustering problem each node in the network should have, if possible, at least $k$ cluster heads within $r$ communication hops away. Assuming that the network topology allows $k$ cluster heads for each

* The research leading to these results has received funding from the Swedish Civil Contingencies Agency (MSB) and has received funding from the European Union Seventh Framework Programme (FP7/2007-2013) under grant agreement n° 257007.

C. Lu, T. Masuzawa, and M. Mosbah (Eds.): OPODIS 2010, LNCS 6490, pp. 79–82, 2010.

node, the set of cluster heads forms a *total (k,r)-dominating set* where the nodes in the set, the cluster heads, also need to have $k$ nodes in the set within $r$ hops. The dominating set problem is well known to be NP-hard in general. Therefore, instead of looking for a global minimum, approximate algorithms are proposed.

Starting from an arbitrary state, self stabilizing algorithms let a system stabilize to, and stay in, a consistent state [3]. There is a multitude of existing clustering algorithms for ad-hoc networks, of which a number is self-stabilizing. A self-stabilizing (1,1)-clustering algorithm that converges fast is presented in [5]. A lot of organizational problems is tackled in a self-stabilizing manner and a self-stabilizing (1,r)-clustering algorithm is presented in [4]. Weighted graphs is taken into account in the self-stabilizing (1,r)-clustering in [1]. Algorithms for the full (k,r)-clustering problem is presented in [7] and [9], but neither is self-stabilizing. The algorithm presented in [8] groups nodes together without assigning cluster heads. It considers malicious nodes inside the network, but is not self-stabilizing.

## 1.1  Our Contribution

We have constructed the first, to the best of our knowledge, self-stabilizing $(k, r)$-clustering algorithm for ad-hoc networks. The algorithm is based on synchronous rounds and makes sure that, within $O(r)$ rounds, all nodes have at least $k$ cluster heads (if the topology permits it) using a deterministic scheme. A randomized scheme complements the deterministic scheme and lets the set of cluster heads stabilize to a local minimum. It stabilizes within $O(g \cdot r \cdot \log n)$ rounds with high probability, where $g$ is a bound on the number of nodes within $2r$ hops, and $n$ is the size of the network.

Our contribution is presented as follows. In section 2 we introduce the system settings. Section 3 describes the algorithm. We discuss multiple paths, proofs and experiments, and conclude in Section 4.

## 2  System Settings

We assume a static network. Changes in the topology are seen as transient faults. We impose no restrictions on the network topology other than that an upper bound, $g$, on the number of nodes within $2r$ hops of any node is known. For a node $p_i$ in the network, we denote the nodes within one hop $N_i$, i.e. the nodes to which it can directly send messages. The nodes within $r$ hops from $p_i$, including $p_i$ itself, is denoted $G_i^r$. We assume undirected communication graphs, i.e. $p_i \in N_j$ iff $p_j \in N_i$. The system is synchronous and progresses in rounds. Each round has two phases. In the receipt phase each node $p_i$ receives messages from all nodes in $N_i$. In the step phase each node $p_i$ broadcasts a message to all nodes $p_j \in N_i$ and that is received reliably in the next receipt phase.

## 3  Self-Stabilizing Algorithm for $(k, r)$-Clustering

The goal of the algorithm is, using as few cluster heads as possible, for each node $p_i$ in the network to have a set of at least $k_i = \min(k, |G_i^r|)$ cluster heads within

```
1  on step phase:
2    increase timer modulo T
3    schedule escape attempt uniformly at random from [0,T-2r-2] if timer = 0
4    if state = HEAD and |heads| > k and it is time to attempt escape then
5       state ← ESCAPING
6       remove i from heads
7    if state = ESCAPING for 2r+1 rounds (all nodes in Gᵢʳ had the chance to veto escape) then
8       state ← SLAVE
9    if |heads| < k then
10      add more nodes in Gᵢʳ to heads, if there are any left to add
11   Broadcast state and heads with a TTL of r to direct neighbors, together with queued forwards
12
13 on receive jstate and jheads originating from node pⱼ
14   if i ∈ jheads and state was not set to ESCAPING within the last 2r rounds then
15      state ← HEAD
16      add i to heads
17   if j ∈ heads and jstate = ESCAPING and |heads| > k then
18      remove j from heads
19   else if jstate = HEAD then
20      add j to heads if not present
21   queue up a forward of jstate and jheads with one lower TTL if TTL is not already 1
```

**Fig. 1.** Simplified pseudocode of the clustering algorithm, for a node $p_i$ with id $i$

its $r$-hop neighborhood $G_i^r$. We achieve a local minimum, i.e. a set from which no cluster head can be removed without violating the aforementioned goal.

Space does not allow us to present the algorithm it its entirety. Full description with all details and everything that is needed to get self-stabilization can be found in [6]. The basic idea of the algorithm is presented in Figure 1. A node can have state HEAD or ESCAPING, in which it is a cluster head, or have state SLAVE, in which it is not a cluster head. In every round each node $p_i$ sends out its state and its cluster heads (line 11). Forwarding (line 21) with a time-to-live (TTL) mechanism makes sure that this information reaches all nodes in $G_i^r$ within $r$ rounds. A node $p_i$ that does not have $k$ cluster heads elects new ones by adding them to *heads* (lines 9-10), which will be broadcasted as a join signal (line 11). A node that receives such a join signal becomes a cluster head if not already so (lines 14-16). A node that gets to know about a new cluster head adds it to *heads* (lines 19-20).

This procedure might overshoot and establish too many cluster heads in the network, i.e. there is some cluster head node $p_j$ for which all nodes that has $p_j$ as a cluster head also have at least $k$ other cluster heads. To reach a local minimum, a cluster head node $p_i$ tries to escape at random points in time. This is done by setting the state to ESCAPING (lines 2-6) and ignoring incoming join signals for $2r$ rounds (line 14). If a node can allow a cluster head to go, it removes it from *heads* (lines 17-18). Otherwise it disallows the escape attempt by continuing to send join signals. If no nodes disallow an escape attempt, the node can become SLAVE (line 7-8). If two cluster heads of which any one, but only one, can be allowed escape, tries to escape concurrently, both might fail. Therefore, the cluster heads repeatedly starts escape attempts at random points in time, within periods of a constant $T$ rounds, to resolve such conflicts. Eventually their attempts will not overlap.

# 4    Discussion and Conclusions

The full algorithm merges information received by different sources and with
different TTL values. This reduces the total amount of information that needs
to be transmitted. The flooding of messages makes sure that if there exist
multiple paths within $r$ hops between a pair of nodes they will all be used. This
provides higher fault tolerance. It can also give us higher security if nodes in the
network can be compromised.

In [6] we prove that enough cluster heads are assigned within $O(r)$ rounds.
Under the assumption of synchronized timers, we prove that the set of cluster heads convergences to a local minimum within $O(g \cdot r \cdot \log n)$ rounds, with
high probability. We also show experimentally that synchronized timers are not
needed and that $g$ does not have to be known very accurately. In addition, we
present some preliminary experimental results on how far from optima our sets
of cluster heads come, and provide more thorough discussions on security and
discussions on complexity versus fault tolerance.

To conclude, we have presented the first self-stabilizing $(k, r)$-clustering algorithm for ad-hoc networks. A deterministic mechanism guarantees that all
nodes, if possible for the given topology, have $k$ cluster heads within $r$ hops. A
randomized mechanism lets the set of cluster heads stabilize to a local minimum.

# References

1. Caron, E., Datta, A.K., Depardon, B., Larmore, L.L.: A self-stabilizing k-clustering
   algorithm using an arbitrary metric. In: Sips, H., Epema, D., Lin, H.-X. (eds.) Euro-
   Par 2009. LNCS, vol. 5704, pp. 602–614. Springer, Heidelberg (2009)
2. Chen, Y.P., Liestman, A.L., Liu, J.: Clustering Algorithms for Ad Hoc Wireless
   Networks, vol. 2, ch. 7, pp. 154–164. Nova Science Publishers, Bombay (2004)
3. Dolev, S.: Self-Stabilization. MIT Press, Cambridge (March 2000)
4. Dolev, S., Tzachar, N.: Empire of colonies: Self-stabilizing and self-organizing distributed algorithm. Theoretical Computer Science 410(6-7), 514–532 (2009)
5. Johnen, C., Nguyen, L.H.: Robust self-stabilizing weight-based clustering algorithm.
   Theoretical Computer Science 410(6-7), 581–594 (2009)
6. Larsson, A., Philippas, T.: Self-stabilizing (k, r)-clustering in wireless ad-hoc networks with multiple paths. Technical report, Chalmers University of Technology
   (2010)
7. Spohn, M.A., Garcia-Luna-Aceves, J.J.: Bounded-distance multi-clusterhead formation in wireless ad hoc networks. Ad Hoc Networks 5(4), 504–530 (2007)
8. Sun, K., Peng, P., Ning, P., Wang, C.: Secure distributed cluster formation in wireless sensor networks. In: ACSAC 2006, Washington, DC, USA, pp. 131–140. IEEE
   Computer Society, Los Alamitos (2006)
9. Wu, Y., Li, Y.: Construction algorithms for k-connected m-dominating sets in wireless sensor networks. In: MobiHoc 2008, pp. 83–90. ACM, New York (2008)

# Self-stabilizing Byzantine Asynchronous Unison[*][**]

Swan Dubois[1], Maria Gradinariu Potop-Butucaru[1],
Mikhail Nesterenko[2], and Sébastien Tixeuil[1]

[1] LIP6 - UMR 7606 Université Pierre et Marie Curie - Paris 6 & INRIA, France
[2] Kent State University, USA

**Abstract.** We explore asynchronous unison in the presence of systemic transient and permanent Byzantine faults in shared memory. We observe that the problem is not solvable under less than strongly fair scheduler or for system topologies with maximum node degree greater than two.

We present a self-stabilizing Byzantine-tolerant solution to asynchronous unison for chain and ring topologies. Our algorithm has minimum possible containment radius and optimal stabilization time.

*Asynchronous unison* [2, 3] requires processors to maintain synchrony between their counters called *clocks*. Specifically, each processor has to increment its clock indefinitely while the clock *drift* from its neighbors should not exceed 1. Asynchronous unison is a fundamental building block for a number of principal tasks in distributed systems such as distributed snapshots [4] and synchronization [5, 6].

A practical large-scale distributed system must counter a variety of transient and permanent faults. A systemic transient fault may perturb the system and leave it in an arbitrary configuration. *Self-stabilization* [7, 8] is a versatile technique for transient fault forward recovery. *Byzantine fault* [9] is the most generic permanent fault model: a faulty processor may behave arbitrarily. However, designing distributed systems that handle both transient and permanent faults proved to be rather difficult [10–12]. Some of the difficulty is due to the inability of the system to counter Byzantine behavior by relying on the information contained in the global system configuration: a transient fault may place the system in an arbitrary configuration.

In the context of the above discussion, considering joint Byzantine and systemic transient fault tolerance for asynchronous unison appears futile. Indeed, the Byzantine processor may keep setting its clock to an arbitrary value while the clocks of the correct processors are completely out of synchrony. Hence, we are happy to report that the problem is solvable in some cases. In this paper we present a shared-memory Byzantine-tolerant self-stabilizing asynchronous unison algorithm that operates on chain and ring system topologies. The algorithm operates under a strongly fair scheduler. We show that the problem is unsolvable

---

[*] A full version of this work is available in [1].
[**] This work was funded in part by ANR projects SHAMAN, ALADDIN, and SPADES.

C. Lu, T. Masuzawa, and M. Mosbah (Eds.): OPODIS 2010, LNCS 6490, pp. 83–86, 2010.
© Springer-Verlag Berlin Heidelberg 2010

for any other topology or under less stringent scheduler. Our algorithm achieves minimal fault-containment radius: each correct processor eventually synchronizes with its correct neighbors. We prove our algorithm correct and demonstrate that its stabilization time is asymptotically optimal.

*Related work.* The impetus of the present research is the result by Dubois et al [13]. They consider joint tolerance to crash faults and systemic transient faults. The key observation that enables this avenue of research is that the adopted definition of asynchronous unison does not preclude the correct processors from decrementing their clocks. This allows the processors to synchronize and maintain unison even while their neighbors may crash or behave arbitrarily.

There are several pure self-stabilizing solutions to the unison problem [2, 3, 14]. None of those tolerate Byzantine faults. Classic Byzantine fault tolerance focuses on masking the fault. There are self-stabilizing Byzantine-tolerant clock synchronization algorithms for completely connected synchronous systems both probabilistic [11, 15] and deterministic [16, 17]. The probabilistic and deterministic solutions tolerate up to one-third and one-fourth of faulty processors respectively.

Another approach to joint transient and Byzantine tolerance is *containment*. For tasks whose correctness can be checked locally, such as vertex coloring, link coloring or dining philosophers, the fault may be isolated within a region of the system. *Strict stabilization* guarantees that there exists a containment radius outside of which the processors are not affected by the fault [12, 18–20]. We say that that an algorithm is $(c, f)$−strictly stabilizing if it is strictly stabilizing with a radius of $c$ and can tolerate up to $f$ Byzantine processors. Yet some problems are not local and do not admit strict stabilization. However, the tolerance requirements may be weakened to *strong-stabilization* [21, 22] which allows the processors outside the containment radius to be affected by Byzantine processors after the convergence of the system. The faulty processors can affect these correct processors only a finite number of times after the convergence of the system. Strong-stabilization enables solution to several problems, such as tree orientation and tree construction.

*Contributions.* The first step is to characterize the necessary assumptions on the system (topology, scheduling, and number of faults) to solve the asynchronous unison in a strictly stabilizing way. Following [13] and the fact that a Byzantine processor may simulate a crashed one, we can state some impossibility results.

**Theorem 1.** *There does not exist a $(f, d)$-strictly stabilizing solution to the asynchronous unison problem in shared memory for any distance $d \geq 0$ if the communication graph of the distributed system contains processors of degree greater than two or if the number of faults is greater than one or if the scheduler is either unfair or weakly fair.*

Then, it remains to explore the case of asynchronous unison on chains and rings under the strongly fair scheduler with at most one Byzantine processor. We provide in the following the intuition of the provided algorithm.

The *drift* between two processors $p$ and $q$ is the absolute value of the difference between their clock values. Two processors $p$ and $q$ are *in unison* if the drift between them is no more than 1. An *island* is a segment of correct processors such that for each processor $p$, if its neighbor $q$ is also in this island, then $p$ and $q$ are in unison. A processor with no in-unison neighbors is assumed to be a single-processor island. The main idea of the algorithm is as follows. Our algorithm forms islands of processors with synchronized clocks. The algorithm is designed such that the clocks of the processors with adjacent islands drift closer to each other and the islands eventually merge. If a faulty processor restricts the drift of one such island, for example by never changing its clock, the other islands still drift and synchronize with the affected island. Then, we prove the following theorem:

**Theorem 2.** *Our algorithm is a $(1, 0)$-strictly stabilizing asynchronous unison under the strongly fair scheduler on ring and chains topology.*

Finally, we explore the time complexity of our algorithm and of the problem. We prove the following result:

**Theorem 3.** *The stabilization complexity of our algorithm is optimal. It stabilizes in $\Theta(L)$ asynchronous rounds where $L$ is the largest drift between correct processors in the initial configuration of the system.*

*Conclusion.* In this paper we explored joint tolerance to Byzantine and systemic transient faults for the asynchronous unison problem in shared memory. Some open problems follow.

Solutions under distributed scheduler, that allows multiple concurrent steps, remain to be explored. Another way to complete these results is to consider bounded clocks. The existence of a solution for shared memory execution model opens another avenue of research. It is interesting to consider the existence of a solution in lower atomicity models such as shared register or message-passing. We conjecture that a solution in such model is more difficult to obtain as the lower atomicity tends to empower faulty processors. Indeed, in the shared-register model a Byzantine processor may report differing clock values to its right and left neighbor. Such behavior makes a single fault ring topology essentially equivalent to two fault chain topology. The latter is proven unsolvable. Hence, we posit that in the lower atomicity models, the only topology that allows a solution to asynchronous unison to the chain.

# References

1. Dubois, S., Potop-Butucaru, M.G., Nesterenko, M., Tixeuil, S.: Self-stabilizing byzantine asynchronous unison. CoRR abs/0912.0134 (2009)
2. Gouda, M.G., Herman, T.: Stabilizing unison. Inf. Process. Lett. 35(4), 171–175 (1990)
3. Couvreur, J.M., Francez, N., Gouda, M.G.: Asynchronous unison (extended abstract). In: ICDCS, pp. 486–493 (1992)
4. Chandy, K.M., Lamport, L.: Distributed snapshots: Determining global states of distributed systems. ACM Trans. Comput. Syst. 3(1), 63–75 (1985)

5. Awerbuch, B.: Complexity of network synchronization. J. ACM 32(4), 804–823 (1985)
6. Awerbuch, B., Kutten, S., Mansour, Y., Patt-Shamir, B., Varghese, G.: A time-optimal self-stabilizing synchronizer using a phase clock. IEEE Trans. Dependable Sec. Comput. 4(3), 180–190 (2007)
7. Dijkstra, E.W.: Self-stabilizing systems in spite of distributed control. ACM Commun. 17(11), 643–644 (1974)
8. Dolev, S.: Self-stabilization. MIT Press, Cambridge (March 2000)
9. Lamport, L., Shostak, R.E., Pease, M.C.: The byzantine generals problem. ACM Trans. Program. Lang. Syst. 4(3), 382–401 (1982)
10. Daliot, A., Dolev, D.: Self-stabilization of byzantine protocols. In: Herman, T., Tixeuil, S. (eds.) SSS 2005. LNCS, vol. 3764, pp. 48–67. Springer, Heidelberg (2005)
11. Dolev, S., Welch, J.L.: Self-stabilizing clock synchronization in the presence of byzantine faults. J. ACM 51(5), 780–799 (2004)
12. Nesterenko, M., Arora, A.: Tolerance to unbounded byzantine faults. In: 21st Symposium on Reliable Distributed Systems (SRDS 2002), p. 22. IEEE Computer Society, Los Alamitos (2002)
13. Dubois, S., Potop-Butucaru, M., Tixeuil, S.: Brief announcement: Dynamic FTSS in Asynchronous Systems: the Case of Unison. In: Keidar, I. (ed.) DISC 2009. LNCS, vol. 5805, pp. 291–293. Springer, Heidelberg (2009)
14. Boulinier, C., Petit, F., Villain, V.: When graph theory helps self-stabilization. In: Chaudhuri, S., Kutten, S. (eds.) PODC, pp. 150–159. ACM, New York (2004)
15. Ben-Or, M., Dolev, D., Hoch, E.N.: Fast self-stabilizing byzantine tolerant digital clock synchronization. In: Bazzi, R.A., Patt-Shamir, B. (eds.) PODC, pp. 385–394. ACM, New York (2008)
16. Dolev, D., Hoch, E.N.: On self-stabilizing synchronous actions despite byzantine attacks. In: Pelc, A. (ed.) DISC 2007. LNCS, vol. 4731, pp. 193–207. Springer, Heidelberg (2007)
17. Hoch, E.N., Dolev, D., Daliot, A.: Self-stabilizing byzantine digital clock synchronization. In: [23], pp. 350–362
18. Masuzawa, T., Tixeuil, S.: Stabilizing link-coloration of arbitrary networks with unbounded byzantine faults. International Journal of Principles and Applications of Information Science and Technology (PAIST) 1(1), 1–13 (2007)
19. Sakurai, Y., Ooshita, F., Masuzawa, T.: A self-stabilizing link-coloring protocol resilient to byzantine faults in tree networks. In: Higashino, T. (ed.) OPODIS 2004. LNCS, vol. 3544, pp. 283–298. Springer, Heidelberg (2005)
20. Dubois, S., Masuzawa, T., Tixeuil, S.: The impact of topology on byzantine containment in stabilization. In: Lynch, N.A., Shvartsman, A.A. (eds.) DISC 2010. LNCS, vol. 6343, pp. 495–509. Springer, Heidelberg (2010)
21. Masuzawa, T., Tixeuil, S.: Bounding the impact of unbounded attacks in stabilization. In: [23], pp. 440–453
22. Dubois, S., Masuzawa, T., Tixeuil, S.: On byzantine containment properties of the min+1 protocol. In: Dolev, S., Cobb, J., Fischer, M., Yung, M. (eds.) SSS 2010. LNCS, vol. 6366, pp. 96–110. Springer, Heidelberg (2010)
23. Datta, A.K., Gradinariu, M. (eds.): SSS 2006. LNCS, vol. 4280. Springer, Heidelberg (2006)

# Reliably Detecting Connectivity Using Local Graph Traits

Alejandro Cornejo and Nancy Lynch

Massachusetts Institute of Technology,
Cambridge MA 02139-4307, USA

**Abstract.** Local distributed algorithms can only gather sufficient information to identify local graph traits, that is, properties that hold within the local neighborhood of each node. However, it is frequently the case that global graph properties (connectivity, diameter, girth, etc) have a large influence on the execution of a distributed algorithm.

This paper studies local graph traits and their relationship with global graph properties. Specifically, we focus on graph $k$-connectivity. First we prove a negative result that shows there does not exist a local graph trait which perfectly captures graph $k$-connectivity. We then present three different local graph traits which can be used to reliably predict the $k$-connectivity of a graph with varying degrees of accuracy.

As a simple application of these results, we present upper and lower bounds for a local distributed algorithm which determines if a graph is $k$-connected. As a more elaborate application of local graph traits, we describe, and prove the correctness of, a local distributed algorithm that preserves $k$-connectivity in mobile ad hoc networks while allowing nodes to move independently whenever possible.

## 1 Introduction

The $t$-neighborhood of a node $u$ of a graph $G$, is the induced subgraph of $G$ consisting of all vertices at distance at most $t$ from $u$, and all edges connecting two such vertices. A graph trait is a pair $(t, T)$ where $t$ is a function from the positive integers to the positive integers, and $T$ is a predicate over a graph. A graph trait $(t, T)$ is satisfied by a graph $G$ on $n$ vertices, if the $t(n)$-neighborhood of every node of $G$ satisfies $T$. A graph trait is local if $t$ is a constant.

Our motivation for studying local graph traits comes from the classical synchronous distributed system model. In this model, each node of an undirected graph $G$ is occupied by a processor. The system progresses in synchronous lock-step rounds, and at each round a process can send a message to its neighbors, receive messages, and perform local computation. Observe that after running for $t$ rounds, the knowledge of a process is limited to learning about all nodes at distance at most $t$, as well as the edges present between these nodes (i.e. its $t$-neighborhood). Since we do not restrict either the amount of local computation or the message size, it follows that after $O(diameter(G))$ rounds, every process can acquire complete knowledge of the graph and can compute any function of

C. Lu, T. Masuzawa, and M. Mosbah (Eds.): OPODIS 2010, LNCS 6490, pp. 87–102, 2010.

$G$. Therefore, distributed algorithms whose runtime is independent of the diameter of the network are especially interesting. Awerbuch et al. [1] defined a *local* algorithm as one whose runtime is significantly smaller than $n$ for any possible diameter of the network[1]. Local distributed algorithms can only learn their local neighborhood, and therefore they are limited to observing local graph traits.

Despite the fact that local distributed algorithms are limited to observe local graph traits, it is often the case that global graph properties have a great influence on the execution of a distributed algorithm. For example, the chromatic number of a graph is a lower bound on the number of rounds required for every node to broadcast once without colliding with its neighbors. Similarly, in algorithms which require coordination, the connectivity of a graph is an upper bound on the fault-tolerance of an algorithm, since higher connectivity implies more nodes can fail without disconnecting the graph.

Given the effects that graph properties, both local and global, have on the execution of distributed algorithms, it is not surprising that studying the relationship between local graph traits and global graph properties is a fruitful direction for proving upper and lower bounds on local distributed algorithms. This was first observed in the seminal work of Linial [11], who used an elegant construction relying on $t$-neighborhood graphs to prove that any distributed algorithm that finds a maximal independent set in a cycle must take at at least $\Omega(\log^* n)$ rounds.

However, the study of the relationship of local graph traits and global graph properties dates further back. In 1983, Wigderson [14] showed that if a graph is locally $k$-chromatic, then it has a chromatic number of $O(\sqrt{kn})$. Even earlier, in 1952, Dirac [6] proved that if $G$ has at least three vertices, and all nodes have degree at least $n/2$, then $G$ is Hamiltonian. In the same vein, we study local graph traits which imply global graph $k$-connectivity.

Paraphrasing the formal definition given in Section 2, the connectivity of a graph $G$, denoted $\kappa(G)$, is the size of the smallest set of vertices whose removal disconnects the graph. Although a complete graph on $n$ vertices cannot be disconnected by removing vertices, by convention its connectivity is $n-1$. We say a graph $G$ is $k$-connected if $\kappa(G) \geq k$. In Section 3, we show that there does not exist a local graph trait that characterizes a $k$-connected graph. More precisely, we prove that for any constant $k > 0$ there does not exist a local graph trait $(t, T)$ such that a graph $G$ satisfies $(t, T)$ if and only if $G$ is $k$-connected. We show a similar result holds even when considering only *simply connected* graphs. Namely, there does not exist a local graph trait $(t, T)$ such that a connected graph $G$ satisfies satisfies $(t, T)$ if and only if $G$ is $k$-connected. These results hold even in the case of unit disk graphs.

Since its not possible to locally characterize the $k$-connectivity of a graph, in Section 4 we turn our attention to local graph traits that when satisfied imply

---

[1] We remark that the algorithms presented in this paper satisfy a more stringent notion of locality, since their runtime is constant and therefore independent of $n$ or the diameter of the network. However, our impossibility results hold for the weaker notion of locality.

$k$-connectivity. Specifically, we describe three different local graph traits which are parametrized by $k$, and when fulfilled imply that the is graph $k$-connected.

As a simple application of these results, in Section 5 we present straightforward algorithmic implementations of the local traits described in Section 4 which yield constant time distributed algorithms to test for $k$-connectivity. We also describe a lower bound for distributed algorithms that reliably predict $k$-connectivity, which is derived directly from the impossibility results described in Section 3. As a more elaborate application, we show how to exploit the local graph traits presented, to extend the algorithm described in [4] to preserve $k$-connectivity in a mobile ad hoc network while allowing the agents of the network to move as freely as possible.

Most of the previous work on $k$-connectivity is in the field of topology control. Jorgic et al. [8] reported the experimental results of three different distributed algorithms to detect $k$-connectivity on random geometric graphs, but the paper lacks any formal guarantees. Czumaj and Zhao [5] presented a greedy centralized algorithm to construct a $k$-connected $t$-spanner with runtime $\tilde{O}(nk)$. Thurimella [13] described a distributed algorithm to identify sparse $k$-connected subgraphs that runs in $O(diameter(G) + \sqrt{n})$ time. Jia et al. [7] described a centralized algorithm to approximate the minimum power assignment while preserving $k$-connectivity. Similarly, Li and Hou [9] describe a distributed algorithm that given a $k$-connected graph finds a $k$-connected spanner. For a list of related work on preserving connectivity see [4] and references therein. A version of the algorithmic counterpart of two of the local graph traits described in Section 4 appeared in [3].

## 2   Model

The communication network is modeled as a undirected graph. We use $G = (V, E)$ to denote an undirected graph, where $V$ is the set of vertices, and $E$ is the set of edges (two-element subsets of $V$). A pair of vertices $u, v \in V$ are neighbors if and only if $\{u, v\} \in E$. For ease of exposition we use the notation $E(G)$ (and $V(G)$) to denote the set of edges (and vertices) of a graph $G$. It is well known that most graph functions cannot be computed in anonymous networks, even for very simple graphs $G$. Hence, we define labeled graphs, denoted by a tuple $(G, id)$, where $id : V \rightarrow I$ is an injective function that maps each vertex to a unique identifier. In mobile ad hoc networks it is often useful to assume processes know their own position. To this end, we consider two-dimensional Euclidean graphs, denoted by a tuple $(G, p)$ (or $(G, p, id)$ when considering labeled two-dimensional Euclidean graphs) where $p : V \rightarrow \mathbb{R}^2$ is a function that maps each vertex to a point in the Euclidean plane. A two-dimensional Euclidean graph $(G, p)$ is a *unit disk graph* if there is an edge between two nodes if and only if they are at distance at most one, that is $E := \{\{u, v\} \mid \|p(u) - p(v)\| \leq 1\}$.

We consider a synchronous network model. Specifically, each node of an undirected graph $G$ is occupied by a process. The system progresses in synchronous lock-step rounds. At each round a process can send a message to its neighbors, receive messages from its neighbors and perform local computation. If the graph is labeled, we assume that at time zero the processor occupying node $v \in V(G)$ knows the identifier $id(v)$ of that node. Similarly, if the graph has an associated embedding (i.e. two-dimensional Euclidean graphs) we assume that at time zero the processor occupying node $v \in V(G)$ knows the position $p(v)$ of that node.

For a positive integer $t$ we denote with $N^t[u]$ the closed $t$-neighbors of $u$, the set of vertices reachable by paths starting at $u$ and of length at most $t$. Let $G^t(u)$ be the $t$-neighborhood of node $u$, the graph induced by the closed $t$-neighbors of $u$ in $G$. When $t = 1$ we simply use $N[u]$ and $G(u)$ to denote the 1-neighbors and 1-neighborhood of node $u$ respectively.

Since this model does not restrict the message size or the amount of local computation, after $t$ rounds a process at vertex $v$ can learn about its $t$-neighbors (including their unique ids and embedding when considering labeled Euclidean graphs), but it cannot learn about any node which is more than $t$ hops away. In particular in unit disk graphs, after $t$ rounds a process at vertex $v$ can learn exactly its $t$-neighborhood $G^t(v)$. In general graphs, after $t$ rounds a node can learn all edges between its $t$-neighbors, except for those edges whose endpoints are at distance exactly $t$. As shown in Section 5, this subtle difference between unit disk graphs and general graphs can be bridged by using an additional communication round to learn the $t$-neighborhood of a node in $t + 1$ communication rounds.

In Section 3 and 4 we e study the relationship between local graph traits and global graph properties. A graph trait is a pair $(t, T)$ where $t$ is a function from the positive integers to the positive integers and $T$ is a predicate over a graph, which (if applicable) can make use of the labeling or embedding of the graph. A graph $G$ on $n$ vertices satisfies a trait $(t, T)$, if the $t(n)$-neighborhood of every vertex $v \in V(G)$ satisfies $T$. A graph trait is local if $t \in O(1)$; a graph trait is weakly-local if $t \in o(n)$.

A graph trait $(t, T)$ implies a graph property $P$ if any graph which satisfies $(t, T)$ also satisfies $P$. Similarly, a graph property $P$ implies a graph trait $(t, T)$ if any graph which satisfies $P$ also satisfies $(t, T)$. A graph trait $(t, T)$ characterizes a graph property $P$ (or alternatively a graph property $P$ is characterized by a graph trait $(t, T)$) if $(t, T)$ implies $P$ and $P$ implies $(t, T)$. Given the graph traits $(t, T)$ and $(t', T')$ which imply a graph property $P$, we say that $(t, T)$ is *more accurate* than $(t', T')$ with respect to $P$ if every graph which satisfies $(t', T')$ also satisfies $(t, T)$, and there exists a graph which satisfies $P$ and $(t, T)$, but not $(t', T')$.

In particular, the global graph property that we are concerned with is graph $k$-connectivity. A *vertex cut* $C$ of a connected graph $G$ is a set of vertices whose removal renders $G$ disconnected. The *size* of a vertex cut $C$ is the number of vertices $|C|$. A vertex cut is said to be a *minimum vertex cut* if it is a vertex cut of smallest size. The connectivity of a graph $G$, denoted by $\kappa(G)$, is the size of a smallest vertex cut of $G$. A complete graph on $n$ vertices has no cuts at all, but by convention its connectivity is $n - 1$. We say a graph $G$ is $k$-connected if $\kappa(G) \geq k$.

# 3   The Impossibility of Locally Characterizing Connectivity

In this section we show that it is impossible to characterize the $k$-connectivity of a graph using (weakly-)local graph traits. The results hold even when restricted to simply connected labeled unit disk graphs. As a warm up, we first show that there does not exist a weakly-local graph trait that characterizes simple connected graphs.

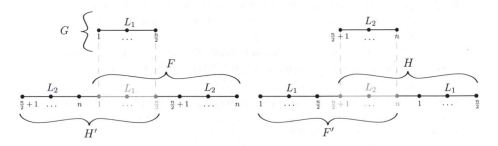

**Fig. 1.** All nodes are embedded in the horizontal axis. Neighboring nodes are 1 unit apart.

**Theorem 1.** *There does not exist a weakly-local graph trait $(t, T)$ that characterizes a simply connected graph.*

*Proof.* Fix any local trait $(t, T)$ which is implied by a simply connected graph. We will show that there exists a disconnected graph $G$ which satisfies $(t, T)$.

Since $t \in o(n)$ there exists a sufficiently large $n$ such that $n > 4t(n)$, we consider graphs over the vertex set $V = \{1, \ldots, n\}$. Throughout the proof we assume all graphs are labeled using the same injective function. We group the vertices into two connected components $L_1$ and $L_2$. Component $L_1$ is a line graph of the first $\frac{n}{2}$ nodes, namely for each $i \in [1, \frac{n}{2} - 1]$ vertex $i$ is connected with vertex $i + 1$. Component $L_2$ is a line graph of the remaining nodes, namely for each $i \in [\frac{n}{2} + 1, n - 1]$ node $i$ is connected with node $i + 1$.

In the rest of the proof we describe how to connect $L_1$ and $L_2$ to produce a disconnected graph $G$ and four connected graphs $F$, $F'$, $H$ and $H'$. We then show that since $(t, T)$ is satisfied by the four connected graphs by assumption, it must be that $G$ also satisfies $(t, T)$.

Specifically, $G$ is the disconnected graph made up of $L_1$ and $L_2$ with no additional edges. The graphs $F$ and $F'$ result from joining $L_1$ and $L_2$ with the edge $\{\frac{n}{2}, \frac{n}{2} + 1\}$, and the graphs $H$ and $H'$ result from joining $L_1$ and $L_2$ with the edge $\{n, 1\}$.

These graphs can be embedded as unit disk graph such that: (i) $L_1$ has the same embedding in $F$, $G$ and $H'$. (ii) $L_2$ has the same embedding in $F'$, $G$ and $H$ (cf. figure 1).

By assumption $T$ is satisfied on the $t(n)$-neighborhood of every node in $F$, $F'$, $H$ and $H'$. To show that $G$ satisfies the local trait $(t, T)$, it suffices to show that every node has the same $t(n)$-neighborhood (including the labeling and embedding of the nodes) in $G$ as it does in $F$, $F'$, $H$ or $H'$. We proceed by a case analysis on $i \in V$.

1. If $i \in [1, \frac{n}{4}]$ the $t(n)$-neighborhood of node $i$ in $G$ is a line graph with the nodes $\max(1, i - t(n)), \ldots, i, \ldots, i + t(n)$, which is the same $t(n)$-neighborhood of node $i$ in $F$.
2. If $i \in [\frac{n}{4} + 1, \frac{n}{2}]$ the $t(n)$-neighborhood of node $i$ in $G$ is a line graph with the nodes $i - t(n), \ldots, i, \ldots, \min(i + t(n), \frac{n}{2})$, which is the same $t(n)$-neighborhood of node $i$ in $H'$.
3. If $i \in [\frac{n}{2} + 1, \frac{3n}{4}]$ the $t(n)$-neighborhood of node $i$ in $G$ is a line graph with the nodes $\max(\frac{n}{2}, i - t(n)), \ldots, i, \ldots, i + t(n)$, which is the same $t(n)$-neighborhood of node $i$ in $H$.
4. If $i \in [\frac{3n}{4} + 1, n]$ the $t(n)$-neighborhood of node $i$ in $G$ is a line graph with the nodes $i - t(n), \ldots, i, \ldots, \min(i + t(n), n)$, which is the same $t(n)$-neighborhood of node $i$ in $F'$.
□

The previous theorem relies on the fact that if $t \in o(n)$ we can construct a large enough disconnected graph where every $t(n)$-neighborhood is indistinguishable from one in a connected graph. The same argument can be extended to show it is possible to construct a large enough disconnected graph whose $t(n)$-neighborhood is indistinguishable from a $k$-connected graph.

However, if we restrict our attention to characterizing the connectivity of *simply connected* graphs, the same argument no longer works. In particular in a graph which is connected but not $k$-connected, there exists a minimum vertex cut $C$ of size $1 \leq |C| < k$. It is conceivable that the $t(n)$-neighborhood of a node $u \in C$ in the cut might not fulfill all the local traits implied by a $k$-connected graph. The following theorem rules out that possibility by showing that even when restricted to connected graphs, there does not exist a local graph trait that characterizes a $k$-connected graph.

**Theorem 2.** *For any constant $k > 0$ there does not exist a weakly-local graph trait $(t, T)$ that characterizes $k$-connectivity of $\frac{k}{2}$-connected graphs.*

*Proof.* Let $k$ be any positive constant, and fix any local trait $(t, T)$ which is implied by $k$-connectivity. We will show that there exists a $\frac{k}{2}$-connected graph $G$ which is not $k$-connected and satisfies $(t, T)$.

Since $k$ is a constant and $t \in o(n)$, then there exists a sufficiently large $n$ such that $n = 2m \cdot k$ where $m > 4t(n)$, we consider graphs over the vertex set $V = \{1, \ldots, n\}$. We assume all graphs are labeled with the same injective function. We partition the vertices $V$ into four sets $V_1, V_2, V_3$ and $V_4$ each of size $mk/2$. Each vertex set $V_i$ is partitioned further into $m$ cliques $K_1^i, \ldots, K_m^i$, each of size $\frac{k}{2}$. In a slight abuse of notation we say cliques $A$ and $B$ are connected if every node in $A$ is connected to every node in $B$. For each vertex set $V_i$ we

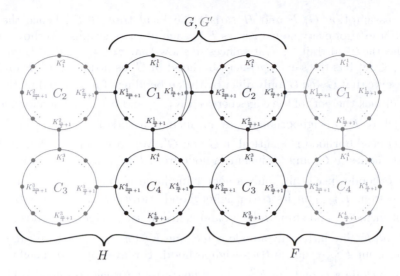

**Fig. 2.** Each point in the graph represents a clique of size $k/2$ embedded at that point, there is a line between cliques $A$ and $B$ if every node in clique $A$ is connected to every node in clique $B$. The clique cycle $C_i$ is formed by arranging the cliques uniformly around a circle at distance 1 from each other. To form $F$, $H$ and $G$ we arrange the clique cycles in a ring, where each clique cycle is at distance 1 from its neighboring clique cycle. To break the links between $C_1$ and $C_2$ in $G'$ we "push" the nodes of $K^1_{\frac{m}{4}-1}$ some $\varepsilon > 0$ towards the center of $C_1$.

consider the clique cycle graph $C_i = \langle V_i, E_i \rangle$ formed by connecting clique $K^i_j$ to clique $K^i_{j+1 \bmod m}$ for each $j \in [1, m]$.

In the rest of the proof we describe how to connect these clique cycles to produce the $k$-connected graphs $F$, $G$ and $H$, and the graph $G'$ with connectivity $\frac{k}{2}$. We then argue that since $(t, T)$ is satisfied by $F$, $G$ and $H$ by assumption, then it must also be satisfied by $G'$, which completes the theorem.

To construct the graphs $F$, $G$ and $H$ we connect the clique cycles $C_1, C_2, C_3$ and $C_4$ in a ring. Specifically in $G$ we connect cliques $K^1_{\frac{m}{4}+1}$ and $K^2_{\frac{3m}{4}+1}$, cliques $K^2_{\frac{m}{2}+1}$ and $K^3_1$, cliques $K^3_{\frac{3m}{4}+1}$ and $K^4_{\frac{m}{4}+1}$, and cliques $K^4_1$ and $K^1_{\frac{m}{2}+1}$. In $F$ and $H$ we connect cliques $K^1_{\frac{3m}{4}+1}$ and $K^2_{\frac{m}{4}+1}$, cliques $K^2_{\frac{m}{2}+1}$ and $K^3_1$, cliques $K^3_{\frac{m}{2}+1}$ and $K^4_{\frac{3m}{4}+1}$, and cliques $K^4_1$ and $K^1_{\frac{m}{2}+1}$. Finally, $G'$ is the graph that results from removing the edges between $C_1$ and $C_2$ in $G$. Observe that to disconnect $F$, $G$ or $H$ we need to remove delete all the nodes of at least two cliques, and since each clique is of size $k/2$, it follows that these graphs are $k$-connected. Similarly, to disconnect $G'$ it is sufficient and necessary to remove all the nodes of a single clique, and since each clique is of size $k/2$, it follows $G'$ has connectivity $\frac{k}{2}$.

These graphs can be embedded as a unit disk graph such that: (i) The embedding of $G$ and $G'$ are identical except for the clique $K^1_{\frac{m}{4}+1}$. (ii) The clique cycles $C_1$ and $C_4$ have the same embedding in $G'$ and $H$. (iii) The clique cycles $C_2$ and $C_3$ have the same embedding in $G'$ and $F$ (cf. figure 2).

By assumption $G$, $F$ and $H$ satisfy the local trait $(t,T)$, hence the $t(n)$-neighborhood of every node satisfies $T$ in each of these graphs. To show that $G'$ satisfies the local trait $(t,T)$ it suffices to show that $\forall i \in [1,4], \forall j \in [1,m]$ every node $v \in K_j^i$ has the same $t(n)$-neighborhood in $G'$ as it does in $G$, $F$ or $H$.

Observe that $G$ and $G'$ only differ by the embedding of $K_{\frac{m}{4}+1}^1$ and by the presence (or lack thereof) of the edges between $K_{\frac{m}{4}+1}^1$ and $K_{\frac{3m}{4}+1}^2$. Therefore, for any node whose $t(n)$-neighborhood does not include a node from $K_{\frac{m}{4}+1}^1$ or $K_{\frac{3m}{4}+1}^2$, its $t(n)$-neighborhood is identical in $G$ and $G'$. Moreover since $t(n) < m/4$ only a "few" nodes in $C_1$ and $C_2$ include a node from $K_{\frac{m}{4}+1}^1$ or $K_{\frac{3m}{4}+1}^2$ in their $t(n)$-neighborhood. Specifically, only a node $v$ in clique $K_j^1$ for $j \in [2, \frac{m}{2}]$ can include a node from $K_{\frac{m}{4}+1}^1$ in its $t(n)$-neighborhood. However, its $t(n)$-neighborhood cannot include a node from $K_{\frac{3m}{4}+1}^1$, and hence its $t(n)$-neighborhood is identical in $G'$ and $F$. Similarly, only a node $v$ in clique $K_k^2$ for $k \in [\frac{m}{2}+2, m]$ can include a node from $K_{\frac{3m}{4}+1}^2$ in its $t(n)$-neighborhood. However, its $t(n)$-neighborhood cannot include a node from $K_{\frac{m}{4}+1}^2$, and hence its $t(n)$-neighborhood is identical in $G'$ and $H$. $\qquad\square$

Given that it is impossible to characterize the connectivity of a graph with local graph traits, in the next section we focus on studying local graph traits which imply $k$-connectivity. In Section 5 we leverage these local graph traits to design local distributed algorithms.

## 4   Local Graph Traits That Imply $k$-Connectivity

We describe three local graph traits which imply graph $k$-connectivity for simply connected graphs. The first graph trait holds for general graphs, while the other two local traits hold only for unit disk graphs.

### 4.1   A Natural Local Trait for $k$-Connectivity

Perhaps the simplest and most intuitive local graph trait for $k$-connectivity is to check if the neighborhood of a vertex is $k$-connected. Specifically, consider the local trait $(c, K(k))$ where $c > 0$ is a positive constant and $K(k)$ is the predicate that checks if the $c$-neighborhood of a vertex is $k$-connected. We now show that this local graph trait implies $k$-connectivity.

**Theorem 3.** *The local graph trait $(c, K(k))$ implies $k$-connectivity for simply connected graphs.*

*Proof.* Suppose by contradiction that a connected graph satisfies the local graph trait $(c, K(k))$ but it is not $k$-connected. Since $G$ satisfies $(c, K(k))$, then $G^c(u)$ must have at least $k+1$ vertices, hence $|V| \geq k+1$. Since by assumption $G$ is not $k$-connected, it has a vertex cut with at most $k-1$ vertices. On the other hand since $G$ is connected, any vertex cut is of size at least 1.

In particular let $C$ denote a minimum vertex cut, and let $P$ and $Q$ be two connected components produced by removing all vertices in $C$. Fix any vertex $u \in C$, we make the following claim (proved later):

*Claim.* There exists vertices $p, q \in N(u)$ such that $p \in P$ and $q \in Q$.

Let $U = N^t[u] \setminus \{p, q\}$, where $p$ and $q$ are fixed as in the claim. Observe that since $G$ satisfies $(c, K(k))$ then $G^c(u)$ is $k$-connected. In particular this means $|N^c[u]| \geq k+1$ and hence $|U| \geq k-1$. Moreover, this also implies that removing any subset of $U$ of size at most $k - 1$ leaves a path from $p$ to $q$ in $G^c(u)$.

However by assumption, removing the set $C \subseteq V$ of size at most $k-1$ produces two connected components $P$ and $Q$. Since removing $C \subseteq V$ disconnects $P$ from $Q$ in $G$, then removing $U \cap C \subseteq U$ has to disconnect $p$ and $q$ in $G^c(u)$.

Finally since $|C| \leq k - 1$ then $|U \cap C| \leq k - 1$, but this contradicts that removing any subset of $U$ of size at most $k - 1$ leaves a path from $p$ to $q$ in $G^c(u)$, which completes the theorem.     $\square$

*Proof.* [of Claim] By assumption, $C$ is a minimum vertex cut that separates $P$ and $Q$. Hence, if we consider the smaller vertex set $C' = C \setminus \{u\}$, it must be that removing the vertices from $C'$ does not separate $P$ and $Q$.

This implies that for any pair of vertices $p' \in P$ and $q' \in Q$ there exists a simple path between $p'$ and $q'$ using only vertices from the set $V - C'$. Since this path does not exist when removing the set $C$, the path must go through $u$.

Follow the path starting at $p' \in P$, and let $p \in P$ be the last vertex in the path that belongs to $P$. It must be that the vertex in the path after $p$ is $u$ (and hence $p \in N(u)$). Otherwise it would contradict that $P$ is a component separated from the rest of the vertices when removing $C$. By following the path starting at $q' \in Q$ the same argument can be used to show there exists a vertex $q \in Q$ such that $q \in N(u)$, which completes the proof.     $\square$

It's not immediate how to improve the accuracy of $(c, K(k))$. To illustrate this difficulty, assume there exists some local graph trait $(c, K'(k))$ which implies $k$-connectivity and is more accurate than $(c, K(k))$. Therefore, there must exist a $k$-connected graph $G$ with a vertex $u \in V$ whose $c$-neighborhood does not satisfy $K(k)$ but does satisfy $K'(k)$. However, this also implies that if we consider the graph $G' = G^c(u)$, then $G'$ is not $k$-connected but $K'(k)$ is satisfied at node $u$.

Using this logic it is tempting to go further and argue that since any local graph trait which implies $k$-connectivity should not be satisfied by a graph which is not $k$-connected, then $K'(k)$ does not imply $k$-connectivity (reaching a contradiction). However, a graph trait $(c, K'(k))$ is only satisfied by a graph, if the $c$-neighborhood of *all* nodes satisfies $K'(k)$.

The next subsection describes a local graph traits which is less accurate than $(c, K)$. However, this local trait will introduce ideas which will inspire a better graph trait presented in the last subsection, one which uses techniques that allow us to use it to preserve $k$-connectivity in Section 5. We remark that up to this point, we have not used either labeled or Euclidean graphs.

## 4.2    Small Edges Increase Connectivity

Given an Euclidean graph $(G, p)$, we define the length of an edge as the Euclidean distance between the embedding of its endpoints. In unit disk graphs, one would expect that moving all nodes closer together (thereby decreasing the length of all the edges) would increase the connectivity of the graph. This observation is exploited by the local graph trait $(c, Small(k))$. Here $c > 0$ is a positive constant and $Small(k)$ is a predicate that checks if the $c$-neighborhood of a node has at least $k + 1$ nodes and has a connected spanning subgraph using only edges of length at most $1/k$. To prove that this local graph trait implies $k$-connectivity we first show the following:

**Lemma 4.** *If a unit disk graph with $n \geq k+1$ vertices has a connected spanning subgraph using edges of length at most $1/k$, then it is $k$-connected.*

*Proof.* Fix any unit disk graph graph $G$ with $n \geq k + 1$ vertices which has a connected spanning graph using edges of length at most $1/k$. If $G$ were a clique then it is $k$-connected, hence we assume $G$ is not a clique and let $C$ be a minimum vertex cut of $G$. We will show that $|C| \geq k$, which implies that $G$ is $k$-connected.

Let $P$ and $Q$ be two connected components produced by the cut $C$. Since $G$ has a connected spanning subgraph using edges of length at most $1/k$, then for any pair of vertices $p \in P$ and $q \in Q$ there exists a simple path $p \rightsquigarrow q$ from $p$ to $q$ in using only edges of length $1/k$. We use the vertices of $C$ to define a *gap* in $p \rightsquigarrow q$, as a maximal set of contiguous vertices in $p \rightsquigarrow q$ that belong to $C$. For each gap $g$ let $g.first$ and $g.last$ be the vertices in the path immediately before and after the gap.

Any gap $g$ is of size at most $|g| \leq |C|$, and the Euclidean distance between $g.first$ and $g.last$ is bounded by $(|g| + 1)/k$. Hence if $|C| \leq k - 1$, then the distance between the $g.first$ and $g.last$ is at most $k/k = 1$. However, since $G$ is a unit disk graph by assumption, there must exist an edge $(g.first, g.last)$ in $G$ which bridges the gap and there would exist a path from $p$ to $q$. Therefore, for $C$ to be a cut, it must be that $|C| \geq k$, and thus $G$ is $k$-connected.    □

We can stitch Lemma 4 with Theorem 3, which showed that $(c, K(k))$ implies $k$-connectivity of connected graphs, to prove the following.

**Theorem 5.** *The local graph trait $(c, Small(k))$ implies $k$-connectivity for simply connected unit disk graphs.*

*Proof.* By Theorem 3 it suffices to show that, for every vertex $u \in V$, if the $c$-neighborhood of $u$ satisfies $Small(k)$ then it also satisfies $K(k)$.

Fix a vertex $u \in V$ which satisfies $Small(k)$, then it follows that $G^c(u)$ has at least $k + 1$ vertices and has a connected spanning subgraph using edges of length at most $1/k$. However, then by Lemma 4 $G^c(u)$ is $k$-connected and it satisfies $K(k)$.    □

In the process of proving Theorem 5 we showed that the graphs which satisfy $(c, Small(k))$ also satisfy $(c, K(k))$. It is not difficult to construct unit disk graphs which satisfy $(c, K(k))$ but where $(c, Small(k))$ does not hold (i.e. a clique with $k + 1$ vertices using only "large" edges). Therefore it follows that by definition $(c, K(k))$ is more accurate than $(c, Small(k))$ for $k$-connectivity.

A natural question, is to ask weather *all* edges in the connected spanning graph need to be small for the connectivity of a graph to increase, or is it sufficient only for *some* edges to be small? We answer this question in the next subsection.

### 4.3   Spanning Trees with Small Edges Imply $k$-Connectivity

Given an Euclidean graph $(G, p)$, let $MST_G$ denote a minimum spanning tree of $G$. Observe that in labeled graphs, ties between edges of the same length can be broken consistently using the unique identifiers associated with each node. Therefore, in Euclidean labeled graphs, we can assume distinct edge lengths, which implies there is a unique minimum spanning tree.

For any positive constant $c > 0$, let $LMST_G^c = (V, F)$ denote the local minimum spanning tree of $G = (V, E)$. The edge set of the local minimum spanning tree is $F := \{\{u, v\} \mid \{u, v\} \in E(MST_{G^c(u)}) \cap E(MST_{G^c(v)})\}$. In other words, $LMST_G^c$ is the intersection of the minimum spanning trees associated with the $c$-neighborhood of every node in $G$. It is known [10] that in graphs $G$ with a unique minimum spanning tree $MST_G$, the local minimum spanning tree contains the minimum spanning tree and is therefore connected. This property suggests an improved local graph trait using the same ideas of $(c, Small(k))$.

Consider the local graph trait $(c, MSTSmall(k))$ where $c > 0$ is any positive constant and $MSTSmall(k)$ is a predicate that checks if the $c$-neighborhood of a node $u$ has at least $k + 1$ nodes and all the edges of the form $\{u, v\}$ in its minimum spanning tree have length at most $1/k$. The next theorem shows that $(c, MSTSmall(k))$ implies $k$-connectivity as a consequence from the properties of $LMST_G^c$ and Lemma 4.

**Theorem 6.** *The local graph trait $(c, MSTSmall(k))$ implies $k$-connectivity for simply connected labeled unit disk graphs.*

*Proof.* Let $G$ be any simply connected labeled unit disk graph. Therefore $G$ has a unique minimum spanning tree $MST_G$. If $G$ satisfies $(c, MSTSmall(k))$ then by definition it follows that all the edges in $LMST_G^c$ are of length $1/k$.

Moreover, since $MST_G \subset LMST_G^c$ and $MST_G$ is a connected spanning graph by definition, then clearly $LMST_G$ is also a connected spanning subgraph of $G$. Finally since $G$ has a connected spanning subgraph with edges of length at most $1/k$ (namely $LMST_G^c$), then Lemma 4 implies it is $k$-connected.       □

A feature which is shared by graphs that satisfy $(c, MSTSmall(k))$ and $(c, Small(k))$, is that they contain connected spanning subgraphs using edges of length at most $1/k$. This will prove to be a valuable property in Section 6.

A well known folklore result is that amongst all spanning trees, a minimum spanning tree minimizes the length of the longest edge. Together with the fact that it is possible to construct a $k$-connected graph where $(c, MSTSmall(k))$

is satisfied, but not $(c, Small(k))$, we can conclude $(c, MSTSmall(k))$ is more accurate than $(c, Small(k))$ with respect to $k$-connectivity.

In fact, it turns out that $(c, MSTSmall(k))$ is satisfied in some $k$-connected graphs where $(c, K(k))$ is *not* satisfied. However, the converse is also true, and therefore the accuracy of $(c, K(k))$ and $(c, MSTSmall(k))$ for $k$-connected graphs is incomparable. Which of them is more useful depends on the characteristics of the graphs being considered.

## 5  Applying Local Graph Traits to Distributed Algorithms

In the first part of the section we warm up by designing a simple constant time distributed algorithm that tests for $k$-connectivity using local graph traits. In the later subsection we consider a more elaborate application that leverages the local graph trait $(c, MSTSmall(k))$ together with a connectivity preserving algorithm presented in [4], to yield a $k$-connectivity preserving algorithm for mobile ad hoc networks.

### 5.1  Testing for $k$-Connectivity

Consider the following constant time procedure (which is the algorithmic counterpart of the local trait $(c, K(k))$). The process running at each node $u \in V(G)$ executes a full information protocol for $c + 1$ communication rounds to recover the $c$-neighborhood of $u$. At the end of the $c + 1$ rounds, the process outputs true if $G^c(u)$ is $k$-connected and outputs false otherwise.

Since $(c, K(k))$ implies $k$-connectivity, then if this procedure outputs true at every node, then $G$ is guaranteed $k$-connected. On the other hand, if $G$ is not $k$-connected we are guaranteed that at least one process will output false. We remark that if $G$ is disconnected, not all processes are guaranteed to output false, just a non empty subset of them, which is enough for most of applications.

This procedure can be used by itself as a constant time distributed algorithm to test for $k$-connectivity, or can be used as a building block to solve other problems. For example in a distributed topology control algorithm, to guarantee $k$-connectivity, every process could run the procedure repeatedly with an increasing power assignment, stopping when the procedure outputs true. If the maximum transmission power is sufficiently large and the graph has at least $k+1$ nodes, this algorithm eventually stops and guarantees a $k$-connected graph. However, it might not stop in the first round when the graph becomes $k$-connected. Moreover, the impossibility result on weakly-local graph traits for $k$-connectivity, implies that any distributed topology control algorithm that finds an optimal solution requires at least $\Omega(n)$ communication rounds.

In deployments where the unit disk graph assumption holds and nodes are equipped with GPS, an algorithmic implementation of $(c, MSTSmall(k))$ requires one less communication round, and might yields better results.

## 5.2 Maintaining $k$-Connectivity of Robot Swarms

We consider a mobile ad hoc network composed of $n$ mobile robots (aka processes). When possible we adhere to the standard synchronous network model described in Section 2. At the beginning of every round, in addition to the usual operations, each robot can query its own position (perhaps using GPS), query its intended target position for the next round (via an existing motion planner) and feed a trajectory to its actuators (for example, a linear trajectory to its intended target). Actuators are imperfect, and hence a robot following a trajectory may stop or slow down abruptly and travel only a fraction, possible none, of this trajectory. We assume the communication graph is a unit disk graph induced by the positions of the robots. For simplicity, we will assume that at the beginning of every round each robot knows its neighbors in the communication graph and their positions, this could be implemented by exchanging hello messages tagged with the position of the robots. This model closely resembles the $\mathcal{FSYNC}$ model introduced in [12], with some subtle yet important differences.

Since robots (as opposed to regular processes) can move and change their position from round to round, we extend our notation to account for this. Let $p(v, r)$ denote the position of the robot occupying node $v$ at round $r$. Similarly, let $G(r) = (V, E(r))$ denote the communication graph induced at round $r$, and let $N[u, r]^t$ be the closed $t$-neighbors of node $u$ at round $r$. We use $N[u, r]$ as short hand notation for the closed 1-neighbors of $u$ at round $r$.

In previous work [2, 4] we addressed the problem of maintaining connectivity ($k = 1$) for robot swarms. Specifically, we described a distributed algorithm that modifies an existing short-term motion plan to ensure connectivity. The algorithm uses only local information, is stateless, does not require a fixed set of neighbors and does not make any assumptions on the current or goal configurations. Moreover, the algorithm is *robust* to the robots' speed changes; if robots travel any fraction of the trajectory (perhaps none) at any speed, connectivity is preserved. The *progress* of the algorithm is defined as the total distance traveled by all robots (summing over all the robots) towards their intended destinations. Let $d$ be the minimal distance each robot intends to move and let $R$ be the communication radius. Assuming that the target configuration of the robots is connected and the motion does not require breaking any cycles, we proved that the algorithm guarantees that the progress is at least $\min(d, R)$. Furthermore, we exhibited a class of configurations where no local algorithm can do better than this bound, and hence under these conditions the bound is tight and the algorithm is asymptotically optimal. Finally we proved that all robots get $\varepsilon$-close to their target within $O(D_0/R + n^2/\varepsilon)$ rounds where $D_0$ is the total initial distance to the targets and $n$ is the number of robots [4].

Starting with a graph which satisfies the local trait $(c, MSTSmall(k))$, we describe how to extend the CONNSERV algorithm presented in [4] to enforce the local graph trait $(c, MSTSmall(k))$ throughout the execution and preserve $k$-connectivity with similar robustness, safety and progress conditions as the original algorithm.

**Connectivity Maintenance Algorithm.** The CONNSERV algorithm [4] is parametrized by a communication radius $R \in \mathbb{R}$ and a neighbor filtering function $f : 2^V \rightarrow 2^V$ which receives a closed set of neighbors $N[u, r]$ and returns a filtered set of neighbors $N'(u, r) \subseteq N[u, r]$. These parameters should satisfy the following properties: $P1$. Any two robots which are at distance $R$ or less can reliably exchange a message (i.e. are connected). $P2$. Filtered neighbors are within distance $R$ $(\forall v \in N[u, r], \|p(v) - p(v)\| \leq R)$. $P3$. Preserving connectivity with the filtered neighbors is sufficient to preserve global graph connectivity. Formally, if $G(r) = (V, E(r))$ is connected, then the spanning subgraph $H = (V, F)$ where $F := \{\{u, v\} \mid u \in N'(v, r) \wedge v \in N'(u, r)\}$ is also connected.

When run by a robot at node $u$ at round $r$, the input of the CONNSERV algorithm is a tuple $(p_u, N_u, t_u)$, where $p_u = p(u, r)$, $N_u = N[u, r]$ and $t_u$ is the intended target position at round $r$. The output of the CONNSERV algorithm is a new target position $t_u^*$. For any parameters which satisfy the properties above, the CONNSERV algorithm was shown to provide the following guarantees [4].

**Safety Theorem.** *If* $u \in N'(v, r)$ *and* $v \in N'(u, r)$, *then* $|t_u^* - t_v^*| \leq R$

In other words, if by the beginning of the next round every robot moves to the target position output by CONNSERV, if $G(r)$ was connected then $G(r + 1)$ will also be connected (this follows from the safety theorem and $P3$).

However, it would be unreasonable to expect all robots to be able to reach the target output by the algorithm by the beginning of the next round. For example, a robot might encounter an obstacle, it might stop or slow down suddenly due to hardware malfunction, or it might be to slow to complete the trajectory. This motivates the next result, which shows that the graph will remain connected even if robots stop or slow down unexpectedly.

**Robustness Theorem.** *If* $u \in N'(v, r)$ *and* $v \in N'(u, r)$, *then for any point $p$ in the linear trajectory from $p(u, r)$ to $t_u^*$, and any point $q$ in the linear trajectory from $p(v, r)$ to $t_v^*$, it holds that* $\|p - q\| \leq R$.

For the algorithm to be useful, it needs to provide a progress guarantee that relates the input and the output target, since a trivial algorithm which forces all robots to remain stationary vacuously satisfies the safety and robustness theorems. On the other hand, it is not possible to guarantee progress unconditionally, since for example, if two robots want to move in opposite directions as to disconnect the graph, guaranteeing any progress would violate the safety and robustness theorems. Therefore, the progress guarantees must be conditioned on the assumption that the intended targets do not require breaking any edges needed for connectivity.

We define the progress of a robot as the distance advanced to the input target assuming it moves to the output target. If at round $r$ a robot at node $u$ has an input target $t_u$, let $d_u = \|p(u, r) - t_u\|$ be the distance from its current position to its input target. If the algorithm CONNSERV outputs a target $t_u^*$ the progress is defined as $\delta_u = d_u - \|t_u - t_u^*\|$.

The progress of the system is then the sum of the progress of each robot, that is $\sum_{u \in V} \delta_u$. In the following $d$ is defined as $d = \min_{u \in V} d_u$.

**Progress Theorem.** *In any configuration where the intended targets do not require breaking edges needed for connectivity, the progress is at least* $\min(d, R)$.

Finally, assuming the robots have the same target for sufficiently many rounds, the following result provides an upper bound on the number of rounds required for every robot to reach their target. Here $D_0 = \sum_{u \in V} d_u$ is the sum of the distances from each robot to its intended long term target.

**Termination Theorem.** *In any configuration where the intended targets do not require breaking edges needed for connectivity, every robot gets $\varepsilon$-close to its target within* $O(D_0/R + n^2/\varepsilon)$ *rounds.*

**$k$-Connectivity Maintenance.** We will argue that if the starting configuration satisfies $(c, MSTSmall(k))$, it is possible to choose parameters for the CONNSERV algorithm so that it preserves $k$-connectivity.

Concretely, we let $R = R_{\min}/k$ where $R_{\min}$ is the smallest distance such that any two robots within distance $R_{\min}$ can exchange messages reliably. For the filtering function $f$, let $S \subseteq N[u, r]$ be the subset of vertices which are at distance less than or equal to $R_{\min}/k$ from $u$. We let $f$ return every vertex $v$ such that $v \in E(MST_S)$, in other words the neighbors of $u$ in the minimum spanning tree involving only vertices in $S$ (i.e. at distance at most $R_{\min}/k$ from $u$). Finally, we assume the communication graph is initially $k$-connected, specifically we assume $G(0)$ satisfies the local graph trait $(c, MSTSmall(k))$.

It is clear that the parameters described satisfy $P1$ and $P2$, but it is not obvious that $P3$ is satisfied or that the resulting algorithm preserves $k$-connectivity.

**Theorem 7.** $\forall r \geq 0$, $f$ *satisfies $P3$ and $G(r)$ is $k$-connected.*

*Proof.* Let $H(r)$ be the graph that results from removing all edges of $G(r)$ which are of length more than $R_{\min}/k$. We make the following claim (proved later).

*Claim.* $H(r)$ is a *connected* spanning subgraph of $G(r)$.

Then it follows that $H(r)$ satisfies $(c, MSTSmall(k))$ and hence $H(r)$ (and therefore $G(r)$) are $k$-connected. Finally since the filtered neighbors returned by $f$ define a local minimum spanning tree over $H(r)$, $f$ satisfies $P3$. □

*Proof.* [of Claim] We proceed by induction on $r$. The base case is trivial since $G(0)$ satisfies $(c, MSTSmall(k))$. Suppose by inductive hypothesis that $H(r)$ is a connected spanning subgraph of $G(r)$. Let $LMST(r)$ be the connected spanning subgraph of $H(r)$ described by the filtered neighbors returned by $f$.

By the safety and robustness theorems, all the edges of $LMST(r)$ are present in $G(r+1)$ with length at most $R = R_{\min}/k$. Therefore the subgraph $H(r+1) \subseteq G(r+1)$ also contains $LMST(r)$, and thus it is a connected spanning subgraph of $G(r+1)$. □

Therefore, since the parameters used in the CONNSERV algorithm satisfy $P1, P2$ and $P3$, the safety and robustness theorems imply the graph is $k$-connected at every time instant even if the robots slow down or stop unexpectedly and only execute some fraction of the trajectory.

Finally, observe that the progress theorem implies that at every round the progress of the system is at least $\min(d, R_{min}/k)$. Similarly the termination theorem implies the system gets $\varepsilon$-close to its targets within $O(D_0 k/R_{min} + n^2/\varepsilon)$. Thus, we conclude that leveraging local graph traits allows us to preserve $k$-connectivity at a cost which is linear in $k$ when compared to preserving simple connectivity.

# References

[1] Awerbuch, B., Luby, M., Goldberg, A.V., Plotkin, S.A.: Network decomposition and locality in distributed computation. In: Proc. of the 30th Annual Symposium on Foundations of Computer Science (1989)

[2] Cornejo, A., Lynch, N.: Connectivity Service for Mobile Ad-Hoc Networks. In: Spatial Computing Workshop (2008)

[3] Cornejo, A., Lynch, N.: Fault-Tolerance Through k-Connectivity. In: Workshop on Network Science and Systems Issues in Multi-Robot Autonomy: ICRA 2010, vol. 2 (2010)

[4] Cornejo, A., Kuhn, F., Ley-Wild, R., Lynch, N.: Keeping mobile robot swarms connected. In: Keidar, I. (ed.) DISC 2009. LNCS, vol. 5805, pp. 496–511. Springer, Heidelberg (2009)

[5] Czumaj, A., Zhao, H.: Fault-tolerant geometric spanners. Discrete and Computational Geometry 32(2), 207–230 (2004)

[6] Dirac, G.A.: Some theorems on abstract graphs. Proc. London Mathematical Society 2 (1952)

[7] Jia, X., Kim, D., Makki, S., Wan, P.J., Yi, C.W.: Power assignment for k-connectivity in wireless ad hoc networks. Journal of Combinatorial Optimization 9(2), 213–222 (2005)

[8] Jorgic, M., Goel, N., Kalaichevan, K., Nayak, A., Stojmenovic, I.: Localized detection of k-connectivity in wireless ad hoc, actuator and sensor networks. In: Proc. 16th ICCCN (2007)

[9] Li, N., Hou, J.C.: FLSS: a fault-tolerant topology control algorithm for wireless networks. In: Proceedings of the 10th Annual International Conference on Mobile Computing and Networking, pp. 275–286. ACM, New York (2004)

[10] Li, N., Hou, J.C., Sha, L.: Design and analysis of an MST-based topology control algorithm. In: INFOCOM, vol. 3, pp. 1702–1712 (2003)

[11] Linial, N.: Distributive graph algorithms Global solutions from local data. In: 28th Annual Symposium on Foundations of Computer Science 1987, pp. 331–335 (1987)

[12] Suzuki, I., Yamashita, M.: Distributed anonymous mobile robots: formation of geometric patterns. SIAM J. Comput. 28(4), 1347–1363 (1999)

[13] Thurimella, R.: Sub-linear distributed algorithms for sparse certificates and bi-connected components. In: PODC, pp. 28–37. ACM, New York (1995)

[14] Wigderson, A.: Improving the performance guarantee for approximate graph coloring. Journal of the ACM (JACM) 30(4), 735 (1983)

# Distributed Game-Theoretic Vertex Coloring*

Ioannis Chatzigiannakis[1,2], Christos Koninis[1,2],
Panagiota N. Panagopoulou[2], and Paul G. Spirakis[1,2]

[1] Computer Engineering and Informatics Department, Patras University
[2] Research Academic Computer Technology Institute, Greece
{ichatz,koninis,panagopp,spirakis}@cti.gr

**Abstract.** We exploit the game-theoretic ideas presented in [12] to study the vertex coloring problem in a distributed setting. The vertices of the graph are seen as players in a suitably defined strategic game, where each player has to choose some color, and the payoff of a vertex is the total number of players that have chosen the same color as its own. We extend here the results of [12] by showing that, if any subset of non-neighboring vertices perform a selfish step (i.e., change their colors in order to increase their payoffs) in parallel, then a (Nash equilibrium) proper coloring, using a number of colors within several known upper bounds on the chromatic number, can still be reached in polynomial time. We also present an implementation of the distributed algorithm in wireless networks of tiny devices and evaluate the performance in simulated and experimental environments. The performance analysis indicates that it is the first practically implementable distributed algorithm.

## 1 Introduction

One of the central optimization problems in Computer Science is the problem of *vertex coloring* of graphs, i.e., the problem of assigning a color to each vertex of the graph so that no pair of adjacent vertices gets the same color (i.e., so that the coloring is *proper*) and so that the total number of distinct colors used is minimized. In this work, we deal with the problem of vertex coloring in a distributed setting, with a focus on distributed implementations applicable to Wireless Sensor Networks (WSNs). Finding a good (with respect to the total number of colors used) coloring of the nodes of a WSN has many practical applications: First, colors may be seen as frequencies, so that a proper coloring of the nodes corresponds to a solution to the frequency assignment problem; Furthermore, a coloring of a WSN actually partitions its nodes into subsets (each corresponding to a color), such that no communication link exists between any pair of nodes in the same subset, and such a partition might be useful when designing sleep/awake protocols in order to save energy or providing secure group communication.

* This work has been partially supported by the ICT Programme of the European Union under contract numbers ICT-2008-215270 (**FRONTS**) and ICT-2008-224460 (**WISEBED**).

C. Lu, T. Masuzawa, and M. Mosbah (Eds.): OPODIS 2010, LNCS 6490, pp. 103–118, 2010.
© Springer-Verlag Berlin Heidelberg 2010

The challenge of designing vertex coloring algorithms applicable in WSNs lies in both the intrinsic difficulty of the original problem of vertex coloring and the particularities of WSNs. More specifically, the global optimum of vertex coloring is NP-hard [7], and the best polynomial time approximation algorithm achieves an approximation ratio of $O(n(\log \log n)^2/(\log n)^3)$ [6] ($n$ being the number of vertices). In the distributed setting of wireless sensor networks the problem of vertex coloring has been studied before; the randomized algorithm presented in [10] needs, with high probability, $O(\Delta \log n)$ time and uses $O(\Delta)$ colors, where $n$ and $\Delta$ are the number of nodes in the network and the maximum degree, respectively. This algorithm requires knowledge of a linear bound on $n$ and $\Delta$. This result was improved in [14], where the coloring problem is solved in $O(\Delta + \log \Delta \log n)$ time, given an estimate of $n$ and $\Delta$, and $O(\Delta + \log^2 n)$ without knowledge of $\Delta$, while it needs $\Delta + 1$ colors.

In this work, our objective is to find a (suboptimal, inevitably) proper coloring, that at least guarantees some bounded maximum total number of colors used. Towards this direction, we try to exploit the ideas presented in [12]: the vertices of the graph are seen as players in a suitably defined strategic game, where each player has to choose some color, and the payoff of a player is the total number of players that have chosen the same color as his own (unless some neighbor has also chosen the same color, in which case the payoff is 0). The vertices are allowed to perform, sequentially, selfish steps, i.e., change their colors in order to increase their payoffs, and in [12] it was shown that this selfish improvement sequence converges, in polynomial time, into a pure Nash equilibrium of the game, which is actually a proper coloring of the vertices of the graph that uses a total number of colors satisfying all known upper bounds on the chromatic number of the graph (that is, the minimum number of colors needed to color the graph).

*Our contribution.* Our objective here is to exploit the game-theoretic ideas presented in [12] so as to deploy an efficient, in terms of both time complexity and number of colors used, distributed algorithm for vertex coloring. The algorithm of [12] is in fact a local search method (the vertices perform local changes, by moving to color classes of higher cardinality, until no further local moves are possible); however, it relies on several assumptions that obstruct its straightforward implementation in a distributed setting. In particular, it requires that global information about the cardinalities of all color classes is at any time available to each vertex, and that only a single vertex at each time is allowed to perform a local move. Here, we propose an implementation of the algorithm that deals with these issues, thus enabling its application to distributed settings, while preserving both the efficiency (in terms of time complexity) and the quality (in terms of number of colors used) of the coloring produced.

The solution we propose relies on an extension of the results of [12] that we give here. Namely, we raise the requirement that only one vertex at a time is allowed to perform a selfish step in order to guarantee polynomial time convergence into a Nash equilibrium coloring. In particular, we show that if any subset of non-neighboring vertices perform a selfish step in parallel, then a Nash equilibrium coloring satisfying the bounds given in [12] can still be reached in polynomial time.

This potentiality of parallelization of selfish steps allows us to derive a distributed implementation of the game-theoretic vertex coloring algorithm, which computes, in polynomial time, a proper coloring of the vertices of a graph using at most

$$k \leq \min \left\{ \Delta_2(G) + 1, \; \frac{n + \omega(G)}{2}, \; n - \alpha(G) + 1, \; \frac{1 + \sqrt{1 + 8m}}{2}, \right.$$
$$\left. \frac{\chi(G) + 1}{2} + \sqrt{m - \frac{(\chi(G) - 1)(\chi(G) + 1)}{4}} \right\}$$

distinct colors, where $n$ is the number of vertices, $m$ is the number of edges, $\Delta_2(G)$ is the maximum degree of a vertex which is connected to a vertex of equal or higher degree, and $\omega(G)$, $\alpha(G)$ and $\chi(G)$ are the clique number, independence number and chromatic number, respectively, of the graph under consideration. To the best of our knowledge, this is the first distributed, polynomial-time implementation of a vertex coloring algorithm that achieves all the above bounds. Our algorithm requires $O(n)$ memory to maintain the local lists with the cardinalities of all color classes. Also, the protocol does not require any initial knowledge on the network (e.g., network size, average node degree etc.) but gathers all necessary information dynamically.

The paper is organized as follows. The next section introduces the definitions and notations needed to solve the problem of vertex coloring. We then present the game-theoretic approach for vertex coloring based on local search. In Section 3 we first prove that a "parallelization" of the local search approach is still possible. Based on this fact, we then present a distributed implementation of the algorithm that converges fast into a pure Nash equilibrium, and thus into a proper coloring. We also describe a self-stabilizing version of the distributed algorithm that recovers from transient faults regardless of its initial state. Section 4 presents an implementation of the distributed algorithm in wireless networks of tiny devices and evaluates the performance in simulated and experimental environments. The performance analysis indicates that our distributed algorithm is the first practically implementable for real networks.

## 2     Background

**Definitions and notation.** For a finite set $A$ we denote by $|A|$ the cardinality of $A$. For a positive integer $n \in \mathbb{N}$ let $[n] = \{1, \ldots, n\}$. Denote $G = (V, E)$ a simple, undirected graph with vertex (node) set $V$ and set of edges $E$. For a vertex $v \in V$ denote $N(v) = \{u \in V : \{u, v\} \in E\}$ the set of its neighbors, and let $\deg(v) = |N(v)|$ denote its degree. Let $\Delta(G) = \max_{v \in V} \deg(v)$ be the maximum degree of $G$. Let

$$\Delta_2(G) = \max_{u \in V} \; \max_{v \in N(u):d(v) \leq d(u)} \; \deg(v)$$

be the maximum degree that a vertex $v$ can have, subject to the condition that $v$ is adjacent to at least one vertex of degree no less than $\deg(v)$. Clearly,

$\Delta_2(G) \leq \Delta(G)$. Let diam$(G)$ be the diameter of $G$, i.e., the maximum length of a shortest path between any pair of vertices of $G$. Let $\omega(G)$ and $\alpha(G)$ denote the clique number and independence number of $G$, i.e. the number of vertices in a maximum clique and a maximum independent set of $G$.

**The Vertex Coloring problem.** One of the central optimization problems in Computer Science is the problem of *vertex coloring* of graphs: given a graph $G = (V, E)$ of $n$ vertices, assign a color to each vertex of $G$ so that no pair of adjacent vertices gets the same color (i.e., so that the coloring is *proper*) and so that the total number of distinct colors used is minimized. The *chromatic number* of $G$, denoted by $\chi(G)$, is the global optimum of vertex coloring, i.e., the minimum number of colors needed to properly color the vertices of $G$. The vertex coloring problem is known to be NP-hard [7], and the chromatic number cannot be approximated to within $\Omega(n^{1-\epsilon})$ for any constant $\epsilon > 0$, unless NP $\subseteq$ co-RP [5].

**The game-theoretic approach.** Panagopoulou and Spirakis [12] proposed an efficient vertex coloring algorithm that is based on *local search*: Starting with an arbitrary proper vertex coloring (e.g. the trivial proper coloring where each vertex is assigned a unique color), each vertex (one at a time) is allowed to move to another color class of higher cardinality, until no further local moves are possible. This local search method is illustrated in [12] via a game-theoretic analysis, because of the natural correspondence of the local optima of the proposed method to the pure Nash equilibria of a suitably defined *strategic game*.

In particular, given a finite, simple, undirected graph $G = (V, E)$ with $|V| = n$ vertices, the *graph coloring game* $\Gamma(G)$ is defined as the game in strategic form where the set of players is the set of vertices $V$, and the action set of each vertex is a set of $n$ colors $[n] = \{1, \ldots, n\}$ (for simplicity, we represent each color by an integer). A *configuration* or *pure strategy profile* $\mathbf{c} = (c_v)_{v \in V} \in [n]^n$ is a combination of actions, one for each vertex. That is, $c_v$ is the color chosen by vertex $v$. For a configuration $\mathbf{c} \in [c]^n$ and a color $x \in [n]$, we denote by $n_x(\mathbf{c})$ the number of vertices that are colored $x$ in $\mathbf{c}$, i.e. $n_x(\mathbf{c}) = |\{v \in V : c_v = x\}|$. The *payoff* that vertex $v \in V$ receives in the configuration $\mathbf{c} \in [n]^n$ is

$$\lambda_v(\mathbf{c}) = \begin{cases} 0 & \text{if } \exists u \in N(v) : c_u = c_v \\ n_{c_v}(\mathbf{c}) & \text{else} \end{cases}.$$

In other words, given a proper coloring, the payoff of a vertex equals the cardinality of the color class it belongs to.

A *pure Nash equilibrium* [11] is a configuration $\mathbf{c} \in [n]^n$ such that no vertex can increase its payoff by unilaterally deviating. Let $(x, \mathbf{c}_{-v})$ denote the configuration resulting from $\mathbf{c}$ if vertex $v$ chooses color $x$ while all the remaining vertices preserve their colors. Then, a configuration $\mathbf{c} \in [n]^n$ of the graph coloring game $\Gamma(G)$ is a pure Nash equilibrium if, for all vertices $v \in V$, $\lambda_v(x, \mathbf{c}_{-v}) \leq \lambda_v(\mathbf{c}) \quad \forall x \in [n]$. We also say that $\mathbf{c}$ is a *Nash equilibrium coloring*.

A vertex $v \in V$ is *unsatisfied* in the configuration $\mathbf{c} \in [n]^n$ if there exists a color $x \neq c_v$ such that $\lambda_v(x, \mathbf{c}_{-v}) > \lambda_v(\mathbf{c})$; else we say that $v$ is *satisfied*. Clearly, a configuration $\mathbf{c}$ is a Nash equilibrium coloring if and only if all vertices are satisfied in $\mathbf{c}$. For an unsatisfied vertex $v \in V$ in the configuration $\mathbf{c}$, we say that $v$ performs a *selfish step* if $v$ unilaterally deviates to some color $x \neq c_v$ such that $\lambda_v(x, \mathbf{c}_{-v}) > \lambda_v(\mathbf{c})$.

The analysis of the graph coloring game given in [12] illustrates that $\Gamma(G)$ has always pure Nash equilibria, and that each pure equilibrium is a proper coloring of $G$. Furthermore, there exists a pure equilibrium that corresponds to an optimum coloring. It is also shown that any pure Nash equilibrium of the game is a proper coloring of $G$ that uses a number of colors, $k$, bounded above by all the general known to us upper bounds on the chromatic number of $G$. In particular, $k$ is proved to be bounded above by $\Delta_2(G)+1$, $\frac{n+\omega(G)}{2}$, $n-\alpha(G)+1$, and $\frac{1+\sqrt{1+8m}}{2}$. In [4], it was further shown that $k \leq \frac{\chi(G)+1}{2} + \sqrt{m - \frac{(\chi(G)-1)(\chi(G)+1)}{4}}$. Therefore, the number of colors used by any pure Nash equilibrium of $\Gamma(G)$ is at most

$$k \leq \min \left\{ \Delta_2(G) + 1, \ \frac{n + \omega(G)}{2}, \ n - \alpha(G) + 1, \ \frac{1 + \sqrt{1 + 8m}}{2}, \right.$$
$$\left. \frac{\chi(G) + 1}{2} + \sqrt{m - \frac{(\chi(G) - 1)(\chi(G) + 1)}{4}} \right\}. \tag{1}$$

Most interestingly, it is proven that any sequence of selfish steps, when started with a proper (e.g., the trivial) coloring, always reaches a pure Nash equilibrium in $O(n \cdot \alpha(G))$ selfish steps. The proof follows from the existence of a *potential function* [9], which is a function defined over the set of pure strategy profiles of a game, and has the property that the difference of the function's value after a player deviates equals the corresponding difference of that player's payoff. In particular, [12] showed that the graph coloring game $\Gamma(G)$ possesses a potential function defined as

$$\Phi(\mathbf{c}) = \frac{1}{2} \sum_{x \in [n]} (n_x(\mathbf{c}))^2 \quad \forall \mathbf{c} \in [n]^n \ ,$$

with the property that, for any proper coloring $\mathbf{c} \in [n]^n$, for any vertex $v \in V$, and for any color $x \in [n]$, it holds that

$$\lambda_v(x, \mathbf{c}_{-v}) - \lambda_v(\mathbf{c}) = \Phi(x, \mathbf{c}_{-v}) - \Phi(\mathbf{c}) \ .$$

Therefore, if any vertex $v$ performs a selfish step then the value of $\Phi$ is increased as much as the payoff of $v$ is increased. Since the payoff of $v$ is increased by at least 1 and the value of $\Phi$ is bounded above by $\frac{n \cdot \alpha(G)}{2}$, it follows that after at most $\frac{n \cdot \alpha(G)}{2}$ selfish steps there will be no vertex that can improve its payoff (because $\Phi$ will have reached a local maximum, which is no more than the global maximum, which is no more than $(n \cdot \alpha(G))/2$, so a pure Nash equilibrium will have been reached. This implies the following simple centralized algorithm $\mathcal{A}$ that computes, in polynomial time, a pure Nash equilibrium of $\Gamma(G)$ – and thus a proper coloring of $G$, satisfying Inequality 1:

> **Algorithm $\mathcal{A}$**
> *Input:* Graph $G$ with vertex set $V = \{v_1, \ldots, v_n\}$
> *Output:* A pure Nash equilibrium $\mathbf{c} = (c_{v_1}, \ldots, c_{v_n}) \in [n]^n$ of $\Gamma(G)$
> Initialization: **for** $i = 1$ **to** $n$ **do** $c_{v_i} = i$
> **repeat**
>     find an unsatisfied vertex $v \in V$ and a color $x \in [n]$
>         such that $\lambda_v(x, \mathbf{c}_{-v}) > \lambda_v(\mathbf{c})$
>     set $c_v = x$
> **until** all vertices are satisfied

## 3 Distributed Vertex Coloring as a Game

The algorithm $\mathcal{A}$ presented in the previous section is actually a local search algorithm: starting with an arbitrary proper vertex coloring (such as the trivial proper coloring), we allow each vertex (one at a time) to move to another color class of higher cardinality, until no further local moves are possible. Our aim is to apply this local search method in a distributed environment, so that we end up with a distributed algorithm that computes a proper coloring of the vertices of a graph using a total number of colors satisfying Inequality 1.

However, there are two main challenges arising in implementing $\mathcal{A}$ in a distributed environment: First, $\mathcal{A}$ requires *mutual exclusion*: only one node at a time can perform a selfish step in order to guarantee convergence into a pure Nash equilibrium coloring. Second, $\mathcal{A}$ requires *global knowledge*: each node needs to be aware of the cardinalities of each color class in order to decide whether to perform a selfish step or not.

In this section, we deal with both the above challenges. First, we prove that a "parallelization" of algorithm $\mathcal{A}$ is possible, in the sense that even if we allow multiple non-neighboring unsatisfied vertices to perform simultaneous selfish steps, then (a slight modification of) the algorithm still converges into a pure Nash equilibrium, and thus into a proper coloring satisfying the bounds of Inequality 1. Then, we propose a method to deal with the global information problem (possibly at the expense of parallelization) so that, after performing a selfish step, a node needs not to inform all the other nodes about the change, but only a small subset of them. Finally, we discuss the ability of our distributed algorithm to self-stabilize.

In the following, we assume that each node has, and is aware of, a unique identification number (id).

### 3.1 Simultaneous Execution of Selfish Steps

We study whether polynomial time convergence into a Nash equilibrium coloring can still be guaranteed if vertices are allowed to perform selfish steps simultaneously. We still assume however that all vertices have access to global information about the cardinality of each color class. Note that, in order to avoid non-proper colorings, we should not allow any pair of neighboring vertices change their colors simultaneously. Towards this direction, we allow an unsatisfied vertex to perform a selfish step only if it has the maximum id among the unsatisfied vertices in the neighborhood.

More specifically, given a configuration $\mathbf{c} \in [n]^n$, we say that the vertices perform a *joint selfish step* if:

1. Each vertex that is unsatisfied in $\mathbf{c}$ sends a message to inform its neighbors.
2. If an unsatisfied vertex $v$ does not receive any message from a vertex of higher id, then it performs a selfish step, i.e., it changes its color to $c'_v$ such that $\lambda_v(c'_v, \mathbf{c}_{-v}) > \lambda_v(\mathbf{c})$.

Clearly, in a joint selfish step, possibly more than one vertices will change their color simultaneously. We will examine whether a sequence of joint selfish steps converges into a Nash equilibrium coloring.

Given a configuration $\mathbf{c} \in [n]^n$, we define vector $\mathbf{\Lambda}(\mathbf{c})$ as the $n$-vector whose $i$th entry, $\Lambda_i(\mathbf{c})$, is the cardinality of the $i$th largest color class. We say that $\mathbf{\Lambda}(\mathbf{c})$ is lexicographically greater than $\mathbf{\Lambda}(\mathbf{c}')$ if $\Lambda_i(\mathbf{c}) > \Lambda_i(\mathbf{c}')$ for some $i \in [n]$ and, if $i > 1$, then $\Lambda_j(\mathbf{c}) = \Lambda_j(\mathbf{c}')$ for all $j \in \{1, \dots, i\}$. If $\Lambda_i(\mathbf{c}) = \Lambda_i(\mathbf{c}')$ for all $i \in [n]$, then $\mathbf{\Lambda}(\mathbf{c})$ is lexicographically equal to $\mathbf{\Lambda}(\mathbf{c}')$.

**Lemma 1.** *Let $\mathbf{c} \in [n]^n$ be any configuration and let $\mathbf{c}' \in [n]^n$ be a configuration that results from $\mathbf{c}$ after a joint selfish step. Then, $\mathbf{\Lambda}(\mathbf{c})$ can not be lexicographically greater than $\mathbf{\Lambda}(\mathbf{c}')$.*

*Proof.* Let $Y \subseteq [n]$ be the subset of colors that correspond to the color classes of maximum cardinality in $\mathbf{c}$. If the cardinality of such a maximum color class increases, then $\Lambda_1(\mathbf{c}) < \Lambda_1(\mathbf{c}')$ and we are done.

Assume now that the cardinality of each maximum color class in $\mathbf{c}$ does not increase. Observe that a vertex colored $x$ in $\mathbf{c}$ for some $x \in Y$ gets maximum payoff, so it could improve it only by moving to a color class of equal (and thus maximum, as well) cardinality. But since the cardinality of each maximum color class in $\mathbf{c}$ does not increase, it must be the case that all these cardinalities remain the same. Therefore $\Lambda_i(\mathbf{c}) = \Lambda_i(\mathbf{c}')$ for all $i \in [k]$. This further implies that no vertex colored $x' \notin Y$ in $\mathbf{c}$ chooses color $x \in Y$ in $\mathbf{c}'$.

Similar arguments apply for the cardinalities of the second to maximum color classes in $\mathbf{c}$ and so on, to conclude that either (i) $\mathbf{\Lambda}(\mathbf{c}')$ is lexicographically greater than $\mathbf{\Lambda}(\mathbf{c}')$ or (ii) $\mathbf{\Lambda}(\mathbf{c}')$ is lexicographically equal to $\mathbf{\Lambda}(\mathbf{c}')$ and each vertex that performs a selfish step moves to a color class of equal cardinality in $\mathbf{c}$. $\square$

**Lemma 2.** *Let $\mathbf{c} \in [n]^n$ be any configuration and let $\mathbf{c}' \in [n]^n$ be a configuration that results from $\mathbf{c}$ after a joint selfish step. If $\mathbf{\Lambda}(\mathbf{c})$ is lexicographically greater than $\mathbf{\Lambda}(\mathbf{c}')$, then $\Phi(\mathbf{c}) > \Phi(\mathbf{c}')$.*

*Proof.* Clearly, for any configuration $\mathbf{c}''$, $\Phi(c'') = \sum_{i=1}^{n} \Lambda_i^2(\mathbf{c}'')$. The proof follows from the fact that $\mathbf{\Lambda}(\mathbf{c}')$ has at least the same number of zero entries as $\mathbf{\Lambda}(\mathbf{c}')$, since no vertex has an incentive to choose a color that is not used by any other vertex. $\square$

In order to avoid loops of the joint selfish steps sequence, we have to deal with the case where $\boldsymbol{\Lambda}(\mathbf{c}')$ is lexicographically equal to $\boldsymbol{\Lambda}(\mathbf{c})$, and hence $\Phi(\mathbf{c}') = \Phi(\mathbf{c})$. To do so, we propose the *probabilistic joint selfish step*, given a configuration $\mathbf{c} \in [n]^n$:

1. The vertices perform a joint selfish step, resulting in configuration $\mathbf{c}'$.
2. Each vertex which performed a selfish step checks if $\Phi(\mathbf{c}) = \Phi(\mathbf{c}')$. If yes, it changes back to its previous color with probability $1/2$ and repeats this step.

We show that the probabilistic joint selfish steps sequence converges into a pure Nash equilibrium in polynomial time, with high probability:

**Theorem 1.** *With high probability, the probabilistic joint selfish step sequence converges in polynomial time into a Nash equilibrium coloring.*

*Proof.* Assume $\Phi(\mathbf{c}') = \Phi(\mathbf{c})$ after a joint selfish step. Then Lemma 1 and Lemma 2 imply that at least $k \geq 2$ vertices with different colors but with the same payoff in $\mathbf{c}$ permuted their colors. If some, but not all, of these vertices return to their original colors, then some payoffs will increase, yielding to a lexicographically greater configuration $\mathbf{c}''$, and thus $\Phi(\mathbf{c}'') > \Phi(\mathbf{c})$. The probability that this does not happen is $2 \cdot 2^{-k}$ (equal to the probability that all change their colors or all preserve their colors). So the probability that Step 2 of the probabilistic joint selfish steps sequence is repeated $t$ times is $2^{-t-k+1}$.    □

## 3.2    Distributing Global Information

The (probabilistic) joint selfish steps sequence converges to a Nash equilibrium coloring provided that all vertices know the cardinalities of all color classes. In order to achieve this in a distributed environment, we let each vertex maintain a local list with the cardinalities of all color classes. Initially, each vertex is assumed to have a unique color (its id) and all these cardinalities equal to 1. An arbitrary vertex, which we call *the Judge* initiates the coloring procedure. It performs a selfish step, if possible, and sends the original list of cardinalities to its neighbors, who become its children. Then these nodes perform a joint selfish step, if possible, and also pass the original list to their neighbors. If a vertex has no parent and receives the list from some vertex, it sets itself as the child of that vertex.

This way, a tree (whose root is the Judge) is constructed, and when the leaves are reached, all unsatisfied vertices will have actually performed a joint selfish step, based on *the same* global information about the cardinalities of color classes. This is assured by the fact that no matter if a vertex performs a selfish step or not, it sends to its children the list of cardinalities that was originally sent by the Judge.

Now, the information about the color changes has to be sent back to the Judge: starting by the leaves, each child sends the information about its new color to its parent and so on, until the root of the tree gets informed about the new cardinalities of all color classes. Then the root sends the updated list to its children and the above procedure is repeated until the Judge gets the same list in two successive steps. In order to avoid loops, if the root observes that the list has changed but the value of the potential function $\Phi$ did not change, a message is forwarded to ask the vertices to change back to their original color with probability $1/2$.

The above procedure, combined with the results of the previous subsection, guarantees that after at most $O(n \cdot a(G))$ phases (each phase starting with the Judge forwarding the list of cardinalities and ending with the Judge receiving the information about all color changes that have occurred), a proper coloring of $G = (V, E)$ will be produced, using a number of colors satisfying the bounds of Inequality 1. To the best of our knowledge, this is the first distributed implementation of a coloring algorithm that achieves all these bounds. We also note that the complexity bound of $O(n \cdot a(G))$ is rather strict, in the sense that it assumes that no more than $o(1)$ vertices perform a selfish step in the same phase. In Section 4 we provide experimental evidence that, in practice, the number of vertices performing a joint selfish step is large, and therefore the coloring procedure converges faster than the $O(n \cdot a(G))$ bound suggests.

### 3.3 Self-stabilization

We now present a self-stabilizing version of the distributed vertex coloring algorithm; that is, starting from an arbitrary state (coloring of vertices), it guarantees to converge to a legitimate state (proper coloring) in finite number of steps and to remain in a legitimate set of states thereafter. Essentially the self-stabilizing version of our distributed algorithm can recover from transient faults.

In the sequel we consider that the system can start in any configuration. That is, the color of each vertex can be corrupted. Note that we do not make any assumption on the bound of the corrupted nodes. In the worst case all the nodes in the system may start in a corrupted configuration. However, we assume that the unique identification numbers of the vertices are stored in a read-only memory segment that cannot be affected by transient faults.

A central modification to the algorithm is to force vertices to periodically broadcast their color and id to their neighbors. By doing so they can detect (a) whether the coloring is not proper and (b) if a vertex is unsatisfied. This simple mechanism allows the vertices to detect if a transient fault has occurred or if the algorithm was initiated from an arbitrary state. The periodic broadcasting also guarantees that after all vertices have been satisfied the algorithm will not terminate; it will continue to check whether a selfish step can be taken indefinitely. Therefore, when a transient failure occurs, the procedure will be re-initiated.

When two or more vertices detect that their color is not proper (i.e., the same color is used by another vertex) they locally resolve the conflict by changing their color. Each vertex chooses as a new color the maximum color used in the neighborhood plus its own id (recall that for simplicity colors are represented as integers). This will increase the number of colors used in the neighborhood but will result in a proper coloring in $O(1)$ steps. After selecting the new colors, the vertices notify the Judge about the conflict resolution by using the tree structure. In order to guarantee that the tree structure will be functional despite the transient failures, we replace the tree construction algorithm used in the previous section with a self-stabilizing tree structure (e.g., see [1]).

When the Judge receives notification of a conflict resolution, it waits for a $O(\text{diam}(G))$ period (where $\text{diam}(G)$ is a predefined upper bound of the diameter of $G$) so that any other conflict is propagated through the tree structure. Then the Judge broadcasts a request message to all vertices to report their color as if they have changed their color during the previous phase. This allows the Judge to recount the color classes and reconstruct the necessary global information. Finally when this reconstruction phase is complete, the new list of colors is broadcast to the vertices to locally check if they wish to conduct a selfish step or not.

The above discussion clearly implies the following:

**Lemma 3.** *The self-stabilizing version of the distributed vertex coloring algorithm assures that:*

1. *Starting from an arbitrary configuration, eventually all vertices are properly colored.*
2. *Starting from an arbitrary configuration, the Judge is informed about the color of each vertex.*
3. *Starting from an arbitrary configuration, eventually all vertices are informed about the cardinality of each color class.*
4. *Starting from an arbitrary configuration, the algorithm returns to a proper configuration (*convergence*).*
5. *Starting from a proper configuration, the algorithm preserves the proper configuration (*closure*).*

## 4    Algorithm Engineering

Most of the times in Computer Science, researchers tend to design an algorithm in an abstract way. This happens because an algorithm should be able to be used in many different situations and it is up to the developer to decide the way it should be turned into code for a real system. Almost every time the developer finds many limitations in the ways she can operate within the given hardware and software specifications. These problems are further augmented when implementing algorithms for wireless sensor networks due to the extremely limited resources and also due to the heterogeneous nature (both in terms of hardware and software). Algorithm development for such networks is complex as it unites the challenges of distributed applications and embedded programming.

As a starting point we implemented a centralized version that follows closely the initial design of [12]. We call this algorithm JColoring (noted as JC in the figures). The algorithm was implemented in a real environment that actually exchanged network messages containing protocol payload. The implementation of this centralized version required two sub-protocols that enabled (i) only a single vertex to perform a selfish step each time (mutual exclusion) and (ii) all vertices to know the cardinalities of all color classes (global knowledge).

In this centralized version, the Judge coordinates the graph coloring game by communicating with each vertex sequentially. The Judge sends a message to the next vertex in the sequence containing the list of color classes along with their cardinalities. Upon receiving this message, the vertex has to decide whether to take a selfish step or not and informs the Judge about its new color. Communication between vertices is implemented using the well-established ad-hoc routing protocol TORA [13]. The particular protocol was selected because of its simplicity and ability to operate adequately in wireless settings.

As expected, the evaluation of this version indicated the very poor scalability of the system due to the single point of coordination. It also revealed another major practical problem: in the hardware platform we used to test the algorithm, the available payload of each packet was about 120 bytes. Therefore in a single message we were able to store about 30 colors. So when the number of vertices grew beyond this number, the messages containing the list with the color cardinalities were fragmented in two (or more) packets thus drastically decreasing the available bandwidth and leading to longer execution times.

After achieving a satisfactory state of the centralized version, the next step was to implement our distributed algorithm as described in Section 3. We call this algorithm PJColoring (noted as PJC in the figures). With the data structures and the message fragmentation mechanism in place we proceeded by implementing (i) the simultaneous execution of selfish steps (thus leveraging the mutual exclusion limitation) and (ii) the distributed global information provisioning. Essentially the tree construction algorithm used for efficiently broadcasting the distributed global information replaced the TORA routing algorithm.

The evaluation of our distributed algorithm revealed yet another interesting Algorithm Engineering issue. The convergecast operation for collecting the information about the color changes resulted in a high number of message exchanges. To resolve this issue we implemented an aggregation mechanism. Each parent, after collecting the new colors of its children, it summarizes the information and propagates a single message to its parent. This aggregation is repeated by each parent until the root of the tree is reached. In many cases this technique led to a reduction of up to 50% of message exchanges. In the figures we note this improved version of the algorithm as PJC (with aggregation).

Successful Algorithm Engineering for tiny artifacts requires the validation and evaluation of algorithms in experiments on real networked embedded devices. We ported our distributed algorithm into the real hardware platform as a final step of validation and evaluation of its performance. This immediately revealed a serious problem related to wireless communication, that of medium congestion. The implementation of the simultaneous execution produced a large number of concurrent exchanges and thus message collisions that were unresolved by the MAC. To reduce the collisions we imposed a short random delay before each transmission; yet, some messages were still not delivered properly. This was inevitable due to the non-deterministic behavior of the wireless medium. To completely overcome these problems we implemented the self-stabilizing version of the algorithm leading to a system that was fully operational in real wireless sensor network deployments.

## 4.1   Implementation Details

We decided to implement our algorithms using Wiselib [2]: a code library, that allows implementations to be OS-independent. It is implemented based on C++ and templates, but without virtual inheritance and exceptions. Algorithm implementations can be recompiled for several platforms and firmwares, without the need to change the code. Wiselib can interface with systems implemented using C (Contiki), C++ (iSense), and nesC (TinyOS).

Furthermore, an important feature of Wiselib is the already implemented algorithms and data structures. Since different kind of hardware uses different ways to store data (due to memory alignment, inability to support dynamic memory, etc.), it is important to use these safe types as much as possible since they have been tested before on most hardware platforms. As of mid 2010, the Wiselib includes about 40 Open Source implementations of standard algorithms, and is scheduled to grow to 150-200 algorithms by the end of 2011. We use Wiselib data structure to implement the maps for counting the cardinalities of the colors and the implementation of TORA algorithm.

Finally, Wiselib also runs on the simulator Shawn [8], hereby easing the transition from simulation to actual devices. This feature allows us to validate the faithfulness of our implementation and also get results concerning the quality of our algorithms without time consuming deployment procedures and harsh debugging environments. Shawn allows repeatability of simulations in an easy way by using only a single configuration file. It provides many options such as packet loss, radius of communication, ways of communicating and even mobility in an abstract way, without needing to provide specific code for every hange.

## 4.2   Performance Evaluation: Simulations

We start by presenting the results of the validation and performance evaluation based on simulated executions of our implementation in the Shawn environment. Using different types of networks and experimenting with the different parameters, we were able to identify that parameters that influence the correctness and performance of our protocol. The first type of network topologies (called "Fixed Diameter"), deploys $n = [25, 250]$ vertices in a way such that the network diameter remains constant as $n$ increases: the resulting networks have $\operatorname{diam}(G) = 6$ and the average vertex degree (which we call "density") is $2 \le \operatorname{avgDeg}(G) \le 40$. The second type of network topologies (called "Fixed Density"), deploys $n = [10, 230]$ in a way such that the average vertex degree remains constant as $n$ increases: the resulting networks have $3 \le \operatorname{diam}(G) \le 15$ and $\operatorname{avgDeg}(G) = 12$.

A very important performance metric is the number of colors used. Figure 1 depicts the number of colors used for each set of topologies. As expected, as the density of the network increases (i.e., the average vertex degree), the chromatic number increases and hence the number of colors used by our algorithm.

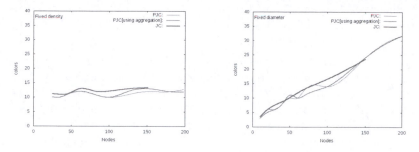

**Fig. 1.** Colors used for both types of network topologies

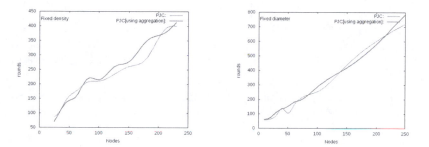

**Fig. 2.** Time complexity for both types of network topologies

When the density remains more or less constant, the chromatic number remains constant, as also does the number of colors used by the algorithm.

The next algorithm property that we evaluate is the time complexity. Figure 2 depicts the number of rounds required to reach a Nash equilibrium. It is evident that as the number of vertices increases, so does the number of rounds required. However, the "Fixed Diameter" topologies seem to be harder to address. This is caused due to the fact that only few vertices can change color in each step of the algorithm in dense environments. As a result the algorithm takes more steps until it reaches an equilibrium. These topologies require about twice the number of rounds than the "Fixed Density" topologies. In these figures we have not included the results for the JColoring algorithm as they cannot be fit properly: in all cases the resulting rounds are about 20 times more than the PJColoring algorithm and thus even for $n = 50$ it becomes totally impractical.

In the sequel we examine the communication complexity of the algorithms. Figure 3 shows the number of message exchanges for each set of topologies considered. The behavior of the algorithms are similar to that observed in the previous figure: (a) as the number of vertices increases, so does the number of messages exchanged, (b) the "Fixed Diameter" topologies seem to be harder to address and (c) the JColoring exchanges much greater messages than the PJColoring algorithms. In these figures it is clear that for $n > 50$ the JColoring algorithm becomes impractical, while PJColoring performs well for all the topologies examined.

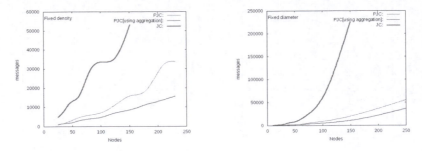

**Fig. 3.** Communication complexity for both types of network topologies

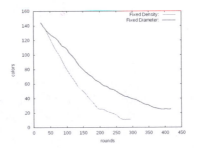

**Fig. 4.** Colors used as the simulation evolves for PJColoring (with aggregation) for different type of topologies of fixed number of vertices ($n = 150$)

We conduct a final set of simulated executions to evaluate the ability of PJ-Coloring (with aggregation) to perform parallel selfish steps. Figure 4 depicts the number of colors used as the network evolves over time. We present the results for both types of topologies when $n = 150$. We observe that in both cases the algorithm reaches the equilibrium fast, although in the denser topology the progress is slower. It is evident that the algorithm succeeds in parallelizing the process of selfish coloring.

## 4.3    Performance Evaluation: Experiments

Although simulations enable both verification and performance evaluation of protocols, they do not take into account read-world effects that may change their behavior. Also, in physical deployment protocols must operate within the given hardware and software specifications (i.e., memory and message payload constrains). Thus we continue the evaluation of the PJColoring (with aggregation) algorithm in our experimental testbed that consists of iSense sensor nodes with a 32 Bit RISC Controller running at 16 MHz, with 96 KB RAM and 128 KB Flash and an IEEE 802.15.4 compliant wireless radio interface. All nodes come with a permanent energy source allowing for an arbitrary duty cycle. Nodes are equipped with a multitude of sensors measuring light, temperature, humidity, acceleration, magnetic-field levels and barometric pressure. For more details on the installation and technical aspects of the experimental testbed see [3]. In the sequel we consider that each node represents a single vertex.

The first set of experiments was conducted in networks where all the nodes were within communication range of each other, thus forming a clique. In Tab. 1 we can see the breakdown of the total execution time of the protocol. The selection of this topology is twofold. Firstly, to determine the effect of network density in the execution of our protocol (i.e., collisions, etc.). We observed that up to networks sizes of eight nodes the results were always correct (i.e., messages were delivered properly), while in larger networks message loses always occurred forcing the non-stabilizing version of the protocol to produce not consistent results. Secondly, to determine the real execution time of a protocol phase. In these topologies the protocol terminates immediately after the initial assignment of colors, i.e., after checking that every available color is taken by a neighbor. For all topologies considered the execution time of a single phase is about $130ms$.

**Table 1.** Real execution times of PJColoring (with aggregation) for clique topology

| Nodes | Tree Construction | PJColoring | Total |
|-------|-------------------|------------|-------|
| 5 | 745 ms | 132 ms | 877 ms |
| 6 | 739 ms | 133 ms | 872 ms |
| 7 | 754 ms | 141 ms | 895 ms |
| 8 | 752 ms | 137 ms | 889 ms |

The second set of experiments was conducted in networks where the nodes where positioned in a line. Tab. 2 lists the total execution time of the protocol. This topology produces networks of increasing diameter thus requiring more time to construct the broadcast tree and complete the convergecast operation. Based on the results, the initialization of the two processes requires about $750ms$, and then for each further hop an additional time of about $300ms$ was required. The initialization time of the two processes is also confirmed by the previous set of experiments that used single-hop networks (see Tab. 1). In these topologies, we observe that the time required by the selfish steps is also increasing with the network size. It seems that for each additional hop an additional time of $600ms$ is required, or, based on the previous set of experiments, about 4 protocol phases. By carefully examining the particular topology, it is evident that the addition of a new node forces all other nodes to take additional selfish steps until a Nash equilibrium is reached.

The performance evaluation we conducted both in simulated and experimental environments indicates interesting aspects of the vertex coloring problem. It also indicates that our solution successfully raises the limitation of [12] where only one vertex at a time is allowed to perform a selfish step in order to guarantee polynomial time convergence into a Nash equilibrium coloring. Our distributed algorithm allows non-neighboring vertices to perform a selfish step in parallel, thus drastically reducing time complexity. The Algorithm Engineering process

**Table 2.** Real execution times of PJColoring (with aggregation) for line topology

| Nodes | Tree Construction | PJColoring | Total |
|-------|-------------------|------------|---------|
| 2 | 736 ms | 138 ms | 1389 ms |
| 3 | 1032 ms | 728 ms | 1760 ms |
| 4 | 1298 ms | 1324 ms | 2622 ms |
| 5 | 1756 ms | 2011 ms | 3767 ms |

led to a practical algorithm that can be executed in real networks and, interestingly, further reduces communication complexity while preserving the time complexity and the number of colors used.

# References

1. Afek, Y., Kutten, S., Yung, M.: Memory-efficient self stabilizing protocols for general networks. In: Toueg, S., Kirousis, L.M., Spirakis, P.G. (eds.) WDAG 1991. LNCS, vol. 579, pp. 15–28. Springer, Heidelberg (1992)
2. Baumgartner, T., Chatzigiannakis, I., Fekete, S., Koninis, C., Kröller, A., Pyrgelis, A.: Wiselib: A Generic Algorithm Library for Heterogeneous Sensor Networks. In: Silva, J.S., Krishnamachari, B., Boavida, F. (eds.) EWSN 2010. LNCS, vol. 5970, pp. 162–177. Springer, Heidelberg (2010)
3. Chatzigiannakis, I., Koninis, C., Mylonas, G., Colesanti, U., Vitaletti, A.: A peer-to-peer framework for globally-available sensor networks and its application in building management. In: DCOSS/IWSNE 2009 (2009)
4. Escoffier, B., Gourves, L., Monnot, J.: Strategic Coloring of a Graph. In: Calamoneri, T., Diaz, J. (eds.) CIAC 2010. LNCS, vol. 6078, pp. 155–166. Springer, Heidelberg (2010)
5. Feige, U., Kilian, J.: Zero knowledge and the chromatic number. Journal of Computer and System Sciences 57(2), 187–199 (1998)
6. Halldórsson, M.: A still better performance guarantee for approximate graph coloring. Information Processing Letters 45, 19–23 (1993)
7. Karp, R.M.: Reducibility among combinatorial problems. In: Complexity of Computer Computations, pp. 85–103. Plenum Press, New York (1972)
8. Kröller, A., Pfisterer, D., Buschmann, C., Fekete, S.P., Fischer, S.S.: A new approach to simulating wireless sensor networks. In: DASD 2005, pp. 117–124 (2005)
9. Monderer, D., Shapley, L.S.: Potential games. Games and Economic Behavior 14, 124–143 (1996)
10. Moscibroda, T., Wattenhofer, R.: Coloring Unstructured Radio Networks. In: SPAA 2005, pp. 39–48 (2005)
11. Nash, J.F.: Non-cooperative games. Annals of Mathematics 54(2), 286–295 (1951)
12. Panagopoulou, P.N., Spirakis, P.G.: A game theoretic approach for efficient graph coloring. In: Hong, S.-H., Nagamochi, H., Fukunaga, T. (eds.) ISAAC 2008. LNCS, vol. 5369, pp. 183–195. Springer, Heidelberg (2008)
13. Park, V.D., Corson, M.S.: A Highly Adaptive Distributed Routing Algorithm for Mobile Wireless Networks. In: INFOCOM 1997, pp. 1405–1413 (1997)
14. Schneider, J., Wattenhofer, R.: Coloring unstructured wireless multi-hop networks. In: PODC 2009, pp. 210–219 (2009)

# Constructing a Map of an Anonymous Graph: Applications of Universal Sequences

Jérémie Chalopin[1,*], Shantanu Das[1], and Adrian Kosowski[2]

[1] LIF, CNRS & Aix-Marseille University, France
jeremie.chalopin@lif.univ-mrs.fr, shantanu.das@acm.org
[2] Gdańsk University of Technology, Poland
and LaBRI, INRIA Bordeaux Sud-Ouest, France
adrian@kaims.pl

**Abstract.** We study the problem of mapping an unknown environment represented as an unlabelled undirected graph. A robot (or automaton) starting at a single vertex of the graph G has to traverse the graph and return to its starting point building a map of the graph in the process. We are interested in the cost of achieving this task (whenever possible) in terms of the number of edge traversal made by the robot. Another optimization criteria is to minimize the amount of information that the robot has to carry when moving from node to node in the graph.

We present efficient algorithms for solving map construction using a robot that is not allowed to mark any vertex of the graph, assuming the knowledge of only an upper bound on the size of the graph. We also give universal algorithms (independent of the size of the graph) for map construction when only the starting location of the robot is marked. Our solutions apply the technique of universal exploration sequences to solve the map construction problem under various constraints. We also show how the solution can be adapted to solve other problems such as the gathering of two identical robots dispersed in an unknown graph.

**Keywords:** Graph Exploration, Map Construction, Anonymous Networks, Mobile Robot, Universal Exploration Sequences.

## 1   Introduction

We consider the problem of exploration and mapping of an unknown unlabelled environment by a mobile entity which we call the agent. The environment is usually modelled as a graph where the agent is initially located at any arbitrary node of the graph. The objective of the agent is to build a map of the graph. The graph is anonymous i.e. the nodes of the graph do not have any identifying labels and thus, all nodes of the same degree look identical to the agent. However, the edges incident to a node are locally ordered with a *port numbering* that allows the agent to deterministically choose an edge and traverse along it. Note that if the agent is allowed to somehow mark the nodes that it visits (such

---

* This author was partially supported by ANR Projects SHAMAN and ECSPER.

C. Lu, T. Masuzawa, and M. Mosbah (Eds.): OPODIS 2010, LNCS 6490, pp. 119–134, 2010.
© Springer-Verlag Berlin Heidelberg 2010

that it can recognize them on future visits), then a simple depth-first search suffices to solve the problem. When the agents do not have the capability to mark nodes it is sometimes difficult to solve the map construction problem. A known technique for traversing unlabelled graphs is to use the so-called universal traversal sequences [18]. A universal traversal sequence is a sequence of port numbers such that if the agent traverses the edges of any graph $G$ according to this sequence it is guaranteed to visit all nodes of $G$ irrespective of the topology of $G$ and the port-numbering on $G$. However, such sequences tend to be very large and thus it is perhaps not the most efficient method of traversing a graph. Moreover, traversing the graph does not necessarily imply that the agent can build a map. In certain cases, when the graph has enough symmetry, it may not be possible to build a map of the complete graph. In this paper, we concerned about the time complexity (or number of moves made by the agent) for building a map in those cases when it is possible to do so. An efficient method for map construction is useful as a basic step for an autonomous agent in solving other tasks in unknown unlabelled environments.

One application of the map construction problem is the task of gathering together two autonomous agents that are dispersed in a unknown environment. This is called the *rendezvous* problem. When the two dispersed entities can not communicate from a distance, solving rendezvous is essential for an exchange of information or for achieving even the simplest form of coordination between the mobile entities. The rendezvous problem belongs to the class of symmetry-breaking problems (e.g. leader election is another such problem) that are central to study of computability in distributed systems. The importance of the problem is evident from the large volume of literature [5,10,13,16,19,11,26] dedicated to solving the problem under various conditions and restrictions.

Even if the agents succeed in building a map of the graph, it may not always be possible to rendezvous. For instance, if the agents are in a ring of even size and they start from diametrically opposite nodes in the ring, then no deterministic algorithm is guaranteed to solve rendezvous in this case. However, if the agents start from any other location (except being opposite to each other) then it is possible to solve rendezvous, as soon as we allow the agents to mark their starting locations [19]. In this paper, we solve rendezvous in anonymous graphs assuming that the starting locations of the agents are marked. However, the agents are not allowed to mark any other vertices during their traversal. Further, the agents may not have any prior information about the graph not even the size of the graph.

**Related Work:** Previous studies on graph exploration have mostly concentrated on *labelled* graphs (or digraphs), with an emphasis on minimizing the cost of exploration in terms of either the number of moves (edge traversals) or the amount of memory used by the agent. Panaite and Pelc [21] gave an algorithm for exploring labelled undirected graphs that uses $m + O(n)$ moves, improving on the standard *Depth-First Search* algorithm that takes $2m$ moves. On the other hand, Deng and Papadimitrou [12] as well as Albers and Henzinger [1] studied the exploration of strongly connected directed graphs under the same conditions.

There have also been some studies on the efficiency of exploration when some prior information about the graph is available with the agent—for instance, when the agent possesses an unlabelled isomorphic map of the graph [22].

Given an unknown, unlabelled (sometime called anonymous) graph, it is not always possible to construct an exact map of the graph (due to the presence of symmetries). There exists characterizations of anonymous graphs where it is possible to solve the problem [25].

For exploring arbitrary anonymous graphs, various methods of marking nodes have been used by different authors. Bender *et al.* [7] proposed the method of dropping a pebble on a node to mark it and showed that any strongly connected directed graph can be explored using just one pebble, if the size of the graph is known and using $O(\log \log n)$ pebbles, otherwise. Dudek *et al.* [14] used a set of distinct markers to explore unlabeled undirected graphs. In [15] the authors focus on minimizing the amount of memory used by the agents for exploration (however, they do not require the agents to construct a map of the graph). Others have studied the exploration of mazes or labyrinths, which have been shown [8] to be easier to explore than graphs, due to the availability of orientation information.

In the absence of any device for marking nodes, unknown anonymous graphs can still be explored using universal traversal/exploration sequences [18]. Aleliunas et al. [2] showed that there exists universal traversal sequences of polynomial size for all connected graphs of a given size $n$. A recent result by Reingold [23] showed that universal exploration sequences can be constructed in logarithmic space. Such sequences have been used for solving the rendezvous problem [10,24] though only in the synchronous setting.

The problems of rendezvous and leader election has been extensively studied as symmetry-breaking problems in unknown anonymous graphs, starting from the work of Angluin [4]. Characterizations of the solvable instances for leader election in *message passing* networks of processors, have been provided by Boldi *et al.* [9] and by Yamashita and Kameda [25] among others. Recently, Fusco and Pelc [17] have shown that leader election can be solved if each process has a memory of $O(\log n)$ bits, matching the lower bound given by Ando et al. [3]. The rendezvous problem has been solved under various different assumptions such as distinct labels for the agents, sense of direction information, or prior knowledge of topology (e.g.[5,13,16,19,26]). In the most general setting of unknown anonymous graph with identical agents, the problem was recently solved in [10], though only for synchronous agents. In the asynchronous case, an almost complete solution using distinct labels has been provided in [11]. The idea of solving rendezvous by marking the starting locations with tokens was first proposed by Baston and Gal [6].

**Our Results:** We study the complexity of map construction in anonymous graphs by a mobile agent that is not allowed to write on the nodes of the graph. We present several polynomial time deterministic algorithms for map construction.

In the model where no vertices of the graph are marked, for the task of map construction to be feasible, the agents must know some bound $n$ on the number of nodes of the graph and some bound $d \leq n$ on its degree. The folklore algorithm based on view construction [25] requires $O(d^n)$ moves by the agent. The recent paper [10] provides more efficient map construction algorithms: a polynomial-time approach (with very high exponent) using small memory, and an $O(n^{10}d^5 \log^2 n)$-time algorithm using $O(n^9 d^4 \log^2 n)$ memory. Herein we put forward two improved algorithms which offer different time/memory tradeoffs:

- a simple algorithm running in $O(n^6 d^2 \log n)$ time, using $O(n^6 d^2 \log n)$ memory (Prop. 6),
- a more advanced algorithm running in $O(n^6 d^3 \log n)$ time, using $O(n^3 d^2 \log n \log d)$ memory (Prop. 7).

In the model in which the agent has no prior knowledge of graph parameters (such as $n$ or $d$), in order to make the problem feasible, we assume that the starting location of the agent is specially marked. In this case, we show how to guess the value of $n$ and thus solve map construction in polynomial time using an optimal memory ($\Theta(\log n)$) algorithm (Prop. 8). We also present another algorithm which requires slightly more agent memory ($O(nd \log n)$) but is much more efficient in terms of time steps, requiring only $O(n^3 d)$ steps (Prop. 9). Finally, in this model we also show how our algorithms can be extended to solve the rendezvous of two mobile agents in anonymous graphs with marked homebases even in the asynchronous case (Prop. 11).

## 2   Model, Definitions and Known Results

### 2.1   Our Model

The environment is represented by a simple undirected connected graph $G = (V(G), E(G))$. The agent starts from a single node of the graph, called the *homebase*. The agent can traverse any edge of the graph incident to its current location. At each node $v \in V(G)$, the edges incident to $v$ are distinguishable to any agent arriving at $v$. There is a bijective function

$$\lambda_v : \{(v, u) \in E(G) : u \in V(G)\} \rightarrow \{0, 1, 2, \ldots d(v) - 1\}$$

which assigns unique labels (port-numbers) to the edges incident at node $v$ (where $d(v)$ is the degree of $v$). An agent at a node $u$ can choose to leave through any incident edge $e = (u, v)$ simply by specifying the port number $\lambda_u(u, v)$ of the edge. On reaching the node $v$, the agent knows the port number $\lambda_v(v, u)$ of the edge through which it arrived. The $i$th successor of a node $u$, denoted by $succ(u, i)$ is the node $v$ reached by taking port number $i$ from node $u$ (where $0 \leq i < deg(u)$). For any edge $(u, v)$, we use $\lambda(u, v)$ to denote the ordered pair of labels $(\lambda_u(u, v), \lambda_v(u, v))$. A path in $G$ is a sequence of nodes $P = (u_0, u_1, \ldots, u_k)$ such that $(u_j, u_{j+1}) \in E(G)$, $\forall j, 0 \leq j < k$ and the label sequence of path $P$ is $\Lambda(P) = (\lambda(u_0, u_1), \ldots \lambda(u_{k-1}, u_k))$.

The nodes of $G$ do not have visible identities by which a visiting agent can identify them. In other words, nodes having the same degree look identical to the agents. The agents have computing and storage capabilities. When an agent moves from one node to another, it carries with its own local memory which consists of two parts. One part is a write-only stable storage which is used to write the output (we assume it is large enough to store a map of $G$). The other part is the agent's private memory which is used for remembering the information obtained in previous moves. Our objective is to minimize the private memory of the agent i.e. the amount of information it needs to remember while moving along the graph. When the agent is located at any node of the graph, it has access to a read-write memory which can be used for local computation (but not for storing information). We are not concerned about the cost of performing local computations at node. We are interested in minimizing the total number of edge traversals (steps) made by the agent in achieving its tasks.

## 2.2   Universal Exploration Sequences

In this paper, we will use the notion of a *Universal Exploration Sequence* (UXS) [18]. Let $(a_1, a_2, \ldots, a_k)$ be a sequence of integers. An *application* of this sequence to a graph $G$ at node $u$ is the sequence of nodes $(u_0, \ldots, u_{k+1})$ obtained as follows: $u_0 = u, u_1 = succ(u_0, 0)$; for any $1 \leq i \leq k$, $u_{i+1} = succ(u_i, (p + a_i) \mod d(u_i))$, where $p$ is the port number at $u_i$ corresponding to the edge $\{u_{i-1}, u_i\}$. A sequence $(a_1, a_2, \ldots, a_k)$ whose application to a graph $G$ at any node $u$ contains all nodes of this graph is called a UXS for this graph. A UXS for a class $\mathcal{G}$ of graphs is a UXS for all graphs in this class.

For all feasible pairs of $N$ and $D$, let $U(N, D)$ be a UXS for the class $\mathcal{G}_{N,D}$ of all graphs with at most $N$ nodes and maximum degree at most $D$. The following important result, based on a reduction from Koucký [18], is due to Reingold [23].

**Proposition 1 ([23]).** *For any positive integer $n$, there exists a UXS $Y(n) = (a_1, a_2, \ldots, a_M)$ for the class $\mathcal{G}_n$ of all graphs with at most $n$ nodes, such that*

- *$M$ is polynomial in $n$,*
- *for any $i \leq M$, the integer $a_i$ can be constructed using $O(\log n)$ bits of memory.*

The above result implies that a (usually non-simple) path $(u_0, \ldots, u_{M+1})$ traversing all nodes can be computed (node by node) in memory $O(\log n)$, for any graph with at most $n$ nodes. Moreover, logarithmic memory suffices to walk back and forth on this path: to walk forward at node $u_i$, port $(p + a_i) \mod d(u_i)$ should be computed when coming by port $p$, to walk backward, port $(p - a_i) \mod d(u_i)$ should be computed.

**Proposition 2 ([2]).** *For any positive integers $n, d, d < n$, there exists a universal exploration sequence of length $O(n^3 d^2 \log n)$ for the family of all graphs with at most $n$ nodes and maximum degree at most $d$.*

Note that the exploration sequences in the proposition above are not constructible in logarithmic memory, while the log-space constructible sequences from Proposition 1 are much longer (though still polynomial in $n$).

## 2.3   The Map Construction Problem

As mentioned before, the problem of reconstructing the topology of a network of processors has been studied before, notably in [25]. That paper introduced the concept of the *view* of a node in a graph, which we restate below:

**Definition 1 ([25]).** *The view $\mathcal{V}_{G,\lambda}(v)$ of node $v$, in a graph $G$ with port-numbering $\lambda$, is an infinite edge-labelled rooted tree $T$, whose root represents the node $v$ and for each neighboring node $u_i$ of $v$, there is a vertex $x_i$ in $T$ and an edge from the root to $x_i$ with the same labels as the edge from $v$ to $u_i$ in $G$. The subtree of $T$ rooted at $x_i$ is again the view $\mathcal{V}_{G,\lambda}(u_i)$ of the node $u_i$.*

We shall drop the subscript $\lambda$ when it is obvious from the context.

**Proposition 3 ([20]).** *Given any simple graph $G$ with $n$ nodes and a port-numbering $\lambda$, two vertices $u, u' \in V(G)$ have the same view (i.e. $\mathcal{V}_G(u) = \mathcal{V}_G(u')$) if and only if the views truncated to a depth of $n$ are equal (i.e. $\mathcal{V}_G^n(u) = \mathcal{V}_G^n(u')$).*

If two nodes of a graph have identical views then these nodes are said to be equivalent to each other. If the nodes of the graph are grouped into classes such that two nodes are put in the same class if and only if they have the same view, then such a classification is an equivalence partition of $V(G)$, where all classes have the same size. Based on this partitioning, the *quotient graph* of $G$ is defined as follows.

**Definition 2 ([25]).** *Given an undirected connected graph $G$ with port-numbering $\lambda$, the quotient graph $H$ is an edge-labelled multigraph such that there exists a homomorphism $\varphi$ from $G$ to $H$ satisfying the following: (i) For any two nodes $u$ and $v$, $\varphi(u) = \varphi(v)$ if and only if $\mathcal{V}_G(u) = \mathcal{V}_G(v)$, (ii) For each edge $(u, v)$ of $G$, there is an edge $(\varphi(u), \varphi(v))$ in $H$ labelled with $\lambda(u, v)$ and (iii) $H$ has no other edges.*

If two graphs $G_1$ and $G_2$ have identical quotient graph then it is not possible to distinguish between them by just traversing them (without making any marks on the graph). Any deterministic algorithm executed on $G_1$ would produce the same output as the same deterministic algorithm executed on $G_2$. Thus, for such graphs, it is not possible to reconstruct an exact copy of the graph. In fact the maximum information that can be obtained by an agent traversing the graph, is represented by the quotient graph.

**Definition 3.** *We define the Map Construction problem as follows. Given an undirected connected graph $G$ with port-numbering $\lambda$, an agent starting at any node of $G$ has to build the (edge-labelled) multigraph $H$ such that $H$ is the quotient graph of $(G, \lambda)$.*

Note that if $G$ has no symmetry (i.e. when all nodes have distinct views) then the quotient graph of $G$ is $G$ itself. Thus for these cases, the maps constructed by our algorithms would be the exact copy of $G$.

Finally we present a well known impossibility result for the rendezvous problem.

**Proposition 4 ([9,25]).** *Given a graph $G$ with a port-numbering $\lambda$, the deterministic rendezvous of two agents is impossible if the starting location of the two agents have the same view.*

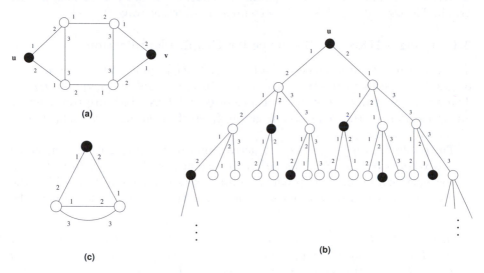

**Fig. 1.** (a) An example graph containing two agents initially at the marked nodes. The view of each agent is shown in (b) while the quotient graph is shown in (c).

## 3    Map Construction with Knowledge of Upper Bound

In this section, we assume that the agent has prior knowledge of $n$, the size of the graph. In fact, in all our algorithms, the value of $n$ can be replaced by any upper bound $N \geq n$.

A polynomial-time approach for solving the map construction problem can be obtained by applying a subroutine from [10], which, for any given starting node, computes an integer in the range $[1, n]$ which is a unique identifier of the node in the quotient graph. In this way, the map exploration problem can be solved by performing a DFS exploration of the graph and computing the identifiers of the endpoints of all the traversed edges. The claim below follows.

**Proposition 5 ([10]).** *The map construction problem can be solved in time $O(nd \cdot T(n,d))$, where $T(n,d)$ is the (polynomial) time complexity of computing the identifier of a node in a graph of order $n$ and degree $d$.*

Using the routines from [10], the operation of computing the identifier is extremely time consuming. When the agent is equipped with only $O(\log n)$ memory, we have $T(n,d) = O(|U(n^2,d)|^2 |U(n,d)|^2)$, where the used exploration sequences need to be logarithmically constructible. It is possible to implement the signature detection routines in $T(n, d) = O(|U(n^2, d)||U(n, d)|)$ steps, but using memory of the same order as the number of steps. Thus, for the best known upper bounds on the length of exploration sequences, this means that the map is constructed in $O(n^{10}d^5 \log^2 n)$ time and $O(n^9 d^4 \log^2 n)$ memory.

In this section we put forward two algorithms which solve the map construction problem more efficiently. The first relies on the intriguing property that the traversal of a sufficiently long exploration sequence is sufficient to identify the graph. The second uses UXS-s in a completely different way.

### 3.1    Using a UXS as a Sequence for Graph Identification

Suppose that a fixed sequence $Y = (a_1, a_2, \ldots, a_M)$ applied at a node $u = u_0$ of graph $G$ results in the traversal of $G$ visiting the nodes $(u_0, u_1, \ldots, u_{M+1})$. The *signature* of node $u$ is the sequence of edge labels which are traversed by an application of the sequence in graph $G$ starting at node $u$: $S_{(Y,G)}(u) = (\lambda(u_0, u_1), \ldots, \lambda(u_M, u_{M+1}))$.

The results in [10] provide a constructive criterion for distinguishing the views of two vertices of a graph $G$ based on the signatures of vertices. In fact, by a minor modification of their proof, we obtain a method for distinguishing the views of vertices in any two (not necessarily identical) graphs, and we have the following result.

**Lemma 1.** *Let $G_1 = (V_1, E_1)$ and $G_2 = (V_2, E_2)$ be graphs with at most $n$ nodes. Then, for any nodes $u_1 \in V_1$, $u_2 \in V_2$, we have $\mathscr{V}_{G_1}(u_1) \neq \mathscr{V}_{G_2}(u_2)$, if and only if $S_{(Y,G_1)}(u_1) \neq S_{(Y,G_2)}(u_2)$, where $Y = U(2n^2, d)$.*

**Proposition 6.** *There exists an algorithm for map construction which runs in $|U(2n^2, d)|$ steps and requires $O(|U(2n^2, d)|)$ memory.*

*Proof.* The algorithm proceeds by performing a traversal of the sequence $Y = U(2n^2, d)$, starting from the agent's homebase $u$. The agent records successive labels encountered during its traversal, thus computing the signature $S_{(Y,G)}(u)$. Based on this, the agent computes its quotient graph (using local computations only), as the smallest graph $G' = (V', E')$ with distinguished node $u' \in V'$ such that $S_{(Y,G')}(u') = S_{(Y,G)}(u)$. Such a graph must exist, since the tested property is by Lemma 1 equivalent to the condition $\mathscr{V}_G(u) = \mathscr{V}_{G'}(u')$, which is satisfied by a non-empty family of graphs, having a unique element which is smallest in terms of the number of nodes. This element is precisely the quotient graph of $G$, which completes the proof.                                                                            □

### 3.2    An Algorithm with Efficient Identification of Nodes

We now present an algorithm to solve map construction more efficiently. Our algorithm uses ideas that are usually used to minimize a deterministic automaton.

Given a graph $G$ and node $u$ of $G$ and a sequence of edge-labels $Y = ((p_1, q_1), (p_2, q_2), \ldots, (p_j, q_j))$, we say that $Y$ is *accepted* from $u$ if there exists a path $P = (u = u_0, u_1, \ldots, u_j)$ in $G$ such that $\Lambda(P) = Y$, i.e. for each $i, 1 \leq i \leq j$, $(p_i, q_i) = \lambda(u_{i-1}, u_i)$. For any $k > 0$, two vertices $u, v$ that have the same view up to depth $k$ are said to be $k$-equivalent; we denote it by $u \sim_k v$. The $k$-class of $u$ is the set of all vertices that are $k$-equivalent to $u$ and this set is denoted by $[u]_k$. Given any two distinct $k$-classes $C, C'$, a $(C, C')$-distinguishing path is a sequence of edge-labels $Y_{C,C'} = ((p_1, q_1), (p_2, q_2), \ldots, (p_j, q_j))$ of length at most $k$ such that $Y_{C,C'}$ is accepted from each node $u \in C$ and it is not accepted from any node $v \in C'$. For any two distinct $k$-classes, there always exists either a $(C, C')$-distinguishing path or a $(C', C)$-distinguishing path.

To compute the quotient graph of $G$, it suffices to visit every node $v$ of $G$ and identify the $n$-class of $v$ and each of its neighbors. Recall from Proposition 3 that $[u]_n = [u]_\infty = [u]$ for any node $u \in G$. Once these equivalence classes are known, one can construct the quotient graph $H$ as follows. The vertices of $H$ are the equivalence classes, and there is an edge labelled by $(p, q)$ from $[u]$ to $[v]$ in $H$ if and only if, $u$ has a neighbor $v' \in [v]$ such that $\lambda_u(u, v') = p$ and $\lambda_v(v', u) = q$.

We present an algorithm (See Algorithm 1) that iterates over $k$, and for each $k$, explores the graph and identifies the $k$-classes of the visited nodes and their neighborhoods. We use a UXS $U(n, d)$ of size $O(n^3 d^2 \log n)$ for the traversal.

For $k = 1$, it is easy to determine the $k$-class of any node $v$ by traversing each edge incident to $v$ and noting the labels. From this information, one can find the distinguishing paths for any pair of 1-classes. For $k \geq 2$, it is possible to identify the $k$-classes and the corresponding distinguishing paths (from knowledge of the $k - 1$ classes) using the properties below.

**Lemma 2.** *For $k \geq 2$, two nodes $u$ and $v$ belong to the same $k$-class, if and only if (i) $u$ and $v$ belong to the same 1-class and (ii) for each $i, 0 \leq i \leq deg_G(u) = deg_G(v)$, the $i$th neighbor $u_i$ of $u$ and the $i$th neighbor $v_i$ of $v$ belong to the same $(k - 1)$-class and $\lambda(u, u_i) = \lambda(v, v_i) = (i, j)$, for some $j \geq 0$.*

**Proposition 7.** *Algorithm 1 solves map construction for any graph of size $n$ in $O(|U(n, d)| \cdot n^3 d)$ moves and requires $O(n^3 \log n + |U(n, d)| \log d)$ memory.*

*Proof.* Let $n_k$ be the number of $k$-classes. During the $k$th iteration, on each node $v$ reached by the UXS, for each neighbor $w$ of $v$, the agent computes the $k - 1$ class of $w$. To do so, it needs to check at most $n_k$ different paths of length $k - 1$. Consequently, for each node $v$, it needs $O(\deg(v) \cdot n_k \cdot k)$ moves to compute the $k$-class of $v$. Thus, during the $k$th iteration of the algorithm, the agent performs $O(d \cdot n_k \cdot k \cdot |U(n, d)|)$ moves, where $d$ is the maximum degree of the graph. Due to Proposition 3 there are at most $n$ iterations, and $n_k \leq n$; so the total number of moves made by the agent is $O(|U(n, d)| \cdot n^3 d)$.

At the end of the $k$th iteration, the agent needs to remember the number $n_k$ of $k$-classes and $n_k(n_k - 1)/2$ distinguishing paths, each of length at most $k$. This can be stored using $O(n^3 \log n)$ bits. During the $k$th iteration, the agent needs to remember for each $v$ and for each neighbor $w$ of $v$, the label of the edge $(v, w)$ and the index of the $(k - 1)$-class of $w$. For each $v$, it needs $O(\deg(v) \cdot \log n)$ bits. However, the agent does not need to remember the $k$-class of each $v_i$, but it is sufficient to identify the distinct $k$-classes that exist in the graph. Thus, since there are at most $n$ different $k$-classes, the agent needs $O(n \cdot d \cdot \log n)$ bits of memory to compute the number of $k$-classes, and to compute the corresponding distinguishing paths using the distinguishing paths for the $(k-1)$-classes. Since the agent can store the UXS using $O(|U(n,d)| \log d)$ bits, the agent can execute this algorithm using $O(n^3 \log n + |U(n,d)| \log d)$ memory.    □

---

**Algorithm 1.** Class-Refinement(n)

---

Let $v_1, v_2, \ldots v_t$ be the sequence of nodes visited by $U(n,d)$, possibly containing duplicate nodes ;
Apply $U(n,d)$ and **for** *each node $v_i$* **do**
    ⌊ Store the labels of each edge incident to $v_i$;
Compute the number of 1-classes and store a distinguishing path for each pair of distinct classes ;
k := 2;
**repeat**
    Apply $U(n,d)$ and **for** *each node $v_i$* **do**
        **for** *each edge $(v_i, w)$ incident to $v_i$* **do**
            Compute the $(k-1)$-class of $w$ (using the distinguishing paths);
            Store the label of $(v_i, w)$ and the index of the $(k-1)$-class of $w$ ;
    Compute the number of $k$-classes and store a distinguishing path for each pair of distinct $k$-classes ;
    Increment k;
**until** *the number of $k$-classes is equal to the number of $(k-1)$-classes* ;
Compute the quotient graph ;

---

# 4    Universal Algorithms for Map Construction

In this section, we assume that the agents do not know the size of the graph $G$ and we are interested in designing universal algorithms that work for graphs of any size. Note that it is not possible to perform exploration with stop in unlabelled graphs of arbitrary size and topology. No terminating algorithm can guarantee to visit all the nodes of an arbitrary connected graph with unmarked nodes. To get around this problem, we assume that the starting location of an agent is specially marked, so that it can be distinguished from the other nodes. This is a much weaker assumption compared to allowing the agent to have a pebble which it can drop at any node and later retrieve it. However this weak assumption is sufficient for obtaining universal algorithms for the map construction problem.

## 4.1    Guessing the Value of $n$

The universal exploration sequences used in the previous section used the order of the graph as input. If this information is not available, we can try to guess a value of an upper bound $N$ on $n$. If the assumed value of $N$ is not big enough, we may not be able to explore the entire graph using $U(N,N)$. The idea is to detect this fact and increase the value of $N$ and try again. Eventually, we would reach a correct upper bound on the size of the graph. In this case, any of the algorithms from the previous section can be applied to solve the map construction problem.

The first of the proposed approaches is implementable in logarithmic space.

**Proposition 8.** *There exists an algorithm for an agent with a marked homebase which computes an upper bound $N \geq n$ on the order of the graph, $N \in \mathrm{poly}(n)$, using $O(\log n)$ memory.*

*Proof.* Let $K$ be a parameter which is initially set as 1 and doubled in successive iterations of the algorithm. The idea of the proof is to detect in each iteration whether the universal exploration sequence $U(K, K)$, starting from the homebase $r$ of the agent has visited all nodes of the graph $G$. The considered UXS is obtained through Reingold's log-space construction [23]. For the smallest value of parameter $K$ such that $U(2K, 2K)$ explores $G$, and $U(K, K)$ does not, we have that $K < n \leq U(2K, 2K)$. Hence, by putting $N = U(2K, 2K)$ we obtain the sought polynomial upper bound on the value of $N$, since the length of the considered UXS is polynomial in $K$.

It remains to describe a subroutine which allows the agent to decide if an exploration sequence $U(K, K)$, starting from homebase $r$, explores the entire graph $G$. Let $S = (r = u_0, u_1, \ldots, u_M)$ be the sequence of vertices visited during the traversal, and $U = \{u_0, u_1, \ldots, u_M\}$. Observe that since $G$ is a connected graph, the considered traversal does not completely explore $G$ if and only if there exists a node $v \in V \setminus U$ which is a neighbor of some node $u \in U$. The algorithm proceeds by visiting the successive vertices $(u_0, u_1, \ldots, u_M)$ of the exploration sequence. At each node $u_i$, the agent makes a detour to explore its neighborhood $Nbd(u_i)$. The agent visits successive nodes of this neighborhood, and for each node $v \in Nbd(u_i)$, $v \neq r$, executes a subroutine to decide if $v \in U$. More precisely, when located at $v$, for successive values of index $j = 1, 2, \ldots, M$, the agent performs a test to decide whether $v = u_j$, and then returns to $v$. Testing the condition $v = u_j$ is performed by traversing a path starting at $v$ and defined through the sequence of port labels which appear in the traversal $(u_j, u_{j-1}, \ldots, u_0)$. In other words, we follow a reversal of the $j$-prefix of the exploration sequence $U(K, K)$, starting by leaving node $v$ through the port by which $u_j$ is entered in sequence $S$. Since each node can be uniquely identified by the sequence of ports appearing on any path leading from the marked homebase $r$ to this node, we have that $v = u_j$ if and only if the traversal of the considered path terminates at the marked node $r$.

We finally note that since navigating the robot along sequence $U(K, K)$, or any prefix or reversal of $U(K, K)$, only requires $O(\log n)$ memory (cf. [18]), the entire algorithm runs using $O(\log n)$ memory. □

## 4.2   More Efficient Map Construction

In this section we consider other methods of exploration rather than using an UXS. The fact that the starting node $r$ of the agent is marked and can be distinguished from other nodes, makes it easier to perform an exploration. The agent can perform a breadth-first traversal building a BFS-tree $T$ rooted at $r$.

During the traversal, whenever the agent explores a new edge and reaches a node $v$, it checks whether $v$ is same as some node $u$ in its tree. This can be done by successively applying the label-sequences for the back-paths from each node $u \in T$ to the root $r$, and checking if one of these hits the marked node. Based on this idea, we have an algorithm for building a map of $G$ starting from the single marked homebase in $G$ (See Algorithm 2). The algorithm maintain a BFS-tree $T$ containing the visited nodes and a data structure called ROOT_PATHS that stores the edge-labelled path $P$ in $T$ from any node $v$ to the homebase $r$. For such a stored path $P$, $\text{Start}(P)$ refers to the node $v$.

**Proposition 9.** *There exists an algorithm for map construction for an agent with a marked homebase which runs in $O(n^3 d)$ steps and uses $O(n \cdot d \log n)$ memory.*

*Proof.* First we show that the $Map$ output by algorithm BFS-Tree-Construction is an exact copy of $G$ and the graph $T$ output by the algorithm is a spanning tree of $G$. Note that the sequence of labels on the path from the homebase $r$ to each node in $T$ is unique. Thus no node appears more than once in $T$. Since the algorithm performs a breadth-first search, every node is reached by the algorithm. If the algorithm does not add a reached node $u$ to $T$ then there is path from $u$ to $r$ which is identically labelled as an exisitng path $P \in$ ROOT_PATHS. Hence by the previous argument $u$ already exists in $T$. It is easy to see that $T$ is connected and every edge in $T$ appears in $G$. Thus, $T$ is a spanning tree of $G$. The $Map$ is a super-graph of $T$ and every edge that is traversed by the algorithm is added to $Map$ (either as tree-edge or as a cross-edge). The algorithm traverses each edge incident to any node in $T$ and thus all edges of $G$ are traversed by the algorithm. Thus we conclude that $Map$ is an isomorphic copy of $G$.

Whenever the algorithm traverses an unexplored edge at a node $v$, it has to check at most $n$ paths in ROOT_PATH, each of length at most $n$. This takes $O(n^2)$ steps for each edge and thus $O(n^3 d)$ steps in total. The agent requires $O(n \cdot d \cdot \log n)$ memory to store $Map$ and $T$. The data-structure $ROOT\_PATHS$ does not need to be stored explicitly and can be obtained from $T$.     □

### 4.3   Solving Rendezvous

We now show the above techniques can be used to solve the rendezvous of two dispersed agents in an unknown graph. Note the algorithm BFS-Tree-Construction from the previous section will fail to build a map if there are more than one agents in the graph. If there are two marked nodes in $G$ and an agent can confuse between these two nodes, as they would look identical to the agent. However, if we execute the algorithm BFS-Tree-Construction in a graph with two marked homebases, the following properties would be satisfied.

**Lemma 3.** *If two agents starting from marked homebases in a connected graph $G$ execute algorithm BFS-Tree-Construction, then the following holds:*
*(i) The graph $T$ constructed by each agent would be an acyclic connected (not necessarily spanning) subgraph of $G$.*
*(ii) If the maps constructed by the two agents are identical then the views from the two homebases are identical.*

---

**Algorithm 2.** BFS-Tree-Construction

---

$Map := T := \{r\}$ ;
Add $r$ to Queue;
ROOT_PATHS $:= \emptyset$;
**while** *Queue is not empty* **do**
  Get next node $v$ from Queue and go to $v$ using $Map$;
  **while** *node $v$ has unexplored edges* **do**
    Traverse the next unexplored edge $e = (v, u)$;
    **for** *each path $P \in ROOT\_PATHS$* **do**
      Apply sequence $\Lambda(P)$ at node $u$ ;
      **if** *successfully reached a marked node* **then**
        Add to $Map$ a cross-edge from $v$ to Start($P$);
        Update the number of explored edges at the node Start($P$);
        Return to node $v$ using $T$ and exit Loop;
      **else**
        Backtrack to node $u$ ;

    **if** *All path sequences failed to reach a marked node* **then**
      Add a new node $u$ to $T$ and $Map$ ;
      Add edge $(v, u)$ to $T$ and $Map$ ;
      Insert $u$ to Queue ;
      ROOT_PATHS $:=$ ROOT_PATHS $\cup$ Path$_T(u, r)$ ;
      Backtrack to node $v$ ;

---

*Proof.* (i) An agent executing Algorithm 2 adds a node $u$ to $T$ only if this node does not exist in $T$ (If the node $u$ already belongs to $T$ the agent can correctly detect this fact). Thus result (i) follows from properties of breadth-first search. (ii) The $Map$ constructed by an agent $a$ consists of a BFS-tree (call it $T_a$) and some cross-edges. The tree $T_a$ is a subgraph of $G$ rooted at the homebase $r_a$ of the agent. If the maps of the two agents are identical then, for every cross-edge $(u, v)$ in the $Map$ of agent $a$, there is a cross-edge $(u', v')$ in the $Map$ of the other agent (say, agent $b$) such that either $(u, v)$ and $(u', v')$ are actual edges in $G$, or $(u, v')$ and $(u', v)$ are edges in $G$. It is possible to build the view of the agent $a$ using the information contained in its $Map$ (and the fact that the two Maps are identical). We replace each cross-edge $(u, v)$ in the $Map$, by an edge $(u, u_v)$ and a new node $u_v$, and plug in a copy of $Map$ re-rooted at $v$ at the new node $u_v$. We can repeat this recursively from the top level down to any depth $N$ until there are no cross-edges up to depth $N$. Finally, for each tree edge $(x, y)$ where $x$ is the parent of $y$, we can add an edge $(y, y_x)$ and a new node $y_x$ and attach a copy of the current $Map$ re-rooted at $x$ at the new node $y_x$. Using this process recursively, one can obtain the view of agent $a$ up to any desired depth $N$. Hence we conclude that the views of the two agents are identical if and only if the Maps obtained by Algorithm 2 are identical.                                □

Due to the above results and Proposition 4, we know that when the maps obtained by the two agents are identical, then rendezvous is not solvable deterministically. So, we only need to consider the case when the maps are distinct. In this case if we could compare the maps of the agents, we can elect one of the agents and the agents could rendezvous at the homebase of the elected agent.

The map constructed by an agent is a rooted edge-labelled graph, where the edge-labelling is a port-numbering on $G$. There exists a total ordering on the family of such graphs. In the following we will use a fixed ordering on this family of graphs and we say $M_1 < M_2$, if $M_1$ is distinct from $M_2$ and appears earlier than $M_2$ in this fixed ordering. We now present an algorithm for rendezvous of the two agents using the algorithm BFS-Tree-Construction as a basic step, followed by comparison of the maps (See Algorithm 3).

---

**Algorithm 3.** Universal-RDV

---

$(T, Map) :=$ BFS-Tree-Construction();
Let ROOTPATHS be the set of paths obtained during the algorithm;
Traverse Map and **for** *each cross-edge* $e = (u, v) \in Map$ **do**
> Apply the sequence $\lambda(u, v)$;
> Apply the sequence for the path $P \in$ ROOTPATHS that starts at $v$;
> // The agent has reached some marked homebase
> $(T_2, Map2) :=$ BFS-Tree-Construction();
> **if** $Map2 < Map$ **then**
>> Traverse tree edges from current node to reach root of $Map2$;
>> **Terminate**;
>
> **else if** $Map2 > Map$ **then**
>> Traverse tree edges from current node to reach node $v$;
>> Apply the sequence $\lambda(v, u)$;
>> Apply the sequence for the path $P \in$ ROOTPATHS that starts at $u$;
>> **Terminate**;

Output: "Rendezvous is not solvable";

---

**Proposition 10.** *Algorithm Universal-RDV solves rendezvous of two agents in any connected graph $G$ with marked homebases, whenever it is deterministically possible and otherwise detects failure.*

*Proof.* The algorithm constructs a map using Algorithm 2 as a sub-procedure and then compares it with the map of the other agent. Since the procedure BFS-Tree-Construction is deterministic, the map of the other agent can be obtained by simply executing algorithm 2 from the homebase of the other agent. So we need to show that the algorithm succeeds in reaching the other homebase. Suppose $T_a$ and $T_b$ be the two trees constructed by the two agents $a$ and $b$ and $r_a$ and $r_b$ be the corresponding homebases. Since each node is included in one of the two trees, there exists a node $v$ in $T_a$ that is adjacent to some node $w$ in $T_b$. When agent $a$ explored the neighborhood of $v$, the neighbor $w$ was not added to $T_a$. This implies that there must be a node $u \in T_a$, such the path from $u$ to $r_a$ is identically labelled as the path from $w$ to $r_b$. In other words there is a cross-edge $(v, u)$ in $Map_a$ that corresponds to an actual edge $(v, w)$ in $G$. Thus, when the agent $a$ traverses this cross-edge and follows the path to the root, it will reach the homebase of the other agent. Note that the agent does not know which path leads to the other homebase, so it must repeat this process for each cross-edge in its Map.

If the maps from the two homebases are distinct, the agents can always agree on a rendezvous location by comparing the maps. The algorithm fails only if the two maps are identical. In that case, we know that rendezvous is not solvable due to Lemma 3 and Proposition 4.                                                      □

**Proposition 11.** *Any execution of Algorithm Universal-RDV on a graph of size $n$ and maximum degree $d$ by two agents, requires $O(n^4 d^2)$ moves by each agent. Each agent requires a private memory of size $O(nd \log n)$.*

*Proof.* If there are $n$ nodes in the graph, then the Map of an agent can contain at most $n$ nodes. The map construction process requires $O(n^3 d)$ steps as before. However the process is repeated for each cross-edge in the Map. Each cross-edge corresponds to a distinct edge in $G$, thus there can be at most $n \cdot d$ cross-edges. Hence the result follows. The agent stores the Map in its memory, which requires $O(nd \log n)$ memory space.                                                   □

Note that the algorithm presented here solves rendezvous with detect in the asynchronous case (in contrast to [10]). In case the agents possess only logarithmic memory, we can use the techniques from Section 4.1 to obtain a log-space algorithm for solving rendezvous with detect, in the same setting.

# References

1. Albers, S., Henzinger, M.R.: Exploring unknown environments. SIAM Journal on Computing 29(4), 1164–1188 (2000)
2. Aleliunas, R., Karp, R.M., Lipton, R.J., Lovász, L., Rackoff, C.: Random walks, universal traversal sequences, and the complexity of maze problems. In: 20th Annual Symposium on Foundations of Computer Science (FOCS 1979), pp. 218–223 (1979)
3. Ando, E., Ono, H., Sadakane, K., Yamashita, M.: The space complexity of leader election in anonymous networks. International Journal of Foundations of Computer Science 21(3), 427–440 (2010)
4. Angluin, D.: Local and global properties in networks of processors. In: 12th Symposium on Theory of Computing (STOC 1980), pp. 82–93 (1980)
5. Barrière, L., Flocchini, P., Fraigniaud, P., Santoro, N.: Rendezvous and election of mobile agents: impact of sense of direction. Theory of Computing Systems 40(2), 143–162 (2007)
6. Baston, V., Gal, S.: Rendezvous search when marks are left at the starting points. Naval Research Logistics 48(8), 722–731 (2001)
7. Bender, M., Fernández, A., Ron, D., Sahai, A., Vadhan, S.: The power of a pebble: Exploring and mapping directed graphs. In: 30th ACM Symposium on Theory of Computing (STOC 1998), pp. 269–278 (1998)
8. Blum, M., Kozen, D.: On the power of the compass (or, why mazes are easier to search than graphs). In: 19th Annual Symposium on Foundations of Computer Science (FOCS 1978), pp. 132–142 (1978)

9. Boldi, P., Vigna, S.: An effective characterization of computability in anonymous networks. In: Welch, J.L. (ed.) DISC 2001. LNCS, vol. 2180, pp. 33–47. Springer, Heidelberg (2001)
10. Czyzowicz, J., Kosowski, A., Pelc, A.: How to meet when you forget: log-space rendezvous in arbitrary graphs. In: 29th Annual ACM Symposium on Principles of Distributed Computing (PODC 2010), pp. 450–459 (2010)
11. Czyzowicz, J., Labourel, A., Pelc, A.: How to meet asynchronously (almost) everywhere. In: 21st Annual ACM-SIAM Symposium on Discrete Algorithms (SODA 2010), pp. 22–30 (2010)
12. Deng, X., Papadimitriou, C.H.: Exploring an unknown graph. Journal of Graph Theory 32(3), 265–297 (1999)
13. Dessmark, A., Fraigniaud, P., Kowalski, D.R., Pelc, A.: Deterministic rendezvous in graphs. Algorithmica 46(1), 69–96 (2006)
14. Dudek, G., Jenkin, M., Milios, E., Wilkes, D.: Robotic exploration as graph construction. Transactions on Robotics and Automation 7(6), 859–865 (1991)
15. Fraigniaud, P., Ilcinkas, D.: Digraphs exploration with little memory. In: Diekert, V., Habib, M. (eds.) STACS 2004. LNCS, vol. 2996, pp. 246–257. Springer, Heidelberg (2004)
16. Fraigniaud, P., Pelc, A.: Deterministic rendezvous in trees with little memory. In: Taubenfeld, G. (ed.) DISC 2008. LNCS, vol. 5218, pp. 242–256. Springer, Heidelberg (2008)
17. Fusco, E.G., Pelc, A.: How much memory is needed for leader election. In: Lynch, N.A., Shvartsman, A.A. (eds.) DISC 2010. LNCS, vol. 6343, pp. 251–266. Springer, Heidelberg (2010)
18. Koucký, M.: Universal traversal sequences with backtracking. Journal of Computer and System Sciences 65(4), 717–726 (2002)
19. Kranakis, E., Krizanc, D., Santoro, N., Sawchuk, C.: Mobile agent rendezvous in a ring. In: 23rd International Conference on Distributed Computing Systems (ICDCS 2003), pp. 592–599 (2003)
20. Norris, N.: Universal covers of graphs: isomorphism to depth n–1 implies isomorphism to all depths. Discrete Applied Mathematics 56(1), 61–74 (1995)
21. Panaite, P., Pelc, A.: Exploring unknown undirected graphs. Journal of Algorithms 33(2), 281–295 (1999)
22. Panaite, P., Pelc, A.: Impact of topographic information on graph exploration efficiency. Networks 36(2), 96–103 (2000)
23. Reingold, O.: Undirected connectivity in log-space. Journal of the ACM 55(4) (2008)
24. Ta-Shma, A., Zwick, U.: Deterministic rendezvous, treasure hunts and strongly universal exploration sequences. In: 18th Annual ACM-SIAM Symposium on Discrete Algorithms (SODA 2007), pp. 599–608 (2007)
25. Yamashita, M., Kameda, T.: Computing on anonymous networks: Part I - characterizing the solvable cases. IEEE Transactions on Parallel and Distributed Systems 7(1), 69–89 (1996)
26. Yu, X., Yung, M.: Agent rendezvous: A dynamic symmetry-breaking problem. In: Meyer auf der Heide, F., Monien, B. (eds.) ICALP 1996. LNCS, vol. 1099, pp. 610–621. Springer, Heidelberg (1996)

# Effect of Fairness in Model Checking of Self-stabilizing Programs*

Jingshu Chen, Fuad Abujarad, and Sandeep Kulkarni

Michigan State University,
3115 Engineering Building, 48824 East Lansing, US
{chenji15,abujarad,sandeep}@cse.msu.edu
http://www.cse.msu.edu/~{chenji15,abujarad,sandeep}

**Abstract.** Existing approaches for verifying self-stabilization with symbolic model checker have relied on the use of weak fairness. We point out that this approach has limited scalability. To overcome this limitation, we show that if self-stabilization is possible without fairness then cost of verifying self-stabilization is substantially lower. The practical meaning of this observation is if the extra effort required to verify self-stabilization under weak fairness is not necessary then the state space reached by model checking of self-stabilizing programs could be substantially larger.

**Keywords:** Self-stabilization, Fairness, Fault-tolerance, Verification, Model checking.

## 1 Motivation

Verification of self-stabilizing programs is a challenging task [6, 1], because it requires us to consider all possible states that could be substantially large [2]. Moreover, due to complex recovery algorithms used in self-stabilizing program, it is desirable to automate the verification of the self-stabilization property.

One approach for automated verification of self-stabilization is to utilize symbolic model checking, a technique to automatically verify whether a given model meets a given property while utilizing boolean encoding to represent state space. Unlike theorem proving approaches (e.g. [7]), the model checking approach does not require the designer to have considerable experience in logic reasoning and hence, it is widely used in verifying the distributed algorithms. Furthermore, if the program does not meet the given property, the process of model checking typically produces a counterexample. Besides, the problem of state space explosion can be reduced with the help of symbolic techniques. Thus, symbolic model checking could be used as a useful tool by the designer while developing self-stabilizing protocols.

Although the work in [1] has demonstrated feasibility of applying symbolic model checking for verifying self-stabilizing programs, it also shows that verification is feasible only for programs with a small number of processes. To overcome this limitation, in this paper, we focus on the bottlenecks involved in the

---

* This work was partially sponsored by NSF CAREER CCR-0092724 and ONR Grant N00014-01-1-0744.

C. Lu, T. Masuzawa, and M. Mosbah (Eds.): OPODIS 2010, LNCS 6490, pp. 135–138, 2010.

verification of self-stabilizing programs. In particular, we show that issue of fairness significantly affects the verification performance for symbolic model checking of self-stabilizing programs.

## 2    Current Results

Existing model checkers have focused on weak fairness in their representation of fairness. We point out that if self-stabilization is possible under unfair computation, verification cost can be significantly lower. In fact, we observe from the following case study where that the cost of verification under weak fairness is more than 1000 times that of the cost under no fairness. The practical meaning of this observation is if the extra effort required to verify self-stabilization under weak fairness is not necessary and hence should be removed, the state space reached by model checking of self-stabilizing programs could be larger.

### 2.1    Case Study: Dijkstra's K-State Program

In this section, we study the K-state token ring program [2]. First, we describe it in terms of guarded commands. Then, we show its modeling in SMV [5]. And, finally, we provide verification results under unfair and weakly fair computation.

The K-state program consists of $N + 1$ processes, numbered from 0 to $N$. The program topology is a unidirectional ring. Each process $p.i$, $0 \leq i \leq N$, has one variable $x.i$ that denotes the current state value. Each variable has the domain $[0, \ldots, K - 1]$.

The program consists of two types of actions. The first type is for process 0. This action is enabled when $x.0$ equals $x.N$. When $p.0$ executes its action, it increments $x.0$ by 1 in modulo K arithmetic. The second type of action is for process $p.i$, $i \neq 0$. This action is enabled when $x.i$ does not equal to $x.(i - 1)$. When $p.i$ executes its action, it copies $x.(i-1)$. Thus, the actions are as follows:

$$K_0:: \quad x.0 = x.N \qquad \longrightarrow \qquad x.0 = (x.0 + 1) \bmod K;$$
$$K_i:: \quad x.i \neq x.(i - 1) \quad \longrightarrow \qquad x.i = x.(i - 1);$$

*Remark 1.* This program is known to be self-stabilizing if $K > N$. In subsequent discussion, we let $K = N + 1$.

**Legitimate states.** The state where $x$ values of all processes is 0 is a legitimate state. In this state, only process 0 is enabled. After process 0 executes, $x.0$ changes to 1 and all other $x$ values are still 0. In this state, only process 1 is enabled. Hence, it can execute and change $x.1$ to 1. Continuing this further, eventually, we reach a state where all $x$ values are 1 where process 0 is the only enabled process and process 0 will increment $x.0$ to 2. The legitimate states of the K-state program are equal to all the states reached in such subsequent execution.

Now, we show how we model the K-state program in SMV under unfair computation. Our illustration is for $K=3$. For unfair computations, we model the K-state program in terms of its transitions. Specifically, we can model action $K_0$ as a set of transitions $(s_0, s_1)$ where the guard of $K_0$, $(x.0=x.N)$, is *true* in $s_0$ and $s_1$ is obtained by executing $x.0=x.0 + 1$ from state $s_0$. In SMV, the action is written as a Boolean formula. In this formula, $x.0$, $x.1$, $x.2$ denote the value of the corresponding variable in the source state, i.e., $s_0$. And, $next(x.0)$, next $(x.1)$, $next(x.2)$ denote the value of the corresponding variable in the target state, i.e., $s_1$. Thus, action $K_0$ can be written as $(x.0 = x.2) \wedge next(x.0) = x.0 + 1 \wedge next(x.1) = x.1 \wedge next(x.2) = x.2$.

To model the fact that the program starts from any initial state, we specify the initial value of the state as: $INIT\ x.i = \{0, 1, 2\}$. Since SMV utilizes Boolean representations of such states, the representation of corresponding formula is *true* and, hence, is compact. (Additional details about modeling are available in [9].)

We verified the K-state program for $3 \leq K \leq 9$ *or* $K = 50$. Table 1 gives the verification time for model checking the K-state program for different values of $K$. $N/A$ in this table means the result was not available within an admissible amount of time (1 hour).

We chose the smaller values of $K$ for comparing the verification time with [1] where only weak fairness is considered. Specifically, in [1], authors have shown the feasibility of verification for upto $K = 8$ . In particular, the time reported in [1] for $K = 8$ is 1836.0s whereas the time for the corresponding verification is 139.1s. Since the underlying tool as well as the program remains the same, this change is due to improved hardware over last few years. However, what this result does show is that in spite of the improved hardware, the ability to verify under weak fairness remains essentially the same. Specifically, if we assume a reasonable time constraint permissible (e.g., one hour) for verification then the change in hardware made it possible to achieve verification for $K = 9$ as opposed to $K = 8$.

By contrast, verification of much larger systems is possible if we consider unfair computations. In particular, it was possible to achieve verification for $K = 50$ in less than 1 hour. And, in this case, the corresponding state space is $10^{85}$. By contrast, verification with weak fairness could not complete when state space was $10^{11}$. We also studied other typical cases in the literature of self-stabilization, including Ghosh's mutual exclusion program [3] and Hoepman's ring-orientation program [4]. The details are discussed in [9]. Our case studies show that scalability of verifying self-stabilization is unlikely to change with improved hardware. However, scalability of verifying self-stabilization can be significantly improved for the case where the program is correct self-stabilizing without fairness.

**Table 1.** Verification Results for the K-state program

| Execution time(s) | | | | | | | | | |
|---|---|---|---|---|---|---|---|---|---|
| | K=3 | K=4 | K=5 | K=6 | K=7 | K=8 | K=9 | K=10 | K=50 |
| unfair | 0 | 0 | 0 | 0 | 0.02 | 0.03 | 0.05 | 0.08 | 3466.30 |
| weakly-fair | 0 | 0.03 | 0.63 | 5.33 | 34.30 | 139.10 | 1276.08 | N/A | N/A |
| results reported in [1] | 0.1 | 0.4 | 4.6 | 43.5 | 285.2 | 1836.0 | N/A | N/A | N/A |
| approximate state space | $10^1$ | $10^2$ | $10^3$ | $10^4$ | $10^5$ | $10^7$ | $10^8$ | $10^{10}$ | $10^{84}$ |

## 2.2 Other Results

For those cases where weak fairness is essential for self-stabilization, we also identify two approaches: (1) manually assist the model checker to make it more effective (e.g., decomposition), or (2) verify a slightly different property (or model) that still provides *good* assurance (e.g., weak stabilization [8]). We show that both these approaches improve the scalability significantly in [9].

## 3 Conclusion

In this paper, we focused on scalable model checking of self-stabilizing algorithms. We point out that while a significant percentage of the literature on self-stabilization routinely assumes weak fairness, where if an action is continuously enabled, it is guaranteed to be executed, verification under such weak fairness is not scalable. Our observation was that in many cases, the assumption of weak fairness is superfluous. And, in these cases, scalable verification of self-stabilization is possible under unfair computation model. To our knowledge, this is the first paper that has shown feasibility of verifying the typical self-stabilizing programs, e.g., K-state program, with large number of processes. Thus, the results in this paper provide several avenues to designers of self-stabilizing programs to verify correctness of their programs or to identify bugs. In future, we intend to provide a simplified tool that will allow designers to specify programs in guarded commands and utilize verification under different levels of fairness.

## References

1. Tsuchiya, T., Nagano, S., Paidi, R.B., Kikuno, T.: Symbolic Model Checking for Self-Stabilizing Algorithms. IEEE Trans. Parallel Distrib. Syst. 12, 81–95 (2001)
2. Dijkstra, E.W.: Self stabilizing systems in spite of distributed control. Communications of the ACM 17(11) (1974)
3. Ghosh, S.: Binary Self-stabilization in Distributed Systems. Information Processing Letter 40(3), 153–159 (1991)
4. Hoepman, J.H.: Uniform Deterministic Self-Stabilizing Ring-Orientation on Odd-Length Rings. In: Tel, G., Vitányi, P.M.B. (eds.) WDAG 1994. LNCS, vol. 857, pp. 265–279. Springer, Heidelberg (1994)
5. McMillan, K.L.: Symbolic Model Checking. Kluwer Academic, Dordrecht (1993)
6. Dolev, S.: Self-Stabilization. MIT Press, Cambridge (2000)
7. Qadeer, S., Shankar, N.: Verifying a self-stabilizing mutual exclusion algorithm. In: Gries, D., Roever, W.-P. (eds.) IFIP International Conference on Programming Concepts and Methods (PROCOMET 1998), Shlter Island, NY, pp. 424–443. Chapman & Hall, Boca Raton (June 1998)
8. Gouda, M.G.: The Theory of Weak Stabilization. In: Datta, A.K., Herman, T. (eds.) WSS 2001. LNCS, vol. 2194, Springer, Heidelberg (2001)
9. Chen, J., Abujarad, F., Kulkarni, S.: Towards Scalable Model Checking of Self-Stabilizing Programs, MSU-CSE-10-25, Computer Science and Engineering, Michigan State University (2010)

# A Formal Framework for Conformance Testing of Distributed Real-Time Systems

Moez Krichen

Research Unit of Development and Control of Distributed Applications
Higher Institute of Computer Science and Multimedia of Sfax
Technopole de Sfax BP 242, 3021, Sfax-Tunisia
moez.krichen@redcad.org

**Abstract.** We extend our previous work on model-based testing [2]. We propose a formal framework for black-box conformance testing for distributed real-time systems. Our framework is based on the model of partially-observable, non-deterministic timed automata. A given distributed system can be modeled either as a single timed automaton or a network of timed automata. Our algorithm for generating test suites is based on an on-the-fly determinization of the specification automaton. Our testing architecture may be either centralized or not.

## 1 Introduction

We are interested in *conformance testing* where the aim is to check conformance of the system under test SUT to a given specification. The SUT is often a black box in the sense that we do not have knowledge about its internals, thus, can only rely on its observable input/output behavior. We focus our attention on real-time systems. These are systems that operate in an environment with strict timing constraints. Distributed real-time systems correspond to a particular class of real-time systems where the system in hand is made of several interacting components. In this case, the real-time constraints to consider may be either local (i.e., constraints on a single component) or global (i.e., constraints on the whole system). However, even though a local time constraint may be defined locally at the level of a single component it will generally depend on the interaction with other components of the system. This makes testing distributed real-time systems a harder problem in general.

## 2 Modelling Distributed Systems Using Timed Automata

We use timed automata [1] with deadlines to model urgency. A *timed automaton over* Act is a tuple $A = (Q, q_0, X, \mathsf{Act}, \mathsf{E})$, where: $Q$ is a finite set of *locations*; $q_0 \in Q$ is the initial location; $X$ is a finite set of *clocks*; E is a finite set of *edges*. Each edge is a tuple $(q, q', \psi, r, d, a)$, where: $q, q' \in Q$ are the source and destination locations; $\psi$ is the *guard*, a conjunction of constraints of the form $x \# c$, where $x \in X$, $c$ is an integer constant and $\# \in \{<, \leq, =, \geq, >\}$; $r \subseteq X$ is a set of clocks to *reset* to zero; $d \in \{\mathsf{lazy}, \mathsf{delayable}, \mathsf{eager}\}$ is the *deadline*; $a \in \mathsf{Act}$ is the action.

C. Lu, T. Masuzawa, and M. Mosbah (Eds.): OPODIS 2010, LNCS 6490, pp. 139–142, 2010.
© Springer-Verlag Berlin Heidelberg 2010

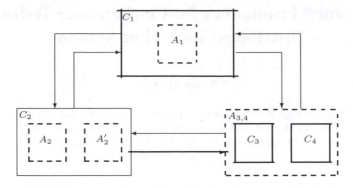

**Fig. 1.** An example of the structure of a real-time distributed system modeled using a network of (four) timed automata

Specifications are usually built in a *compositional* way, from many components. This greatly simplifies modeling. Notice that a compositional specification does not require the SUT to be implemented following the same structure. Composition is merely a way of modeling the specification. Thus, a given distributed system can be modeled either as a single timed automaton or as a network of timed automata. To model our system, we may associate one timed automaton per component or not. Moreover a single timed automaton may be used to model the behavior of the composition of several interacting components. Thus, the total number of timed automata used for modeling and the number of components of the system to model may not be the same.

For instance, the distributed system shown in Figure 1 consists of four components, namely $C_1$, $C_2$, $C_3$ and $C_4$. The behavior of component $C_1$ (solid line box) is modeled by TAIO $A_1$ (dotted box). The behavior of $C_2$ is modeled by TAIO $A_2 \| A_2'$ obtained by the parallel composition of the two TAIO $A_2$ and $A_2'$. Finally, the subsystem made of the two components $C_3$ and $C_4$ is modeled by TAIO $A_{3,4}$. That is we do not dispose, in this case, of the model of each of the components $C_3$ or $C_4$ separately. Rather, we dispose of a model of the subsystem they make.

## 3   Testing Architecture

The role of a tester consists in interacting with the SUT in order to execute the available test cases and then to observe the response of the SUT due to this excitation. In case of monitoring, the role of the tester is limited to observe the behavior of the SUT and to decide whether the generated behavior is accepted or not. In case of on-line testing, the tester may be also in charge of generating test cases (i.e., deriving them from the specification) on-the-fly while testing.

A given tester may be global or local. We may either associate only one tester with the whole system under test (e.g., Figure 2 - (a)) or associate one tester with each component of the system (e.g., Figure 2 - (b)). These are two extreme situations for a possible testing architecture which may be referred to as *global-tester based architecture* and *local-tester based architecture*, respectively.[1]

---

[1] Also referred to as "centralized" and "decentralized" testing architectures.

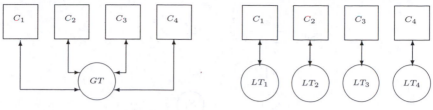

(a) Global-tester based architecture             (b) Local-tester based architecture

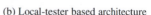

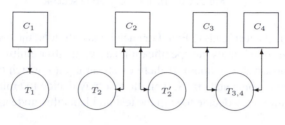

(c) Hybrid architecture

**Fig. 2.** Different possible testing architectures for distributed real-time systems

In this work, we propose a more general testing architecture named *hybrid architecture* which allows both extreme situations. That is in our framework, a given component of the SUT may be connected to more than one testers simultaneously. On the other hand, a given tester may be associated with one or several components of the SUT as well. For instance, a possible hybrid testing architecture for the distributed real-time system of Figure 1 is depicted in Figure 2 - (c). Accordingly, one tester $T_1$ is associated with component $C_1$; two testers $T_2$ and $T_2'$ are associated with $C_2$; and a same tester $T_{3,4}$ is associated with the two components $C_3$ and $C_4$.

It is worth noting that the testing architecture proposed for this example is not unique. Also note that there is no correlation between the way the SUT is modeled and the testing architecture to adopt. In other terms, it is not necessary to associate a tester with each TAIO appearing in the model of the SUT. For instance, the number of testers to use may exceed the number of TAIO appearing in the model and vice versa. In our example, we have chosen to test component $C_2$ using two testers namely $T_2$ and $T_2'$. This may be justifiable by the fact that in this case we are interested in checking the conformance of $C_2$ with respect to two distinct and independent properties (one encoded by $A_2$ and the other one by $A_2'$ for instance). So each tester has a separate job to achieve independently from the other tester.

## 4   Test Generation Principle

We adapt the untimed test generation algorithm of [3]. Roughly speaking, the algorithm builds a test in the form of a tree. A node in the tree is a set of states $S$ of the specification and represents the "knowledge" of the tester at the current test state. The algorithm extends the test by adding successors to a leaf node, as illustrated in Figure 3. For all

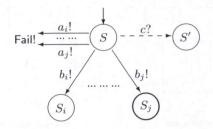

**Fig. 3.** Generic test-generation scheme

*illegal* outputs $a_i$ the test leads to Fail. For each legal output $b_i$, the test proceeds to node $S_i$, which is the set of states the specification can be in after emitting $b_i$ (and possibly performing unobservable actions). If there exists an input $c$ which can be accepted by the specification at some state in $S$, then the test may decide to emit this input. At any node, the algorithm may decide to stop the test and label this node as Pass.

## 5   Conclusion and Future Work

An important contribution in this work was trying to make decorrelation between the number of components of the system under test, the number of entities used to model this system and finally the number of testers used for test.

Many extensions are possible. For instance, we need selection techniques -based on coverage criteria- to guide test generation and to reduce the number of generated tests. As a future work direction, we are intending to implement our testing methodology in the context of distributed software architectures.

## References

1. Alur, R., Dill, D.: A theory of timed automata. Theoretical Computer Science 126, 183–235 (1994)
2. Krichen, M., Tripakis, S.: Conformance testing for real-time systems. Formal Methods in System Design 34(3), 238–304 (2009)
3. Tretmans, J.: Testing concurrent systems: A formal approach. In: Baeten, J.C.M., Mauw, S. (eds.) CONCUR 1999. LNCS, vol. 1664, pp. 46–65. Springer, Heidelberg (1999)

# Signature-Free Broadcast-Based Intrusion Tolerance: Never Decide a Byzantine Value

Achour Mostéfaoui and Michel Raynal

IRISA, Université de Rennes 1, 35042 Rennes, France
{achour,raynal}@irisa.fr

**Abstract.** Providing application processes with strong agreement guarantees despite failures is a fundamental problem of fault-tolerant distributed computing. Correct processes have not to be "polluted" by the erroneous behavior of faulty processes. This paper considers the consensus agreement problem in a setting where some processes can behave arbitrarily (Byzantine behavior). In such a context it is possible that Byzantine processes collude to direct the correct processes to decide on a "bad" value (a value proposed only by faulty processes).

The paper has several contributions. It presents a family of consensus algorithms in which no bad value is ever decided by correct processes. These processes always decide a value they have proposed (and this is always the case when they all propose the same value) or a default value $\perp$. These algorithms are called *intrusion-free* consensus algorithms. To that end, each consensus algorithm is based on an appropriate underlying broadcast algorithm. One of these abstractions, called *validated broadcast* is new and allows the design of a resilience-optimal consensus algorithm (i.e., it copes with up to $t < n/3$ faulty processes where $n$ is the total number of processes). All proposed consensus algorithms assume the underlying system is enriched with additional computational power provided by a binary Byzantine consensus algorithm. The paper presents also a resilience-optimal randomized binary consensus algorithm based on the validated broadcast abstraction. An important feature of all these algorithms lies in the fact that they are signature-free (and hence particularly efficient).

**Keywords:** Asynchronous message-passing system, Broadcast abstraction, Byzantine process, Consensus problem, Fault-tolerance, Intrusion-tolerance, Reliable broadcast, Resilience, Signature-free algorithm, Time-free algorithm.

## 1 Introduction

*Asynchronous Byzantine consensus.* A process has a *Byzantine* behavior when it behaves arbitrarily. This bad behavior can be intentional (malicious behavior, e.g., due to intrusion) or simply the result of a transient fault that altered the local state of a process, thereby modifying its behavior in an unpredictable way. We are interested here is solving the *consensus* problem in asynchronous distributed systems prone to Byzantine process failures whatever their origin.

In a classical crash failure setting, the consensus problem is defined as follows: every process proposes a value and the non-faulty processes have to decide (termination property) on the same value (agreement property), that has to be one of the proposed

C. Lu, T. Masuzawa, and M. Mosbah (Eds.): OPODIS 2010, LNCS 6490, pp. 143–158, 2010.

values (validity). In a Byzantine failure setting, the notion of "value proposed by a faulty process" is not well-defined. Hence, the validity property is weakened and usually replaced by the following: if all non-faulty processes propose the same value, that value is decided.

*Aim of the paper.* Unfortunately, the previous validity property leaves open the possibility for the non-faulty processes to decide an arbitrary value when all of them do not propose the same value, and such a "bad" value can "pollute" their behavior. Hence, the idea to introduce an additional validity property, that we call *non-intrusion*, to prevent this type of behavior, namely, a value proposed only by faulty processes cannot be decided by non-faulty processes. Said in another way, the non-faulty processes are required to decide the value proposed by one of them (and this has to be always the case when they all propose the same value), or a default value (denoted $\perp$) when they are not enough to propose the very same value.

The paper presents a family of asynchronous Byzantine multivalued consensus algorithms that satisfy the previous property. We call them *intrusion-free* consensus algorithms. Of course, as consensus cannot be solved in asynchronous system in which even only one process may crash [12], the underlying system has to be enriched with additional computational power in order for the consensus to be solved despite the net effect of asynchrony and Byzantine failures. We consider here that this additional power is given by an underlying binary Byzantine consensus algorithm (e.g., [20,27]).

*Content of the paper.* All the multivalued Byzantine consensus algorithms presented in the paper are signature-free (no underlying cryptography mechanism is assumed). Each algorithm relies on an appropriate underlying broadcast operation (that can be implemented despite asynchrony and up to $t$ Byzantine processes, where $t$ is constrained by a function on the total number $n$ of processes). These broadcast abstractions are the classical unreliable broadcast (that requires $t < n$), the "echo" broadcast introduced in [3] (that we call *no-duplicity* broadcast), the reliable broadcast introduced in [4], plus a novel all-to-all broadcast abstraction that can be interesting by itself, that we call *validated broadcast*. All these broadcast abstractions (but unreliable broadcast) require $n > 3t$ to be implemented in an asynchronous system prone to Byzantine failures. They differ in the number of consecutive communication steps they need.

As we will see, the new validated broadcast abstraction is particularly interesting in the context of Byzantine processes. This is because it allows a correct process to deliver a message only if that message has been validated by at least one correct process. Said differently, validated broadcast eliminates the "noise" introduced by "bad" values (i.e., values proposed only by Byzantine processes).

A resulting multivalued Byzantine consensus algorithm is then characterized by its underlying broadcast algorithm that has a particular cost counted by the number of communication steps, the size of control additional information messages have to carry, and their messages. As we will see, the proposed algorithms are highly modular and exhibit a tradeoff relating their time efficiency (the weaker the underlying broadcast abstraction, the more efficient the algorithm), and the constraint on $t$ they need (the weaker the underlying broadcast abstraction, the stronger the constraint on $t$).

The paper also presents a binary Byzantine consensus algorithm (which can provide the previous algorithms with the required additional computing power). This algorithm is signature-free, requires $t < n/3$ (and is consequently optimal with respect to resilience), and needs six communication steps per round.

*Related work.* Numerous Asynchronous Byzantine algorithms have been proposed (e.g., [3,4,5,13,14,16,17,24,26] to cite a few; see also [21] for a short survey). To our knowledge, the only algorithm that considers the non-intrusion property is the one described in [10] (this algorithm requires messages to carry a vector of proposed values which, as shown here, is not necessary).

The idea to direct the processes to decide $\perp$ in "bad scenarios" (i.e., when they cannot decide a value they have proposed) is different but in the same spirit as the idea developed in the notion of *abortable* objects [1]. In that case, the "bad scenarios" are when there is concurrency among operations. In a concurrency context, operations can return $\perp$, while an operation has always to return a non-trivial result when executed in a concurrency-free context.

*Road map.* The paper is made up of 8 sections. Section 2 presents the computation model, the different broadcast abstractions, and an algorithm implementing the validated broadcast abstraction. Section 3 presents the intrusion-free Byzantine consensus problem. Then, Sections 4 and 5 present a suite of intrusion-free multivalued Byzantine consensus algorithms that differ mainly in the underlying broadcast abstraction they use. Section 6 discusses the previous algorithms. Section 7 presents a randomized binary consensus algorithm based on the validated broadcast abstraction. Finally, Section 8 concludes the paper. Due to page limitations, will find all proofs in [19]).

## 2   Computation System Model

### 2.1   Base Model

*Asynchronous processes.* The system is made up of a finite set of $n > 1$ processes denoted $p_1, \ldots, p_n$ that communicate by exchanging messages through a communication network. Each process proceeds to its own speed, which means that processes are asynchronous.

*Multiset.* All algorithms presented in the paper use multisets. A *multiset* (sometimes also called *bag*) differs from a set in that it can contain several copies of the same value. Given a multiset $rec_i$, $\#_v(rec_i)$ denotes the occurrence number of $v$ in $rec_i$.

*Failure model.* Up to $t$ processes can exhibit a *Byzantine* behavior. A Byzantine process is a process that behaves arbitrarily: it can crash, fail to send or receive messages, send arbitrary messages, start in an arbitrary state, perform arbitrary state transition, etc. Moreover, Byzantine processes can collude to "pollute" the computation. Yet, it is assumed that they do not control the network. This means that they cannot corrupt the messages sent by non-Byzantine processes, and the schedule of message delivery is uncorrelated to Byzantine behavior. A process that exhibits a Byzantine behavior is called *faulty*. Otherwise, it is *correct* or *non-faulty*. Given an execution, $\mathcal{C}$ denotes the set of processes that are correct in that execution.

*Notation.* This process model is denoted $\mathcal{BZ\_AS}_{n,t}[\emptyset]$. In the following, this model is enriched with a constraint on $t$ and a specific broadcast abstraction. As an example, $\mathcal{BZ\_AS}_{n,t}[n > 5t, \mathsf{WB}]$ is $\mathcal{BZ\_AS}_{n,t}[\emptyset]$ in which fewer than $n/5$ processes are faulty and processes communicate using the operations of the WB broadcast abstraction (see below).

## 2.2 Asynchronous Communication Network

*Base communication network.* Each pair of processes is connected by a channel (which means that when a process receives a message, it knows which is the sender of the message). Each channel is asynchronous (no bound on message transfer delay, except it is finite), and reliable (no loss, creation or corruption of messages). Hence, the network is asynchronous.

A process $p_i$ sends a message to a process $p_j$ by invoking the primitive "send TAG$(m)$ to $p_j$", where TAG is the type of the message and $m$ its content. To simplify the presentation, it is assumed that a process can send messages to itself. A process receives a message by executing the primitive "receive()".

In the following, several types of broadcast operations are defined. They all can be implemented from the base send and receive primitives, which means that, while they provide us with distinct communication abstraction levels, they do not provide the processes with additional computing power.

When considering the broadcast abstraction XX (where XX stands for WB, NDB, VB or RB, see below), we say that a process "XX-broadcasts" or "XX-delivers" a message.

*Unreliable broadcast.* The pair of operations denoted WB_broadcast() and WB_deliver() are used to denote a simple unreliable broadcast. WB_broadcast TAG$(m)$ is used as a shortcut for

**for each** $j \in \{1, \ldots, n\}$ send TAG$(m)$ to $p_j$ **end for**,

and WB_deliver() is synonym with receive(). This means that a message broadcast by a correct process is delivered to all correct processes. Differently, while it is assumed to send the same message to all processes, a faulty process can actually send different messages to distinct processes and no message to others.

Trivially, an invocation of WB_broadcast TAG$(m)$ costs one communication step and $O(n)$ messages (more precisely, $n - 1$ messages). This communication abstraction is called WB, and the corresponding system model is denoted $\mathcal{BZ\_AS}_{n,t}[\mathsf{WB}]$.

*Remark.* When measuring the cost of a broadcast abstraction we do not take into account the size of the "data message" that is broadcast. This is because this size is independent of the way the broadcast is implemented. We only consider the size of the additional control information required by the corresponding broadcast implementation.

*No-duplicity broadcast.* This communication abstraction, denoted NDB, is defined by the operations NDB_broadcast() and NDB_deliver() that provide the processes with a higher abstraction level than WB. Considering an instance where NDB_broadcast() is invoked by process $p_i$, this broadcast abstraction is defined by the following properties.

- NDB-No-duplicity. No two correct processes NDB-deliver distinct messages from $p_i$.

– NDB-Termination. If the sender is correct, all correct processes eventually NDB-deliver its message.

The corresponding system model is denoted $\mathcal{BZ\_AS}_{n,t}[\text{NDB}]$. Let us observe that, if the sender $p_i$ is faulty, it is possible that some correct processes deliver a message from $p_i$ while others do not. The no duplicity property prevents correct processes from delivering different messages from a faulty sender. (When considering the less severe crash failure model, no-duplicity broadcast and weak broadcast are equivalent.)

This broadcast primitive has been defined by Toueg [26]. It can be built on top of the base send/receive primitives in systems where $t < n/3$. Such an implementation uses two consecutive communication steps and $O(n^2)$ underlying messages ($n - 1$ in the first communication step, and $n(n - 1)$ in the second one). The size of the control information added to a message is $\log_2 n$ (sender identity).

*Reliable broadcast.* The reliable broadcast abstraction, denoted RB, has been proposed by Bracha [4]. Strictly stronger than the no-duplicity broadcast, it provides processes with the operations RB_broadcast() and RB_deliver() defined by the following properties.

– RB-No-duplicity. No two correct processes RB-deliver distinct messages from $p_i$.
– RB-Termination. If the sender is correct, all correct processes eventually RB-deliver its message.
– RB-Uniformity. If a correct process RB-delivers a message from $p_i$ (possibly faulty) then all correct processes eventually RB-deliver a message from $p_i$.

It has been proved in [4] that $n > 3t$ is a necessary requirement to implement this operation. If the sender is correct, only three communication steps and $O(n^2)$ messages whose size is $O(\log_2 n)$ bits are necessary.

*Validated broadcast.* This last communication abstraction, denoted VB, is defined by the operations VB_broadcast() and VB_deliver() described below. It is a new abstraction that provides the processes with a communication level higher than no-duplicity broadcast. More precisely, validated broadcast is an *all-to-all* reliable broadcast with a notion of message *validation*, namely, a message has to be validated by enough processes in order to be VB-delivered, otherwise the default value $\perp$ is VB-delivered instead of it.

As it is an all-to-all broadcast abstraction, VB requires that all correct processes invoke VB_broadcast(). The idea is that a value $v$ is valid if there is at least one correct process that broadcasts that value. As no process knows if it is itself correct or faulty (e.g., a process can correctly execute its algorithm and then crash), a value broadcast by a process is required to be validated by $n - 2t \geq t + 1$ processes to be valid. As already indicated, if a message value is not validated, $\perp$ is delivered instead of it.

More precisely, assuming a broadcast instance in which every correct process invokes VB_broadcast(), let us consider the invocation of a particular process $p_i$ that invokes VB_broadcast($m$). VB is defined by the following properties.

– VB-No-duplicity. No two correct processes VB-deliver distinct messages from $p_i$ (the message that is VB-delivered can be a non-$\perp$ value or the default value $\perp$).
– VB-Termination. If the sender is correct and VB-broadcast $m$, all correct processes eventually VB-deliver the same message $m'$ where $m'$ is $m$ or $\perp$.

- VB-Uniformity. If a correct process VB-delivers a message from $p_i$ (possibly faulty), all correct processes eventually VB-deliver a message from $p_i$.
- VB-Validation. If $\perp$ is VB-delivered, there is at least one correct process that does not validate the message VB-broadcast by $p_i$. If $m' \neq \perp$ is VB-delivered, $m'$ has been validated by at least one correct process.

Let us remark that a notion of validated broadcast based on local knowledge (and not on distributed knoweledge as here) has been proposed in [8].

**Table 1.** Cost and constraint of the different broadcast abstractions

| broadcast | x-to-y | # comm. steps | message size | # msgs | constraint on $t$ |
|---|---|---|---|---|---|
| WB | 1-to-$n$ | 1 | constant | $n - 1$ | $n > t$ |
| NDB | 1-to-$n$ | 2 | $\log_2 n$ | $0(n^2)$ | $n > 3t$ |
| RB | 1-to-$n$ | 3 | $\log_2 n$ | $0(n^2)$ | $n > 3t$ |
| VB | $n$-to-$n$ | 6 | $\log_2 n$ | $n \times 0(n^2)$ | $n > 3t$ |

*Comparing the broadcast abstractions.* Table 1 compares the costs of the three previous broadcast abstractions. Considering one broadcast instance, the second column indicates the broadcast type (1-to-$n$ or $n$-to-$n$). The third column indicates the number of (sequential) communication steps that are needed. The fourth column presents the size of additional control information that an implementation message has to carry (the $\log_2 n$ comes from the fact that the identity of the process that broadcasts a message has to be sent together with it when forwarded by another process). The fifth column indicates the number of implementation messages that are needed. Finally, the last column states the constraint on $t$ required to implement the corresponding abstraction in $\mathcal{BZ\_AS}_{n,t}[\emptyset]$.

## 2.3   An Implementation of the Validated Broadcast Abstraction

Algorithm 1 implements the all-to-all validated broadcast. Let us recall that all-to-all means here that all correct processes are assumed to invoke VB_broadcast(). This means that a process VB-delivers at least $n - t$ messages. This implementation uses two consecutive RB-broadcast invocations. Its cost is consequently, $2 \times 3 = 6$ communication steps and $O(n^3)$ messages of size $O(\log_2 n)$ bits. The implementation of a VB-broadcast instance is made up of two parts.

- The first part is made up of two consecutive RB-broadcasts. More precisely, a process $p_i$ first invokes RB_broadcast INIT($v_i$) and waits until it has RB-delivered messages from at least $n - t$ processes (lines 01-03). The values RB-delivered are deposited in a multiset denoted $rec_i$.
  Then, if value $v_i$ has been RB-delivered from at least $n - 2t \geq t + 1$ processes (which means that it has been RB-broadcast by at least one correct process), $p_i$ validates it by assigning $yes$ to $aux_i$. Otherwise $p_i$ sets $aux_i$ to $no$ at line 04 (in that case it does not validate $v_i$). Then, $p_i$ issues a second RB-broadcast (line 05) to disseminate $aux_i$ (that is equal to $yes$ or $no$) to all processes.

---

**operation** VB_broadcast$(v_i)$
(01)   RB_broadcast INIT$(v)$;
(02)   **let** $rec_i$ **let be** the multiset of values RB_delivered to $p_i$;
(03)   **wait until** $(|rec_i| \geq n - t)$;
(04)   **if** $(\#_{v_i}(rec_i) \geq n - 2t)$ **then** $aux_i \leftarrow yes$ **else** $aux_i \leftarrow no$ **end if**;
(05)   RB_broadcast VALID$(aux_i)$.

**for** $1 \leq j \leq n$ **VB-delivery task** $T_i[j]$:
(06)   **wait until** $\bigl($VALID$(x)$ and INIT$(v)$ are RB_delivered from $p_j\bigr)$;
(07)   **if** $(x = yes)$ **then wait until** $(\#_v(rec_i) \geq n - 2t)$; $d \leftarrow v$
(08)        **else   wait until** $(\#_{v',v'',\ldots\neq v}(rec_i) \geq t + 1)$; $d \leftarrow \perp$
(09)   **end if**;
(10)   VB_deliver$(d)$ at $p_i$ as the value VB-broadcast by $p_j$.

---

**Algorithm 1.** A reliable-broadcast-based implementation of VB-broadcast

– The second part is made up of $n$ tasks. The task $T_i[j]$ starts by the wait statement for both the value $v$ RB-broadcast by $p_j$ and the boolean $x$ RB-broadcast by $p_j$ to say whether its value $v$ has been validated or not. Note that the value $v$ can be delivered either at line 03 or at line 06. (Let us remind that each time a message INIT$(v)$ is RB-delivered to $p_i$, the value $v$ is added to $rec_i$, which means that, after the predicate $|rec_i| \geq n - t$ has become true at line 03, the set $rec_i$ still keeps on being updated when new messages INIT$()$ are RB-delivered to $p_i$.)
  If $x = yes$, as $p_j$ can be Byzantine, $v$ has not necessarily been validated. Hence, $p_i$ has to check it. To that end, $p_i$ waits until the predicate $\#_v(rec_i) \geq n - 2t$ becomes true (line 07). When this predicate $\#_v(rec_i) \geq n - 2t \geq t + 1$ becomes true (if ever it does, line 07) we have $\#_v(rec_i) \geq t + 1$ and, consequently, $v$ is VB-delivered to $p_i$ as being the value VB-broadcast by $p_j$.
  Differently, if $x = no$, $p_i$ waits until $rec_i$ contains more than $t$ values different from $v$ (the value RB-delivered from $p_j$). When this occurs (if ever it does, line 07) $p_i$ VB-delivers $\perp$ as the value VB-broadcast by $p_j$.

**Theorem 1.** *Algorithm 1 implements the validated broadcast abstraction in the system model* $\mathcal{BZ\_AS}_{n,t}[t < n/3, \text{RB}]$.

## 3   Intrusion-Tolerant Byzantine Consensus and the Enriched Model

### 3.1   Byzantine Consensus

*Byzantine consensus.* The consensus problem has been informally stated in the Introduction. Assuming that at least each correct process proposes a value, each of them has to decide on a value in such a way that the following properties are satisfied.

– C-Termination. Every correct process eventually decides on a value.
– C-Agreement. No two correct processes decide on different values.
– C-Obligation (validity). If all correct processes propose the same value $v$, then $v$ is decided.

*Intrusion-tolerant Byzantine (ITB) consensus.* In Byzantine consensus, if not all correct processes propose the same value, any value can be decided. As indicated in the Introduction, we are interested in a stronger version of the consensus problem in which a value proposed only by faulty processes can never be decided. This consensus problem instance is defined by the termination, agreement and obligation properties stated above plus the following validity property.

- C-Non-intrusion (validity). A decided value is a value proposed by a correct process or $\perp$.

The fact that no value proposed only by faulty processes can be decided gives its name (namely *intrusion-tolerant*) to that consensus problem instance.

*Binary consensus.* The consensus is *binary* when only two values (e.g., 0 and 1) can be proposed. When more than two values can be proposed, consensus is *multivalued*.

Interestingly, the fact that only two values can be proposed to a binary Byzantine consensus algorithm provides it with an interesting property, namely, if all correct processes propose the same value $b \in \{0, 1\}$, it follows from the obligation property that they decide $b$, whatever the value ($b$ or $\bar{b} = 1 - b$) proposed by the faulty processes. Hence, we have the following property (that is no longer true for multivalued consensus).

*Property 1.* Any binary Byzantine consensus algorithm that satisfies the obligation property, satisfies also the non-intrusion property. Moreover, $\perp$ is never decided.

### 3.2   Enriched Model for Multivalued ITB Consensus

*Additional power is required.* It is well-known that Byzantine consensus cannot be solved when $t \leq n/3$ in synchronous systems [15,22]. Moreover, consensus cannot be solved in asynchronous systems as soon as even only one process may crash [12], which means that Byzantine consensus cannot be solved either as soon as one process can be faulty. Said another way, additional computational power is needed if one wants to solve Byzantine consensus in an asynchronous system.

Such an additional power can be obtained by randomization (e.g., [3,10,14,23,26]), failure detectors (e.g., [14,16]), additional synchrony assumptions (e.g., [11,17]), or even the assumption that there is a binary consensus algorithm that is given for free by the underlying system (e.g., [6,10,20,24,27]).

*Enriched model for multivalued ITB consensus.* In the following, $BBC$ denotes any algorithm that solves the binary Byzantine consensus problem. (Such algorithms are described in [4,10,14,26]. See also Section 7). Let $\mathcal{BZ\_AS}_{n,t}[\text{XX}, BBC]$ denote the system model $\mathcal{BZ\_AS}_{n,t}[\emptyset]$ enriched with $BBC$ (computational power) and the broadcast abstraction XX.

As announced in the Introduction, the aim is to design a generic multivalued ITB consensus algorithm on top of $\mathcal{BZ\_AS}_{n,t}[\text{XX}, BBC]$.

# 4   Generic Consensus Based on the WB or NDB Abstraction

This section presents a generic multivalued ITB consensus algorithm that can be instantiated with either WB or NDB instances. There is one instance of WB or NDB per process. Moreover, all these instances are independent the ones from the others. The algorithm uses two rounds for each process to compute a value it proposes to the underlying binary consensus. The instantiation with WB requires $n > 5t$, while the one with NDB requires $n > 4t$.

*Principles and description of the algorithm.* When considering algorithm 2, a process invokes propose$(v_i)$ where $v_i$ is the value it proposes to the consensus. It terminates when it executes the return() statement (line 14) that supplies it with the decided value. (In order to prevent confusion, the operation of the underlying binary consensus that is built is denoted bin_propose().)

In order to reduce the Byzantine consensus problem to its binary counterpart to benefit from $BBC$, the processes first exchange the values they propose. If a process sees that a value $v$ has been proposed "enough" times, it proposes 1 to $BBC$, otherwise it proposes 0. Then, if 1 is decided from $BBC$, the correct processes decide the value $v$ that has been proposed "enough" times, otherwise they decide $\perp$ (lines 09-14). For this to work, two things are necessary:

- (a) A value has to appear as if it has been proposed by enough processes.
- (b) If a process $p_i$ proposes 1 to $BBC$ because it has seen enough copies of a value $v$, it must be sure that any other correct process $p_j$ will be able to decide $v$ even if it has proposed 0 to $BBC$ (because it has not seen enough copies of $v$).

---

**operation** propose$(v_i)$
(01)    XX_broadcast EST1$(v_i)$;
(02)    **wait until** $\big($EST1$(-)$ messages XX_delivered from $(n - t)$ processes$\big)$;
(03)    **let** $rec1_i$ = multiset of values XX_delivered and carried by EST1 messages;
(04)    **if** $(\exists v :\ \#_v(rec1_i) \geq n - 2t)$ **then** $aux_i \leftarrow v$ **else** $aux_i \leftarrow \perp$ **end if**;
(05)    XX_broadcast EST2$(aux_i)$;
(06)    **wait until** $\big($EST2$(-)$ messages XX_delivered from $(n - t)$ processes$\big)$;
(07)    **let** $rec2_i$ = multiset of values XX_delivered and carried by EST2 messages;
(08)    **if** $(\exists v \neq \perp :\ \#_v(rec2_i) \geq n - 2t)$ **then** $bp_i \leftarrow 1$ **else** $bp_i \leftarrow 0$ **end if**;
(09)    **if** $(\exists v \neq \perp : v \in rec2_i)$ **then let** $v$ = most frequent non-$\perp$ value in $rec2_i$;
(10)                                    $res_i \leftarrow v$
(11)                              **else** $res_i \leftarrow \perp$
(12)    **end if**;
(13)    $b\_dec_i \leftarrow$ bin_propose$(bp_i)$;       % underlying binary consensus %
(14)    **if** $(b\_dec_i = 1)$ **then** return$(res_i)$ **else** return$(\perp)$ **end if**.

---

**Algorithm 2.** A generic intrusion-tolerant Byzantine consensus algorithm

These two issues are solved by two asynchronous rounds executed before invoking the underlying $BBC$ algorithm (lines 01-12). The messages of the first round and the second round are tagged EST1 and EST2, respectively. Interestingly, we will state below two properties $PR1$ and $PR2$ that are the same as the properties used in [18] to solve consensus on top of an asynchronous system enriched with any of Chandra and Toueg's failure detectors [9].

It is important to remark that, at the abstraction level of the consensus algorithm, a message carries only a tag (EST1 or EST2) and a proposed value or $\perp$. Hence, considering that proposed values have constant size, the size of the messages used by the algorithm is $O(1)$ (no message is required to carry array-like data structures whose size would depend on $n$).

*First round.* The aim of this round (lines 01-04) is to direct each process $p_i$ to define a "new" proposed value $aux_i$ in such a way that the values $aux_i$ of the correct processes satisfy the following property:

$$PR1 \equiv \big[\forall i, j \in C : \big((aux_i \neq \perp) \wedge (aux_j \neq \perp)\big) \Rightarrow$$
$$(aux_i = aux_j = v) \wedge (v \text{ has been proposed by a correct process})\big].$$

Hence this round replaces (for the correct processes) the set of values they propose by a non-empty set including at most two values (namely, a value $v$ proposed by a correct process and $\perp$).

From an operational point of view, this is obtained as follows. The processes first exchange (with the help of the underlying broadcast facility) the values they propose (lines 01-02). The values delivered at $p_i$ are kept in the multiset $rec1_i$. Then, if there is a value $v$ in $rec1_i$ such that $\#_v(rec1_i) \geq n - 2t$, $v$ is assigned to $aux_i$. Otherwise $aux_i = \perp$.

*Second round.* The aim of the second round (lines 05-12) is to establish the following property denoted $PR2$ in order the result of the underlying $BBC$ algorithm can be safely exploited as described previously (lines 13-14). The local variable $bp_i$ contains the value proposed by $p_i$ to the underlying $BBC$ algorithm, and $res_j$ contains the non-$\perp$ value that any correct process $p_j$ will decide if $\perp$ is not decided.

$$PR2 \equiv \big[(\exists i \in C : bp_i = 1) \Rightarrow (\forall j \in C : res_j = res_i = v \neq \perp)\big].$$

Operationally, this is obtained as follows. With the help of the underlying broadcast abstraction the correct processes exchange the values of their $aux_i$ variables. The values delivered at $p_i$ are saved in the multiset $rec2_i$. (This multiset contains $n - t$ values, and, due to $PR1$, those can be $\perp$, a non-$\perp$ value $v$ proposed by a correct process, and at most $t$ arbitrary values sent by faulty processes.)

If there is a non-$\perp$ value $v$ such that $\#_v(rec2_i) \geq n - 2t$, $p_i$ proposes $bp_i = 1$ to the binary consensus. Otherwise, $p_i$ has not seen enough copies of a value $v \neq \perp$ and consequently proposes $bp_i = 0$. In all cases, $p_i$ defines $res_i$ as the most frequent non-$\perp$ value it has received. As the proof shows [19], if a correct process $p_i$ invokes bin_propose(1), all correct processes will have the same non-$\perp$ value in their $res_j$ variables.

**Theorem 2.** *Algorithm 2 solves the multivalued consensus problem (as defined by the C-termination, C-agreement, C-obligation and C-non-intrusion properties) in both system models* $BZ\_AS_{n,t}[t < n/5, \mathsf{WB}, BBC]$ *and* $BZ\_AS_{n,t}[t < n/4, \mathsf{NDB}, BBC]$.

# 5   Consensus Based on the Validated Broadcast Abstraction

This section presents an intrusion-free Byzantine consensus algorithm based on validated broadcast. This algorithm requires $t < n/3$ and has consequently an optimal resilience. It requires a single round (instead of 2 as in Figure 2) but, as it uses a validated broadcast, this round requires four communication steps.

---

**operation** propose($v_i$)
(01)     VB_broadcast EST1($v_i$);
(02)     **wait until** $(\text{EST}(-)$ messages VB_delivered from $(n - t)$ processes$)$;
(03)     **let** $rec_i$ = multiset of the values $v$ such that $\text{EST}(v)$ is VB_delivered to $p_i$;
(04)     **if** $(\exists v \neq \bot : \#_v(rec_i) \geq n - 2t) \wedge (rec_i$ contains a single non-$\bot$ value$)$
(05)         **then** $bp_i \leftarrow 1$ **else** $bp_i \leftarrow 0$
(06)     **end if**;
(07)     $b\_dec_i \leftarrow$ bin_propose($bp_i$);       % underlying binary consensus %
(08)     **if** $(b\_dec_i = 1)$
(09)         **then wait until** $(\exists v \neq \bot$ such that $\text{EST}(v)$ VB_delivered from $(n - 2t)$ processes$)$;
(10)             return($v$)
(11)         **else** return($\bot$)
(12)     **end if**.

---

**Algorithm 3.** A validated-broadcast-based intrusion-tolerant Byzantine consensus algorithm

In Algorithm 3, after having VB-broadcast its value, a process $p_i$ waits for $\text{EST}()$ messages from $n - t$ processes and deposits the corresponding values in the multiset $rec_i$. Let us notice that all the values that are VB-delivered satisfy the VB-validation property.

Then, $p_i$ checks if (in addition to $\bot$) it has VB-delivered exactly one non-$\bot$ value $v$ and that value has been VB-broadcast by at least $n - 2t$ processes (line 04). If there is such a value, $p_i$ proposes 1 to the underlying binary consensus, otherwise it proposes 0 (line 05).

Finally, $p_i$ decides $\bot$ if the underlying binary consensus returns 0 (lines 08 and 11). Differently, if 1 is returned, $p_i$ waits until it has VB-delivered $n - 2t$ $\text{EST}()$ carrying the very same value $v$ (line 09) and then decides that value (line 10). Let us notice that, among these $n - 2t$ messages, some have been already VB-delivered at line 02. The important point is (as shown in the proof) that the net effect of (a) the validated broadcast, (b) the predicate used at line 04, and (c) the predicate used in the wait statement at line 09, ensures that if a correct process invokes bin_propose(1), then all correct processes eventually VB-deliver $n - 2t$ times the very same value $v$ and decide it.

**Theorem 3.** *Algorithm 3 solves the multivalued consensus problem (defined by the C-termination, C-agreement, C-obligation and C-non-intrusion properties) in the system model $\mathcal{BZ\_AS}_{n,t}[t < n/3, \text{VB}, BBC]$.*

## 6  Discussion

*An additional property of the previous ITB consensus algorithms.* Let $v$ be the most proposed value (it is possible that several values are equally most proposed, in that case any of them is chosen), and let $\#_v$ be the number of processes that propose it. The previous algorithms have the following noteworthy property. (This follows from properties $PR1$ and $PR1$, Theorem 2 for the instances obtained from the generic Algorithm 2, and from Theorem 3 for Algorithm 3 based on a validated broadcast.)

- If $\#_v \geq n - t$, then $v$ is always decided by the correct processes (let us observe that, in that case, there is a single most proposed value).
- If $\#_v < n - 2t$, then $\perp$ is always decided by the correct processes.
- If $n - 2t \leq \#_v < n - t$, then which value ($v$ or $\perp$) is decided by the correct processes depends on the behavior of the Byzantine processes.

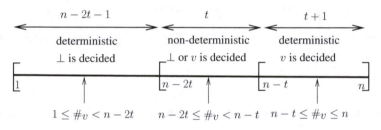

**Fig. 1.** Deterministic vs non-deterministic scenarios

Let us consider an omniscient observer that would know which are the proposed values. In the first and the second cases, this omniscient observer can compute the result in a deterministic way. Differently, in the last case it cannot. The value that is decided depends actually on the behavior of Byzantine processes (that can favor the most proposed value, or entail a $\perp$ decision). These different possibilities are depicted on Figure 1. Of course, a value proposed only by Byzantine processes is necessarily proposed by fewer than $n - 2t$ processes as $n > 3t$ and hence cannot be the decision value.

*Comparing the previous signature-free multivalued ITB algorithms.* Table 2 presents a summary of the cost and the constraint on $t$ associated with the previous signature-free multivalued ITB consensus algorithms. As they can all use the same underlying $BCC$ algorithm, the comparison does not take it into account.

**Table 2.** Cost of the ITB consensus algorithms

| Consensus algorithm instantiated with | # communication steps | message size at send/receive level | # msgs at send/receive level | constraint on $t$ |
|---|---|---|---|---|
| Algorithm 2 with WB | $2 \times 1 = 2$ | constant | $O(n^2)$ | $n > 5t$ |
| Algorithm 2 with NDB | $2 \times 2 = 4$ | $\log_2 n$ | $O(n^3)$ | $n > 4t$ |
| Algorithm 3 with VB | $1 \times 6 = 6$ | $\log_2 n$ | $O(n^3)$ | $n > 3t$ |

It is easy to see that, due the weaker constraint on $t$, Algorithm 3 instantiated with VB outperforms Algorithm 1 instantiated with NDB. On another side, in a system where the number of Byzantine processes remains small, Algorithm 1 instantiated with WB is the most efficient.

## 7    A Randomized VB-Based Byzantine Binary Consensus Algorithm

This section presents a randomized Byzantine binary consensus algorithm (that can be used as the underlying $BBC$ algorithm). The additional power needed to solve consensus is given here by random coins. In addition to being optimal from a resilience point of view ($t < n/3$), this algorithm has two noteworthy features:

– It is based on the validated broadcast abstraction, and
– Each round requires 6 communication steps (a single VB-broadcast instance).

When looking at Byzantine consensus algorithms that are optimal from a resilience point of view (i.e., algorithms able to cope with up to $\lfloor(n-1)/3\rfloor$ faulty processes), the best consensus algorithm we are aware of has rounds made up of three communication steps [7]. Moreover, this algorithm is based on signatures (public key cryptography). As far as signature-free algorithms are concerned, the best resilience-optimal algorithm, that uses control information whose size is only $O(\log_2 n)$ we are aware of, is the one described in [26,25], which requires five communication steps per round. Algorithm 4 that is presented in this section is signature-free and requires six communication steps per round. The fifth step in [25] is necessary to ensure errorless termination as explained in the original paper [26].

*Common coin.* The asynchronous system is equipped with a *common coin* as defined by Rabin [23] and improved in [7] in order to get rid of the trusted dealer. Such an oracle is denoted $CC$, hence the system model is $\mathcal{BZ\_AS}_{n,t}[t < n/3, CC]$. A common coin can be seen as a global entity that delivers a sequence of random bits $b_1, b_2, \ldots, b_r, \ldots$ to processes (each bit $b_r$ has the value 0 or 1, with probability $1/2$).

More precisely, this oracle provides the processes with a primitive denoted random() that returns a bit each time it is called by a process. In addition to being random, this bit has the following global property: the $r$th invocation of random() by any correct process $p_i$ returns it the bit $b_r$. This means the same random bit $b_r$ is returned to each correct process as the result of its $r$th invocation of random(). It is important to notice that the adversary (which determines the scheduling of processes and messages) has no access to the common coin, which corresponds to the *oblivious scheduler* model [2]. (The reader interested in the implementation of a common coin can consult [2,7].)

*On randomized consensus.* When using additional computing power provided by random coins, the consensus termination property can no longer be deterministic. The *Randomized Consensus* problem is defined by C-validity (Obligation), C-agreement plus the following termination property. [3,23]:

– Proba-C-Termination: Every correct process decides with probability 1.

*Underlying principles and description of the algorithm.* In Algorithm 4, a process $p_i$ invokes the function bin_propose$(v_i)$ where $v_i$ is the value it proposes. It decides when it executes the statement decide$(v)$ (line 07). The design of Algorithm 4 is close to an algorithm we have proposed in [14]. Its fundamental difference is that it is resilience-optimal ($t < n/3$), while the one described in [14] requires $t < n/5$.

---

**operation** bin_propose$(v_i)$
$est_i \leftarrow v_i; r_i \leftarrow 0;$
**repeat forever**
(01)    $r_i \leftarrow r_i + 1;$
(02)    VB_broadcast EST$(r_i, est_i);$
(03)    **let** $rec_i$ = multiset of values $est$ such that EST$(r_i, est)$ has been VB_delivered to $p_i;$
(04)    **wait until** $(|rec_i| \geq n - t);$
(05)    $s_i \leftarrow$ random$();$
(06)    **if** $(\exists v \neq \bot : \#_v(rec_i) \geq n - 2t) \wedge (rec_i$ contains a single non-$\bot$ value)
(07)        **then** $est_i \leftarrow v;$
(08)            **if** $(v = s) \wedge (p_i$ has not yet decided) **then** decide$(s_i)$ **end if**
(09)        **else** $est_i \leftarrow s_i$
(10)    **end if**
**end repeat**.

---

**Algorithm 4.** A binary Byzantine consensus algorithm based on VB-broadcast

The local variable $est_i$ of process $p_i$ keeps its current estimate of the decision value (initially, $est_i = v_i$). The processes proceed by consecutive asynchronous rounds. Thus, the pair $(r_i, est_i)$ of a correct process $p_i$ describes its current state ($r_i$ is $p_i$'s current round number). The first part of Algorithm 4 consists of lines 01-04 that describes the communication of the current round. The second part, made up of lines 05-10, defines the management of the local estimate $est_i$ and the decision rule. More precisely, we have the following.

- At every round $r_i$, each correct process $p_i$ VB-broadcasts EST$(r_i, est_i)$, and waits until it has VB-delivered EST$(r_i, -)$ from at least $n - t$ processes (lines 02-04).
- In the second part, $p_i$ first computes the random number $s$ associated with the current round $r_i$ (line 05). Then, $p_i$ checks if it has received a non-$\bot$ value $v$ from at least $n - 2t$ different processes, and $v$ is the only non-$\bot$ value in $rec_i$ (predicate at line 06). If this predicate holds, $p_i$ adopts $v$ as new estimate (line 07) and decides the random value $s$ if $v = s$ (line 08). If the predicate is false, $p_i$ updates its estimate $est_i$ to the random value $s$. In all cases, $p_i$ starts a new asynchronous round.

The statement decide$()$ allows the invoking process to decide but does not stop its execution. Hence, a process executes rounds forever. This facilitates the description of the algorithm. Using techniques such as the one developed in [14] allows a process to both decide and stop.

**Remark.** It is possible to add the following test after line 04:
        **if** $(\exists v : \#_v(rec_i) \geq n - t)$ **then** decide$(v)$ **end if**.

This allows the algorithm to always terminate in a single round whatever the value of the common coin when all correct processes propose the same value and no process exhibits a Byzantine behavior. This scenario is very likely to happen in actual executions.

**Theorem 4.** *Algorithm 4 solves the randomized binary consensus problem in the system* $\mathcal{BZ\_AS}_{n,t}[t < n/3, \mathsf{VB}, CC]$.

**Theorem 5.** *Let* $n > 3t$. *The expected decision time of Algorithm 4 is constant.*

## 8  Conclusion

The paper has presented a family of multivalued intrusion-free Byzantine consensus algorithms. The intrusion-freedom property means that no value proposed only by Byzantine processes can ever be decided. These consensus algorithms are built on top of appropriate broadcast abstractions. One of these abstractions, called validated broadcast, is new (and can be interesting by itself to solve other problems than consensus). Moreover, all proposed algorithms are signature-free (hence efficient).

The intrusion-free consensus algorithm based on the validated broadcast abstraction has several noteworthy features: it is optimal from a resilience point of view ($t < n/3$) and each round requires a single validated broadcast.

The paper has also presented a novel randomized binary Byzantine consensus algorithm that is resilient-optimal and, in a very interesting way, is also based on the validated broadcast abstraction.

## References

1. Aguilera, M.K., Frolund, S., Hadzilacos, V., Horn, S., Toueg, S.: Abortable and Query-abortable Objects and their Efficient Implementation. In: Proc. 26th Annual ACM Symposium on Principles of Distributed Computing (PODC 2007), pp. 23–32 (2007)
2. Aspnes, J.: Lower Bounds for Distributed Coin Flipping and Randomized Consensus. JACM 45(3), 420–450 (1998)
3. Ben-Or, M.: Another Advantage of Free Choice: Completely Asynchronous Agreement Protocols. In: Proc. 2nd ACM Symposium on Principles of Distributed Computing (PODC 1983), pp. 27–30. ACM Press, New York (1983)
4. Bracha, G.: Asynchronous Byzantine Agreement Protocols. Information & Computation 75(2), 130–143 (1987)
5. Bracha, G., Toueg, S.: Asynchronous Consensus and Broadcast Protocols. Journal of the ACM 32(4), 824–840 (1985)
6. Brasileiro, F., Greve, F., Mostéfaoui, A., Raynal, M.: Consensus in One Communication Step. In: Malyshkin, V.E. (ed.) PaCT 2001. LNCS, vol. 2127, pp. 42–50. Springer, Heidelberg (2001)
7. Cachin, C., Kursawe, K., Shoup, V.: Random Oracles in Constantinople: Practical Asynchronous Byzantine Agreement using Cryptography. In: Proc. 19th Annual ACM Symp. on Principles of Distributed Computing (PODC 2000), pp. 123–132 (2000)
8. Cachin, C., Kursawe, K., Petzold, F., Shoup, V.: Secure and efficient asynchronous broadcast protocols. In: Kilian, J. (ed.) CRYPTO 2001. LNCS, vol. 2139, pp. 524–541. Springer, Heidelberg (2001)

9. Chandra, T., Toueg, S.: Unreliable Failure Detectors for Reliable Distributed Systems. JACM 43(2), 225–267 (1996)
10. Correia, M., Ferreira Neves, N., Verissimo, P.: From Consensus to Atomic Broadcast: Time-free Byzantine-Resistant Protocols without Signatures. The Computer J. 49(1), 82–96 (2006)
11. Dwork, C., Lynch, N., Stockmeyer, L.: Consensus in the Presence of Partial Synchrony. JACM 35(2), 288–323 (1988)
12. Fischer, M.J., Lynch, N.A., Paterson, M.S.: Impossibility of Distributed Consensus with One Faulty Process. Journal of the ACM 32(2), 374–382 (1985)
13. Friedman, R., Mostéfaoui, A., Raynal, M.: $\Diamond \mathcal{P}_{mute}$-Based Consensus for Asynchronous Byzantine Systems. Parallel Processing Letters 20(1-2), 162–182 (2005)
14. Friedman, R., Mostéfaoui, A., Raynal, M.: Simple and Efficient Oracle-Based Consensus Protocols for Asynchronous Byzantine Systems. IEEE Transactions on Dependable and Secure Computing 2(1), 46–56 (2005)
15. Lamport, L., Shostack, R., Pease, M.: The Byzantine Generals Problem. ACM Transactions on Programming Languages and Systems 4(3), 382–401 (1982)
16. Kihlstrom, K.P., Moser, L.E., Melliar-Smith, P.M.: Byzantine Fault Detectors for Solving Consensus. The Computer Journal 46(1), 16–35 (2003)
17. Martin, J.-P., Alvizi, L.: Fast Byzantine Consensus. IEEE Transactions on Dependable and Secure Computing 3(3), 202–220 (2006)
18. Mostéfaoui, A., Raynal, M.: Solving Consensus Using Chandra-Toueg's Unreliable Failure Detectors: a General Quorum-Based Approach. In: Jayanti, P. (ed.) DISC 1999. LNCS, vol. 1693, pp. 49–63. Springer, Heidelberg (1999)
19. Mostéfaoui, A., Raynal, M.: Signature-Free Broadcast-Based Intrusion Tolerance: Never Decide a Byzantine Value. Tech. Report #1954, IRISA, Université de Rennes, France (2010)
20. Mostéfaoui, A., Raynal, M., Tronel, F.: From Binary Consensus to Multivalued Consensus in Asynchronous Message-Passing Systems. Inf. Processing Letters 73, 207–213 (2000)
21. Okun, M.: Byzantine Agreement. In: Springer Encyclopedia of Algorithms, pp. 116–119 (2008)
22. Pease, M., Shostak, R., Lamport, L.: Reaching Agreement in the Presence of Faults. JACM 27, 228–234 (1980)
23. Rabin, M.: Randomized Byzantine Generals. In: Proc. 24th IEEE Symposium on Foundations of Computer Science (FOCS 1983), pp. 116–124. IEEE Computer Society Press, Los Alamitos (1983)
24. Song, Y.J., van Renesse, R.: Bosco: One-Step Byzantine Asynchronous Consensus. In: Taubenfeld, G. (ed.) DISC 2008. LNCS, vol. 5218, pp. 438–450. Springer, Heidelberg (2008)
25. Srikanth, T.K., Toueg, S.: Simulating Authenticated Broadcasts to Derive Simple Fault-tolerant Algorithms. Distributed Computing 2, 80–94 (1987)
26. Toueg, S.: Randomized Byzantine Agreement. In: Proc. 3rd Annual ACM Symposium on Principles of Distributed Computing (PODC 1984), pp. 163–178 (1984)
27. Turpin, R., Coan, B.A.: Extending Binary Byzantine Agreement to Multivalued Byzantine Agreement. Information Processing Letters 18, 73–76 (1984)

# A Fault Avoidance Strategy Improving the Reliability of the EGI Production Grid Infrastructure

Francesco Palmieri[1], Silvio Pardi[2], and Paolo Veronesi[3]

[1] Università degli studi di Napoli Federico II
Via Cinthia, 5 – 80126, Napoli, Italy
francesco.palmieri@unina.it
[2] INFN Sezione di Napoli and INDAM
Via Cinthia, 5 – 80126, Napoli, Italy
spardi@na.infn.it
[3] INFN CNAF
Viale Berti Pichat 6/2 - 40127 Bologna, Italy
paolo.veronesi@cnaf.infn.it

**Abstract.** Reliability is a crucial issue for the development of stable and effective production grid infrastructures. That is, grid users must be able to trust upon the runtime service they request and receive from the underlying grid. Many runtime services and capabilities offered by modern Grid infrastructures are not available in advance to the application developers and dynamically bound only at the execution time, leading to an increased incidence of interaction faults. In this work we propose, implement and evaluate a novel low-impact fault-avoidance scheme, specifically conceived to improve the grid reliability from the user/application point of view, by providing proper service status information to the workload management system. In particular, starting from the EGEE experience, we designed a strategy inhibiting the use of some specific runtime capabilities on the available resources as soon as the monitoring system detect any anomalous behavior associated to these capabilities and re-integrating them when they restart to correctly work again. The results of a significant set of tests ran on the production EGEE infrastructure, have been presented to show the effectiveness of our approach.

**Keywords:** Reliability, Fault Avoidance, Monitoring, Resource Management.

## 1 Introduction

Grid computing is evolving very rapidly, and the amount of heterogeneous resources and services made available from the participating sites, distributed all-over the world, is increasing at an astonishing pace. Unfortunately, Grid systems are highly dependent upon one another and their provided services become more and more important for several fields in the modern society so that in a few years they will be viewed as critical infrastructures. In such a complex and dynamic environment the possible fault conditions grow together with the available resources and services. This

C. Lu, T. Masuzawa, and M. Mosbah (Eds.): OPODIS 2010, LNCS 6490, pp. 159–172, 2010.

makes troubleshooting an extremely difficult task and poses a severe limit on the overall model scalability. In more detail, the shared resources made available on a grid include supercomputers, computing clusters, storage systems, and associated hardware, together with all the needed managing and monitoring software. These resources may employ different processor architectures and operating systems and belong to many different organizations having little knowledge of each other. As such, they are likely to be managed in different administrative domains, with very dissimilar access, maintenance and security policies. All the above circumstances can undoubtedly lead to an increased incidence of interaction faults, hence, from the application's point of view, there may be no guarantee that resources available in a grid will be reliable. This is especially true if the application consists of multiple tasks organized into a complex workflow or features a large amount of services. In the latter case a fault in a single service can be amplified as the effects of the fault move further down the chain. The likelihood of hardware and software errors interesting single or multiple devices is exacerbated by the fact that several applications may perform long tasks that may require the use of many resources for several days of computation. Furthermore, the great heterogeneity of the connected resources, together with their dynamic behavior in entering and leaving the grid, will likely lead to component interactions that may result in communication problems, which adversely affect the execution of user applications. Furthermore, the need for managing long-lasting high-bandwidth data connections to transport large data sets requires the concurrent use of a large number of network components. This in turn increases the chance of failures that necessitate rerouting connections through alternative devices/paths. For these reasons, ensuring reliable transport and runtime services in large-scale grid infrastructures is an extremely critical issue. Consequently, the involved technical support teams invest huge efforts in troubleshooting operations, fixing middleware bugs, repairing hardware and software faults and excluding the faulty or out-of-service resources from the information system. In order to promote the pervasive adoption of grid technologies within the emerging high-level application scenarios, future grid infrastructures must guarantee an adequate degree of service reliability and quality. The information system plays a fundamental role in associating the right resources to the applications and the presence of obsolete or incorrect information can cause scheduling error, frequent resubmission and, finally, job faults. For this sake we propose a low-impact application-driven fault avoidance strategy, easily implementable within the European Grid Infrastructure (EGI). This strategy is based on the introduction of a verification/feedback mechanism within the grid information system implemented through on-line service-monitoring checks periodically handled by several independent monitoring systems, The results of these checks are used to modify in real-time the information providing specific hints to the workload management system to drive resource discovery operations according to a fault-free resource scheduling plan.

This solution, whose main goal is avoiding as many runtime failures as possible, to minimize the job execution time, demonstrated to be effective in incrementing both the user perceivable quality and the overall grid performance.

## 2 The EGI Grid: Technology Backgrounds

EGI is the Europe's flagship Research infrastructure, one of the greatest Grid experiences worldwide, developed within the *Enabling Grid for E-sciencE* project (EGEE) context, and collecting several grid communities whose cooperation is based on the gLite middleware [1] stack, that combines components developed in various related projects, such as Condor [2], Globus [3] and LCG [4]. Such middleware provides high level services for scheduling and running computational jobs, accessing and moving data, and obtaining information on the Grid infrastructure as well as Grid applications, all embedded into a consistent security infrastructure. The *Workload Management System* (WMS) is the gLite meta-scheduling engine that manages the assignment of the available computing and storage resources to the submitted jobs according to several performance objectives, user preferences and specific status information gathered from the grid information system. It also keeps track of the jobs it manages in a consistent way via the logging and bookkeeping service. The EGEE system for information and monitoring is based on the *Top Berkeley Database Information Index* (BDII). A specific top BDII management service gathers in a database (updated every couple of minutes) the information from all site Information Systems, thus having almost real time information from all the sites, their availability and operational and status details on their resources. The WMS takes decisions based on the information stored in the top BDII's. There is in general one BDII for every WMS instance in the current EGI/LCG production infrastructure. The monitoring subsystem plays a central role in supporting the EGI grid operation and management activities, and in particular for troubleshooting problems on the individual resources. It is divided into two main service classes, corresponding at two different views of the Grid: the internal and the external ones. Essentially, internal monitoring is a way used by the resources to directly report information about their operating status. Typical examples of the information reported are the memory and CPU occupation, end-to-end network performance measurements and so on. The most common internal monitoring tools used in EGI are GSTAT 2.0, GridView, GridMap. On the other hand, the external monitoring subsystem realizes a Grid snapshoot from the user perspective, that is, showing which services of resources are really available to the users. It is composed by a set of tools performing probes, tests and measurements directly at the middleware level, by simulating the user behavior. The main external monitoring service in the EGI infrastructure is the *Service Availability Monitoring* (SAM) Framework, a centralized service, periodically executing and checking a set of typical user actions such as job submission, data access, authentication etc. In recent times the system evolved according to a distributed probing scheme. In particular the Nagios-based centralized service availability monitor is a re-engineering of the SAM framework providing several regional Monitoring facilities based on the Nagios tool, instead of a centralized one, responsible to periodically perform the probes/tests on a more limited set of resource falling within their competence region. Through the SAM system the EGI collaboration calculates for each site offering services to the grid, an availability and reliability index. All the faults detected by SAM can be due to any kind of fault at the network, middleware or local operating system level, so that each error really represents a set of possible faults in the chain from the user to the local resource. For this reason SAM is not an easy instrument for troubleshooting the grid but at the EGI state-of-the-art it is the most powerful available system to know its real behavior. However, the effectiveness of the monitoring subsystem alone in ensuring grid reliability is extremely limited since it can be only viewed as an almost passive fault reporting facility. More effective reliability improvements can be obtained by allowing the use of monitoring information to condition and change in real-time the top-level scheduling policies according to an end-user visibility of the really working services. We started from the consideration that the computing resources participating to a Grid usually don't publish any online information about the runtime operational status of their service portfolio or about the possible faults occurred on it (e. g. an host certificate is expired or some crucial service is down for some reason). Moreover, due to the strong interdependency among the Grid resources, we considered that the best monitoring approach is performing periodical sanity checks on the services offered on the individual resources and providing such information to the users through the Grid information system.

## 3   Improving Grid Reliability through Fault Avoidance

One of the best strategies available for ensuring reliability in complex systems is potentially preventing the occurrence of faults (aka fault avoidance), instead of recovering from failure situations that have already took place, with the consequent significant gains in terms of execution times and resource usage. In fact, preventing the selection of unreliable, flawed or incorrectly configured resources within the execution context of specific grid tasks may be extremely desirable to avoid handling error conditions, exceptions and re-runs adversely conditioning the individual task completion time and success chances, together with the overall grid performance. However, the main problems to be faced when designing a fault avoidance strategy arise from the necessity of ensuring a reasonable trade-off between scalability and the main objectives of availability and reliability. Accordingly, we propose a fault avoidance strategy operating at the information system layer and based on a common view of the different resource and status monitoring systems available on the grid sites. Such strategy has been specifically conceived to ensure the scalability needed from large production grid infrastructures, and at the same time its implementation is sufficiently simple and flexible to be introduced seamlessly within the most common existing middleware frameworks. Furthermore, it is not exclusive in its application and can be easily associated to other available fault tolerance and management techniques such as data replication, checkpointing, process migration and so on. The main idea behind our proposal is blacklisting at the collective Information Index level specific resources or services capabilities that are suspected to operate incorrectly, as soon as the monitoring system, by performing periodic sanity checks on them, detects any anomalous behavior, and re-putting them back in service when they restart working properly. Clearly, the exact knowledge of the real operational status of the various services offered by the available Grid resources is the crucial information needed to support such a fault avoidance strategy. This can be achieved by introducing several advanced on-line service status monitoring objects within the grid information system. These objects, that are associated to online operational information periodically gathered from the individual resource capabilities and continuously kept up-to-date by the grid monitoring system, can be used, under the control of client applications, as specific hints to the workload management system to condition its resource discovery and selection decisions according to a scheduling plan that ensures the occurrence of no (avoidable) faults. Thus, the Grid's resource scheduling system becomes the key element for handling these runtime faults, being able through its selection policies and choices to greatly reduce or exacerbate the above phenomena. The above methodology allows the users or applications to avoid the resources that doesn't work correctly early at the submission time. The greatest strength of the approach is that it has minimal impacts on the existing grid middleware framework since we don't need any change to be introduced in both local site middleware infrastructure and data model, and we can re-use the already existing resource monitoring information system (producing and consuming the above monitoring objects) to correctly drive resource selection in such a way that no (or minimal) runtime fault will occur. Moreover, only specific resource capabilities (i.e. the one that are in fault for some reason) can be selectively excluded from selection, without putting out-of-service, entire resources by ensuring that all the other services that are provided correctly will still be available to the grid. Such strategy allows us to guarantee at each time the greatest possible level of resource and services availability, simultaneously taking advantage from maximum reliability. The most significant expected benefits come from the ability of following the highly dynamic behavior of the grid model in excluding and reintegrating at the right time the specific runtime capabilities offered from the available resources.

## 4 Introducing Fault-Avoidance in the EGI Production Infrastructure

To implement a really effective fault avoidance solution in a wide and complex environment such as the EGI production infrastructure, we at first need to identify and categorize the most common runtime faults that may occur in EGI. Some immediate examples may include the incorrect or missing support of *Message Passing Interface* (MPI) or other library services. Such information is needed in order to provide clues to the underlying scheduling system (the gLite's Workload Management System) for preventing as possible failure in applications due to their occurrence, with the consequent job resubmission, damaging both the individual job success and the overall grid performance. To benefit from these clues, the resource scheduler should enable autonomous and intelligent coupling of monitoring data and operational meta-data in a framework able to condition its selection/admission decisions. Such association can be achieved by defining specific *Job Description Language* (JDL) [5] parameters, associated to job requests, to be checked by the WMS during the first stage of the resource brokering process. These parameters are based on Condor classified advertisements (*ClassAds*) and structured as attribute-value pairs published on the Grid top-level information system database (the top-BDII). In such a way, we can easily embed some specific service operational status request into the scheduling strategy in order to improve the efficiency and the utilization of the Grid and avoiding, as possible, execution failures. ClassAds comparisons are performed in the matchmaking phase starting from JDL attributes in an evaluation environment that maps any attribute accessed to the match candidate under consideration. In detail, the users submitting a job to the Grid, can describe in advance their service status requirements by referencing the proper monitoring metric, by using JDL expressions containing a *Requirements* and a *Rank* ClassAds expression. This drives the discovery of grid-resources that are "certified" to support the needed services. "Requirements" is an expression defining the conditions to be met by the resource. It can include the check for extended attributes or specific runtime environment variables that can be used for referencing the proper service support capabilities on the involved node. "Rank" specifies the preference criteria to be adopted to select one of the resource instances that satisfy the Requirements. In addition, authorization checks are also performed against the Virtual Organization attributes. This lead to the formulation of a schedule plan conditioned, as possible, from the above requirements. In the following script (fig. 1) we show a JDL example with the requirements of discovering grid-resources that support the MPICH library service. Specifically, we are looking for computational resources within the "*matisse*" Virtual Organization that publish the MPICH variable in their runtime environment. Such variable indicates that the MPICH library is configured and the Nagios checks are enabled, so that the involved job can run only on sites/nodes supporting MPICH services. Thus, if the MPI library does not work correctly on specific nodes we can prevent job submission on them by deleting the MPICH label from the corresponding resources on the information system. Accordingly, the presence of the MPICH suffix guarantees that the MPI libraries are correctly working, so that the users don't need to change their working conventions, that is a crucial issue in production infrastructures.

```
Type = "Job";
JobType = "MPICH";
CpuNumber = 16;
Executable = "mpi-start-wrapper.sh";
Arguments = "cpi MPICH";
StdOutput = "mpich-test.out";
StdError = "mpich-test.err";
InputSandbox = {"mpi-start-wrapper.sh",
"mpi-hooks.sh","cpi.c"};
OutputSandbox = {"mpich-test.err","mpich-test.out"};
RetryCount = 1;
Requirements = Member("MPICH",
other.GlueHostApplicationSoftwareRunTimeEnvironment);
```

**Fig. 1.** Simple JDL for MPICH test

To support the above mechanisms we introduced additional information coming from the Nagios monitoring system on the top level information index, the top-BDII, that can be used in WMS "requirement" expressions matching. Thus the top-BDII service will realize the additional function of aggregating at the grid collective services layer all the resources and services status update information provided from the centralized monitoring information system through properly crafted incremental *ldif* structures. An example entry in the ldif file used to inhibit the MPICH support for the Taiwan-LCG2 site is reported in the following figure.

```
# SITE Taiwan-LCG2, CE quanta.grid.sinica.edu.tw
# CHECK org.sam.WN-MPI, RETRIEVED FROM samnag012.cern.ch
# SUMMARY: node147: CRITICAL: MPI Status: ERROR
# EXECUTION TIME: 2010-04-15T12:20:22Z
dn:
GlueSubClusterUniqueID=quanta.grid.sinica.edu.tw,
GlueClusterUniqueID=quanta.grid.sinica.edu.tw,Mds-Vo-name=
Taiwan-LCG2,Mds-Vo-name=local,o=grid
changetype:modify
delete: GlueHostApplicationSoftwareRunTimeEnvironment
GlueHostApplicationSoftwareRunTimeEnvironment:MPICH
delete: GlueHostApplicationSoftwareRunTimeEnvironment
GlueHostApplicationSoftwareRunTimeEnvironment:MPICH-1_2-7
delete: GlueHostApplicationSoftwareRunTimeEnvironment
GlueHostApplicationSoftwareRunTimeEnvironment:MPICH2
```

**Fig. 2.** An example ldif entry

In more detail, we identified a set of typical fault scenarios that are crucial for the success of Grid jobs and data-management activities and developed the corresponding runtime functional probes within the Nagios-Based centralized Service Availability Monitoring system. The Nagios system periodically runs the above runtime probes and sanity check jobs on all the available resources and collects the results through an Apache ActiveMQ messaging and Integration Patterns provider, exposing a STOMP (Streaming Text Orientated Messaging Protocol) interface that allows clients to subscribe for some specific event of interest. We created a new *Python* plug-in module interfacing the above system by subscribing through the STOMP protocol for the occurrence of each specific error event of interest (e. g. a fault in MPI availability). These custom "sensors" simulate typical grid client behavior by using required grid service interface to perform a specific operation to be probed on a node (e.g. submit job, transfer file). In particular, the result of each sensor check is represented by a string containing the probed status. For example, if the probed service does not respond, the sensor returns status Unknown. If the service responds, the corresponding sensor analyzes the output and if it is correct, the returned status is Ok, otherwise it is set to Failure. Hence, when a new failure event results from a resource check, the associated Python script updates the corresponding information on the top-BDII index by adding a new entry containing the needed directives in an ldif file to be periodically passed through HTTP to the new top level information index itself. In such a way all the information indexes locally available on each participating sites remains unmodified while only the top-BDII index used for requirements matching by the WMS should be reconfigured. The whole operating scenario is depicted in fig. 3 below.

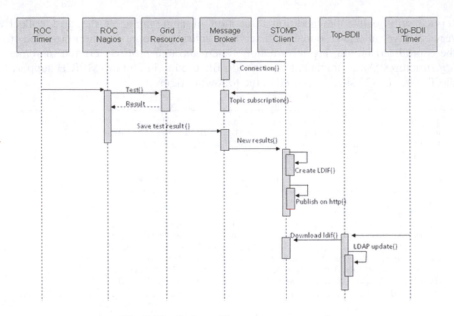

**Fig. 3.** The fault-avoidance strategy scenario

In the above sequence diagram the ROC Nagios service performs the sanity tests and sends the results to the Message Broker, that in turns reroutes them to the Stomp subsystem, if the involved topic has been previously subscribed by the Stomp client. When a fault is detected the Stomp client updates the *ldif* file. Periodically the TOP BDII downloads via http the latest *ldif* file and modifies its database accordingly.

## 5   Performance Evaluation and Results Analysis

In this section we present some performance evaluation experiences, based on real use cases within the EGI/EGEE community, aiming at demonstrating the effectiveness of the implemented architecture in providing reliability improvements from the user perspective.

### 5.1   Studying the Impact of Fault Avoidance for MPI

In the following tests, performed on the live EGI grid infrastructure, to avoid impacts on the production top-BDII, and to have a reference for comparison, we created a new LDAP collective information index server, with the same scope and visibility within the grid and following the same standard glue-schema (describing the Grid resources information stored in the information system), to be used in alternative to the production one for WMS "requirement" expressions matching. Then, in order to demonstrate the effectiveness of the proposed approach, we compared, by analyzing some specific status reporting objects, the content of the standard top-BDII egee-bdii.cnaf.infn.it implemented at the CNAF site in Bologna with the information provided by the alternative top-BDII service create in the EGI Naples site, grisu002.dsf.unina.it, taking into account the monitoring information to be used in our fault-avoidance strategy. We specifically observed the presence of potential faults in Message Passing Interface (MPI) library services. MPI is the most popular programming library for parallel computing available in the C, C++, and Fortran programming environments. Its diffusion within the European gLite-based Grids has greatly increased in the last years to support the emerging needs of scientific applications featuring a large degree of parallelism. The setup and configuration of MPI services on the generic worker nodes is a complex duty, especially when the site administrator has to manage different library implementation flavors, different programming environments and different underlying network fabrics. Moreover, the support of different MPI configurations is a requirement for many scientific communities and the online verification of a correctly working configuration is fundamental to avoid job aborts or abnormal behaviors. To do that, a site-level Nagios server in Naples periodically runs a proper plug-in comparing the number of sites that correctly provide the MPICH and OPENMPI capabilities in the two information systems. In the picture below (fig. 4) we show a 1-week time interval in which the red line represents the trend of standard top-BDII whereas the green one corresponds to the observations taken from the alternate top-BDII influenced by our strategy. As expected the number of sites supporting MPICH in the alternative top-BDII is always below the red line. In particular during the observation period our approach can successfully inhibit the selection of all the faulty resources from the information system.

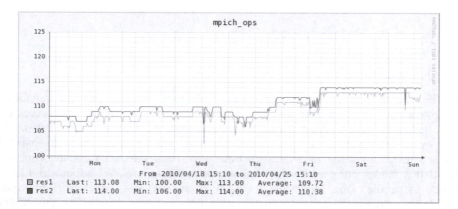

**Fig. 4.** Comparison between the number of MPI resources in the production top-BDII (red line) and on the modified top-BDII (green line)

A monthly observation showed that the number of sites, which support the different MPI flavours, is an extremely dynamic value, with an increasing trend. During the observation period we detected a maximum of 9 nodes that failed the MPICH test at the same time and a mean of 2 nodes in fault (1.8 theoretical value) with a standard deviation of 1.3. The main outcome of our analysis is that in the 98.2% of the time, we find in the standard top-BDII at least one resources that supports in some flavours of MPI but is in fault, and our approach is able to exclude these resources in real-time to improve the reliability from the user point of view, without excluding the resources from the information system. So the reliability is maintained without decrease the availability. In the monthly histogram in fig. 5 below we show the frequency of resources experiencing failures with the MPI, in particular we note that all it is working fine just in the 1.8% of case and we have at least 1 resource in fault in the 76.5% of the time, until more than 2 resource in fault is present in the 2.7% of the time.

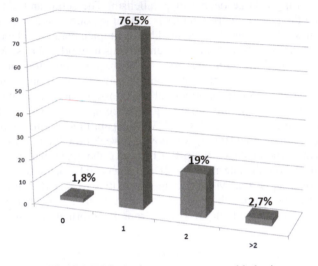

**Fig. 5.** MPI faults frequency on a monthly basis

## 5.2  Job Submission Test

In the second part of our performance evaluation we analyzed the reliability improvements from the user point of view, achieved in terms of jobs successful executed on the grid. In order to match a real use case, the evaluation tests, ran on the test virtual organization *matisse*, sharing about 10.000 CPUs dedicated to material science research on the EGI infrastructure, have been selected from the common simulation framework used by the local scientific community. We submitted a large bunch of parallel jobs, requesting the MPI library services, which perform basic simulation activities by using 16 processors at a time. Currently MPICH, OPENMPI and MVAPICH are the most popular MPI flavors, and whereas in the EGI production infrastructure more than 30% of the working sites publish the support for these libraries, a consistent part of them does not pass the basic Nagios sanity check.

We create an alternative WMS services that point to the modified top-BDII that delete in real-time the MPICH tag from the resource that fault the test. To analyze the improvements on the runtime stability of the grid directly perceivable by the users we at first calculated an approximation of the worst-case job execution failure rate when no fault avoidance strategy is implemented. By querying the production Top BDII services (the collective informative system) we discovered 49 resources publishing the MPICH support available within the *matisse* Virtual Organization. We then verified that only 36 resources between the above ones passed the MPICH Nagios test at job execution time, so that we determined the worst-case failure rate value according to the following formula:

$$\frac{\#\text{of MPI-enabled resources that failed the MPICH Probe}}{\#\text{of MPI-enabled resources available}} = \frac{13}{49} = 0.265$$

Starting from the above considerations we ran two sequences of tests needed to evaluate the introduced performance improvements on a significant number of jobs:

1.  Submission and analysis of 550 jobs on the Grid through the production WMS.
2.  Submission and analysis of 550 jobs on the Grid through by avoiding the failure resource by using the modified top-BDII.

During the first test sequence, we obtained a total of 144 job faults, with a failure rate equal 0.261, that seems to be in accordance with the calculated value. So we can argue that, in absence of any specific Requirements or Ranking tag in the JDL, an increasing load on the Grid produces a uniform usage of the available resources.

The aim of the second test block is quantifying the effective performance improvements introduced by our fault-avoidance strategy within the grid runtime system. First, by using the standard gLite matchmaking client (glite-wms-job-list-match) we observed that the Requirements expression in the JDL script worked correctly forcing the WMS to filter out the 13 resources, which failed the Nagios MPICH probe. After the complete execution of all the 550 jobs enforcing the MPI runtime status check through the JDL requirements mechanism, we detected a total of only 4 aborted jobs with a failure rate equal to 0.007, two order less than the previous case. The chart in figure 6 shows the percentage of jobs successfully completed when the fault-avoidance strategy is enforced, compared with the standard WMS behaviour.

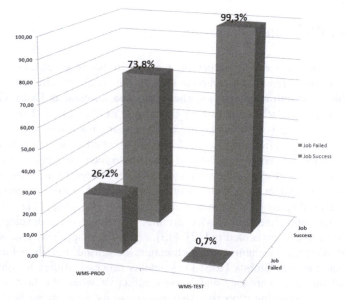

**Fig. 6.** Job execution histogram

The following picture (fig. 7) shows the failure behavior vs. the number of sent job compared in the two test series. The great difference that can be appreciated when our framework is operational gives some significant information about the improvements obtained in term of reliability and fault frequency.

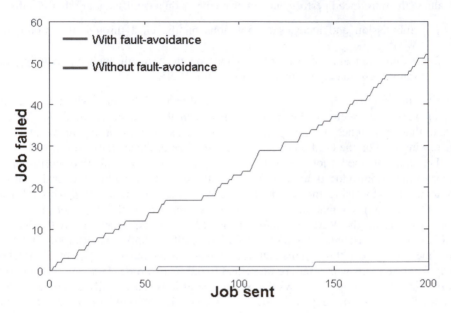

**Fig. 7.** The job failure rate

## 6   Related Work and Discussion

The interest in making Grid infrastructures fault-aware has received a certain attention in literature. In [6] is reported a complete state of the art about reliability support and strategies for Grid infrastructure together with the associated Open Grid Forum (OGF) efforts, particularly emphasizing the importance of the end user/application point of view that is at base of our approach. The same issue is also presented in [7] where failures in large grid infrastructure are demonstrated to be frequent phenomena. The aforementioned work proposes a fault-tolerance strategy based on the introduction of a novel meta-scheduler able to detect several Grid element failures, but does not implement any fault avoidance strategies. Other interesting works concerning reliability in Grid infrastructure are   [8], [9], [10], [11]. In particular, several specific fault detection service architectures have been developed for grid computing systems, as presented in [12], [13], [14]. Other applications adopted some ad hoc fault-tolerance mechanisms, which cannot be reused, nor shared among them [15]. For example, the Globus HBM [13] detects process failures by observing the received and missing heartbeats and is only effective when the host and network connections are functioning properly. Also, the above mechanism has been developed under the assumption that both the grid generic server and the heartbeat monitor run reliably. In any case all the above approaches require the introduction of new complex services and significant modifications to several crucial components of the grid architecture (i.e. the resource broker). On the other side, the approach presented in this paper is based on the specialized use of already available features and mechanisms with no impact on strategic middleware services. We only require the introduction of some simple, flexible and modular monitoring plug-ins each dedicated to specific sanity checks. These features make our fault avoidance solution implementable in a seamless and straightforward way in existing production grids. It can also be easily extended, by adding new plug-ins specialized for checking new services, as they are added to the grid middleware platform.

## 7 Conclusions

Grid reliability encompasses the concept of guaranteeing that all applications to be executed on the grid will fully succeed with no errors or exception conditions. Unfortunately, as grid applications scale to take advantage from the huge quantity of available resources, the probability of runtime faults due becomes no longer negligible and must be taken into account in all the grid operations, management and planning/evolution activities. The introduction of monitoring metrics on the Grid information system that can be used to easily implement a user-driven fault avoidance strategy opens up new interesting scenarios to increase the grid stability and facilitate operations in a production infrastructure. The proposed fault avoidance schema represents a good compromise in reducing the likelihood of erroneous or unsatisfactory results being received by the grid users, and hence enhancing the overall perceived and effective runtime performance, without introducing significant impacts on the various middleware flavors co-operating within the European grid production architecture. By implementing the above strategy in the EGI environment we obtained interesting performance improvements particularly perceivable from the user's point of view. This effect has been evidenced through a set of preliminary tests that showed the positive impact of the proposed approach in terms of number of jobs successfully completed. In future works we plan to deploy this approach in a large scale Grid and obtain further feedbacks in order to detect the best metrics to publish, to improve the data model and maximize the improvements on the infrastructure's stability and reliability. Finally, we plan to investigate the introduction of a site reputation concept as a new metric to be used during the resource discovery process.

## References

1. Laure, E., Hemmer, F., et al.: Middleware for the Next Generation Grid Infrastructure. In: Computing in High Energy and Nuclear Physics (CHEP), Interlaken, Switzerland (September 2004)
2. Frey, J., Tannenbaum, T., Livny, M., Foster, I., Tuecke, S.: Condor-G: A Computation Management Agent for Multi-Institutional Grids. Cluster Computing 5(3), 237–246 (2002)
3. Foster, I., Kesselman, C., Tuecke, S.: The Anatomy of the Grid. The International Journal of High Performance Computing Applications 15(3), 200–222 (2001)
4. The LCG Project, http://cern.ch/lcg

5. Pacini, F.: Job Description Language HowTo (2003),
   `http://server11.infn.it/workload-grid/docs/`
   `DataGrid-01-TEN-0142-02.pdf`
6. Dabrowski, C.: Reliability in grid computing systems, Concurrency and computation: pratice and experience, Published online in Wiley Inter Science (2009) doi:10.1002/cpe.1410, `http://www.interscience.wiley.com`
7. Huedo, E., Montero, R., Llorente, I.: Evaluating the reliability of computational Grids from the end user's point of view. Journal of Systems Architecture 52(12), 727–736 (2006)
8. Abawajy, H., Dandamudi, S.P.: Fault-tolerant grid resource management infrastructure. Journal of Neural, Parallel and Scientific Computations 12, 208–220 (2004)
9. Hwang, S., Kesselman, C.: Gridworkflow: A flexible failure handling framework for the grid. In: 12th International Symposium on High-Performance Distributed Computing (HPDC-12 2003), Seattle, WA, USA, pp. 126–137. IEEE Computer Society, Los Alamitos (2003)
10. Weissman, J.B.: Fault tolerant computing on the grid: What are my options? In: Proceedings of the The Eighth IEEE International Symposium on High Performance Distributed Computing, HPDC 1999, p. 26. IEEE Computer Society, Los Alamitos (1999)
11. Weissman, J.B.: Fault-tolerant wide area parallel computation. In: Proceedings of IDDPS 2000 Workshops, pp. 1214–1225 (2000)
12. Abawajy, J.H.: Fault detection service architecture for grid computing systems. In: Laganá, A., Gavrilova, M.L., Kumar, V., Mun, Y., Tan, C.J.K., Gervasi, O. (eds.) ICCSA 2004. LNCS, vol. 3044, pp. 107–115. Springer, Heidelberg (2004)
13. Stelling, P., Foster, I., Kesselman, C., von Laszewski, G., Lee, C.: A fault detection service for wide area distributed computations. In: Proc. 7th Symposium on High Performance Computing (HPDC), pp. 268–278 (1998)
14. Tierney, B., Crowley, B., Gunter, D., Holding, M., Lee, J., Thompson, M.: A monitoring sensor management system for grid environments. In: HPDC, pp. 97–104 (2000)
15. Soonwook, H.: A Generic Failure Detection Service for the Grid, Ph.D. thesis, University of Southern California (2003)

# Failure Detectors Encapsulate Fairness*

Scott M. Pike, Srikanth Sastry, and Jennifer L. Welch

Dept. of Computer Science and Engineering
Texas A&M University
College Station TX-77843 USA
{pike,sastry,welch}@cse.tamu.edu

**Abstract.** Failure detectors are commonly viewed as abstractions for
the synchronism present in distributed system models. However, inves-
tigations into the exact amount of synchronism encapsulated by a given
failure detector have met with limited success. The reason for this is that
traditionally, models of partial synchrony are specified with respect to
real time, but failure detectors do not encapsulate real time. Instead, we
argue that failure detectors encapsulate the *fairness* in computation and
communication. Fairness is a measure of the number of steps executed
by one process relative either to the number of steps taken by another
process or relative to the duration for which a message is in transit. We
argue that oracles are substitutable for the fairness properties (rather
than real-time properties) of partially synchronous systems. We propose
four fairness-based models of partial synchrony and demonstrate that
they are, in fact, the 'weakest systems models' to implement the canonical failure detectors from the Chandra-Toueg hierarchy.

## 1  Introduction

The inability to distinguish a crashed process from a slow process makes it
impossible to solve several classic problems in distributed computing in crash-
prone asynchronous systems [9]. Efforts to circumvent this impossibility have
spawned two complementary approaches. The first approach, called *partial synchrony* [8,7], focuses on assuming explicit temporal guarantees on computation
and communication to enable crash detection. The second approach focuses on
augmenting asynchronous systems with oracles, called *failure detectors* [3], that
provide potentially incorrect information about process crashes in the system.

It has long been held that failure detectors *encapsulate* partial synchrony. More
precisely, a failure detector $\mathcal{D}$ *encapsulates* a partially-synchronous system model
$M$ if and only if the following two conditions hold: (1) $\mathcal{D}$ can be implemented
in $M$, and (2) every problem $P$ that is solvable in system model $M$ is also
solvable in an asynchronous system augmented with $\mathcal{D}$. Alternatively (and more
informally), the notion of *encapsulation* by a failure detector may be viewed
synonymously with the notion of *mutual reducibility*; that is, a failure detector

* This work was supported in part by NSF grant 0964696 and Texas Higher Education
Coordinating Board grant NHARP 000512-0130-2007.

C. Lu, T. Masuzawa, and M. Mosbah (Eds.): OPODIS 2010, LNCS 6490, pp. 173–188, 2010.

$\mathcal{D}$ encapsulates a system model $M$ if and only if $M$ can implement $\mathcal{D}$, and $\mathcal{D}$ can implement $M$. As such, if $\mathcal{D}$ encapsulates $M$, then $\mathcal{D}$ is substitutable for $M$ because any problem solvable in $M$ is also solvable in asynchrony augmented with $\mathcal{D}$.

**Partial Synchrony.** A system model is partially synchronous [8] if it provides temporal bounds on computational and/or communicational quantities such as message delays and process speeds. The knowledge of such bounds may be incomplete or unknown. Despite such uncertainty, partial synchrony is useful for solving problems in crash-prone distributed systems, and several such models have been proposed in the literature(*e.g.*, [8,7,12,21,22,11,20]). These models vary in the information they provide about these bounds, and consequently they have different crash detection capabilities. One way to formalize this notion of crash detection capability is with failure detectors.

**Failure Detectors.** Informally, a failure detector [3] can be viewed as a system service (or oracle) that can be queried for (potentially unreliable) information about process crashes. The unreliable outputs of such oracles can be false positives (suspecting live processes) or false negatives (not suspecting crashed processes). From an empirical standpoint, most fault-tolerant problems in distributed computing that are otherwise unsolvable in crash-prone asynchronous systems can be solved by either (1) assuming adequate degrees of partial synchrony [8], or (2) augmenting asynchronous systems with sufficiently powerful oracles [15]. This observation suggests that the axiomatic properties of oracles might *encapsulate* the temporal properties of (suitably defined) models of partial synchrony. Accordingly, this conjecture has led to the pursuit of 'weakest system models' to implement various classes of oracles.

Current work on the weakest system models for oracles (see Sect. 2) has met with limited success partly because the proposed system models assume real-time bounds on communication (and possibly computation too). Unfortunately, failure detectors do not preserve such real-time bounds. To find such weakest system models, we need to address a more fundamental question: what precisely about partial synchrony do failure detectors preserve?

**Results.** We answer the foregoing question by demonstrating that failure detectors (at least when restricted to the Chandra-Toueg hierarchy [3]) encapsulate *fairness*: a measure of the number of steps executed by a process relative to other events in the system. We argue that oracles are substitutable for the fairness properties (rather than real-time properties) of partially synchronous systems. We propose four fairness-based models of partial synchrony and demonstrate that they are, in fact, the 'weakest systems models' to implement the canonical failure detectors from the Chandra-Toueg hierarchy in the presence of arbitrary number of crash faults.

**Significance.** Our results further the shift in the direction of oracular research away from real-time notions of partial synchrony (which have traditionally been understood with respect to events that are essentially external to the system) and towards fairness-based partial synchrony (which can be understood solely with respect to other events that are internal to the system). In fact, our results suggest that fairness is the currency for crash tolerance and research on weaker real-time bounds for crash tolerance should focus on enforcing appropriate fairness constraints on empirical systems relative to which known oracles can implemented.

**Organization.** We present related work in Sec. 2. Sect. 3 provides specifications for the asynchronous system model, the four failure detectors, and the four fairness-based partially-synchronous systems that we consider. Sects. 4–6 present the four equivalences between the failure detectors and the fairness-based partially-synchronous systems. We conclude with a discussion in Sect. 7.

# 2   Related Work

**The Chandra-Toueg Hierarchy.** Chandra and Toueg [3] introduced the following four popular oracles: (1) the *perfect failure detector* $\mathcal{P}$, which never suspects any process before the process crashes, after some (unknown) time permanently suspects all the crashed processes, and never transitions from suspecting a process to not suspecting that process; (2) the *eventually perfect failure detector* $\Diamond\mathcal{P}$, which after some (unknown) time stops suspecting correct processes and begins to permanently suspect all crashed processes; (3) the *strong failure detector* $\mathcal{S}$, which never suspects some correct process, after some (unknown) time permanently suspects all the crashed processes; (4) the *eventually strong failure detector* $\Diamond\mathcal{S}$, which after some (unknown) time stops suspecting some correct process and begins to permanently suspects all the crashed processes.

**Chasing the Weakest Model.** Among the aforementioned four Chandra-Toueg oracles, a significant amount of work focuses on $\Diamond\mathcal{P}$ and $\Diamond\mathcal{S}$. A line of work has focused on identifying the weakest system model assumptions that suffice for implementing these oracles. One approach is to weaken real-time constraints on synchrony, while another approach is to dispense with real-time altogether and instead constrain the relative ordering of certain events.

Under the first approach, the weakest real-time based message-passing model known to date that is sufficient to implement $\Diamond\mathcal{P}$ with arbitrary number of crashes guarantees that relative process speeds are bounded (while absolute speeds may remain unbounded above and below) [22] and that there exists an upper bound on the average delay over subsets of messages that are separated by bounded bursts of messages that may experience unbounded (or infinite) delay [21]. Similarly, the weakest message-passing model known to date that is sufficient to implement $\Diamond\mathcal{S}$ in the presence of up to $f$ process crashes guarantees that computation is synchronous and some correct process has $f$ timely outgoing links, although the set of timely links can vary over time [12].

Under the second approach, the weakest fairness-based message passing model known to date for implementing $\Diamond\mathcal{P}$ in environments with at least two correct processes are the $\Theta$-model [11] and the ABC model [20]. The $\Theta$-model bounds the ratio of the end-to-end communication delay of messages that are simultaneously in transit, while the ABC model imposes a restriction on the ratio of the number of messages that can be exchanged between pairs of processes in certain "relevant" segments of an asynchronous execution.

Similarly, for implementing $\Diamond S$, the weakest fairness-based message-passing model known to date was recently proposed by Raynal et al. [1] for systems consisting of $n$ processes with at most $f$ crash faults in which executions progress in "rounds" (the notion of a round is local to each process, not global), and processes send messages to all other processes in each round. A round terminates at a process when the process has received messages from $n - f$ processes for that round. The model guarantees that there exists some correct process $i$ such that eventually some subset of $f$ processes receive a message from $i$ in each of their rounds. Furthermore, this subset of $f$ processes can vary over time, but at all times such a subset exists.

An approach intermediate between the real-time-based and fairness-based approaches is presented by Biely et al. [2]. They prove equivalence (with respect to solvability of some problems) between some models and a set of oracles including $\Diamond S$ and $\Diamond P$. Although the transformations presented in [2] do not preserve bounds on real-time message delay, the authors claim that these bounds are preserved in a "relativistic" sense, but they do not expound on the interpretation of the term "relativistic". Our work formalizes the "relativistic" message delay as a form of *communicational fairness*.

Rajsbaum et al. [18,19] have tackled the problem of finding the weakest read-write shared memory model for implementing various kinds of oracles. They have shown that the power of so-called *limited scope* oracles can be expressed as restrictions on the number of read and write operations by each process in every round. These results are similar to ours in that they identify the power of oracles with some kind of "fairness". Our results differ from theirs in two ways. First, unlike [18,19] we investigate the exact synchronism in perpetually accurate oracles $P$ and $S$ (we are the first to do so). Second, we consider message-passing systems instead of shared-memory systems. In fact, message-passing systems merit separate investigation because results regarding oracles in shared-memory models in general do not carry over to message-passing models. For instance, the weakest oracle for solving wait-free consensus in asynchronous shared-memory is not the same as that for asynchronous message-passing [14,5].

## 3   Definitions

### 3.1   Asynchronous System Model

The asynchronous system [10] consists of a finite set of processes $\Pi$ which can communicate with each other by reliable communication channels. We consider the standard asynchronous system model [10], but with *correct-reliable* channels, and we assume that an arbitrary number of processes can crash. A concise description follows. The detailed description of the system model specifications is available in the full version of the paper at [16] and has been omitted here due to space limitations.

While reliable channels deliver every message sent without duplication or corruption, we consider a weaker form of channels called *correct-reliable* channels which are guaranteed to be reliable while both the sender and the receiver are live. Therefore, if the sender and the receiver are correct, then the channel connecting them is guaranteed to be reliable. Otherwise, the channel is guaranteed to behave like a reliable channel until either the sender or the receiver crashes. Afterwards, the channel is allowed to drop the messages in transit (but not corrupt any messages).

We posit the existence of a discrete global time base whose range of values is the natural numbers $\mathbb{N}$. Global time is used to mark or count the events that occur in the system, and it is not used to measure the real-time duration between two events. Therefore, the real-time duration between consecutive ticks of the global time may be arbitrary, but finite. In the remainder of this paper, 'time' will refer to global time unless explicitly stated otherwise.

Processes execute actions in *atomic steps*. In an atomic step, a process receives at most one message from each process, makes a state transition, and sends at most one message to each process. A *run* consists of an infinite sequence of steps taken by processes while executing an algorithm.

We consider only crash faults. That is, a process can fail only by *crashing*, which occurs when a process ceases execution without warning; a crashed process never resumes execution. Any process that is not crashed is considered to be *live*. In each run, processes are either *correct* or *faulty*. Correct processes execute actions according to their algorithm specification, and never fail, whereas *faulty* processes fail after finite time. In all runs, a fault process takes only finitely many steps whereas correct processes take infinitely many steps.

In order to demonstrate our results, we consider two variations of the asynchronous system. In the first variation, we assume that the asynchronous system is augmented with a failure detector. In such systems, the state transition function of each process also considers the output of its local failure detector module before determining the new state of the process and the set of messages to be sent. Sect. 3.2 describes the failure detectors considered in this paper.

In the second variation, we assume that there are certain constraints on the relative ordering of the atomic steps by different processes in the systems. Such constraints determine the *fairness* properties satisfied by runs of these systems. These constraints are described in Sect. 3.3.

## 3.2   Failure Detectors

The formal definitions of the Chandra-Toueg failure detectors are provided in [3]. Informally, failure detectors are characterized by the kind and degree of unreliability of their output which is a set of suspected processes. Here, we consider four classes of failure detectors from the original Chandra-Toueg hierarchy [3]: *Perfect failure detector* (denoted $\mathcal{P}$), *Strong failure detector* (denoted $\mathcal{S}$), *Eventually Perfect failure detector* (denoted $\Diamond\mathcal{P}$), and *Eventually Strong failure detector* (denoted $\Diamond \mathcal{S}$). All four aforementioned failure detectors guarantee that *eventually every faulty process is permanently suspected by every correct process.* Additionally, the four failure detectors satisfy the following properties:

- $\mathcal{P}$ ensures that no process suspects any live process.
- $\mathcal{S}$ ensures that *some* correct process is never suspected.
- $\Diamond \mathcal{P}$ ensures that correct processes are eventually never suspected.
- $\Diamond \mathcal{S}$ ensures that *some* correct process eventually is never suspected.

### 3.3   Fairness Constraints

We claim that Chandra-Toueg failure detectors encapsulate *fairness* guarantees of the underlying system. Such fairness is of two kinds: *computational* and *communicational*. *Computational fairness* restricts the number of steps executed by processes relative to each other. *Communicational fairness* restricts the number of steps executed by the recipient of a message while that message is in transit.

**Computational Fairness.** A common specification for computational fairness is *bounded relative process speeds* [8] which states that the system has a bound $\Phi$ on relative process speeds if in all intervals where some process $i$ takes $\Phi+1$ steps, then all the processes not crashed in that interval are guaranteed to take at least 1 step. Note that this fairness property is symmetric; that is, if $i$'s process speed is bounded relative to $j$'s process speed, then *vice versa* is true as well. However, it is possible to define computational fairness properties that are asymmetric.

Consider our definition of proc-fairness. A process $i$ is said to be *k-proc-fair* (where $k$ is a non-negative integer) in an infinite suffix $\gamma$ of a run $\alpha$, if, for all processes $j \in \Pi$, in every segment of $\gamma$ in which $j$ takes $k + 1$ steps, either (1) $i$ takes at least one step, or (2) $i$ is crashed. Note that $i$ being $k$-proc-fair with respect to $j$ does not imply $j$ being $k$-proc-fair with respect to $i$. As such, proc-fairness is an asymmetric fairness property.

We extend this notion of proc-fairness as follows:

- *k-proc-distinguished*: A process $i$ is said to be *k-proc-distinguished* in run $\alpha$ if $i$ is $k$-proc-fair in all suffixes of $\alpha$
- *Eventually k-proc-distinguished*: A process $i$ is said to be *eventually k-proc-distinguished* in $\alpha$ if there exists a prefix of $\alpha$ such that, in the infinite suffix of $\alpha$ that follows, $i$ is $k$-proc-fair.

Like proc-fairness, the property of being proc-distinguished in asymmetric as well. While other processes may be 'fair' with respect to a proc-distinguished process $i$, process $i$ need not be fair with respect to other processes; *i.e.*, a proc-distinguished process may take an unbounded number of steps in the duration between a non-proc-distinguished process' two consecutive steps. This is an important distinction between computational fairness and bounded relative process speeds defined in [8,7]. Bounded relative process speeds may be viewed as a special case where every process is (eventually) $k$-proc-distinguished.

**Communicational Fairness.** Specifying temporal bounds on communication delay in terms of fairness is not straightforward. For a process $i$ to satisfy communicational fairness, it is necessary that $i$ not take 'too many steps' while a message $m$ is en route to $i$. However, there is one exception: if the sender of $m$ crashes while $m$ is in transit to $i$, then $i$ can take an arbitrary number of steps before $m$ is delivered. In fact, $m$ may even be dropped.

We capture the above intuition through the following definition for a *com-fair* process. A process $i$ is said to be *d-com-fair* (where $d$ is a non-negative integer) in a suffix $\gamma$ of a run $\alpha$, if, for all processes $j \in \Pi$, for each message $m$ sent from $i$ to $j$ in $\gamma$, during the segment of $\gamma$ starting from the configuration in which $m$ is sent and ending with the configuration in which $m$ is received, either (1) $j$ takes no more than $d$ steps, or (2) $i$ is crashed.

We extend this notion of com-fairness as follows:

- *d-com-distinguished*: A process $i$ is said to be *d-com-distinguished* in run $\alpha$ if $i$ is $d$-com-fair in all suffixes of $\alpha$.
- *Eventually d-com-distinguished*: A process $i$ is said to be *eventually d-com-distinguished* in run $\alpha$ if there exists a prefix of $\alpha$ such that, in the infinite suffix of $\alpha$ that follows, $i$ is $d$-com-fair.

Recall that in traditional partially-synchronous models [8,7] the bounds on message delay are measured as the number of steps taken by the *sender*. In contrast, our bounds on communicational fairness are measured as the number of steps taken by the receiver, for the following reason. Since these traditional models assume that relative process speeds are bounded, if some live process takes a bounded number of steps while a message is in transit, then all processes take a bounded number of steps while that message is in transit. Hence, asserting the existence of a bound on the number of steps by the sender is equivalent to asserting the existence of a bound on the number of steps by the recipient in the same time interval. In our case, since computational fairness is not a symmetric property, a bound on the number of steps by the sender need not translate to a bound on the number of steps by the receiver in the same time interval. Consequently, we denominate communicational fairness as the number of steps taken by the recipient.

Furthermore, we bound the number of steps taken by the recipient only while the sender is live for the following reason. While the sender is not crashed, it can successfully maintain an operational communication link with the recipient, and the link can ensure that messages are delivered before the recipient takes 'too many steps'. However, if the sender crashes, the link is no longer guaranteed to stay operational, and no guarantees can be provided on message delay and delivery.

### 3.4 Fairness-Based Partially-Synchronous System Models

In this subsection we present four fairness-based partially-synchronous systems models that represent the fairness encapsulated by the four Chandra-Toueg failure detectors specified in Sect. 3.2.

1. *All Fair* $(\mathcal{AF})$ is an asynchronous system model where: in every run, all processes are both $k$-proc-distinguished and $d$-com-distinguished, for known $k$ and $d$.

2. *Some Fair* ($\mathcal{SF}$) is an asynchronous system model where: in every run, some correct process $i$ is both $k$-proc-distinguished and $d$-com-distinguished, for known $k$ and $d$.

3. *Eventually All Fair* ($\Diamond\mathcal{AF}$) is an asynchronous system model where: for each run, there exists a (potentially unknown) time after which the system behaves like $\mathcal{AF}$.

4. *Eventually Some Fair* ($\Diamond\mathcal{SF}$) is an asynchronous system model where: for each run, there exists a (potentially unknown) time after which the system behaves like $\mathcal{SF}$.

## 4     Methodology

We claim that the Chandra-Toueg oracles encapsulate fairness (and not real-time) properties of the underlying system. We will show that the amount of fairness encapsulated by these oracles is specified by the aforedescribed fairness-based system models. In a precise sense, $\mathcal{AF}$, $\mathcal{SF}$, $\Diamond\mathcal{AF}$, and $\Diamond\mathcal{SF}$ specify the exact amount of fairness encapsulated by $\mathcal{P}$, $\mathcal{S}$, $\Diamond\mathcal{P}$, and $\Diamond\mathcal{S}$, respectively. Alternatively, it can be said that in environments where an arbitrary number of processes may crash, $\mathcal{AF}$, $\mathcal{SF}$, $\Diamond\mathcal{AF}$, and $\Diamond\mathcal{SF}$ are the 'weakest' system models to implement $\mathcal{P}$, $\mathcal{S}$, $\Diamond\mathcal{P}$, and $\Diamond\mathcal{S}$, respectively.

The methodology used to establish the above equivalence is as follows. First, we present a construction (described in Sect. 5) that uses a Chandra-Toueg oracle in an otherwise asynchronous system to schedule distributed applications such that each process executes its application steps 'fairly' with respect to other processes (and messages). The fairness properties guaranteed by the scheduler depend on the available failure detector. By employing $\mathcal{P}$, $\mathcal{S}$, $\Diamond\mathcal{P}$, or $\Diamond\mathcal{S}$, the scheduler provides fairness guarantees specified by $\mathcal{AF}$, $\mathcal{SF}$, $\Diamond\mathcal{AF}$, or $\Diamond\mathcal{SF}$, respectively. This shows that the failure detectors encapsulate at least as much fairness as is specified in the corresponding fairness-based system models. Next, we present an algorithm (described in Sect. 6) which implements a Chandra-Toueg oracle on top of these fairness-based systems. When this algorithm is deployed in $\mathcal{AF}$, $\mathcal{SF}$, $\Diamond\mathcal{AF}$, or $\Diamond\mathcal{SF}$, it implements $\mathcal{P}$, $\mathcal{S}$, $\Diamond\mathcal{P}$, or $\Diamond\mathcal{S}$, respectively. Thus, we show that these failure detectors encapsulate no more guarantees on fairness than what is provided by the corresponding fairness-based systems.

## 5     Extracting Fairness

In this section, we present a distributed scheduler that 'extracts' the fairness encapsulated by the Chandra-Toueg failure detectors. The algorithm presented is a universal construction for the Chandra-Toueg hierarchy in the sense that depending on the failure detector used by the algorithm, the appropriate fairness guarantees are provided by the distributed scheduler. For simplicity, we assume that the application at each process always has some enabled step that it can take. Therefore, the local scheduler module is always in one of two states:

*waiting* and *active*. When the scheduler module is *waiting*, the associated application module is not enabled to take steps. Upon becoming *active*, the scheduler module enables the associated application module to execute a single step and then the scheduler goes back to *waiting*. Additionally, the distributed scheduler 'intercepts' and forwards all the communication among the application modules.

The properties to be satisfied by the distributed scheduler are *local progress* and *fairness*. *Local progress* states that every correct process must be scheduled to execute its application steps infinitely often, regardless of process crashes in the system. *Fairness* properties are as follows:

If the distributed scheduler uses:

 - $\mathcal{P}$, then the distributed scheduler implements the $\mathcal{AF}$ system model.
 - $\mathcal{S}$, then the distributed scheduler implements the $\mathcal{SF}$ system model.
 - $\Diamond\mathcal{P}$, then the distributed scheduler implements the $\Diamond\mathcal{AF}$ system model.
 - $\Diamond\mathcal{S}$, then the distributed scheduler implements the $\Diamond\mathcal{SF}$ system model.

### 5.1   Interface between Scheduler and Application

The scheduler provides three interfaces for an application module to interact with the local scheduler module and the application modules at other processes: *executeAPP()*, *receiveAPP()*, and *sendAPP()*. These interfaces are specified in Alg. 1.2 and described below.

The scheduler enables the application to take a step by invoking *executeAPP()* and in response, the application takes a single atomic step. If multiple actions of the application are enabled to be executed, then the scheduler is assumed to make a non-deterministic choice among the enabled actions subject to the constraint of weak fairness (which states that a continuously enabled action is eventually executed).

The application receives messages sent by other processes by invoking *receiveAPP()*. The scheduler at each process $i$ takes all the messages destined for the application module at $i$ and stores them locally in a *receive buffer*. When the application invokes *receiveAPP()*, the scheduler returns the contents of the *local receive buffer* to the application.

The application sends messages by invoking the *sendAPP()* interface. While taking a step, if the application at process $i$ invokes *sendAPP()*, the scheduler at $i$ stores all the messages that the application wants to send to all the processes in a *local send buffer*. The scheduler module at $i$ then sends the messages to destination process where they are stored in the *receive buffers* of the corresponding scheduler modules (in Actions 5 and 6 of Alg. 1.1 as described in Sect. 5.2). These messages are then received by the respective recipient processes when the latter invoke *receiveAPP()*.

### 5.2   Algorithm Description

The algorithm in Algs. 1.1 and 1.2 implements a distributed scheduler with dynamic heights (or priorities) and permits. Alg. 1.1 shows the actions of the

enum $\{waiting, active\}$ : $s_i$.state $\leftarrow$ waiting    *State variable is initially set to waiting*
**integer** $s_i$.ht $\leftarrow 0$                                              *The height of process i*
$\forall j \in \Pi - \{i\}$ :
  **boolean** $s_i$.permit$_j$ $\leftarrow$ $(i.id > j.id)$    *Process with higher id holds the shared permit*
  **boolean** $s_i$.req$_j$ $\leftarrow$ $(i.id < j.id)$  *Process with lower id holds the shared request token*
  **integer** $s_i$.ht$_j$ $\leftarrow 0$                          *Process i's view of the height of process j*
  **integer** $s_i$.seq$_j$ $\leftarrow 0$    *Generates a new sequence number to solicit messages from j*
  **integer** $s_i$.maxAck$_j$ $\leftarrow 0$    *The highest seq. no. among the messages received from j*
  **set** $s_i$.send_buffer$_j$        *The send buffer through which apps. at i send messages to j*
  **set** $s_i$.receive_buffer$_j$    *The receive buffer from which apps. at i receive msgs. from j*
**boolean** $s_i$.permit$_j$ $\leftarrow$ **true**        *Process always holds its own the shared permit*

---

 1 :  $\{s_i$.state $= waiting\}$ $\longrightarrow$                                              *Action 1*
 2 :      $\forall j \in \Pi - \{i\}$ **where** $s_i$.req$_j \wedge \neg s_i$.permit$_j$ **do**        *Request permit*
 3 :          **send** $\langle request, s_i$.ht$\rangle$ **to** $s_j$; $s_i$.req$_j \leftarrow false$

---

 4 :  $\{$**upon receive** $\langle request, ht\rangle$ **from** $s_j\}$ $\longrightarrow$                  *Action 2*
 5 :      $s_i$.req$_j \leftarrow true$                                *Send permit if $s_i$ is waiting*
 6 :      $s_i$.ht$_j \leftarrow ht$                                  *and $s_j$ has higher priority*
 7 :      **if** $(s_i$.permit$_j \wedge (s_i$.state $= waiting) \wedge ((ht > s_i$.ht$) \vee ((ht = s_i$.ht$) \wedge (i < j))))$
 8 :          **send** $\langle permit, s_i$.ht$\rangle$ **to** $s_j$; $s_i$.permit$_j \leftarrow false$

---

 9 :  $\{$**upon receive** $\langle permit, ht\rangle$ **from** $s_j\}$ $\longrightarrow$                  *Action 3*
10 :      $s_i$.permit $\leftarrow true$                                *Send permit if $s_i$ is waiting*
11 :      $s_i$.ht$_j \leftarrow ht$                                  *and $s_j$ has higher priority*
12 :      **if** $(s_i$.req$_j \wedge (s_i$.state $= waiting) \wedge ((ht > s_i$.ht$) \vee ((ht = s_i$.ht$) \wedge (i < j))))$
13 :          **send** $\langle permit, s_i$.ht$\rangle$ **to** $s_j$; $s_i$.permit$_j \leftarrow false$

---

14 : $\{(s_i$.state $= waiting) \wedge (\forall j \notin \mathcal{D} :: s_i$.permit$_j)\}$ $\longrightarrow$ *Action 4 (Note: $\mathcal{D}$ is queried)*
15 :      $s_i$.state $\leftarrow active$        *Active upon holding permits from trusted processes*
16 :  **foreach** $j$ **in** $\Pi - \{i\}$
17 :          increment $s_i$.seq$_j$ by 1    *Generate a new seq. no. to tag a request message*
18 :          **send** $\langle reqMsg, s_i$.seq$_j\rangle$ **to** $s_j$    *Send a request message to all processes*

---

19 : $\{$**upon receive** $\langle reqMsg, num\rangle$ **from** $s_j\}$ $\longrightarrow$                  *Action 5*
20 :      $msgSet \leftarrow s_i$.send_buffer$_j$                    *Received a mesg request.*
21 :      $s_i$.send_buffer$_j \leftarrow \emptyset$
22 :  **send** $\langle msgSet, num\rangle$ **to** $s_j$        *Send the contents of the local send buffer*

---

23 : $\{($**upon receive** $\langle msgSet', num\rangle$ **from** $s_j)\}$ $\longrightarrow$                  *Action 6*
24 :      $s_i$.receive_buffer$_j \leftarrow s_i$.receive_buffer$_j \cup msgSet'$    *Add to local receive buffer*
25 :      $s_i$.maxAck$_j \leftarrow max(num, s_i$.maxAck$_j)$        *Update max. ack receive so far.*

---

26 : $\{(s_i$.state $= active) \wedge (\forall j \in \Pi - \{i\} :: ((s_i$.maxAck$_j = s_i$.seq$_j) \vee (j \in \mathcal{D})))\}$ $\longrightarrow$
                                *Action 7 (Note that the failure detector $\mathcal{D}$ is queried)*
27 :  $executeAPP()$      *Execute an app. step; executeAPP() is specified in Alg. 1.2*
28 :  $s_i$.ht $\leftarrow \min(\forall j \in \Pi - \{i\} :: s_i$.ht$_j, s_i$.ht$) - 1$    *Reduce height below all neighbors*
29 :  $\forall j \in \Pi - \{i\}$ **where** $(s_i$.permit$_j)$                *do whose height is known.*
30 :      **send** $\langle permit, s_i$.ht$\rangle$ **to** $s_j$; $s_i$.permit$_j \leftarrow false$    *Send all held permits*
31 :  $s_i$.state $\leftarrow waiting$        *Exit the active state after executing an app. step*

**Alg. 1.1.** Actions for scheduler at process $i$

---

**procedure** $executeAPP()$
   execute an enabled application action in which
      application action invokes $receiveAPP()$ to receive messages
      application action invokes $sendAPP(m, j)$ to send message $m$ to process $j$

---

**procedure** $receiveAPP()$
   $returnValue \leftarrow \cup_{\forall j \in \Pi - \{i\}} \{(s_i.\mathsf{receive\_buffer}_j, j)\}$
   $\forall j \in \Pi - \{i\}$ **do** $s_i.\mathsf{receive\_buffer}_j \leftarrow \emptyset$
   **return** $returnValue$

---

**procedure** $sendAPP(m, j)$
   $s_i.\mathsf{send\_buffer}_j \leftarrow s_i.\mathsf{send\_buffer}_j \cup \{m\}$

**Alg. 1.2.** Interaction between the scheduler and the application

scheduler and Alg. 1.2 shows the interface between the scheduler and the scheduled application. The idea of dynamic heights and permits (also called *forks*) is borrowed from the algorithms to solve the dining philosophers problem in [17]. All the processes are assigned a static id and all the ids are known to all the processes in the system.

In Alg. 1.1 each process $i$ has the following variables: $s_i.\mathsf{state}$ which determines if the process is *waiting* or *active*. The height of a process is stored in the variable $s_i.\mathsf{ht}$ which is initially 0. For each process $j$ in the system, $i$ maintains the variables: (a) $s_i.\mathsf{permit}_j$ to determine if the permit shared with $j$ is currently held by $i$, (b) $s_i.\mathsf{req}_j$ to determine if the request token to request a permit from $j$ is currently at $i$, and (c) $s_i.\mathsf{ht}_j$ which stores the last received value of $j$'s height (in permits and request messages).

All processes start in the *waiting* state with the permits at higher-id processes and request tokens at lower-id processes. For a *waiting* process to become *active*, it must collect all its shared permits. A *waiting* process requests missing permits in Action 1. Upon receiving such a request in Action 2, the process determines if the request should be honored based on the following condition: if the process is *waiting*, holds the shared permit, and the requesting process has greater height (or equal height and higher process-id), then the process relinquishes the permit. Otherwise the process simply holds the token and defers sending the permit if the permit is present.

Upon receiving a permit in Action 3, the process again determines if the permit should be kept (to be sent later) or sent based on the same condition mentioned previously.

When a *waiting* process (say) $i$ receives shared permits from all the processes not suspected by the failure detector $\mathcal{D}$, $i$ becomes *active* in Action 4. Upon becoming *active*, $i$ sends an application-message request (denoted $\langle reqMsg \rangle$) with a new sequence number ($s_i.\mathsf{seq}_j$) to each process $j$ in the system. Upon receiving such a message in Action 5, process $j$ sends the contents of its *local send buffer* appended with the sequence number in response. Process $i$ receives such a message sent by $j$ in Action 6; process $i$ adds the contents of the received message to its *local receive buffer* and updates its local state to reflect the latest sequence number for which $i$ has received a response from $j$ (stored in $s_i$.macAck$_j$). Upon receiving responses from all trusted processes for the $\langle reqMsg \rangle$ messages sent with the latest sequence number (that is, $s_i$.seq$_j = s_i$.maxAck$_j$ for all $j$ trusted by $i$), Action 7 is enabled at process $i$. In Action 7, process $i$ invokes $executeAPP()$ to execute an application step before exiting.

This mechanism of receiving application messages before invoking $execute$ $APP()$ ensures that an active process $i$ 'waits on' all the messages sent by a correct and trusted process $j$, thus guaranteeing that a correct and trusted process $j$ is also a com-distinguished process.

When a process executes an application step, the application invokes $receive$ $APP()$ described in Alg. 1.2 to receive all the messages in the *local receive buffer*, and the application action sends messages by invoking $sendAPP()$ described in Alg. 1.2 which simply adds the message to the *local send buffer*.

Eventually, the process exits its *active* state by reducing it height below all processes (whose shared permits it holds), sends all the permits away and transits to *waiting* in Action 7.

Relinquishing the shared permits before *waiting* ensures that a correct and trusted process receives permits from other processes every time the other processes take an application step. The reduction in height ensures that the process does not 'steal' the permits (by sending a request token with greater height) after relinquishing them. This allows a correct and trusted process to become a proc-distinguished process as well.

The proof of correctness is available in the full version of the paper at [16] and has been omitted here due to space limitations.

## 6 Extracting Chandra-Toueg Failure Detectors from Fairness-Based Systems

In this section we present an algorithm that implements the failure detectors $\mathcal{P}$, $\mathcal{S}$, $\Diamond\mathcal{P}$, and $\Diamond\mathcal{S}$ in the system models $\mathcal{AF}$, $\mathcal{SF}$, $\Diamond\mathcal{AF}$, and $\Diamond\mathcal{SF}$, respectively. This result combined with the result in Sect. 5 shows that $\mathcal{AF}$, $\mathcal{SF}$, $\Diamond\mathcal{AF}$, and $\Diamond\mathcal{SF}$ have the minimal synchronism necessary to implement $\mathcal{P}$, $\mathcal{S}$, $\Diamond\mathcal{P}$, and $\Diamond\mathcal{S}$, respectively. The algorithm is as follows:

The failure detector module at each process $i$ maintains a timer timerValue$_j$ for each process $j$ in the system which counts down from $k + d$ to 0, where the bounds on fairness in the system models of Sect. 3.4 are specified by the existence of $k$-proc-distinguished and $d$-com-distinguished processes. Every time process $i$ takes a step, it receives zero or more messages from all other processes, decrements the value of timerValue$_j$ by 1, and sends a heartbeat to each process $j$ in the system. If $i$ receives a heartbeat from $j$, then $i$ trusts $j$ and resets the value of timerValue$_j$ to $k + d$. If timerValue$_j$ is decremented to 0, then $i$ suspects $j$. The pseudo-code for the algorithm is given in Alg. 1.3.

The proof of correctness is available in the full version of the paper at [16] and has been omitted here due to space limitations.

constant timeOut $\leftarrow k + d$
set suspectList $\leftarrow \emptyset$
$\forall j \in \Pi - \{i\}$ :
   integer timerValue$_j$ $\leftarrow$ timeOut

| | | |
|---|---|---|
| 1 : | $\{true\} \longrightarrow$ | *Action 1* |
| 2 : | **receive** $\langle msgSet \rangle$ | *Receives zero or more messages from each process* |
| 3 : | $\forall j \in \Pi - \{i\}$ **do** | |
| 4 : | **send** $\langle HB \rangle$ **to** $j$ | *Send a heartbeat to each process* |
| 5 : | **if** $(\langle HB, j \rangle \in msgSet)$ | |
| 6 : | suspectList $\leftarrow$ suspectList $- \{j\}$ | *Trust upon receiving a heartbeat* |
| 7 : | timerValue$_j$ $\leftarrow$ timeOut | *Reset timer* |
| 8 : | **if** (timerValue$_j = 0$) | |
| 9 : | suspectList $\leftarrow$ suspectList $\cup \{j\}$ | *Suspect upon timer expiry* |
| 10 : | timerValue$_j$ $\leftarrow$ max(timerValue$_j - 1, 0$) | *Decrement timer for each process* |

**Alg. 1.3.** Implementing Chandra-Toueg Oracles In System Models Where (Some) Processes are $k$-Proc-Distinguished and $d$-Com-Distinguished

## 7  Discussion

**Complete Synchrony and $\mathcal{P}$.** It was first noted in [4] that there exist time-free problems solvable in synchronous systems that are unsolvable with $\mathcal{P}$. This points to a 'gap in the synchronism' between $\mathcal{P}$ and the synchronous system. The following corollary of our results explains this gap.

$\mathcal{AF}$ — the weakest system model to implement $\mathcal{P}$ — is extremely similar to the synchronous system model with message delay being denominated in recipient's steps in the former and in real time in the latter. However, there is one significant difference. $\mathcal{AF}$ ensures full synchrony for all messages as long as the senders are live. When a sender crashes, $\mathcal{AF}$ 'loses synchronism' for all the sender's messages that are still in transit. On the other hand, synchronous systems ensure the synchronism for these messages as well. This difference in the behavior between $\mathcal{AF}$ and synchronous systems is the 'gap in synchronism' between the perfect failure detector $\mathcal{P}$ and synchronous systems. To our knowledge, we are the first to characterize this gap.

**On Solving Consensus.** It is well known that $\Diamond \mathcal{S}$ is the weakest failure detector to solve consensus in asynchronous systems with a majority of correct processes [14], and we have shown that $\Diamond \mathcal{SF}$ is the weakest fairness-based system model to implement $\Diamond \mathcal{S}$. Does that mean $\Diamond \mathcal{SF}$ is the weakest system model to solve consensus? The answer is *no*. While $\Diamond \mathcal{S}$ is the weakest to solve consensus only in majority-correct environments, $\Diamond \mathcal{SF}$ is the weakest to implement $\Diamond \mathcal{S}$ in all environments. This obervationsuggests that there is a weaker system model which can implement $\Diamond \mathcal{S}$ in majority-correct environments, but not in all environments.

**Open Questions.** We have argued that several oracles encapsulate fairness in executions and provided evidence by demonstrating that the oracles in the

Chandra-Toueg hierarchy encapsulate such fairness constraints. This opens a larger question: *do all oracles encapsulate fairness?* The answer is arguably *no*. Notable candidates for counterexamples include the failure detectors proposed in [13] whose output can be arbitrary and need not provide semantic information about process crashes alone. This presents another question: *what set of oracles do encapsulate fairness?* This question is open even when restricted to the extended Chandra-Toueg hierarchy (which include oracles like $\mathcal{T}$ [6], and other parametric oracles like the ones in [2,19]). If it turns out that all oracles that output process ids do encapsulate fairness, then it provides us with a clean hierarchy of fairness-based system models that mirrors the extended Chandra-Toueg hierarchy. On the other hand, if we discover that there exist oracles within the extended Chandra-Toueg hierarchy that do not encapsulate fairness, then the implication is that these oracles encapsulate something other than fairness. Knowledge of this other encapsulated information could help in designing crash tolerant systems.

Another consequence of oracles encapsulating fairness is that fault environments might encapsulate fairness as well. Recall that the weakest oracles sufficient to solve problems in distributed systems vary depending on the number of processes that may crash. For instance, consider fault-tolerant consensus. Recall that $\Diamond\mathcal{S}$ is the weakest to solve the problem only in majority-correct environments [14]. In environments where an arbitrary number of processes may crash, the weakest failure detector for the problem is a stronger oracle $(\Diamond\mathcal{S}, \Sigma)$ [5]. Given that $\Diamond\mathcal{S}$ encapsulates some fairness constraints, and $\Sigma$ can be implemented in an asynchronous system with majority correct, we conjecture that $\Sigma$ and majority-correct encapsulate equivalent fairness constraints in the system. Furthermore, this implies that fairness is also encapsulated by constraints on the number of processes that may crash in the system. Based on the above observations and arguments, consider the following question: Is fairness a more general primitive to understand crash fault tolerance in distributed systems? That is, can fairness unify the different weakest failure detector results for the same problem in different fault environments?

Much effort is spent pursuing the 'weakest' real-time-based models to implement certain oracles (like $\Omega$, $\Diamond\mathcal{P}$, and such) for two reasons: (1) bounds in many empirical distributed systems are specified with respect to real time, and (2) these oracles are known to be the weakest to solve many problems in distributed computing. However, given the dependence of the weakest-oracle results on the fault environment, and the conjecture that fault environments themselves could encapsulate fairness, it is perhaps beneficial to investigate the 'weakest' real-time-based models to guarantee appropriate fairness constraints (rather than oracles) so that these constraints can then be encapsulated by various combinations of oracles and fault environments.

**Acknowledgment.** We would like to thank the anonymous reviewers and Michel Raynal for their valuable comments and suggestions to improve this paper.

# References

1. Anta, A.F., Raynal, M.: From an asynchronous intermittent rotating star to an eventual leader. IEEE Transactions on Parallel and Distributed Systems 21(9), 1290–1303 (2010), http://dx.doi.org/10.1109/TPDS.2009.163
2. Biely, M., Hutle, M., Penso, L.D., Widder, J.: Relating stabilizing timing assumptions to stabilizing failure detectors regarding solvability and efficiency. In: Masuzawa, T., Tixeuil, S. (eds.) SSS 2007. LNCS, vol. 4838, pp. 4–20. Springer, Heidelberg (2007), http://dx.doi.org/10.1007/978-3-540-76627-8_4
3. Chandra, T.D., Toueg, S.: Unreliable failure detectors for reliable distributed systems. Journal of the ACM 43(2), 225–267 (1996), http://dx.doi.org/10.1145/226643.226647
4. Charron-Bost, B., Guerraoui, R., Schiper, A.: Synchronous system and perfect failure detector: solvability and efficiency issues. In: International Conference on Dependable Systems and Networks, pp. 523–532 (2000), http://doi.ieeecomputersociety.org/10.1109/ICDSN.2000.857585
5. Delporte-Gallet, C., Fauconnier, H., Guerraoui, R., Hadzilacos, V., Kouznetsov, P., Toueg, S.: The weakest failure detectors to solve certain fundamental problems in distributed computing. In: Proceedings of the 23rd ACM symposium on Principles of Distributed Computing (PODC), pp. 338–346 (2004), http://dx.doi.org/10.1145/1011767.1011818
6. Delporte-Gallet, C., Fauconnier, H., Guerraoui, R., Kouznetsov, P.: Mutual exclusion in asynchronous systems with failure detectors. Journal of Parallel and Distributed Computing 65(4), 492–505 (2005), http://dx.doi.org/10.1016/j.jpdc.2004.11.008
7. Dolev, D., Dwork, C., Stockmeyer, L.: On the minimal synchronism needed for distributed consensus. Journal of the ACM 34(1), 77–97 (1987), http://doi.acm.org/10.1145/7531.7533
8. Dwork, C., Lynch, N.A., Stockmeyer, L.: Consensus in the presence of partial synchrony. J. ACM 35(2), 288–323 (1988), http://doi.acm.org/10.1145/42282.42283
9. Fich, F., Ruppert, E.: Hundreds of impossibility results for distributed computing. Distributed Computing 16(2-3), 121–163 (2003), http://dx.doi.org/10.1007/s00446-003-0091-y
10. Fischer, M.J., Lynch, N.A., Paterson, M.S.: Impossibility of distributed consensus with one faulty process. Journal of the ACM 32(2), 374–382 (1985), http://doi.acm.org/10.1145/3149.214121
11. Hermant, J.F., Widder, J.: Implementing reliable distributed real-time systems with the $\Theta$-model. In: Anderson, J.H., Prencipe, G., Wattenhofer, R. (eds.) OPODIS 2005. LNCS, vol. 3974, pp. 334–350. Springer, Heidelberg (2006), http://dx.doi.org/10.1007/11795490_26
12. Hutle, M., Malkhi, D., Schmid, U., Zhou, L.: Chasing the weakest system model for implementing $\Omega$ and consensus. IEEE Transactions on Dependable and Secure Computing 6(4), 269–281 (2009), http://dx.doi.org/10.1109/TDSC.2008.24
13. Jayanti, P., Toueg, S.: Every problem has a weakest failure detector. In: Proceedings of the 27th ACM Symposium on Principles of Distributed Computing, pp. 75–84 (2008), http://doi.acm.org/10.1145/1400751.1400763
14. Lo, W.K., Hadzilacos, V.: Using failure detectors to solve consensus in asynchronous shared-memory systems (extended abstract). In: Tel, G., Vitányi, P.M.B. (eds.) WDAG 1994. LNCS, vol. 857, pp. 280–295. Springer, Heidelberg (1994), http://dx.doi.org/10.1007/BFb0020440

15. Mostefaoui, A., Mourgaya, E., Raynal, M.: An introduction to oracles for asynchronous distributed systems. Future Gener. Comput. Syst. 18(6), 757–767 (2002), http://dx.doi.org/10.1016/S0167-739X0200048-1

16. Pike, S.M., Sastry, S., Welch, J.L.: Failure detectors encapsulate fairness. Tech. Rep. 2010-7-1, Department of Computer Science and Engineering, Texas A&M University (2010), http://www.cse.tamu.edu/academics/tr/2010-7-1

17. Pike, S.M., Song, Y., Sastry, S.: Wait-free dining under eventual weak exclusion. In: Rao, S., Chatterjee, M., Jayanti, P., Murthy, C.S.R., Saha, S.K. (eds.) ICDCN 2008. LNCS, vol. 4904, pp. 135–146. Springer, Heidelberg (2008), http://dx.doi.org/10.1007/978-3-540-77444-0_11

18. Rajsbaum, S., Raynal, M., Travers, C.: Failure detectors as schedulers (an algorithmically-reasoned characterization). Tech. Rep. 1838, IRISA, Université de Rennes, France (2007), http://hal.inria.fr/inria-00139317/PDF/PI-1838.pdf

19. Rajsbaum, S., Raynal, M., Travers, C.: The iterated restricted immediate snapshot model. In: Hu, X., Wang, J. (eds.) COCOON 2008. LNCS, vol. 5092, pp. 487–497. Springer, Heidelberg (2008),
http://dx.doi.org/10.1007/978-3-540-69733-6_48

20. Robinson, P., Schmid, U.: The Asynchronous Bounded-Cycle Model. In: Kulkarni, S., Schiper, A. (eds.) SSS 2008. LNCS, vol. 5340, pp. 246–262. Springer, Heidelberg (2008), http://dx.doi.org/10.1007/978-3-540-89335-6_20

21. Sastry, S., Pike, S.M.: Eventually perfect failure detection using ADD channels. In: Stojmenovic, I., Thulasiram, R.K., Yang, L.T., Jia, W., Guo, M., de Mello, R.F. (eds.) ISPA 2007. LNCS, vol. 4742, pp. 483–496. Springer, Heidelberg (2007), http://dx.doi.org/10.1007/978-3-540-74742-0_44

22. Sastry, S., Pike, S.M., Welch, J.L.: Crash fault detection in celerating environments. In: Procceding of the 23rd IEEE International Parallel and Distributed Processing Symposium, pp. 1–12 (2009), http://dx.doi.org/10.1109/IPDPS.2009.5161050

# $(\text{anti}-\Omega^x \times \Sigma_z)$-Based $k$-Set Agreement Algorithms

Zohir Bouzid[1],[*] and Corentin Travers[2],[**]

[1] University Pierre et Marie Curie - Paris 6, LIP6-CNRS 7606, France
zohir.bouzid@lip6.fr
[2] LaBRI University Bordeaux 1
travers@labri.Fr

**Abstract.** This paper considers the $k$-set agreement problem in a crash-prone asynchronous message passing system enriched with failure detectors. Two classes of failure detectors have been previously identified as necessary to solve asynchronous $k$-set agreement: the class anti-leader $\text{anti}-\Omega^k$ and the weak-quorum class $\Sigma_k$. The paper investigates the families of failure detector $(\text{anti}-\Omega^x)_{1 \le x \le n}$ and $(\Sigma_z)_{1 \le z \le n}$. It characterizes in an $n$ processes system equipped with failure detectors $\text{anti}-\Omega^x$ and $\Sigma_z$ for which values of $k, x$ and $z$ $k$-set-agreement can be solved. While doing so, the paper (1) disproves previous conjunctures about the weakest failure detector to solve $k$-set-agreement in the asynchronous message passing model and, (2) introduces the first indulgent algorithm that tolerates a majority of processes failures.

**Keywords:** Set-agreement, asynchrony, failure detectors, indulgent algorithms.

## 1 Introduction

*The $k$-set-agreement problem $k$-set-agreement* [10] is one of the fundamental problem in fault tolerant distributed computing. In this problem, $n$ processes starting each with an initial private value are required to agree on at most $k$ values chosen among their initial values. The problem generalizes the *consensus* problem, which corresponds to the case where $k = 1$. In an *asynchronous* system, it is well known that 1-set-agreement is impossible as soon as at least one process may fail by crashing [17], whereas the case $k = n$ does not require any coordination at all. For intermediate values of $k$ ($1 < k < n$), asynchronous $k$-set agreement tolerating $t$ crash failures is possible if and only if $k > t$ [6,25,30].

*Failure detectors.* A *failure detector* is a distributed oracle that provides processes with possibly unreliable information on failures [9]. According to the quality

---

[*] Supported by DIGITEO project PACTOLE and by the ANR projects SHAMAN and ALADDIN.
[**] On leave from University Pierre et Marie Curie - Paris 6. Supported by the ANR project SPREADS and by the INRIA project REGAL.

C. Lu, T. Masuzawa, and M. Mosbah (Eds.): OPODIS 2010, LNCS 6490, pp. 189–204, 2010.

of the information, several classes of failure detectors can be defined. Starting with [27,31], the failure detector approach has been investigated to alleviate the $k$-set-agreement impossibility in asynchronous systems. An algorithm that tolerates unreliable failure detection is said to be *indulgent* towards its failure detector [19,21]. Informally, an indulgent algorithm is always *safe*: it never violates the safety part of the problem it is supposed to solve, even when the underlying failure detector gives false information about failures.

*The quest for the weakest failure detector for $k$-set-agreement.* Given a distributed problem $P$, a natural question is to determine the *weakest failure detector* for $P$, that is a failure detector $D$ which is both *sufficient* to solve the problem – there is an asynchronous algorithm based on $D$ that solves $P$ – and *necessary*, in the sense that any failure detector $D'$ that allows solving $P$ can be used to emulate $D$.

The question of the weakest failure detector class for $k$-set agreement ($1 < k < n$) has been first stated in [29]. This line of research [11,12,20,24] culminated with the work of Zieliński who established that the failure detector class anti$-\Omega^{n-1}$ is the weakest to solve $(n-1)$-set-agreement in the wait-free shared memory model [32]. This has later been generalized to any $k$, $1 \leq k < n$ by three independent groups [2,15,18]. Informally, a failure detector anti$-\Omega^k$ outputs sets of $n - k$ process ids such that some non faulty process id eventually never appear in the outputs.

The situation is different in the message passing model where the answer is known only for the two boundaries cases, i.e., $k = 1$ (consensus) and $k = n - 1$ [14]. For consensus ($k = 1$), it has been shown that the class of *eventual leader* failure detector $\Omega = $ anti$-\Omega^1$ is the weakest failure detector in the asynchronous message passing model in which a majority of processes are non-faulty ($t < \frac{n}{2}$) [8]. This result is generalized to the *wait-free* environment in [13] where it is shown that $\Omega \times \Sigma$ is the weakest failure detector class for consensus when $t < n$. Intuitively, failure detector $\Sigma$ provides a reliable quorum system: when queried, a failure detector of the class $\Sigma$ returns a sets of processes ids, such that (1) any two sets intersect and (2) eventually, every set contains only ids of correct processes. Actually, $\Sigma$ is the weakest failure detector to implement a register in the message passing model [5,13].

Recently, the failure detector family $(\Sigma_k \times \Omega^k)_{1 \leq k < n}$ has been conjunctured to be the weakest failure detector classes for $k$-set-agreement [4]. Failure detector $\Sigma_k$ and $\Omega^k$ generalizes the classes $\Sigma$ and $\Omega$ respectively. Intuitively, a failure detector $\Sigma_k$ allows up to $k$ partitions: any collection of $k + 1$ sets outputs by the failure detector contain at least two intersecting sets. $\Omega^k$, which has been introduced by Neiger [28], outputs sets of $k$ ids that eventually converge to a set including the id of a non-faulty process. It is shown in [4] that $\Sigma_{n-1} \times \Omega^{n-1}$ is equivalent to the loneliness failure detector $\mathcal{L}$ which is the weakest failure detector class for $(n-1)$-set-agreement [14]. Before this paper, nothing specific was known about the power of $\Sigma_x \times \Omega^x$ to solve $k$-set-agreement, for $1 < x < n - 1$.

*Content of the paper.* The paper investigates in the message passing model the computational power of the failure detector families $(\Sigma_x)_{1 \leq x \leq n}$ and (anti$-\Omega^z)_{1 \leq z \leq n}$ as far as $k$-set-agreement is concerned. Its main contributions are the following:

1. It has been shown that $\Sigma_k$ is necessary to solve $k$-set-agreement, for each $k, 1 \leq k \leq n - 1$ [4]. Moreover, for $k = 1$, $\Sigma_1 = \Sigma$ alone is not powerful enough to solve consensus whereas $\Sigma_{n-1}$ is sufficient to solve $(n - 1)$-set-agreement [4,14]. We give necessary and sufficient conditions on the values of $k$, $x$ and $n$ in order to $k$-set-agreement to be solvable in an $n$ processes message passing system enriched with $\Sigma_x$ (Theorem 1, section 3). Roughly speaking, we show that $\Sigma_x$ allows to eliminate at most $\lfloor \frac{n}{x+1} \rfloor$ initial values, thereby generalizing prior results for the cases $k = 1$ [12] and $k = n - 1$ [14].
2. The paper then investigates the combined power of $\Sigma_x$ and anti$-\Omega^z$. For $k \geq xz$, we present a $k$-set-agreement algorithm that tolerates any number of failures (Section 5).

    To ensure safety, namely that no more than $x$ values are decided, we design a non-trivial generalization of the alpha abstraction which is at the core of indulgent consensus [22]. Our abstraction (called alpha$_x$, section 4) can be seen as an obstruction-free object that allows processes to store and retrieve at most $x$ distinct values. Its implementation relies solely on a failure detector of the class $\Sigma_x$. Of note, as $\Sigma_x$ can be simulated in an asynchronous message passing system when $t < \frac{xn}{x+1}$, we obtain a $xz$-set-agreement algorithm which is indulgent (towards the underlying failure detector of the class anti$-\Omega^z$) and tolerates $t < \frac{xn}{x+1}$ failures. To our knowledge, every prior indulgent algorithm assumes a majority of correct processes ($t < n/2$) or relies on a strong failure detector (e.g., $\Sigma$) that cannot be implemented in the asynchronous message passing model when a majority of processes may fail ($t \geq n/2$).
3. Finally, we show that for large enough values of $n$, there is no $k$-set-agreement algorithm based on $\Sigma_x \times \Omega^z$ if $k < xz$ (Theorem 2, section 5). This last result has two noteworthy corollaries. First, as anti$-\Omega^z$ can easily be simulated using the output of $\Omega^z$, it implies that the previous algorithm is optimal. Second, it rules out $\Pi_k = \Sigma_k \times \Omega^k$ as a weakest failure candidate for $k$-set-agreement, thus disproving Bonnet and Raynal's conjuncture [4].

*Roadmap.* The paper is made up of 6 sections. Section 2 describes the computing model and the families of failure detector we are interested in. Section 3 investigates the power of $\Sigma_x$ with respect to the solvability of $k$-set agreement. The alpha$_k$ abstraction is introduced in section 4, which presents also an $\Sigma_k$-based implementation. Section 5 then describes an indulgent $k$-set agreement algorithm that relies on the previous abstraction and a failure detector of the class anti$-\Omega^x$. A matching impossibility result is also presented. Finally, section 6 provides some concluding remarks. Due to space limitations, some proofs are presented in a companion technical report [7].

## 2   System Model and Failures Detectors

*Asynchronous message passing system with process crash failures.* The system consists in a set of $n$ processes denoted $\Pi = \{p_1, \ldots, p_n\}$. Processes are asynchronous and may fail by crashing. Processes communicate via sending and receiving messages over an asynchronous network. Each pair of processes is connected by a bi-directional channel. The channels are asynchronous but reliable. Reliable means that there is no creation, alteration or loss of messages whereas asynchronous means that message transfer delays are finite but unbounded.

Processes may fail by *crashing*, i.e., prematurely stop executing their code. A process is *correct* in an execution if it never crashes in this execution; otherwise it is *faulty*. $t(1 \leq t < n)$ denotes an upper bound on the number of processes that can crash in a run. Given an execution, *Correct* denotes the set of correct processes.

*Notation.* As in [26], $\mathcal{MP}_{n,t}$ denotes the asynchronous distributed system made of $n$ processes, among which at most $t$ may crash in any run. $\mathcal{MP}_{n,t}[X]$ denotes a system enriched with a failure detector of a class $X$.

*The k-set agreement problem.* In the $k$-set agreement problem, each process proposes a value and has to decide a value such that the following properties are satisfied: *(Validity)* A decided value is a proposed value; *(Termination)* Every correct process eventually decides a value; *(Agreement)* The number of distinct decided values is at most $k$.

*Families of failure detector classes.* For process $p_i$, $\mathrm{FD}_i^\tau$ is the value output by the failure detector at time $\tau$.

- *The eventual leader family* $(\Omega^k)_{1 \leq k \leq n}$. This family has been introduced in [28] to generalizes the class of failure detectors $\Omega$ defined in [8], with $\Omega^1 = \Omega$. A failure detector of the class $\Omega^k$ maintains at each process $p_i$ a set of processes of size at most $k$ (denoted LEADER$_i$) that satisfies the following property:
  - (Eventual multiple leadership). There is a time after which the sets LEADER$_i$ contains forever the same set of processes and at least one process of this set is correct.
- *The quorum family* $(\Sigma_k)_{1 \leq k \leq n}$ [4]. A failure detector of the class $\Sigma_k$ maintains at each process $p_i$ a variable TRUSTED$_i$ that contains a set of processes. The family generalizes the "quorum" failure detector $\Sigma = \Sigma_1$ introduced in [13]. The sets output by a failure detector of the class $\Sigma_z$ satisfy:
  - (Completeness) There is a time after which every set TRUSTED$_i$ contains only correct processes.
  - (Intersection) For every set $\mathcal{Q} = \{Q_1, \ldots, Q_{k+1}\}$ of $k+1$ sets output by the failure detector, there exists $Q_i, Q_j \in \mathcal{Q}, i \neq j$ such that $Q_i \cap Q_j \neq \emptyset$.
Of note, a failure detector $\Sigma_k$ can be implemented in $\mathcal{MP}_{n,t}$ provided that $\frac{kn}{k+1} > t$. To simulate a failure detector query, a process sends a REQUEST message to all processes and waits for matching RESPONSES. The set $X$

made of the ids of the senders of the first $n - t$ responses received defines the result of the query. It is easy to see that completeness is ensured: eventually, only correct processes send responses. The intersection property follows from the fact that each simulated query returns a set of $n - t \geq \lfloor \frac{n}{k+1} \rfloor + 1$ identities. Hence, any collection of $k + 1$ such sets contains at least two intersecting sets.

– *The anti-$\Omega$ family* (anti$-\Omega^k)_{1 \leq k \leq n}$ [32]. A failure detector of the class anti$-\Omega^k$ outputs at each process $p_i$ a set ANTI-LEADER$_i$ of $n - k$ processes ids. anti$-\Omega^1$ is equivalent to $\Omega$. In every run, there is a correct process such that eventually each set output by the failure detector does not contain the identity of this process.

  - (Anti-leadership) $\exists p_c \in Correct, \exists \tau$ such that $\forall \tau' \geq \tau, \forall p_i \in \Pi, c \notin$ ANTI-LEADER$_i^{\tau'}$.

## 3   $\Sigma_z$ and $k$-Set-Agreement

Among other results, [12] shows that there is a $k$-set-agreement algorithm based on $\Sigma_1$ if $k > n/2$. On the other side ($k = n - 1$), in [14] a $(n - 1)$-set agreement message passing algorithm is presented. The algorithm relies on a failure detector called $\mathcal{L}$, which has been proved in [4] to be equivalent to $\Sigma_{n-1}$. Actually, it is also shown in [14] that failure detector $\mathcal{L}$ is the weakest failure detector for $(n - 1)$-set-agreement in the wait-free message passing model ($t = n - 1$). We generalize these boundary results to the entire family $(\Sigma_z)_{1 \leq z \leq n}$. Specifically, we present a $k$-set-agreement algorithm based on $\Sigma_z$, provided that $k \geq n - \lfloor \frac{n}{z+1} \rfloor$. A simple matching impossibility result is also presented.

**Theorem 1.** *The $k$-set-agreement problem can be solved in $\mathcal{MP}_{n,n-1}[\Sigma_z]$ if and only if $k \geq n - \lfloor \frac{n}{z+1} \rfloor$*

*Solving $k$-set-agreement with $\Sigma_z$.* The algorithm combines ideas borrowed from the $(n - 1)$-set-agreement protocol based on failure detector $\mathcal{L}$ presented in [14] and a $k$-set-agreement protocol based on $\sigma_{2k}$ [12]. In short, a failure detector of the class $\sigma_{2k}$ provides the properties of the class $\Sigma$ only to a subset of size $2k$ of the system. The algorithm is described in Figure 1.

Let $A_1, \ldots, A_{z+1}$ be a partition of the set of processes such that $\forall i, 1 \leq i \leq z, |A_i| = \lfloor \frac{n}{z+1} \rfloor$ and $|A_{z+1}| = \lfloor \frac{n}{z+1} \rfloor + (n \bmod (z + 1))$. Each process in set $A_i$ tries to decide the proposal of some process that belongs to some partition $A_j, j < i$. To that end, each process $p \in A_i$ first sends its proposal to all processes in "higher" partitions, i.e., the processes that belong to the sets $A_{i+1}, \ldots, A_{z+1}$ (line 1). When a process receives a value $w$ from a "lower" partition, it decides that value after broadcasting a $DEC$ message carrying that value (line 5). A process that has not yet decided also decides $w$ when it receives such a message $DEC(w)$ (Task T3). Note that the initial values of the processes in the "highest" partition ($A_{z+1}$) cannot be decided using this mechanism. Hence at most $n - |A_{z+1}| = z \lfloor \frac{n}{z+1} \rfloor$ are decided in that way.

The mechanism sketched above allows every correct process to eventually decide as soon as at least two partitions contain correct processes. However, it may happen that all correct processes are contained in a single partition $A_i$. We notice that in that case, the failure detector output at each process is eventually contained in $A_i$ (by the completeness property of the class $\Sigma_z$). Henceforth, to prevent processes from waiting for values forever, each process $p_i$ periodically checks its failure detector output; If the current set of trusted processes is contained in $p_i$'s partition, $p_i$ is allowed to decide its initial value (task T2, lines 6-8). The proof shows (Lemma 1) that the total number of decided values is at most $k = z\lfloor\frac{n}{z+1}\rfloor + (n \mod (z+1))$.

---

**init**  $A_1, \ldots, A_{z+1}$ sets of processes such that $\forall i, j, i \neq j, A_i \cap A_j = \emptyset; \bigcup A_i = \Pi$;
$\forall i \in [1..z] |A_i| = \lfloor\frac{n}{z+1}\rfloor; |A_{z+1}| = \lfloor\frac{n}{z+1}\rfloor + n \mod (z+1)$

**propose**(v) % code for process $p \in A_i$
(1)  **foreach** $q \in \bigcup_{j>i} A_j$ **do** send VAL(v) to q **endfor**
(2)  start tasks T1, T2, T3

(3) Task T1: **when** VAL(w) is received **do**
(4)  **foreach** $q \in \Pi$ **do** send DEC(w) to q **enddo**
(5)  decide w; return

(6) Task T2: **repeat** $X \leftarrow \Sigma_z\text{-QUERY}()$ **until** $X \subseteq A_i$
(7)  **foreach** $q \in \Pi$ **do** send DEC(v) to q **enddo**
(8)  decide v; return

(9) Task T3: **when** DEC(w) is received
(10)  **foreach** $q \in \Pi$ **do** send DEC(w) to q **enddo**
(11)  decide(w); return

---

**Fig. 1.** $k$-set agreement algorithm in $\mathcal{MP}_{n,n-1}[\Sigma_z]$, $k = z\lfloor\frac{n}{z+1}\rfloor + (n \mod (z+1))$

**Lemma 1.** *The protocol described in the figure 1 solves $k$-set agreement in $\mathcal{MP}_{n,n-1}[\Sigma_z]$ for $k \geq n - \lfloor\frac{n}{z+1}\rfloor$*

*An impossibility result.* Together with Lemma 1, the following lemma completes the proof of Theorem 1.

**Lemma 2.** *$\forall n, k, z$ such that $k < n - \lfloor\frac{n}{z+1}\rfloor$, there is no $k$-set-agreement algorithm in $\mathcal{MP}_{n,n-1}[\Sigma_z]$*

## 4   The Alpha$_k$ Abstraction

This section presents the Alpha$_k$ abstraction that generalizes the Alpha abstraction introduced by Guerraoui and Raynal in [22] to capture the safety part of indulgent consensus[1]. In the very same way, the abstraction Alpha$_k$ captures the safety part of eventual failure detector based $k$-set-agreement algorithms. In short, the Alpha$_k$ abstraction can be viewed as a shared object intended to store at most $k$ values. A process accesses the object via the operation propose($\cdot$)

---

[1] Another generalization has been introduced in [29]. The implementation presented there relies on atomic registers which are not available in our settings.

with as parameter a value it is willing to store and gets back one of the values actually stored in the object. However, in case of concurrent accesses, propose($\cdot$) operations may not store any value and return the special value $\bot$, which is the object initial value.

More precisely, an alpha$_k$ object exports one operation propose($v, r$) with input parameters a value $v$ and a round number $r$. As in [22], distinct processes must input distinct round numbers and each process must use strictly increasing round number. The Alpha$_k$ abstraction is specified by the following properties, where $\bot$ is a special value that cannot be proposed:

- *Termination.* Every invocation of propose($\cdot$) by any non-faulty process returns.
- *Validity.* If the invocation propose($v, r$) returns $v' \neq \bot$, then propose($v', r'$) with $r' \leq r$ has been invoked.
- *$k$-Quasi-Agreement.* Let $V$ be the set of non-$\bot$ values that are returned by propose($\cdot$) invocations. $|V| \leq k$.
- *Conditional non-$\bot$convergence.* Let $I =$ propose($\_, r$) be a terminating invocation. If for every invocation $I' =$ propose($\_, r'$) that starts before $I$ returns, we have $r' < r$, $I$ returns a non-$\bot$ value.

## 4.1  Implementing Alpha$_k$ with $\Sigma_k$

The algorithm implementing Alpha$_k$ in an asynchronous message passing system is described in Figure 2. The algorithm relies on an underlying failure detector of the class $\Sigma_k$. It tolerates any number of failures.

*Algorithm principles.* At any time, each process $p_i$ has a value $v$ (initially $\bot$) stored in the local variable $val_i$ and a pair of integers $\langle r, \rho \rangle$ stored in the variables $\langle lre_i, pos_i \rangle$. The pair $\langle r, \rho \rangle$ can be seen as the *priority* of value $v$ from $p_i$'s point view. As in [22], $lre$ stands for *last round entered.* $r$ is the highest round number passed as a parameter of a propose(.) operation so far, as far as $p_i$ knows. Furthermore, each round $r$ is associated with a sequence of *positions* numbered from 1 to $2^r$. When $\langle lre_i, pos_i, val_i \rangle = \langle r, \rho, v \rangle$, we say that *value $v$ has reached position $\rho$ in round $r$*. Also, based on its position $\rho$ at round $r$, value $v$ logically *occupies* a position $\rho'$ at round $r + \delta$, for each $\delta > 0$. $\rho'$ is defined by the following function $g$:

$$g(\rho, \delta) = 2^\delta(\rho - 1) + 1$$

Any pair of triplets $\langle r, \rho, v \rangle$, $\langle r', \rho', v' \rangle$, $r \leq r'$ can be compared via the function $g$: $\langle r, \rho, v \rangle \prec \langle r', \rho', v' \rangle$, i.e., $v$ has a priority lower than $v'$ iff $g(\rho, r' - r) < \rho'^2$.

An operation propose($v, r$) returns a value $v' \neq \bot$ (possibly $v' \neq v$) only if $v'$ has obtained a priority high enough so that no more that $k - 1$ values $\neq v'$ can be awarded higher priority. Operationally, a process $p_i$ that invokes propose($v, r$) proceeds as follows:

---

[2] When $g(\rho, r' - r) = \rho'$, $v$ has a lower priority if $v < v'$. One can check that the $\prec$ relation is transitive, so $g$ induces a total order on triplets $\langle r, \rho, v \rangle$.

- In the first phase (lines 1-7), process $p_i$ broadcasts the message REQ_R($r$) in order (1) to inform other processes that it has entered round $r$ and (2) to collect triplets $\langle round, position, value \rangle$ held by other processes.

  When a process $p_j$ receives a message REQ_R($r$), it first updates its round and the position of its value (using the function $g$) if $r > lre_j$. It then sends back the current value of its variables $\langle lre_i, pos_i, val_i \rangle$ in a response message RSP_R (lines 17-18).

  $p_i$ is done collecting $\langle round, position, value \rangle$ triplets when it has received such values from each process $p_j$ in a *quorum*, that is a set of processes returned by a query to the underlying failure detector $\Sigma_k$. If $p$ discovers that another propose($\cdot$) operation with input $r' > r$ has already started, it returns $\perp$ (line 5). Note that this does not violate the conditional convergence property. Otherwise, $p_i$ selects among the values received the triplet with the highest priority, and updates its $\langle lre_i, pos_i, val_i \rangle$ accordingly (lines 6). In the case no triplets contain a value $\neq \perp$, $p_i$ selects its own value with position 0 (line 7).
- The second phase (lines 8-16) consists in a repeat loop. In each iteration of the loop, $p_i$ tries to increment the position of the value currently stored in $val_i$. To that end, it first broadcasts a request message REQ_W that carries $p_i$'s current value together with its position and the current round $r$ (lines 9). Process $p_j$ that has learned that a round $> r$ has been started ignores the content of the messages REQ_W($\langle r, \rho, v \rangle$) it receives. Otherwise, $p_j$ updates its round number and the position of its value. In addition, it adopts the received value if it has higher priority (lines 19-24). Finally, $p_j$ answers with a message RSP_W that carries the updated values of its variables $\langle lre_i, pos_i, val_i \rangle$ (lines 25).

  As in the first phase, $p_i$ stops collecting responses matching its request when a response message RSP_W($\cdot$) has been received from each process $p_j$ in a quorum $Q$. Similarly, if one of the response carries a round number $> r$, $p_i$ returns $\perp$. If this not the case, $p_i$ adopts among the values received the triplet with the highest priority, and updates its $\langle lre_i, pos_i, val_i \rangle$ variables accordingly (lines 14). Since $p_i$ always receives a response from itself, the value of $pos_i$ at the end of the iteration is greater that the value of this variable at the end of the previous iteration. Finally, if the current value $v$ of $p_i$ reaches the last position associated with round $r$, $v$ is returned (lines 15-16).

$k$-*Quasi agreement.* The main difficulty is to guarantee that propose($\cdot$) invocations return collectively no more than $k$ non-$\perp$ values. Value $v_1$ is returned at round $r_1$ if it reaches position $\rho_1 = 2^{r_1}$ and it has been adopted by a quorum $Q_1$. This means that for each process $q \in Q_1$, there is a point in time $\tau_q$ at which we have $\langle lre_q, pos_q, val_q \rangle = \langle r_1, \rho_1, v_1 \rangle$. However, because quorums may not intersect, another value $v' \neq v_1$ may reach an arbitrary high position and consequently replaces the value $v_1$ at each process $q \in Q_1$. For example, this might happen if the quorums output by the failure detector during propose($r'$, _) invocations with $r' > r_1$ do not intersect with $Q_1$. In these invocations, $v'$ may

```
init lre_i ← 0; val_i ← ⊥; pos_i ← 0;

function propose(r, v)
(1)   for each j ∈ Π send REQ_R(r) end for;
(2)   repeat Q ← Σ_k-QUERY()
(3)   until  (∀p_j ∈ Q ∪ {p_i} : RSP_R(r, ⟨lre_j, pos_j, val_j⟩) has been received from p_j)
(4)   let RCV = {⟨lre_j, pos_j, val_j⟩ : RSP_R(r, ⟨lre_j, pos_j, val_j⟩) has been received };
(5)   if (∃lre : ⟨lre, _, _⟩ ∈ RCV: lre > lre_i) then return (⊥) endif
(6)   let pos_M = max{pos : ⟨r, pos, v⟩} ∈ RCV;
      val_i ← max{v : ⟨r, pos_M, v⟩ ∈ RCV}; pos_i ← pos_M;
(7)   if val_i = ⊥ then val_i ← v endif
(8)   repeat pos_i ← pos_i + 1;
(9)       for each p_j ∈ Π send REQ_W(⟨r, pos_i, val_i⟩) to p_j end for
(10)      repeat Q ← Σ_k-QUERY()
(11)      until  (∀p_j ∈ Q ∪ {p_i} : RSP_W(r, pos_i, ⟨lre_j, pos_j, val_j⟩) has been received from p_j)
(12)      let RCV = {⟨lre_j, pos_j, val_j⟩ : RSP_W(r, pos_i⟨lre_j, pos_j, val_j⟩) has been received };
(13)      if (∃lre_j, ⟨lre_j, _, _⟩ ∈ RCV : lre_j > r) then return (⊥) end if
(14)      let pos_M = max{pos : ⟨r, pos, v⟩ ∈ RCV};
          val_i ← max{v : ⟨r, pos_M, v⟩ ∈ RCV}; pos_i ← pos_M;
(15)  until (pos_i = 2^r)
(16)  return (val_i)

when REQ_R(rd) is received from p_j
(17) if rd > lre_i then pos_i ← g(pos_i, rd − lre_i); lre_i ← rd end if
(18) send RESP_R(rd, ⟨lre_i, pos_i, val_i⟩) to p_j

when REQ_W(⟨rd, pos_j, val_j⟩) is received from p_j
(19) if (rd ≥ lre_i) then pos_i ← g(pos_i, rd − lre_i); lre_i ← rd
(20)     case pos_j > pos_i then val_i ← val_j; pos_i ← pos_j
(21)          pos_i = pos_j then val_i ← max(v_i, v_j)
(22)          pos_j < pos_i then nop
(23)     end case
(24) end if
(25) send RSP_W(rd, pos_j, ⟨lre_i, pos_i, val_i⟩) to p_j
```

**Fig. 2.** Implementing $\mathsf{Alpha}_k$ with $\Sigma_k$ (code for $p_i$)

be selected at the end of the first phase and its position can be increased in the second phase. In that case, $v'$ has an higher priority than $v_1$, i.e., a process $q \in Q_1$ that receives $\langle r', \rho', v' \rangle$ will adopts $v'$.

The key idea of the algorithm is a as follows. Fix some round $r' > r_1$. In order to value $v'$ to "overtake" value $v_1$ in round $r'$, $v'$ has to be adopted by a quorum $Q'$ that does *not intersect* with $Q_1$. Consider the positions associated with round $r'$. At the beginning of round $r'$, an odd position $x$ might be logically occupied by a value $v$. This is the case if for some process $p$ and some round $r < r'$, we have $\langle lre_p, pos_p, val_p \rangle = \langle r, \rho, v \rangle$ and $g(\rho, r' - r) = x$. Differently, by definition of $g$, each even position is initially free. Let $x'$ and $x_1 = g(\rho_1, r' - r_1)$ be the positions logically occupied by values $v'$ and $v_1$ respectively at the beginning of round $r'$. Observe that positions are increased by step of 1 and $x' + 2 \leq x_1$. So, to reach position $x_1$ value $v'$ must first successfully go through position $x_1 - 1$. This can only happen if there is quorum $Q'$ that adopts $\langle r', x_1 - 1, v' \rangle$. For each process $q \in Q_1$, the value $v_1$ held by $q$ has an higher priority, since it logically occupies position $x_1$. So $q$ cannot adopt $\langle r', x_1 - 1, v' \rangle$, hence $Q' \cap Q_1 \neq \emptyset$.

The rationale above can be extended to a chain of values $v_1, \ldots, v_\ell$ that each reaches higher and higher priorities to imply the existence of $\ell$ pairwise disjoint quorums. As any collection of $k + 1$ quorums contains at least two intersecting quorums, the length of such a chain is at most $k$. In particular, this implies that at most $k$ distinct values are returned – see the second part of the proof for more details (Lemmas 6–10).

*Remark.* The algorithm is generic in the sense that the parameter $k$ is never explicitly used in the code. In order to implement an $\mathsf{Alpha}_{k'}$ abstraction, it is sufficient to replace the underlying failure detector by a failure detector in the class $\Sigma_{k'}$. On the other hand, the algorithm uses $2^r$ positions per round. We have also developed along the same principles an algorithm that uses $O(r^{k-1})$ positions per round. However, determining which is the round $r'$ position corresponding to a round $r < r'$ position, i.e., defining the equivalent of the $g$ function, is more involved. As a result, the correctness proof is more intricate.

## 4.2   Proof

Consider a *well-formed* execution, in which processes execute the algorithm described in Figure 2 when $\mathsf{propose}(\cdot)$ is invoked. An execution is well-formed if the following conditions are fulfilled: (1) Only round number $r > 0$ are used as input parameters; (2) For any invocations $\mathsf{propose}(\_, r)$ and $\mathsf{propose}(\_, r')$ performed by processes $p$ and $p'$ respectively, if $p \neq p'$ then $r \neq r'$ and, if $p = p'$ and $\mathsf{propose}(\_, r)$ is invoked before $\mathsf{propose}(\_, r')$ then $r < r'$.

**Lemma 3 (Termination).** *Every invocation of* $\mathsf{propose}(\cdot)$ *by a correct process terminates.*

**Lemma 4 (Validity).** *Suppose that the invocation* $\mathsf{propose}(r, v)$ *returns* $v' \neq \perp$. *Then* $\mathsf{propose}(r', v')$ *with* $r' \leq r$ *has been invoked by some process.*

**Lemma 5 (Conditional non-$\perp$convergence).** *Let* $I = \mathsf{propose}(r, \_)$ *be a terminating invocation. If for every invocation* $I' = \mathsf{propose}(r', \_)$ *that starts before* $I$ *returns we have* $r' < r$, $I$ *returns a non-$\perp$ value.*

*$k$-quasi agreement.* The next lemma is central in the proof of the $k$-quasi agreement property. In the following, a quorum is a set of processes returned by a query to the underlying $\Sigma_k$ failure detector.

**Lemma 6.** *Let* $V$ *be the set non-$\perp$ values that are returned by the* $\mathsf{propose}(\cdot)$ *invocations.* $|V| = x \Rightarrow \exists x$ *quorums* $Q_1, \ldots, Q_x, \forall 1 \leq i < j \leq x, Q_i \cap Q_j = \emptyset$.

The $k$-quasi agreement property then follows easily from Lemma 6.

**Lemma 7 ($k$-quasi agreement).** *Suppose that the protocol described in Figure 2 is instantiated with a failure detector of the class* $\Sigma_k$. *The total number of non-$\perp$ values that are returned by the* $\mathsf{propose}(\cdot)$ *invocations is at most* $k$.

*Proof.* Assume for contradiction that $x > k$ non-$\perp$ values are returned. It then follows from Lemma 6 that at least $k + 1$ disjoint quorums are output by the underlying failure detector $\Sigma_k$. This contradicts the intersection property of the class $\Sigma_k$. $\qquad\Box$

In order to prove Lemma 6, we define a sequence $S = s_1, \ldots, s_i = \langle r_i, \rho_i, v_i \rangle, \ldots$ where for each $i$, $r_i$ is a round number, $\rho_i$ a position associated to round $r_i$, and $v_i$ a value. The sequence $S$ is defined inductively as follows:

- $r_1$ is the smallest round $r$ such that the invocation propose$(\_, r)$ returns a non-$\perp$ value, if any. $\rho_1 = 2^{r_1}$ and $v_1$ is the value returned by that invocation.
- Suppose that $s_1, \ldots, s_{i-1}$ have been defined. $r_i$ is the first round $r > r_{i-1}$ during which a value $v \neq \{v_1, \ldots, v_{i-1}\}$ reaches a position $\geq g(\rho_{i-1}, r - r_{i-1})$ (if such a round exists), i.e., $r_i = \min \{r : r > r_{i-1}, \exists p_x, \exists v \notin \{v_1, \ldots, v_{i-1}\}, \langle lre_x, pos_x, val_x \rangle = \langle r, g(\rho_{i-1}, r - r_{i-1}), v \rangle\}$. $v_i$ is then this value, and we define $\rho_i = g(\rho_{i-1}, r_i - r_{i-1}) - 1$.

In the next lemma we give a formula for computing values $\rho_i$.

**Lemma 8.** *Suppose that $|S| \geq \ell$. $\forall i, 2 \leq i \leq \ell$, $\rho_i = 2^{r_i}(1 - \frac{1}{2^{r_1}} - \ldots - \frac{1}{2^{r_{i-1}}})$*

Suppose that value $v$ reaches position $\rho$ in round $r$, i.e., there exists a process $p_i$ for which we have $\langle lre_i, pos_i, val_i \rangle = \langle r, \rho, v \rangle$ at some time. For every round $r' \geq r$, value $v$ then logically occupies round $r$ position $g(\rho, r' - r)$. Indeed, if process $p_i$ later receives a read or write request carrying round $r' \geq r$, $pos_i$ is updated to the value $g(\rho, r' - r)$ (at line 17 or line 19). Given a round $r$, we can then define the highest position logically occupied by value $v$ as follows:

**Definition 1.** *Given a value $v$ and a round number $r$, let* mpos$(v, r)$ *denotes the maximal position logically occupied by value $v$ at the beginning of round $r$. Formally,* mpos$(v, r) = \max\{g(\rho', r - r') : \exists p_j, r' < r$ *and a time at which* $\langle lre_j, pos_j, val_j \rangle = \langle r', \rho', v \rangle\}$*; if no invocation* propose$(r', v)$ *with* $r' < r$ *occurs,* mpos$(v, r) = 0$.

**Lemma 9.** *Suppose that $|S| \geq \ell$. Let $\langle r, \rho, v \rangle$ be the value of process $p_i$ variables $\langle lre_i, pos_i, val_i \rangle$ at some time. If $r \leq r_\ell$ and $v \notin \{v_1, \ldots, v_\ell\}$, $g(\rho, r_\ell - r) < \rho_\ell$.*

**Lemma 10.** *Let $V$ be the set non-$\perp$ values that are returned by the* propose$(\cdot)$ *invocations. If $|V| = x$, $s_1, \ldots, s_x$ are well defined.*

We are now ready to prove Lemma 6. To do so we associate to each $s_i \in S$ a quorum $Q_i$. Intuitively, the processes in $Q_i$ are those processes that allow value $v_i$ to reach position $\rho_i$ during round $r_i$. Each process $q \in Q_i$ hence holds the triplet $\langle r_i, \rho_i, v_i \rangle$ at some time. Note that, after that time, the round $r$ and position $\rho$ are always such that $r \geq r_i$ and $\rho \geq g(\rho_i, r - r_i)$. The crucial observation is that $q$ cannot allow any value $v_j \neq v_i$ to reach position $\rho_j$, essentially because either $r_i > r_j$ (in the case $i > j$) or $g(\rho_i, r_j - r_i) > \rho_j$ (if $j > i$).

*Proof of Lemma 6.* Suppose that $|V| = x$. Let $\ell, 1 \leq \ell \leq x$. We first bound mpos$(v_\ell, r_\ell)$. Suppose that $\langle r, \rho, v_\ell \rangle$ are stored by some process $p$, with $r \leq r_\ell$. There are two cases:

- $1 \leq r \leq r_{\ell-1}$. Since $v_\ell \notin \{v_1, \ldots, v_{\ell-1}\}$, it follows from Lemma 9 that $g(\rho, r_{\ell-1} - r) < \rho_{\ell-1}$. Hence, $g(g(\rho, r_{\ell-1} - r), r_\ell - r_{\ell-1}) < g(\rho_{\ell-1}, r_\ell - r_{\ell-1})$ from which we have $g(\rho, r_\ell - r) < g(\rho_{\ell-1}, r_\ell - r_{\ell-1})$.

- $r_{\ell-1} < r < r_\ell$. By definition of $s_\ell$, we have $\rho < g(\rho_{\ell-1}, r - r_{\ell-1})$. Therefore $g(\rho, r_\ell - r) < g(g(\rho_{\ell-1}, r - r_{\ell-1}), r_\ell - r)$ which implies $g(\rho, r_\ell - r) < g(\rho_{\ell-1}, r_\ell - r_{\ell-1})$.

We conclude that $mpos(v_\ell, r_\ell) < g(\rho_{\ell-1}, r_\ell - r_{\ell-1}) = \rho_\ell + 1$. By definition of $g(\cdot)$, $\rho + 1$ is odd. Similarly, as there exists $r' < r_\ell, \rho'$ such that $mpos(v_\ell, r_\ell) = g(\rho', r_\ell - r')$, $mpos(v_\ell, r_\ell)$ is odd. Consequently $mpos(v_\ell, r_\ell) < \rho_\ell$.

We now define a quorum $Q_\ell$ associated with the triplet $\langle r_\ell, \rho_\ell, v_\ell \rangle$. By definition of $s_\ell$, $r_\ell$ is the first round during which value $v_\ell$ reaches a position $\geq \rho_\ell + 1$. There is a (unique) process $p_\ell$ that invokes propose($\cdot$) with input parameter $r_\ell$. Otherwise, round $r_\ell$ is never entered and value $v_\ell$ cannot reach position $g(\rho_{\ell-1}, r_\ell - r_{\ell-1}) = \rho_\ell + 1$ in round $r_\ell$.

Note that (1) value $v_\ell$ reaches a position $\geq \rho_\ell + 1$ in round $r_\ell$, (2) the highest position logically occupied by $v_\ell$ at the beginning of round $r_\ell$ is $< \rho_\ell$. Moreover, (3) only process $p_\ell$ increases positions in round $r_\ell$, and (4) $p_\ell$ tries to move at most one value from position $\phi$ to position $\phi + 1$, for every position $\phi$. It then follows that $p_\ell$ successfully moves value $v_\ell$ from position $\rho_\ell - 1$ to position $\rho_\ell + 1$. In more details, this means that the variable $pos_\ell$ successively contains the values $\rho_\ell - 1, \rho_\ell, \rho_\ell + 1$ while the variables $\langle lre_\ell, val_\ell \rangle$ keep the values $\langle r_\ell, v_\ell \rangle$.

In particular, let us consider the iteration of the repeat loop (lines 8-15) in which $pos_\ell = \rho_\ell$. Let $Q_\ell$ be the quorum that allows the inner repeat loop to terminate (lines 10-11). Observe that $Q_\ell$ is a set of process returned by a query to failure detector $\Sigma_k$. For each $q \in Q_\ell$, the message RSP_W received from $q$ must carry the triplet $\langle r_\ell, \rho_\ell, v_\ell \rangle$. If not, $p_\ell$ either picks another pair $\langle \rho, v \rangle$ with $v \neq v_\ell$ and $\rho \geq \rho_\ell$ or returns $\perp$. In both case, $p$ stops moving value $v_\ell$. It cannot move $v_\ell$ later in the same round, as the highest position occupied by $v_\ell$ is $\rho_\ell$, and in subsequent iterations, only values located at position $> \rho_\ell$ can be moved.

Consequently, it follows that $\forall p_i \in Q_\ell$ there exists a time $\tau_i^\ell$ at which we have $\langle lre_i, pos_i, val_i \rangle = \langle r_\ell, \rho_\ell, v_\ell \rangle$.

Finally, we establish that $\forall i, j, 1 \leq i < j \leq \ell, Q_i \cap Q_j$. Observe that if $\langle r_1, \rho_1, v_1 \rangle$ and $\langle r_2, \rho_2, v_2 \rangle$ are the values of the same process variables $\langle lre, pos, val \rangle$ at times $\tau_1 < \tau_2$ respectively, $(r_1 = r_2 \wedge \rho_1 \leq \rho_2) \vee (r_1 < r_2 \wedge g(\rho_1, r_2 - r1) \leq \rho_2)$ $(\star\star\star)$.

Assume for contradiction that $\exists \ell, m, 1 \leq \ell < m \leq x$ such that $Q_\ell \cap Q_m \neq \emptyset$. Let $p_i \in Q_\ell \cap Q_m$. There are two cases:

- $\tau_i^\ell < \tau_i^m$. In that case, $p_\ell$ sends first a message RSP_W carrying $\langle r_\ell, \rho_\ell, v_\ell \rangle$ and later a message RSP_W carrying $\langle r_m, \rho_m, v_m \rangle$. Note that the two triplets are the values at times $\tau_i^\ell$ and $\tau_i^m$ respectively of the variables $\langle lre_i, pos_i, val_i \rangle$. We have:

$$g(\rho_\ell, r_m - r_\ell) = 2^{r_m}(1 - \sum_{j=1}^{\ell} \frac{1}{2^{r_j}}) + 1 \quad \text{and} \quad \rho_m = 2^{r_m}(1 - \sum_{j=1}^{m-1} \frac{1}{2^{r_j}})$$

  from which we obtain $\rho_m < g(\rho_\ell, r_m - r_\ell)$, contradicting observation $(\star\star\star)$.
- $\tau_i^\ell > \tau_i^m$. This implies that $lre_i$ first contains $r_m$ and later $r_\ell < r_m$, which is impossible according to observation $(\star\star\star)$. $\qquad \square$

# 5   A $k$-Set Agreement Algorithm

This section presents an (anti$-\Omega^x \times \Sigma_z$)-based $k$-set-agreement protocol, and a matching impossibility result on solving $k$-set agreement in the family of systems $(\mathcal{MP}_{n,n-1}[\text{anti}-\Omega^x, \Sigma_z])_{1 \le x,z \le n}$. The main results of this section are summarized by the following theorem:

**Theorem 2.** *The $k$-set-agreement problem can be solved in $\mathcal{MP}_{n,n-1}[\text{anti}-\Omega^x, \Sigma_z]$ if $k \ge xz$. Moreover, if $2xz \le n$, the $k$-set-agreement problem cannot be solved in $\mathcal{MP}_{n,n-1}[\Omega^x, \Sigma_z]$ if $k < xz$.*

## 5.1   Solving $k$-Set Agreement with anti$-\Omega^x$ and $\Sigma_z$

For the system $\mathcal{MP}_{n,n-1}[\text{anti}-\Omega^x, \Sigma_z]$, we describe a $k$-set agreement algorithm that requires $k \ge xz$. From a computability point of view, our algorithm is optimal if $n$ is large enough: we later establish that if $k < xz$ and $n \ge 2xz$ there is no $k$-set agreement algorithm in $\mathcal{MP}_{n,n-1}[\text{anti}-\Omega^x, \Sigma_z]$ (Corollary 1).

*Using $\Omega$ and $\Sigma_z$ to solve $k$-set agreement for $k \ge z$.* The algorithm is a simple adaptation of the generic $\Omega$-based consensus algorithm presented in [22], in which an $\mathsf{Alpha}_k$ object is used in place of an $\mathsf{Alpha}$ object. For completeness, the algorithm is described in Figure 3. The fact that any decided value has been returned by an invocation of $\mathsf{Alpha}_k.\mathsf{propose}(\cdot)$ guarantees validity and agreement. Because eventually a unique correct process considers itself the leader, there is a time after which only this process invokes $\mathsf{Alpha}_k.\mathsf{propose}(\cdot)$. Hence, by the conditional convergence property of the object, there is an invocation that returns a non-$\perp$ value. This value is then broadcast, allowing every non-faulty process to decide, therefore ensuring termination.

---

```
SA_propose(v)
(1)   dec_i ← ⊥; r_i ← i;
(2)   while (dec_i = ⊥) do
(3)     if Ω-QUERY() = i then dec_i ← Alpha_k.propose(r_i, v)
(4)                         r_i ← r_i + n end if end do
(5)   for each p_j ∈ Π do send DECIDE(dec_i) to p_j end do

when DECIDE(w) is received do
(6)   for each p_j ∈ Π do send DEC(w) to p_j end do
(7)   decide w; return
```

**Fig. 3.** $k$-set agreement algorithm in $\mathcal{MP}_{n,n-1}[\Omega, \Sigma_k]$, code for $p_i$

---

*Using anti$-\Omega^x$ and $\Sigma_z$ to solve $k$-set agreement for $k \ge xz$* Our algorithm is based on a failure detector vector$-\Omega^x$ [32]. A failure detector of the class vector$-\Omega^x$ is a vector of $x$ sub-detectors, $\Omega_1, \ldots, \Omega_x$, such that at least one $\Omega_i$ is a failure detector of the class $\Omega$. When $k = n - 1$, the vector$-\Omega$ failure detector proposed in [32] is obtained. It was shown there how vector$-\Omega$ can be implemented from anti$-\Omega^{n-1}$ in the wait-free asynchronous shared memory model, and how it can be used to solve $(n-1)$-set agreement. The failure detector vector$-\Omega^x$ was also presented in [32]. It is claimed there that the algorithm to transform anti$-\Omega^{n-1}$ into vector$-\Omega^{n-1}$ (Figure 1 in [32], see also [3]), can be generalized to transform anti$-\Omega^x$ into vector$-\Omega^x$. A close look at the transformation algorithm reveals that it can be easily adapted to the message passing case if a reliable broadcast primitive is available. As reliable broadcast can be implemented in an asynchronous message passing system in which any number of processes may fail [23], vector$-\Omega^x$ can be implemented in $\mathcal{MP}_{n,n-1}[\text{anti}-\Omega^x]$.

To solve $k$-set agreement in $\mathcal{MP}_{n,n-1}[\text{anti}-\Omega^x, \Sigma_z]$, processes simulate outputs of a failure detector vector$-\Omega^x$. We associate to each sub-detector $\Omega_i, 1 \leq i \leq x$ an instance of the $(\Omega, \Sigma_z)$-based $z$-set agreement algorithm described in Figure 3. Each processes participates simultaneously in each of the $x$ instances, and terminates as soon as it decides in one instance.

It follows from the fact that at least one sub detector $\Omega_i$ is a failure detector of the class $\Omega$ that at least one instance terminates. Moreover, since at most $z$ values are decided in each instance, the total number of decided value is upper bounded by $xz$. Therefore,

**Lemma 11.** *Let* $1 \leq k, x, z \leq n$. *There is a $k$-set agreement algorithm in* $\mathcal{MP}_{n,n-1}[\text{anti}-\Omega^x, \Sigma_z]$ *if* $k \geq xz$.

## 5.2   An Impossibility Result

This section investigates $k$-set-agreement solvability when the system is enriched with failure detectors of both classes $\Omega^y$ and $\Sigma_z$. The main result is Lemma 13 which establishes that there is no $k$-set agreement algorithm in the wait-free environment $(t = n - 1)$ where failure detectors $\Omega^y$ and $\Sigma_z$ are provided if $k < yz$.

**Lemma 12.** *Let* $k, 1 \leq k \leq n$ *and* $x, 1 \leq 2x \leq n$. *If* $k < x$, *there is no $k$-set agreement algorithm in* $\mathcal{MP}_{n,n-1}[\Omega^x, \Sigma]$.

**Lemma 13.** *Let* $k, 1 \leq k \leq$ *and* $x, z, 1 \leq 2xz \leq n$. *If* $k < xz$, *there is no $k$-set agreement algorithm in* $\mathcal{MP}_{n,n-1}[\Omega^x, \Sigma_z]$.

Given a failure detector $\Omega^x$, it is easy to simulate a anti$-\Omega^x$ failure detector by outputting the complement of the sets leader output by $\Omega^x$. Therefore,

**Corollary 1.** *Let* $k, 1 \leq k \leq$ *and* $x, z, 1 \leq 2xz \leq n$. *If* $k < xz$, *there is no $k$-set agreement algorithm in* $\mathcal{MP}_{n,n-1}[\text{anti}-\Omega^x, \Sigma_z]$.

Bonnet and Raynal introduce in [4] the failure detector class $\Pi_k$ as a weakest failure detector candidate for message passing $k$-set-agreement. Next corollary disproves this conjuncture.

**Corollary 2.** *Let* $k, n : 1 < k < n-1$ *and* $2k^2 \leq n$. *There is no $k$-set agreement algorithm in* $\mathcal{MP}_{n,n-1}[\Pi_k]$.

*Proof.* [4] proves that $\Pi_k$ is equivalent to $\Sigma_k \times \Omega^k$. The corollary then directly follows from Lemma 13                                                                □

# 6   Concluding Remarks

The paper has investigated the computational power of the failure detector classes $\Sigma_x$ and anti$-\Omega^z$ as far as $k$-set-agreement is concerned in the $n$-processes message passing asynchronous model. The main result is that for large enough values of $n$, namely $n > 2kz$, $k$-set agreement is possible if and only if $k \geq xz$.

The main open question is the weakest failure detector for message passing $k$-set-agreement, for $1 < k < n-1$. Our $xz$-set agreement algorithm may help to demonstrate the sufficiency of weakest failure detector candidate. Another interesting avenue for future research is the complexity of $k$-set-agreement tolerating $t > n/2$ failures. When a majority of processes does not fail, it has been shown that the price of indulgence is constant [1,16]. Is it still true when a majority of processes failures has to be tolerated?

# References

1. Alistarh, D., Gilbert, S., Guerraoui, R., Travers, C.: Of choices, failures and asynchrony: The many faces of set agreement. In: Dong, Y., Du, D.-Z., Ibarra, O. (eds.) ISAAC 2009. LNCS, vol. 5878, pp. 943–953. Springer, Heidelberg (2009)
2. Anta, A.F., Rajsbaum, S., Travers, C.: Brief announcement: weakest failure detectors via an egg-laying simulation. In: PODC 2009, pp. 290–291. ACM Press, New York (2009)
3. Anta, A.F., Rajsbaum, S., Travers, C.: Weakest failure detectors via an egg-laying simulation (preliminary version). Technical report, Universidad Rey Juan Carlos. Reports on Systems and Communications, vol. IX(2) (January 2009), http://gsyc.es/tr-docs/RoSaC-2009-2.pdf
4. Bonnet, F., Raynal, M.: Looking for the weakest failure detector for k-set agreement in message-passing systems: Is $\pi_k$ the end of the road? In: Guerraoui, R., Petit, F. (eds.) SSS 2009. LNCS, vol. 5873, pp. 149–164. Springer, Heidelberg (2009)
5. Bonnet, F., Raynal, M.: A simple proof of the necessity of the failure detector sigma to implement an atomic register in asynchronous message-passing systems. Information Processing Letters 110(4), 153–157 (2010)
6. Borowsky, E., Gafni, E.: Generalized FLP impossibility result for t-resilient asynchronous computations. In: STOC 1993, pp. 91–100. ACM Press, New York (1993)
7. Bouzid, Z., Travers, C.: (anti$-\Omega^x \times \Sigma_z$)-based k-set Agreement Algorithms. Technical report, http://hal.archives-ouvertes.fr/inria-00519606/en/
8. Chandra, T.D., Hadzilacos, V., Toueg, S.: The weakest failure detector for solving consensus. J. ACM 43(4), 685–722 (1996)
9. Chandra, T.D., Toueg, S.: Unreliable failure detectors for reliable distributed systems. J. ACM 43(2), 225–267 (1996)
10. Chaudhuri, S.: More choices allow more faults: Set consensus problems in totally asynchronous systems. Information and Computation 105(1), 132–158 (1993)
11. Chen, W., Zhang, J., Chen, Y., Liu, X.: Weakening failure detectors for k-set agreement via the partition approach. In: Pelc, A. (ed.) DISC 2007. LNCS, vol. 4731, pp. 123–138. Springer, Heidelberg (2007)
12. Delporte-Gallet, C., Fauconnier, H., Guerraoui, R.: Sharing is harder than agreeing. In: PODC 2008, pp. 85–94. ACM Press, New York (2008)

13. Delporte-Gallet, C., Fauconnier, H., Guerraoui, R., Hadzilacos, V., Kouznetsov, P., Toueg, S.: The weakest failure detectors to solve certain fundamental problems in distributed computing. In: PODC 2004, pp. 338–346. ACM Press, New York (2004)
14. Delporte-Gallet, C., Fauconnier, H., Guerraoui, R., Tielmann, A.: The weakest failure detector for message passing set-agreement. In: Taubenfeld, G. (ed.) DISC 2008. LNCS, vol. 5218, pp. 109–120. Springer, Heidelberg (2008)
15. Delporte-Gallet, C., Fauconnier, H., Guerraoui, R., Tielmann, A.: The disagreement power of an adversary: extended abstract. In: PODC 2009, pp. 288–289. ACM Press, New York (2009)
16. Dutta, P., Guerraoui, R.: The inherent price of indulgence. Distributed Computing 18(1), 85–98 (2005)
17. Fischer, M.J., Lynch, N.A., Paterson, M.: Impossibility of distributed consensus with one faulty process. J. ACM 32(2), 374–382 (1985)
18. Gafni, E., Kuznetsov, P.: The weakest failure detector for solving k-set agreement. In: PODC 2009, pp. 83–91. ACM Press, New York (2009)
19. Guerraoui, R.: Indulgent algorithms (preliminary version). In: PODC 2000, pp. 289–297. ACM Press, New York (2000)
20. Guerraoui, R., Herlihy, M., Kuznetsov, P., Lynch, N.A., Newport, C.C.: On the weakest failure detector ever. Distributed Computing 21(5), 353–366 (2009)
21. Guerraoui, R., Lynch, N.A.: A general characterization of indulgence. Transactions on Autonomous and Adaptive Systems 3(4) (2008)
22. Guerraoui, R., Raynal, M.: The alpha of indulgent consensus. The Computer Journal 50(1), 53–67 (2007)
23. Hadzilacos, V., Toueg, S.: Reliable broadcast and related problems. In: Distributed Systems, pp. 97–145 (1993)
24. Herlihy, M., Penso, L.D.: Tight bounds for k-set agreement with limited-scope failure detectors. Distributed Computing 18(2), 157–166 (2005)
25. Herlihy, M., Shavit, N.: The topological structure of asynchronous computability. J. ACM 46(6), 858–923 (1999)
26. Mostéfaoui, A., Rajsbaum, S., Raynal, M., Travers, C.: On the computability power and the robustness of set agreement-oriented failure detector classes. Distributed Computing 21(3), 201–222 (2008)
27. Mostéfaoui, A., Raynal, M.: k-set agreement with limited accuracy failure detectors. In: PODC 2000, pp. 143–152. ACM Press, New York (2000)
28. Neiger, G.: Failure detectors and the wait-free hierarchy. In: PODC 1995, pp. 100–109. ACM Press, New York (1995)
29. Raynal, M., Travers, C.: In search of the holy grail: Looking for the weakest failure detector for wait-free set agreement. In: Shvartsman, M.M.A.A. (ed.) OPODIS 2006. LNCS, vol. 4305, pp. 3–19. Springer, Heidelberg (2006)
30. Saks, M.E., Zaharoglou, F.: Wait-free k-set agreement is impossible: The topology of public knowledge. SIAM J. Comput. 29(5), 1449–1483 (2000)
31. Yang, J., Neiger, G., Gafni, E.: Structured derivations of consensus algorithms for failure detectors. In: PODC 1998, pp. 297–306. ACM Press, New York (1998)
32. Zieliński, P.: Anti-omega: the weakest failure detector for set agreement. In: PODC 2008, pp. 55–64. ACM Press, New York (2008)

# Distributed Programming with Tasks

Eli Gafni[1] and Sergio Rajsbaum[2,*]

[1] University of California, Los Angeles,
Computer Science Department,
Los Angeles, CA 90095
eli@ucla.edu

[2] Instituto de Matemáticas, Universidad Nacional Autónoma de México
Ciudad Universitaria, D.F. 04510
Mexico
rajsbaum@math.unam.mx

**Abstract.** In round-by-round models of distributed computing processes run in a sequence of (synchronous or asynchronous) rounds. The advantage of the round-by-round approach is that invariants established in the first round are preserved in later rounds. An elegant asynchronous round-by-round shared memory model, is the *iterated snapshots model* (IS). Instead of the *snapshots model* where processes share an array $m[\cdot]$ that can be accessed any number of times, indexed by process ID, where $P_i$ writes to $m[i]$ and can take a snapshot of the entire array, we have processes share a two-dimensional array $m[\cdot, \cdot]$, indexed by iteration number and by process ID, where $P_i$ in iteration $r$ writes once to $m[r, i]$ and takes one snapshot of row $r$, $m[r, \cdot]$. The IS model lends itself more easily to combinatorial analysis. However, to show that whenever a task is impossible in the IS model the task is impossible in the snapshots model, a simulation is needed. Such a simulation was presented by Borowsky and Gafni in PODC97; namely, it was shown how to take a wait-free protocol for the snapshots model, and transform it into a protocol for the IS model, solving the same task.

In this paper we present a new simulation from the snapshots model to the IS model, and show that it can be extended to work with models stronger that wait-free. The main contribution is to show that the simulation can work with models that have access to certain communication objects, called 01-tasks. This extends the result of Gafni, Rajsbaum and Herlihy in DISC'2006 stating that renaming is strictly weaker than set agreement from the IS model to the usual non-iterated wait-free read/write shared memory model.

We also show that our simulation works with $t$-resilient models and the more general dependent process failure model of Junqueira and Marzullo. This version of the simulation extends previous results by Herlihy and Rajsbaum in PODC'2010 and DISC'2010 about the topological connectivity of a protocol complex in an iterated dependent process failure model, to the corresponding non-iterated model.

---

* Supported by UNAM-PAPIIT.

C. Lu, T. Masuzawa, and M. Mosbah (Eds.): OPODIS 2010, LNCS 6490, pp. 205–218, 2010.

# 1    Introduction

In round-by-round models of distributed computing processes run in a sequence of (synchronous or asynchronous) rounds. The advantage of the round-by-round approach is that invariants established in the first round are preserved in later rounds. The round-by-round approach has been used successfully in previous papers e.g. [2, 12, 14, 17, 20, 25, 29–31, 37–40] for message-passing as well as for shared-memory models, to prove lower bounds and impossibility results. Also, most of the distributed protocols for solving a *task* are structured in rounds. When solving a task, processes start with private input values, communicate with each other, and decide private output values that should satisfy the task's input/output specification.

The *iterated snapshots model* (IS) [9] is an elegant asynchronous round-by-round shared memory model. Instead of the *snapshots model* where processes share an array $m[\cdot]$ that can be accessed any number of times, indexed by process ID, where $P_i$ writes to $m[i]$ and can take a snapshot [1] of the entire array, we have processes share a two-dimensional array $m[\cdot, \cdot]$, indexed by iteration number and by process ID, where $P_i$ writes once to $m[r, i]$ in iteration $r$, and takes a snapshot of row $r$, $m[r, \cdot]$. Notice that the snapshots model where processes share a single array $m[i]$ is equivalent to the usual wait-free single-writer/multi-reader (SWMR) shared memory model where any number of SWMR arrays are available to the processes [1], for computability (we are not interested in complexity issues in this paper).

The IS model is more restrictive than the snapshots model: processes can write or snapshot each shared array $m[r, \cdot]$ only once, and they access the shared arrays in the same order, by iteration number. However, these restrictions induce two advantages. First, an inductive reasoning is facilitated, that greatly simplifies the understanding of distributed protocols e.g. [24]. Second, it is easier to analyze this structured set of runs using topology e.g. [9, 29, 30, 32], than the much larger set of runs of the snapshots model. But, is the IS model equivalent to the snapshots model? Obviously any task solvable in the IS model is solvable in the snapshots model, but is the converse true? In other words, to show that whenever a task is impossible in the IS model the task is impossible in the snapshots model, a simulation is needed. Such a simulation was introduced in [9]; namely, it was shown how to take a wait-free protocol for the snapshots model, and transform it into a protocol for the IS model, solving the same task.

*Results.* This paper presents a new, elegant simulation from the snapshots model to the IS model. It shows that the simulation works with $t$-resilient models, and with models that have access to 01-exclusion tasks. This is a large family of tasks with power between set agreement and read/write memory [21]. Two important consequences of the simulation are the following.

The simulation presented works with $t$-resilient models and the more general dependent process failure model of Junqueira and Marzullo [34]. Thus, a contribution of this paper is to show that the IS model is equivalent to the snapshots model, in the *t*-resilient case, and more generally, for an adversary that can fail processes even in a non-independent way. As a consequence, the results of [29, 30], that were proved in an IS model, are extended to the corresponding non-iterated model. These results characterized the power of an adversary, in terms of topological connectivity.

The so-called "subconsensus" tasks are tasks too weak to solve consensus for two processes. After a substantial effort, it was discovered that two tasks, set agreement [15] and renaming [4], which are both subconsensus tasks, cannot be implemented in read/write memory [8, 13, 33, 41]. It follows that subconsensus tasks have a fine structure, inaccessible by consensus-based analysis. Further light on this fine structure was shed in [25] with the surprising result that renaming is strictly weaker than set agreement. However, this result was shown in a certain IS model of computation, and it was left open the question weather this result extends to the usual, non-iterated wait-free model. Our simulation finally closes this question, by showing that the IS model used in [25] is indeed equivalent to its non-iterated counterpart.

*Discussion of the results and techniques.* Our main result is the prove that the snapshots model and the IS model are equivalent when enriched with 01-exclusion tasks. Obtaining this result proved harder than we originally thought it would be. It necessitated the discovery of a novel simulation of the snapshots model by the IS model. We needed a simulation with the property that if in the snapshots model a process obtained a snapshot of size $n$, then in the IS simulating model, all processes which obtained a simulated snapshot of size $n$ obtained it at the same iteration. Furthermore, it necessitated the surprising use of the notion of a "dual" of a task. This notion was introduced in [21] for the 01-exclusion family of tasks. The main difficulty when simulating the snapshots model with 01-exclusion tasks, is that in this model processes are not required to invoke 01-exclusion tasks in the same order. By "untangling" a program we mean that two processes should invoke any two tasks always in the same order, and only once.

We stress that extending the simulation to work with 01-exclusion tasks is sufficient to prove the renaming vs. set agreement conjecture of [25]. Nevertheless, we claim that the simulation can be extended to apply to any task that is solvable by set agreement. The main difficulty is how to define the dual for any such task $T$. We leave an exploration of this issue for future work.

Characterizing and unifying models of distributed computing has been a central concern of the Theory of Distributed Computing research. When two models turn out to be equivalent one can examine a problem in either one. Some questions are natural and easy to answer in one model, and some in the other. An example is the IS model enriched with failure detectors [40]. The glaring success example is the equivalence between message-passing and shared-memory [3]. This paper takes another step in the quest of showing that an iterated model and its non-iterated version are equivalent under a wide variety of circumstances.

## 2 Model

We consider a set of $n$ sequential and deterministic *processes* that communicate by reading and writing a shared memory. Processes are *asynchronous*: there is no bound on relative process speed. Processes can *fail* by halting. One process cannot determine whether an unresponsive process has failed or is just slow.

Classical models implicitly assume that processes fail independently. In a distributed system, however, failures may be correlated. To model these situations, it is natural to extend the classical failure models to encompass an *adversary* scheduler $\mathcal{A}$ that can cause certain subsets of processes to fail, perhaps in a non-uniform way.

Following Junqueira and Marzullo [34] we specify an adversary $\mathcal{A}$ as follows. A *core* $C$ for $\mathcal{A}$ is a set of processes such that (1) in every run, some process in $C$ does not fail, and (2) $C$ is minimal: for every proper subset $C'$ of $C$, there is a run in which every process in $C'$ fails. A *survivor set* $S$ for $\mathcal{A}$ is a set of processes such that (1) in some run, the set of processes that do not fail is exactly $S$, and (2) $S$ is minimal: for every proper subset $S'$ of $S$, there is no run in which $S'$ is the set of processes that do not fail. For the wait-free adversary, the entire set of processes is the only core, and for the $t$-faulty adversary, any set of $t+1$ processes is a core. Notice that cores and survivor sets are dual notions: cores determine survivor sets and vice-versa. For the $t$-faulty adversary, any set of $n-t$ processes is a survivor set. Delporte-Gallet et al. [16] used a more general notion to study the computational power of adversaries in asynchronous shared memory.

A *task* [7] is a coordination problem in which each process starts with a private input value taken from a finite set, communicates with the other processes, and eventually decides on a private output value, also taken from a finite set. Examples of tasks include *consensus* [19], *renaming* [4], *set agreement* [15], *committee decision* [26], and *musical benches* [23]. The $k$-*set agreement* task generalizes consensus. Here up to $k$ different values can be decided. In this paper we only consider $(n-1)$-set agreement, and call it simply *set agreement*. We consider only *bounded* decision tasks, whose number of input vectors is finite.

Of special interest to this paper is the *01-exclusion* family of tasks [21], inspired by the *weak symmetry breaking* task [13, 25, 33]. In a 01-exclusion task processes have to decide either 0 or 1, but in runs where all participate, at least one should decide 0 and at least one should decide 1. Each 01-exclusion task is defined by a binary vector $b_1, \ldots, b_{n-1}$, specifying that in a run where $k$ processes participate (and the others fail before taking any steps), not all can decide $b_k$. It is known that when $b_k = 1$ for all $k$, the resulting 01-exclusion task is equivalent to set agreement [8].

In the *snapshots model* processes share an $n$-element array $m[\cdot]$ where the $i$-th process can write element $m[i]$, and can take an atomic snapshot [1] of the entire array. We consider *full-information* protocols: initially, each process's state is its input. Each process alternates between writing its state to memory, taking a snapshot of the memory, and using that snapshot as its new state. A protocol solves a task against adversary $\mathcal{A}$ if after a finite number of steps, the protocol *decides* a value based on its current state, in every run that is *fair* with respect to

$\mathcal{A}$, called an $\mathcal{A}$-run. Namely, runs where each time a process takes a snapshot, it reads new values by a set of processes that include a survivor set (to implement this property a process can repeatedly read the shared memory until it sees that new values have been written by a survivor set).

We also consider a round-by-round version of this snapshots model, called the *iterated snapshots* (IS) model. We refer to the processes of the snapshots model as $\pi = \{p_1, \ldots, p_n\}$, and of the IS model as $\Pi = \{P_1, \ldots, P_n\}$. Instead of having processes share an array $m[\cdot]$, we have processes share a *two-dimensional* array $m[\cdot, \cdot]$, indexed by iteration number and by process ID, where $P_i$ in iteration $r$ writes to $m[r, i]$ and can take a snapshot of row $r$. In iteration $r$, each process $P_i$ writes its state to $m[r, i]$, and takes a snapshot of that row reading values by a set of processes that include a survivor set; i.e., we consider only $\mathcal{A}$-runs (a process waits until the set of processes that have written to row $r$ includes a survivor set, and then takes a snapshot of that row).

The snapshots model and the IS model are equivalent with respect to wait-free task solvability [9]. Furthermore, as we shall see, the two models are equivalent with respect to any adversary (Section 3), and also when we enrich both models with more powerful shared objects (Section 4).

## 3 Basic Protocol: Equivalence of Snapshots Model and IS Model

Consider an adversary $\mathcal{A}$. This section describes the protocol SIMULATION of Figure 3 showing that the IS model, is equivalent to the snapshots memory model described above.

Consider the snapshots model with processes $p_0, \ldots, p_n$ and adversary $\mathcal{A}$. In a full-information protocol, to solve a task, a process eventually decides a value based on its current state $s$. Thus, a protocol defines a predicate $undecided(s)$, and a decision function $\delta(s)$. The function $\delta$ is defined on any state $s$ such that $undecided(s)$ is false. Once a protocol decides, $undecided(s)$ is forever false. Let the $k$-th value read by $p_i$ be $read(k, i)$, and $write(k, i)$ the $k$-th value $p_i$ writes. In protocol SIMULATION lines 7 and 9 describe how $read(k, i)$ and $write(k, i)$ get their values during the simulation; these lines are inside comment brackets, as they do not affect the actual operation of the protocol.

Consider the IS memory model with processes $P_0, \ldots, P_n$, and adversary $\mathcal{A}$. In iteration $r$, each process $P_i$ writes a value $c$ to $m[r, i]$, waits until the set of process that have written to row $r$ includes a survivor set, and then takes a snapshot of that row. A shorthand for this sequence of operations is $IS_i(c, r)$. Thus, $IS_i(c, r)$ returns a snapshot of $m[r, \cdot]$.

The protocol SIMULATION of Figure 3 implies the following result. The converse is trivial. The theorem is proved in the rest of this section.

**Theorem 1.** *Given a protocol that solves a task in the snapshots model with adversary $\mathcal{A}$, protocol SIMULATION solves the same task in the IS model with $\mathcal{A}$.*

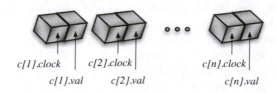

$c[1].clock \qquad c[2].clock \qquad\qquad c[n].clock$

$\qquad c[1].val \qquad\quad c[2].val \qquad\qquad\quad c[n].val$

**Fig. 1.** The $c$ data structure

In the code of Figure 3, all variables are process-local; access to shared variables is encapsulated in the $IS_i(c, r)$ operation. Each $P_i$ maintains a vector $c$ with one entry $c[j]$ per process $P_j$, that consists of two values, $c[j].clock$ and $c[j].val$. See Figure 1.

When $P_i$ writes $k$ in $c[i].clock$ we say "$P_i$ invokes the $k$-th write operation of $p_i$." Intuitively, when $P_i$ has $c[j].clock = k$, $c[j].val$ contains the value of the last (simulated) write by $p_j$ known to $P_i$, the $k$-th write. We describe below when this write is linearized in the simulated run.

Each iteration of the loop $P_i$ accesses a new row of the array, via the $IS$ invocation in line 4. It stores in $view$ the value returned by the $IS_i(c, r)$ operation. Thus, $view[j]$ will contain either a vector which was written by $P_j$, or $\perp$ (if $P_j$'s execution of $IS_j(c, r)$ was not performed concurrently or before the one of $P_i$). See Figure 2.

We use the following partial order on vectors: $c \leq c'$ means $c[i].clock \leq c'[i].clock$ for every $i$, and we write $c < c'$ if $c \leq c'$ and in addition, for some $j$, $c[j].clock < c'[j].clock$. The operation $|c|$ of line 6 adds up the $clock$ entries of $c$. The operation $top(c)$ of line 5 computes the entry-wise maximum of each vector the process has seen, as in the example of Figure 2.

Formally,

$$|c| = \sum_{1 \leq i \leq n} c[i].clock,$$

$$top(view)[i] = view[j][i], \text{s.t.} \max_{1 \leq j \leq n} view[j][i].clock, \text{for each } i$$

where $\perp$ entries count as 0 for the max operation.

In line 6, $P_i$ checks if $|c| = r$, and if additionally $p_i$ is undecided, increments $c[i].clock$ by one, to issue a new simulated write. Let $c.val$ be the vector without the clock values: $c.val[j] = c[j].val$. That is, $c.val$ contains a simulated snapshot, in line 7. Else, the first time $undecided(c.val)$ is false, the decision for $p_i$ is produced, with $\delta(c.val)$.

The correctness of the simulation follows from the next lemmas. In line 6, when $|c| = r$, we say "$P_i$ completes the $k$-th operation of $p_i$," where $k = c[i].clock$. If $undecided(c.val)$ holds, $P_i$ is trying to execute the $k + 1$-st operation of $p_i$. We assume that once $undecided(c.val)$ is false, it remains so, and hence from then on $P_i$ never again increases $c[i].clock$. (It is necessary that $P_i$ keeps on running after deciding, otherwise it could block other processes when they are waiting to hear from a survivor set. But $P_i$ should stop increasing its $clock$, see Lemma 5)

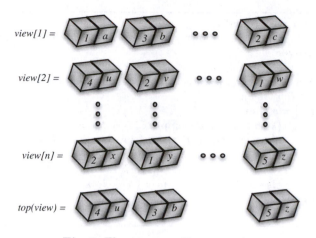

**Fig. 2.** The *top* operation on a *view*

Consider the value of $c$ just before line 5, and suppose it becomes $c'$ after executing line 5. Then, $c' \geq c$, and we have the following simple invariant:

$$r \leq |c|, \tag{1}$$

where once a process $P_i$ decides, the invariant may not hold for $P_i$. Notice that we assume decided processes do not stop in the snapshots model, and we simulate them as if they never fail.

**Lemma 1 (simultaneity).** *All processes that complete an operation in an iteration $r$, do so with the same vector $c$, $|c| = r$. If an undecided process does not complete an operation in that iteration, then its vector $c'$ satisfies $c' > c$.*

*Proof.* Consider an iteration $r$, and the snapshots returned by the $IS_i(c, r)$ operations (of undecided processes). These snapshots can be ordered by containment, $S_1 \subset S_2 \subset \cdots \subset S_k$. The processes that get back snapshot $S_i$ get back the same same value for $c$ in line 4, and hence they all update their $c$ variables in line 5 identically. It follows that the value $|c|$ for each process that returns $S_i$ is the same, let's denote it by $|S_i|$. Thus, $|S_i| \leq |S_{i+1}|$. By invariant (1), $r \leq |S_i|$ (for undecided processes). Also, by the snapshot ordering of operations, if $c$ is the value for some snapshot $S_i$, and $c'$ for some later one $S_{i+1}$, then $c \leq c'$.

The simulated run is fair for $\mathcal{A}$.

**Lemma 2 (fair progress).** *If a process $P_i$ completes an operation in iteration $r$ with a vector $c$, and the next iteration it completes an operation is at iteration $r + k$, with vector $c'$, then there are at least $|S|$ entries $j$, for some survivor set $S$, such that either $c[j].clock < c'[j].clock$ or $P_j$ is decided.*

*Proof.* By Lemma 1, every process that completes an operation in iteration $r$ does so with the same vector $c$ of clock values, $|c| = r$. And (undecided) processes

```
Protocol SIMULATION(input);
 (1)   init r ← 0; c[j].clock ← 0 for j ≠ i; c[i].clock ← 1; c[i].val ← input;
 (2)   loop forever
 (3)        r ← r + 1;
 (4)        view ← IS_i(c, r); {write c, snapshot (including survivor set) row r}
 (5)        c ← top(view);
 (6)        if |c| = r then
 (7)              {k-th snap by p_i completed: read(k, i) ← c.val, k = c[i].clock }
 (8)              if undecided(c.val) then { invoke c[i].clock + 1-th write }
 (9)                   { k-th write of p_i is write(k, i) ← c.val, k = c[i].clock + 1}
(10)                   c[i].val ← c.val; c[i].clock ← c[i].clock + 1;
(11)              else decide δ(c.val) once; { c[i].clock fixed from now on }
(12) end loop.
```

**Fig. 3.** The iterated simulation for adversary $\mathcal{A}$ (code for $P_i$)

that do not complete an operation have vectors larger than $c$. Thus, all these processes, they enter iteration $r + 1$ with vectors $c$, $|c| \geq r + 1$, each one invoking a write operation that has not been executed, i.e., with a new clock value in its $c$ entry. As process $P_i$ completes an operation in iteration $r$, and waits until it sees writes by a survivor set in iteration $r + 1$, its value for $|c|$ in iteration $r + 1$, at line 5, will be incremented by at least $|S|$ (minus the number of processes that have decided).

The value of $r$ is incremented in every iteration. Hence invariant (1) implies the following.

**Lemma 3 (non-blocking progress).** *A correct undecided process eventually completes an operation.*

*Proof.* First notice that an $IS$ operation (line 4) always completes, because adversary $\mathcal{A}$ guarantees that some survivor set will eventually write values to the $r$-th row. For undecided processes, invariant (1) says that the smallest value of $|c|$ in an iteration is lower bounded by $r$, as $r$ increases every iteration, as long as there are correct undecided processes, eventually one will complete an operation.

To complete the proof of Theorem 1, we need to show that the simulated run is a valid run of the snapshot model. A snapshot is linearized when a process completes an operation, and a write is linearized just before the first snapshot that includes it. And it is fair for $\mathcal{A}$, by Lemma 2. Thus, a correct simulated process $p_i$ eventually decides, and then $P_i$ stops incrementing its clock.

More precisely, by Lemma 1, we can consider the sequence of vectors obtained each time an operation is completed, $c^0, c^1, \ldots, c^N$, where $c^0$ is the all 0's vector, and for $i > 0$, $c^i$ is the vector corresponding to the $i$-th completed operation; namely, the operation completed in iteration $|c^i|$. Also, the lemma implies $c^i < c^{i+1}$. Let $r\text{-}proc(c^i)$ be the set of processes that complete an operation in iteration

$|c^i|$, and $w$-$proc(c^i)$ be the set of processes $j$, such that $c^{i-1}[j] < c^i[j]$. Then, a snapshot by each processes in $r$-$proc(c^i)$ is linearized in iteration $|c^i|$ (in arbitrary order), with $c_i.val$ as snapshot value. And just before these snapshots, a write by each process in $w$-$proc(c^i)$ is linearized (in arbitrary order); namely the $c^i[j]$-th write operation by $p_j$. The wait-free termination then follows from Lemma 3. Thus, we have the following lemma, where, abusing notation, $r$-$proc(c^i)$, and $w$-$proc(c^i)$ include which value a process wrote and which value it read.

**Lemma 4 (Linearization).** *For a $\mathcal{A}$ run of the IS model simulating protocol A with input $I$, consider the sequence of vectors obtained each time an operation is completed, $c^0, c^1, \ldots, c^N$. Then,*

$$\text{w-proc}(c^1), \text{r-proc}(c^1), \text{w-proc}(c^2), \text{r-proc}(c^2), \ldots, \text{w-proc}(c^N), \text{r-proc}(c^N),$$

*is an $\mathcal{A}$ run of A with input $I$ in the snapshots model.*

Thus, the simulated run is a valid run of the snapshots model, and it is fair for $\mathcal{A}$. The correctness of the protocol in the snapshots model implies that a correct simulated process $p_i$ eventually decides, and then $P_i$ stops incrementing its clock.

**Lemma 5 (Termination).** *Every correct process eventually decides.*

*Proof.* By Lemma 3, as long as there are correct undecided processes, eventually one will complete an operation, and by Lemma 4 the reads and writes are correct for an $\mathcal{A}$ simulated run. As $r$ increments by 1 every iteration of the **loop**, there must be one process, say $p_1$, that eventually simulates enough operations for line 8, and decides. (because the read and write operations are valid for an $\mathcal{A}$ simulated run) Each time such a process decides, the entry 1 stops increasing in all vectors $c$ of other processes. Hence, invariant (1) implies that some other process will eventually decide (by Lemma 4), say $p_2$, with $c[2].val = dec_2$. Repeating this argument each time a process decides, we have that all must eventually decide.

# 4    Extended Protocol for 01-Exclusion Tasks

We are now ready to extend the read/write simulation protocol of Section 3 with additional tasks. Consider a protocol A for the snapshots model, extended with 01-exclusion tasks. That is, the local algorithm $A_i$ for each process $p_i$, in addition to executing write and snapshot operations, it can invoke 01-exclusion tasks. Similarly, in the IS model, when executing iteration $r$, process $P_i$ writes to $m[r, i]$, may invoke a 01-exclusion task $T$, and then takes snapshots of $m[r, \cdot]$. The algorithm specifies which task $T$ to invoke, provided it invokes the same 01-exclusion task at most once. For the simulation, it is convenient to assume that we insert in the algorithm an additional write and snapshot operation before each task invocation. In this write $p_i$ announces which task is going to invoke, say $T$, and in the snapshot it reads how many processes, $q$, plan to invoke the same task $T$ at this moment, and it also finds out if processes have executed

$T$ earlier. Lets call $q'$ be the total number of processes that invoke $T$ (now or earlier) as known by $c$. Also, let $b$ a value known to have been obtained from $T$, or 0 if no such value exists. Recall that there is a write and snapshot operation around $T$'s invocation; as if $p_i$ writes "in the process of invoking $T$".

For a 01-exclusion task $T$, specified by a vector $b_1^T, \ldots, b_{n-1}^T$, the *dual task* $D(T)$ is the 01-exclusion task with vector $b_1^{D(T)}, \ldots, b_{n-1}^{D(T)}$, where $b_k^{D(T)} = 1 - b_{n-k}^T$. Obviously, $D(D(T)) = T$. The correctness of our simulation is based on the following properties of a 01-task [21].

**Lemma 6.** *Let $T$ be a member of the 01-exclusion family of tasks.*

1. *An $n$-vector is an output vector for $T$ if and only if it is an output vector for $D(T)$.*
2. *$T$ and $D(T)$ are wait-free equivalent.*
3. *Let $R, NR$ be a partition of the $n$ processes into two non-empty sets. Let $O_R$ be an output vector from $T$ when processes $R$ participate, and $O_{NR}$ an output vector from $D(T)$ when processes $NR$ participate. The vector obtained by combining $O_R$ and $O_{NR}$ is an output vector for $T$.*

Property (1) means that when all $n$ processes participate, any output vector is valid for both $T$ and $D(T)$. This is obvious: when all participate, not all should decide 0 and not all should decide 1, in either $T$ or $D(T)$. Property (2) says that given a black box for $T$, there is a wait-free protocol that solves $D(T)$, and vice-versa. Thus, for our simulation, we may assume that instead of $T$, processes can invoke a solution for $D(T)$ (as we already know how to simulate a wait-free protocol in the IS model). Property (3) implies that some processes can produce outputs for $T$ in one iteration, and others for $D(T)$ in another, and the combined outputs are valid for $T$.

The algorithm $A_i$ is simulated in the IS with 01-exclusion tasks with a simple modification of the simulation protocol of Figure 3 as follows. Each time a write or snapshot operation is to be executed as specified by $A_i$, these operations are simulated as explained in Section 3. Consider the moment a task $T$ is to be invoked, as specified by $A_i$. In the previous iteration of $A_i$, $p_i$ announced it will invoke $T$ (with a write operation). Assume $p_i$ completes this snapshot operation with value $c$, in iteration $r$. So, $p_i$ last written value to the shared memory (and hence appears in $c[i]$), specifies that it is about to invoke $T$. We denote this fact by $task(c[i]) = T$. If a process $p_j$ does not want to invoke a task at this moment (according to $A_j$), then $task(c[j]) = \bot$. Notice that is possible that $task(c[j]) = T'$, for $T \neq T'$. Then, $P_i$ computes an output value for $T$ according to three cases, by executing the following code once $|c| = r$ is detected in line 3. That is, process $P_i$ checks which of the following three cases hold for $c$, assuming $P_i$ wants to invoke $T$ in $c$.

**Invoke$_i$($T, c$):**

1. $P_i$ sees in $c$ that already some process returned from $T$ some value, $b$; in this case, it returns $1 - b$ as the output of its invocation to $T$.

2. $P_i$ sees that all processes want to invoke the same task $T$ in $c$, ie $q = n$; in this case it invokes $D(T)$, and returns as simulated output for the invocation, the value it gets back from $D(T)$.

3. Otherwise, it returns as the simulated output for the invocation $1 - b_k^T$, where $k = q'$ is the number of processes that want to invoke $T$, in $c$ or earlier (as seen in $c$).

Once this code is executed, an output for $T$ is obtained, that should be appended to $c[i].val$ in line 10 of the SIMULATION code in Figure 3, so that the next simulated write of $p_i$ will include it.

Notice that the only case when $P_i$ actually invokes a 01-exclusion task is case (2) of the previous lemma, and the task invoked is actually $D(T)$. As mentioned above, we can assume $D(T)$ is available to the processes of the IS model, by Lemma 6(2).

**Lemma 7.** *If all $n$ processes invoke the same 01-task $D(T)$, they invoke it in the same iteration.*

*Proof.* Consider the first iteration, $r$, that a process $P_i$ invokes $D(T)$. Assume $P_i$ sees some vector $c$ when it completes an operation, with $|c| = r$. Then, any other process $P_j$ that does not complete and operation in $r$, sees a vector $c'$, with $c' > c$, by Lemma 1. Thus, when $P_j$ completes an operation in a later iteration, it will be with a vector different from $c$, in particular where more operations have been completed, and not all processes want to invoke the task $T$, and hence $P_j$ with not invoke $D(T)$ in an iteration after $r$.

**Lemma 8.** *The simulation of each task $T$ is correct.*

*Proof.* The code **Invoke**$_i(T, c)$ has three cases. Case (1) is easy: when $P_i$ sees in $c$ that already some process returned from $T$ some value, $b$. In this case it is clear that $P_i$ can return $1 - b$ as the output of its invocation to $T$, because any output vector that contains two different values is a valid output vector for $T$. Case (2) is taken care of by Lemma 7, because if all invoke $D(T)$ in the same iteration, the invocation is to the same instance of the task. And by Lemma 6(1), the output vector returned by $D(T)$ is valid for $T$. Finally, case (3) is when $P_i$ observes in $c$ that a set of processes $R$, with $|R|$ less than $n$, want to invoke $T$. Then $P_i$ returns as the simulated output for the invocation $1 - b_k^T$, where $k = q'$ is the number of processes that want to invoke $T$, in $c$ or earlier. In this case it is safe to return $1 - b_k^T$, because it is possible that these are the only processes that will invoke $T$. And in case the other processes, $NR$ want to invoke $T$ in a later iteration, and they are not aware that some processes $R$ already produced output values for $T$, and invoke $D(T)$ (as in the previous case), then the combined output vector of $R$ and $NR$ is valid for $T$, by Lemma 6(3).

Finally, using Lemma 8, together with the correctness properties for protocol SIMULATION in Figure 3, we get the main theorem.

**Theorem 2.** *If a task is wait-free solvable in the standard model with 01-tasks, then it is wait-free solvable in the EIS model with the same 01-tasks.*

# 5    Conclusion

We have considered two asynchronous shared memory models of distributed computing. The snapshots model, which is equivalent to the usual single-writer, multi-reader shared memory model, and the iterated snapshots (IS) model, where processes access a sequence of snapshots objects, each one only once and in the same order. We have shown that both models are equivalent with respect to task solvability, both in the wait-free case, and in the case where processes can fail, perhaps in a non-uniform way, as modeled by an adversary in the style of [34]. Such an adversary includes the $t$-resilient model, where at most $t$ processes can fail. Our main result is that if processes have access to 01-exclusion tasks, again the snapshots model and the IS model are equivalent. This shows that the renaming task is strictly weaker than the set agreement task, in the snapshots model, something that was known to hold only in the IS model. Our results are based on a new, elegant simulation from the snapshots model to the IS model.

Iterated models are central to the theory of distributed computing [38]. They have proved very useful for lower bound and impossibility results. Many protocols are designed with a round-by-round structure. This paper takes another step in the quest of showing that an iterated model and its non-iterated version are equivalent under a wide variety of circumstances. We believe that we have laid a good foundation to prove that similar equivalences can be proved, when we allow processes to communicate with any task solvable by set agreement. The main difficulty is to define the dual of any such task, and then generalize the properties about 01-exclusion tasks we used in our simulation.

*Acknowledgments.* We thank Michel Raynal and the anonymous referees for their comments on an earlier version of this paper.

# References

1. Afek, Y., Attiya, H., Dolev, D., Gafni, E., Merrit, M., Shavit, N.: Atomic Snapshots of Shared Memory. J. ACM 40(4), 873–890 (1993)
2. Aguilera, M.K., Toueg, S.: A Simple Bivalency Proof that $t$-Resilient Consensus Requires $t + 1$ Rounds. Inf. Process. Lett. 71(3-4), 155–158 (1999)
3. Attiya, H., Bar-Noy, A., Dolev, D.: Sharing Memory Robustly in Message-Passing Systems. J. ACM 42(1), 124–142 (1995)
4. Attiya, H., Bar-Noy, A., Dolev, D., Peleg, D., Reischuk, R.: Renaming in an Asynchronous Environment. J. ACM 37(3), 524–548 (1990)
5. Afek, Y., Gafni, E., Rajsbaum, S., Raynal, M., Travers, C.: Simultaneous Consensus Tasks: A Tighter Characterization of Set-Consensus. In: Chaudhuri, S., Das, S.R., Paul, H.S., Tirthapura, S. (eds.) ICDCN 2006. LNCS, vol. 4308, pp. 331–334. Springer, Heidelberg (2006)
6. Baldoni, R., Raynal, M.: Fundamentals of Distributed Computing: A Practical Tour of Vector Clock Systems. IEEE Distributed Systems Online 3(2) (2002)
7. Biran, O., Moran, S., Zaks, S.: A Combinatorial Characterization of the Distributed 1-Solvable Tasks. J. Algorithms 11(3), 420–440 (1990)

8. Borowsky, E., Gafni, E.: Generalized FLP Impossibility Results for $t$-Resilient Asynchronous Computations. In: Proc. 25th ACM Symposium on the Theory of Computing (STOC 1993), pp. 91–100. ACM Press, New York (1993)

9. Borowsky, E., Gafni, E.: A Simple Algorithmically Reasoned Characterization of Wait-Free Computations (Extended Abstract). In: Proc. 16th ACM Symposium on Principles of Distributed Computing (PODC 1997), pp. 189–198. ACM Press, New York (August 1997)

10. Borowsky, E., Gafni, E., Lynch, N., Rajsbaum, S.: The BG Distributed Simulation Algorithm. Distributed Computing 14(3), 127–146 (2001)

11. Brit, H., Moran, S.: Wait-Freedom vs. Bounded-Freedom in Public Data Structures. J. UCS 2(1), 2–19 (1996)

12. Chou, C.-T., Gafni, E.: Understanding and Verifying Distributed Algorithms Using Stratified Decomposition. In: Proc. 7th ACM Symposium on Principles of Distributed Computing (PODC), Toronto, Ontario, Canada, August 15-17, pp. 44–65. ACM Press, New York (1988)

13. Castañeda, A., Rajsbaum, S.: New combinatorial topology upper and lower bounds for renaming. In: Proc. 27th ACM Symposium on Principles of Distributed Computing (PODC), Toronto, Canada, August 18-21, pp. 295–304 (2008)

14. Charron-Bost, B., Schiper, A.: The Heard-Of model: computing in distributed systems with benign faults. Distributed Computing 22(1), 49–71 (2009)

15. Chaudhuri, S.: More Choices Allow More Faults: Set Consensus Problems in Totally Asynchronous Systems. Information and Computation 105, 132–158 (1993)

16. Delporte-Gallet, C., Fauconnier, H., Guerraoui, R., Tielmann, A.: The Disagreement Power of an Adversary. In: Keidar, I. (ed.) DISC 2009. LNCS, vol. 5805, pp. 8–21. Springer, Heidelberg (2009)

17. Elrad, T., Francez, N.: Decomposition of Distributed Programs into Communication-Closed Layers. Sci. Comput. Program. 2(3), 155–173 (1982)

18. Fidge, C.J.: Logical Time in Distributed Computing Systems. IEEE Computer 24(8), 28–33 (1991)

19. Fischer, M.J., Lynch, N.A., Paterson, M.S.: Impossibility of Distributed Consensus with One Faulty Process. Journal of the ACM 32(2), 374–382 (1985)

20. Gafni, E.: Round-by-Round Fault Detectors: Unifying Synchrony and Asynchrony (Extended Abstract). In: Proc. 17nth ACM Symposium on Principles of Distributed Computing (PODC), Puerto Vallarta, Mexico, June 28-July 2, pp. 143–152 (1998)

21. Gafni, E.: The 0-1-Exclusion Families of Tasks. In: Baker, T.P., Bui, A., Tixeuil, S. (eds.) OPODIS 2008. LNCS, vol. 5401, pp. 246–258. Springer, Heidelberg (2008)

22. Gafni, E., Merritt, M., Taubenfeld, G.: The Concurrency Hierarchy, and Algorithms for Unbounded Concurrency. In: Proc. 21st ACM Symposium on Principles of Distributed Computing (PODC 2001), pp. 161–169. ACM Press, New York (2001)

23. Gafni, E., Rajsbaum, S.: Musical Benches. In: Fraigniaud, P. (ed.) DISC 2005. LNCS, vol. 3724, pp. 63–77. Springer, Heidelberg (2005)

24. Gafni, E., Rajsbaum, S.: Recursion in distributed computing. In: Dolev, S., Cobb, J., Fischer, M., Yung, M. (eds.) SSS 2010. LNCS, vol. 6366, pp. 362–376. Springer, Heidelberg (2010)

25. Gafni, E., Rajsbaum, S., Herlihy, M.: Subconsensus Tasks: Renaming Is Weaker Than Set Agreement. In: Dolev, S. (ed.) DISC 2006. LNCS, vol. 4167, pp. 329–338. Springer, Heidelberg (2006)

26. Gafni, E., Rajsbaum, S., Raynal, M., Travers, C.: The Committee Decision Problem. In: Correa, J.R., Hevia, A., Kiwi, M. (eds.) LATIN 2006. LNCS, vol. 3887, pp. 502–514. Springer, Heidelberg (2006)
27. Herlihy, M.: Wait-Free Synchronization. ACM Transactions on Programming Languages and Systems 11(1), 124–149 (1991)
28. Herlihy, M.: Linearizability. In: Kao, M.-Y. (ed.) Encyclopedia of Algorithms, Springer, Heidelberg (2008)
29. Herlihy, M., Rajsbaum, S.: The Topology of Shared-Memory Adversaries. In: Proc. of the 29th ACM Symposium on Principles of Distributed Computing (PODC), Zurich, Switzerland, July 25-28, pp. 105–113 (2010)
30. Herlihy, M., Rajsbaum, S.: Concurrent Computing and Shellable Complexes. In: Lynch, N., Shvartsman, A. (eds.) DISC 2010. LNCS, vol. 6343, pp. 109–123. Springer, Heidelberg (2010)
31. Herlihy, M., Rajsbaum, S., Tuttle, M.R.: Unifying Synchronous and Asynchronous Message-Passing Models. In: Proc. 17nth ACM Symposium on Principles of Distributed Computing (PODC), Puerto Vallarta, Mexico, June 28-July 2, pp. 133–142 (1998)
32. Herlihy, M., Shavit, N.: A simple constructive computability theorem for wait-free computation. In: Proc. 26th ACM Symposium on Theory of Computing (STOC), Montreal, Quebec, Canada, May 23-25, pp. 243–252 (1994)
33. Herlihy, M., Shavit, N.: The Topological Structure of Asynchronous Computability. Journal of the ACM 46(6), 858–923 (1999)
34. Junqueira, F.P., Marzullo, K.: Designing Algorithms for Dependent Process Failures. In: Schiper, A., Shvartsman, M.M.A.A., Weatherspoon, H., Zhao, B.Y. (eds.) Future Directions in Distributed Computing. LNCS, vol. 2584, pp. 24–28. Springer, Heidelberg (2003)
35. Lamport, L.: Time, Clocks, and the Ordering of Events in a Distributed System. ACM Commun. 21(7), 558–565 (1978)
36. Mattern, F.: Virtual Time and Global States of Distributed Systems. In: Proc. Parallel and Distributed Algorithms Conf., pp. 215–226. Elsevier Science, Amsterdam (1988)
37. Moses, Y., Rajsbaum, S.: A Layered Analysis of Consensus. SIAM J. Comput. 31(4), 989–1021 (2002)
38. Rajsbaum, S.: Iterated Shared Memory Models. In: López-Ortiz, A. (ed.) LATIN 2010. LNCS, vol. 6034, pp. 407–416. Springer, Heidelberg (2010)
39. Rajsbaum, S., Raynal, M., Travers, C.: An impossibility about failure detectors in the iterated immediate snapshot model. Inf. Process. Lett. 108(3), 160–164 (2008)
40. Rajsbaum, S., Raynal, M., Travers, C.: The Iterated Restricted Immediate Snapshot Model. In: Hu, X., Wang, J. (eds.) COCOON 2008. LNCS, vol. 5092, pp. 487–497. Springer, Heidelberg (2008)
41. Saks, M.E., Zaharoglou, F.: Wait-Free k-Set Agreement is Impossible: The Topology of Public Knowledge. SIAM Journal on Computing 29(5), 1449–1483 (2000)
42. Zielinski, P.: Anti-Omega: the weakest failure detector for set agreement. In: Bazzi, R.A., Patt-Shamir, B. (eds.) Proc. 27nth ACM Symp. Principles of Distributed Computing (PODC), Toronto, Canada, August 18-21, pp. 55–64 (2008)

# SkewCCC+: A Heterogeneous Distributed Hash Table*

Marcin Bienkowski[1], André Brinkmann[2],
Marek Klonowski[3], and Miroslaw Korzeniowski[3]

[1] Institute of Computer Science, University of Wrocław, Poland
[2] Paderborn Center for Parallel Computing PC[2], University of Paderborn, Germany
[3] Inst. of Mathematics and Computer Science, Wrocław Univ. of Technology, Poland

**Abstract.** Distributed Hash Tables (DHTs) enable fully distributed Peer-to-Peer network construction and maintenance with name-driven routing. There exist very few DHT approaches that consider heterogeneity of nodes inside the construction process or properly serve data of different load. To our best knowledge, there is no construction which smoothly addresses both these issues.

We propose a Peer-to-Peer construction that explicitly uses heterogeneity to simplify the routing and maintenance process even in the presence of an adaptive adversary. Using a hypercube and cube connected cycles networks as a backbone, we show how to cope with two types of heterogeneity: one for nodes and one for data.

## 1 Introduction

Peer-to-peer (P2P) networks are successfully used to store and retrieve data in scale-out environments like the Internet. Structured P2P networks are typically based on distributed hash tables (DHTs)[15,22], in which data items are mapped to a key space (e.g., a $[0, 1)$ interval), whereas network nodes (called *peers*) are mapped to disjoint subsets of this space. The mapping is performed by a hash function known by all nodes. Peers are responsible for storing information about the data items that are mapped to their subsets. The (logical) network connections between peers are also created on the basis of this mapping: for example, nodes responsible for adjacent subsets are usually directly connected. Such constructions allow for a fully distributed network maintenance and provide a distributed lookup service. Hence, the performance of the DHT depends on two key factors. The first one is the ability to evenly balance the load: each node should be responsible for approximately the same number of data items, which — provided the even distribution of data items in the key space — means that each node should be mapped to a space subset of approximately the same size. The second factor covers the routing parameters of the induced network, such as dilation or congestion.

* Supported by MNiSW grants number N206 257335, 2008-2011, and PBZ/MNiSW/ 07/2006/46.

C. Lu, T. Masuzawa, and M. Mosbah (Eds.): OPODIS 2010, LNCS 6490, pp. 219–234, 2010.

In most previously proposed P2P models, peers are originally treated as being homogeneous, neglecting their different capabilities. Heterogeneity is then added as an additional layer of abstraction, by simply representing each peer by several independent virtual peers and without exploiting the fact that they constitute one entity. There are a few approaches which consider heterogeneity of nodes in the construction process. However, these approaches divide the P2P network into different layers and assign peers to layers according to their strength. Furthermore, to our best knowledge, no previous work handles two aspects of heterogeneity simultaneously, namely the heterogeneous peers and data items of different sizes.

## 1.1  Our Contribution

The main idea and contribution of this paper is a construction of a P2P network, where heterogeneity of nodes not only does not hinder the network maintenance, but allows the network to be more efficient. We base our network, *SkewCCC+*, on our previous solution, SkewCCC [8]. Heterogeneity of nodes is respected from the very beginning: when joining our network, a peer declares its *strength* (which affects its parameters such as the desired node degree), and the network is constructed to meet this requirement. We show how to effectively maintain even load of the nodes. We provide two alternative solutions for checking the imbalance: the first one is based on local information, the second one on global sampling. Moreover, our approach works well even in the presence of an *adaptive* adversary. We note that the previous solution neither handles heterogeneity nor is it able to cope with an adaptive adversary [8].

**Used techniques.** From the perspective of mapping peers to the DHT key space, our approach is based on a classic approach of partitioning trees, which directly map nodes to subsets of the key space. However, in our network, this embedding is not static as for example in the CAN network [21]. Instead, to enable quick and cheap reconfigurations (e.g., due to rebalancing after adversarial changes to the network), our embedding can be dynamically adapted.

The topology of our network resembles a distorted variant of the cube connected cycles (CCC) network, i.e., roughly speaking it is similar to a hypercubic network. Due to its distributed nature, an exact hypercubic topology would be too expensive to maintain and therefore the assumed hypercube dimension is not the same for all the nodes. We model stronger peers as complete hypercubic networks, which are inserted as sub-hypercubes of the main hypercube. This preserves the locality, as opposed to solutions in which such peers are represented by several virtual nodes acting independently in random parts of the network. We avoid the introduction of super-peers, so stronger peers can execute the same algorithms as regular peers.

Finally, we show that our solution allows for data items of different size without hindering the parameters of our construction. We use a second partitioning tree, which spans the same address space to distribute heterogeneous data items.

**Parameters of our network.** In the description below, $n$ denotes the number of nodes in the system and $m$ is the sum of their strengths, i.e., $m = n$ for homogeneous nodes. The parameters below are obtained with high probability.

First, we address network efficiency. The degree of regular peers (those of strength 1) is constant. Inserting or removing a peer of an arbitrary strength takes time and amortized communication cost of $O(\log^2 m)$. Note that bounding the costs in a single step is not feasible as the adversary is able to induce changes in a substantial part of the network. Searching and routing in a stable network takes $O(\log m)$ time and communication cost for nodes of constant strength, $O(\log n)$ for nodes of average strength and $O(1)$ for nodes of strength $\Theta(m)$.

One of our goals is to achieve even load balance among the nodes, which is defined as the maximum deviation from the average key-space size, for which a single node is responsible. We provide two approaches: a version that is based on locally available information achieves a load balance of $O(\sqrt{m})$, whereas the approach using global sampling achieves a load balance of $O(1)$.

Finally, to protect against peer failures, we present a simple algorithm to store $\Theta(\log m)$ copies of each data item in $\Theta(\log m)$ different physical peers.

## 1.2   Related Work

There are many DHT-based P2P solutions suitable for building overlay networks consisting of homogeneous peers; examples are Chord [22], SkipNets [14], Skip-Graphs [3], Hyperrings [5], Viceroy [19], and Distance Halving [20]. They usually require nodes of constant or logarithmic degree and have a logarithmic dilation and induce logarithmic costs for join and leave operations. The P2P networks mentioned above assume that each peer inside the network has the same properties concerning bandwidth or storage capacity. Different peers' capabilities are modeled by introducing virtual peers, where each big peer of strength $k$ is responsible for $k$ virtual peers of strength 1. Each of these virtual peers acts independently of its relatives. This concept hinders the usage of synergies between virtual peers of the same physical peer and even enables intruders to be more effective.

The $Y_0$ approach from Godfrey and Stoica clusters the virtual peers in a restricted range of the Chord network to reduce the out-degree of each peer [12]. The construction of their approach is similar to the Share strategy, which divides the ID space fairly among a set of nodes [9].

Another approach to integrate heterogeneous nodes is the simple distinction between strong and weak nodes, where strong nodes act as super-peers in a super-layer and weak nodes simply connect to these stronger nodes as leaf-layer nodes (see, e.g., [18,24]). Message flooding is only conducted within super-layer and all leaf peers are represented by their corresponding super peers [23]. Of course, this distinction is very coarse grained and the potential of each node can only be again exploited by using virtual nodes.

The Pagoda network is the first P2P network that explicitly uses the strength of a peer inside the join process [7]. The Pagoda network has a constant degree, a logarithmic diameter, and a 1/logarithmic expansion, while its construction is completely deterministic. Pagoda is a leveled network, where every peer is just associated with a single node, and every parent of a peer must have a bandwidth that is at least as large as the bandwidth of that peer. The lowest degree of a peer in Pagoda is constant but not the minimum possible 3 as in our construction. Also, Pagoda is not designed to work against an adaptive adversary ruling the dynamics of the system.

Our approach and analysis is based on a partitioning tree, which has been introduced for P2P networks by Ratnasamy et al. [21] (see Section 2.1 for details on the partitioning tree). The maximum height difference between two leaves inside the partitioning tree can be used to characterize the load imbalance of the P2P network. Adler et al. [2] built a hypercubic network on the top of such tree and showed that it is possible to keep the height difference constant for homogeneous nodes if no nodes may be removed from the network. Our construction works for heterogeneous nodes and serves both insertion and removal operations in an adversarial setting.

Dynamics of the system (so called churn) are specific to Peer-to-Peer designs, however a fully adversarial model is practically never assumed. One exception is a paper by Kuhn et. al [16], where a hyper-cube based network is constructed for adversarial joins and leaves of peers. A somehow perpendicular approach is considered by Awerbuch and Scheideler in [4] where an adversary has full control (not only dynamics of joins and leaves but also other behavior) over a small fraction of all peers. Neither of these approaches consideres heterogeneous peers.

Also more practically oriented results are known for treating with dynamics of the system. An important result is by Guerraoui et. al [13] which shows a distributed overlay construction based on gossiping especially designed to cope with dynamics. This result is experimentally evaluated in both static and dynamic scenarios using traces of a real Peer-to-Peer system.

Some data-oriented P2P networks try to keep semantic relationships between the keys stored in the network and to support multi-attribute queries. Mercury uses the concept of hubs, where each hub is organized as a ring and is responsible for one attribute [6]. The P2P network Oscar tries to improve the routing performance compared to Mercury for skewed key spaces by learning certain regions of the distribution [11]. The approach of P-Grid of building a trie to keep a lexicographic ordering is in some sense similar to the approach presented in this paper [1]. Nevertheless, our approach is based on a random hash-function inside each node of our search tree, which strongly reduces update costs in adversarial (and realistic) environments. In general it is difficult to compare approaches that keep lexicographic orders with our unstructured approach, as keeping this order is always costly.

## 1.3   Organization

The rest of this paper is organized as follows. In Section 2, we define the desired properties of our network. In Section 3, we show how to achieve these properties, i.e., we show algorithms used for searching in the network, joining and leaving it and balancing operations performed on it. In Section 4, we formally analyze our construction (due to space limitations formal proofs can be found in the full version of this paper). In Section 5, we explain how to treat data of different sizes and we conclude in Section 6.

## 2  Construction

In our description inside this section, we concentrate on network construction for heterogeneous peers and unit-sized data items. We show how to extend the given algorithms to storing data of different sizes in Section 5.

We call participants of the network *peers*. Each peer $u$ has a parameter $c(u)$, called its *strength* and reflecting its capabilities (i.e., bandwidth); we assume that each peer of strength $c(u)$ is able to communicate with $3 \cdot c(u)$ neighbors. For simplicity, we assume that the strength is a power of 2 and equals at least 4. We also define the dimension of each peer $u$ as $\dim(u) = \log(c(u))$. Peers whose strength is not a power of two, can be treated in two ways. Either we cut their strength to the highest possible power of 2, thus losing at most a constant fraction of network bandwidth, or we decompose them into independent peers whose strengths are powers of 2, thus adding another logarithmic factor to the cost of basic operations. A trade-off between these two approaches is also possible. Except the total number of peers $n$, we also use the parameter $m = \sum_u c(u)$, called *network capacity*. Naturally, both $n$ and $m$ may change over time and are unknown locally by any peer.

Each peer has a name, name$(u)$, which can be for example its IP address. For simplicity of the description, we assume that each *real* peer is represented in the network by $c(u)$ virtual peers (called *v-peers* for short), each with the same name as the original one. Later, we show how to use the fact that in reality they constitute one entity.

We describe our network design in three levels of abstraction. In the first one, we partition v-peers into groups using a tree (Section 2.1). In the second one, we show which virtual connections are established between these groups (Section 2.2). In the third one, we describe how and by which v-peers these virtual connections are realized (Section 2.3).

### 2.1  Partitioning Tree

A partitioning tree (see Fig. 1a) represents a hierarchical partitioning of the set of all v-peers. We emphasize that this tree is a purely virtual construct, which is used to map the names of virtual peers to addresses in the SkewCCC+ network. Furthermore, it is used for labeling groups of v-peers, and therefore to simplify peers' address management.

**Definition 1 (Partitioning tree).** *A partitioning tree is a proper binary tree. Each v-peer is mapped to a single leaf. We say that a leaf contains all v-peers mapped to it. Moreover, an internal node $s$ contains all peers mapped to leaves from the subtree rooted in $s$. Thus, the root contains all v-peers.*

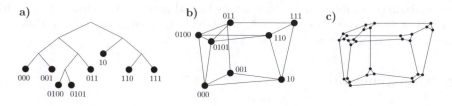

**Fig. 1.** SkewCCC+ construction. Parts a), b), and c) represent the partitioning tree, virtual hypercubic connections and the real SkewCCC+ network. Large dots represent groups of v-peers, small ones v-peers themselves. Only core v-peers are depicted.

**Definition 2 (Edge and node labeling).** *In a partitioning tree: (i) Each internal node has one child connected by an edge labeled* 0 *and one connected by an edge labeled* 1. *(ii) Each node is labeled by the path used to reach it from the root. The root is labeled by an empty string* $\varepsilon$. *(iii) The level of a node is its distance from the root node. The root node is on level* 0. *(iv) For any node* $s$, *$s$-subtree denotes the subtree rooted at* $s$. *(v) At any moment, a hash function* $h_s$, *mapping names of peers to* $\{0,1\}$, *is assigned to each internal node* $s$; *these functions can change over time. The function* $h_s$ *is known within the whole $s$-subtree (by all v-peers belonging to this subtree).*

*Hash functions.* In the analysis of the protocol we treat a hash function as a random oracle, i.e., if we compute the value of a hash function $y = h_s(u)$ for the first time for $u$, the result $y$ is a random bit-string that is stochastically independent of all previously computed values. In practice, the function $h_s(x)$ can be implemented as $h_s(u) = \text{HASH}(s\|u)$, where HASH is a standard hash function like SHA256.

Note that some subprotocols described below need a family of independent hash functions for any fixed $s$. This can be realized by setting $h_s(u)=\text{HASH}(s\|u\|t)$, where $t$ is an integer. Every time the function is required to be changed, $t$ is chosen uniformly at random in some range of integers. This range has to be sufficiently large for $t$ to never occur twice in independent choices and sufficiently small to allow broadcasting of $t$ in a part of the network. In practice, we propose $t$ to be a 160-bit number. For the sake of clarity, we skip $t$ in further notation and we talk about the "current" values of hash functions.

*Choosing peer's position.* Any peer $u$ is mapped to a node of the tree, not necessarily a leaf. First, we describe this mapping in a case when $c(u) = 1$, i.e., $u$ is a single v-peer. In this case $u$ is mapped to a leaf $s$, which is chosen on the basis of name($u$) and the current hash functions for the nodes of the tree. The label of $s$ is fixed bit by bit starting in the root of the tree and proceeding to a leaf: $s_1 = h_\varepsilon(\text{name}(u))$, $s_2 = h_{(s_1)}(\text{name}(u))$, ..., $s_\ell = h_{(s_1,s_2,...,s_{\ell-1})}(\text{name}(u))$.

Choosing a position of a real peer $u$ of strength greater than 1 is performed analogously, but the bit fixing procedure described above may stop in an internal node $s$ of the tree. We first calculate a level $\ell$ on which $u$ is going to be inserted. Starting from the root of the virtual tree, we fix the label of a node on level

$\ell$ bit by bit in the same way as above until level $\ell$ is reached. All v-peers of $u$ are inserted in the $s$-subtree and they are distributed evenly in this subtree, i.e., $c(u)/2$ v-peers are mapped to the left subtree of $s$ and $c(u)/2$ to the right one. This division proceeds recursively to the leaves. We require that the level of $s$ is chosen so that each leaf of the $s$-subtree contains at least 1 of these v-peers and at most 4. For this to be possible, the leaves of the $s$-subtree have to be on levels $\ell + \dim(u)$, $\ell + \dim(u) - 1$ or $\ell + \dim(u) - 2$; later we describe how this condition is ensured. Finally, for any strength of the peer $u$, we say that its label, denoted label$(u)$, is the label of the node $s = (s_1, \ldots, s_\ell)$, to which it is mapped.

## 2.2 Virtual Connections

We now describe, which leaves of the partitioning tree are (virtually) connected. These connections are also depicted in Fig. 1b. Virtual connections create a hypercubic structure. Assume first that the tree is completely balanced, i.e., all leaves are on the same level $\ell$. This means that the labels of all leaves are binary strings of length $\ell$. Then, there are $2^\ell$ leaves and there is a connection between two of them if the Hamming distance of their labels is exactly 1, i.e., their labels differ on exactly 1 bit.

As it is too costly to keep the tree perfectly balanced, we need to relax this condition, in a way similar to that of [8]. Choose any two virtual tree nodes $u$ and $u'$, both on level $\ell$, with Hamming distance 1, such that at least one of them, say $u$, is a leaf. Then we require that either $u'$ is a leaf as well, or $u'$ has height 1, i.e., its both sons are leaves. In the former case, $u$ is connected with $u'$, in the latter one $u$ is connected to both sons of $u'$. We call these virtual connections *hypercubic links across dimension $k$*, where $k$ is the bit on which the labels of $u$ and $u'$ differ.

## 2.3 SkewCCC+

In order to show how virtual connections between tree leaves are actually realized by v-peers, we use the following variation of the hypercubic network. Recall that a standard $d$-dimensional CCC (Cube Connected Cycles) network can be created out of a hypercube [17], where each corner of the hypercube is replaced by a $d$-node ring, and each node becomes responsible for one of the corner's $d$ links. Moreover, $d$ nodes on a ring are sorted by dimension of the hypercubic link they are responsible for (which bit they change in corner label). This way the network remains regular and its degree is reduced from $d$ to 3, whereas almost all of the network properties are only slightly changed. In particular, it is possible to route a packet between any two nodes in $2d$ steps.

In our network, called *SkewCCC+*, we make a similar trick of replacing corners by rings, but we start from the hypercubic structure described in the previous subsection. Assume first that a leaf having $r$ hypercubic connections contains exactly $r$ v-peers. Then, they are connected into a ring in the same sorted manner as described above. In particular, if a leaf has two hypercubic links across the same dimension, then the respective v-peers are adjacent in the ring. Those v-peers which are responsible for hypercubic connections are called *core*; a core v-peer has number $i$ if it is responsible for a hypercubic link across dimension $i$; for each $i$ there are either 1 or 2 v-peers responsible for it. This assignment is depicted in Fig. 1c.

If a leaf contains more than $r$ v-peers, the excessing ones are called *spare*. Spare v-peers are inserted into the ring evenly between core peers (if there are $r'$ v-peers at this leaf, the path between two consecutive core peers should be of length $\lfloor r'/r \rfloor$ or $\lceil r'/r \rceil$). Recall that a leaf at level $d$ has between $d$ and $2d$ virtual connections to other leaves. To ensure that each hypercubic link can be handled by a v-peer, we require that such a leaf contains at least $2d$ v-peers. As routing is slowed down by the number of spare v-peers between any two core v-peers, we also require that a ring contains $O(d)$ v-peers.

# 3   Algorithms

In this section, we show how basic operations such as searching, joining, and leaving the network are implemented. We also describe two rebalancing operations and define what triggers them. The algorithms described below preserve the network structure described in Section 2. Our algorithms use parameters $\alpha$ and $\beta$, meaning that in a ring on level $\ell$ there should be at least $\alpha \cdot \ell$ and at most $\beta \cdot \ell$ v-peers. Specific constraints on these parameters are given in Section 4.

## 3.1   Searching

Searching (look-up) in our structure is performed as searching in a CCC network and inherits its properties; we describe it very shortly for completeness. Assume that a v-peer belonging to a ring $s$ wants to search for a resource with label $k = (k_1, k_2, \ldots)$. For simplicity of the description, we assume that this label is of infinite length. In practice, 160-bit labels are sufficient. The resource associated with label $k$ is stored in the ring $t = (t_1, t_2, ..., t_{d'})$, such that $t_1 = k_1, ..., t_{d'} = k_{d'}$. If the sending v-peer belongs to a peer of dimension $d$, the latter is represented not only in $s$ but also in at least $2^{d-2}$ rings and we can start from any of them. We choose the one whose label agrees with $k$ on the maximal number of bits. We proceed in $d' - (d - 2)$ iterations; in iteration $i$, we find a ring whose name prefix label agrees with $k$ on the first $i$ positions. In one iteration, a request is passed to the next core v-peer on the ring and along a hypercubic link.

## 3.2   Joining and Leaving

When a peer $u$ of strength 1 (i.e., $\dim(u) = 0$) joins the network, it first resolves its corresponding address based on the partitioning tree described in Section 2.1. It finds the ring to which it shall belong, i.e., it looks up its own name in the network using the search operation. Then it joins this ring as a spare v-peer, the ring is split into two rings, if necessary and possible, and it is checked if a rebalancing operation is needed. We treat peers of strength smaller than 4 as if they had strength 1. When $c(u) \geq 4$, i.e., $\dim(u) \geq 2$, the situation is more complicated. The peer checks to which internal node $s$ of the tree it should be mapped in the following manner. First, it searches the tree, behaving as a peer of strength 1 and finds a leaf at level $h$. Let $s$ be the ancestor of this leaf at level $\ell = \max\{h - \dim(u) + 1, 0\}$. Then, $u$ inserts its v-peers in all leaves of the subtree rooted at $s$, distributed as described in Section 2.1. Thus, $u$ should be represented by 1, 2, or 4 v-peers in the leaves of the $s$-subtree. If this is not the case, rebalancing procedures are called as described in Section 3.3.

If the ring that $u$ joins is on level $l$ and already has $\beta \cdot l$ v-peers, a split operation is necessary. If all neighboring rings are of dimension $l$ or $l+1$ the ring splits into two level $l + 1$ rings. If there exists a neighboring ring of dimension $l - 1$, a rebalancing operation is needed.

There is one special situation to be considered. If $u$ is rooted in the root of the virtual tree, it means its dimension (and strength) differs only by a constant factor from the dimension (or total strength) of the whole network. In such case, for robustness, we limit the dimension of $u$ so that it is represented in the initially found ring by 2 v-peers.

A leave operation is performed in a manner mirroring the join operation. Before a peer leaves the network, it removes all its v-peers after which some merge and rebalance operations might be necessary. If a ring at level $l$, from which a v-peer is to be removed, has $\alpha \cdot l$ v-peers, it needs to merge with its brother in the virtual tree. If it still has neighbors at level $l + 1$, a rebalancing is needed.

## 3.3   Rebalancing

We use two procedures for rebalancing. The first one, REBALANCE($s$), is called with a tree node $s$ at level $l$ as an argument and its purpose is to rebuild the whole $s$-subtree. We assume that a concrete v-peer from the $s$-subtree initializes REBALANCE($s$). First, we estimate the number of v-peers in the $s$-subtree by using a procedure for approximating (with a constant factor error) the number of nodes in a peer-to-peer network. The method shown in [10] can be directly used or methods from [4] can be adapted. Based on the returned lower bound $m'_s$ on the number $m_s$ of v-peers in the $s$-subtree, we calculate its new height $h$, so that if we inserted $m'_s$ v-peers to an initially empty network, and they had perfectly balanced labels, we would get a tree with height $h + 1$. Then we construct a tree of height $h$. Information about $h$ is broadcasted to all peers in $s$-subtree and each peer of dimension $d \geq 2$ calculates its new rooting level of $l + h - (d - 1)$, i.e., such that it is represented by 2 v-peers in each node on level $l + h$.

First, new hash functions are chosen in all nodes in the $s$-subtree. This operation is initiated by the peer which started the rebalancing operation. It chooses a random number $t$ which is used as one of the parameters (next to the name of a peer and the address of a virtual tree node) for hash functions in nodes of the virtual tree. Then the chosen $t$ is broadcasted to the whole $s$-subtree and the new hash functions are chosen in this subtree. Then all v-peers in the subtree change their position according to these new hash functions. This change is not immediate. First, the spare v-peers migrate, while the core v-peers still provide an $h$-dimensional hypercubic structure during rebalancing. They are allowed to migrate, when sufficiently many spare peers join a ring according to the new hash functions and can at least temporarily overtake their job. Rings with current heights larger than $h$ are merged to height $h$ and rings with heights smaller than $h$ wait for sufficiently many migrating v-peers to split to dimension $h$ as soon as possible. This is possible, if only the number of v-peers in each ring is large enough, i.e., for sufficiently large constant $\alpha$.

As shall be proven in Section 4, all leaves in the resulting s-subtree are on level $h$, with high probability. We can employ another broadcast to check if this procedure was successful and repeat rebalancing if necessary, thus changing a Monte Carlo randomized algorithm into a Las Vegas one.

The second rebalancing procedure, RECURSIVEREBALANCE$(s, s')$, balances the subtrees $s$ and $s'$, so that their heights differ by no more than one after the procedure. It goes up the tree and finishes at latest in their least common ancestor. From $s$ and $s'$ the one on higher level, say $s$, is chosen. Then REBALANCE$(s)$ is called and if after this operation leaves of the subtree rooted in $s$ are on a level differing by at most 1 from the level of leaves of $s'$, the procedure successfully finishes. Otherwise, the parent of $s$ takes the role of $s$ and the procedure is called anew.

In the following, we describe when rebalancing operations are necessary and at which nodes of the virtual tree they have to be performed. Recall that one of our goals is to keep a difference of at most 1 between levels of any two rings in the network. If we detect two rings which differ in levels by more than 1, we start a rebalancing procedure in which these two rings are parameters. Another goal is for heterogeneous peers to be represented by 1, 2, or 4 v-peers in a ring. If we detect a peer which violates it, we repair it. These two basic conditions are checked formally in a set of rules given in the full version of this paper. Also local and global methods for checking imbalance are described shortly in the full version.

## 4   Analysis

In this section, we provide an analysis for the given model and algorithms. In the analysis, we assume that the situation is stable, i.e., that there are no pending RECURSIVEREBALANCE or REBALANCE operations. We divide the analysis in the following blocks. First, we prove that if neither any RECURSIVEREBALANCE or any REBALANCE is needed anywhere in the network, then the network is globally balanced. Second, we show that rebalancing procedures have linear times polylogarithmic cost and take polylogarithmic time in a subtree in which they are called. Then, we prove that they are called only every linear number of insert/remove operations, so their amortized cost is polylogarithmic. Lastly, we show that the imbalance detection subroutines detect any imbalance in the network in a deterministic way in case of the local mode and with high probability in case of the global mode. Due to space constraints, proofs have been moved to the full version of this paper.

## 4.1   The Adversary

Before analyzing our construction we firstly define what we mean by an adaptive adversary. At any moment, the adaptive adversary has full information about all peers present in the network, about all current hash functions for all nodes of the virtual tree, and about all possible values of these hash functions for all peers. As long as no REBALANCE operations are needed, all hash functions are fixed and the adversary can insert peers whose names are chosen by her, and thus she can insert these peers into any node of the virtual tree. She also has the power of removing any peer, basing on her current knowledge.

The only weakness of the adversary lies in the replacement of hash functions. If the current hash function $h_{s,t}$ for a virtual tree node $s$ was changed in a deterministic way during the rebalancing process (e.g. always to $h_{s,t+1}$), then the adversary would be able to insert peers with such names that their hashes for some number of next steps would be actually chosen by her. In contrast, our construction ensures that the space of possible replacements of $h_{s,t}$ is sufficiently large; this makes it impossible to tamper with by the adversary, and thus the newly chosen hash functions can indeed be treated as random functions for the set of currently present peers.

We stress that our model differs from models sometimes assumed in security analysis (for example in [4]). In such context, it is assumed that the adversary has total control over some fraction of peers. Our model is somehow perpendicular: the adversary has full information about all peers, and fully controls their dynamics but not other behavior. Notice that if she controlled some peers, she would be able to choose future hash functions as in our construction a single peer decides about new hash functions for a whole rebalanced subtree of the virtual tree.

## 4.2   Imbalance in a Correct Stable State

First, we show that the number of levels on which rings (i.e., leaves in the virtual tree) appear is bounded even if our algorithm is only allowed to communicate locally (any imbalance conditions are checked only between neighboring rings).

**Lemma 1.** *The maximal dimension in the network is $\left\lceil \frac{3}{2} \cdot d \right\rceil$ if the minimal is $d$.*

We use the above lemma to strictly bound the number of levels at which rings appear by $\frac{1}{2}\log m$, which yields a worst-case imbalance (the maximal ratio of key-space parts assigned to rings) of $\frac{\beta}{\alpha} \cdot 2^{\frac{\log m}{2}} = O(\sqrt{m})$.

**Lemma 2.** *The level of each ring is between $\frac{2}{3}\log m - O(\log \log m)$ and $\frac{3}{2}\log m$. The difference in levels between the highest- and the lowest-dimensional ring is at most $\frac{1}{2}\log m$.*

## 4.3    Correctness and Cost of Balancing

Below we prove the quality of our rebalancing procedures, assuming that the applied approximation procedure (taken e.g. from any of [10,4]) lower bounds the number of v-peers in a s-subtree and overestimates it by at most a constant factor $\gamma$. Note that by using an additional maximum computation through rumor spreading, it is possible to have the same approximation $m'_s$ in each participating peer. We use the following formulation.

**Theorem 1 (based on [10]).** *For a s-subtree containing $m_s$ v-peers it is possible to calculate $m'_s$ in a distributed fashion in time $O(\log m_s)$, so that, with high probability, $m_s \leq m'_s \leq \gamma \cdot m_s$, where $\gamma$ is constant.*

We assume that the approximation guarantee is $\gamma = 5/4$. If we prefer a weaker approximation procedure (better due to other costs), our construction is still possible, but we have to weaken imbalance conditions: leaves have to be allowed to differ in levels by more than 1. We take $\gamma = 5/4$ for simplicity of description.

**Lemma 3.** *Assume that the chosen height of an s-subtree is $h$ after a rebalancing operation* REBALANCE($s$) *for node $s$ on level $k$. Then each leaf in the s-subtree receives an expected number $EX$ of v-peers, where $2\alpha(k + h) \leq EX \leq 5\alpha(k + h)$.*

At this point we fix $\beta = 10 \cdot \alpha$, as this yields that a ring wants to change its dimension after a linear (in the size of the ring) number of joins and/or leaves in this ring.

**Lemma 4.** *Assume that the chosen height of an s-subtree is $h$ after a rebalancing operation* REBALANCE($s$) *for node $s$ on level $k$. Then with high probability the number $X$ of v-peers in each leaf satisfies $|X - EX| \leq \frac{1}{4}\alpha(k + h)$.*

**Lemma 5.** *For a subtree rooted in an internal node $s$, containing $m_s$ v-peers,* REBALANCE($s$) *takes time $O(\log m)$ and its communication cost is $O(m_s \cdot \log m)$.*

## 4.4    Amortizing the Cost of Balancing

In order to bound how often the REBALANCE procedure is called in different nodes in the virtual tree, we use Lemma 6 stated below. Together with Lemma 3 (which shows that the tree should be balanced on expectation after performing a REBALANCE procedure), Lemma 4 (which proves that with high probability the tree is indeed balanced), and Lemma 5 (which shows how expensive a single rebalancing is), we will be able to state the total amortized cost of balancing.

**Lemma 6.** *Let $s$ be the last node for which* REBALANCE($s$) *was called before reaching a stable situation and $m_s$ the number of v-peers in the s-subtree at that time. Then the total number of v-peers inserted or removed from the s-subtree since the previous* REBALANCE($s$) *is $\Omega(m_s)$.*

In the following theorem we gather the above results. Assuming that after each REBALANCE operation, the tree on which it was performed is indeed balanced, which Lemma 4 guarantees with high probability, we multiply the cost of rebalancing obtained from Lemma 5 together with its frequency obtained from Lemma 6.

**Theorem 2.** *The expected and holding with high probability amortized cost of inserting or removing a v-peer is $O(\log^2 m)$. Each operation takes time $O(\log m)$.*

## 4.5  Detecting Imbalance

In this section we consider the setting where imbalance is checked in a global fashion, i.e., through sampling random locations whenever a peer is inserted or removed. We initiate a rebalancing procedures if we find a ring differing in dimension by more than 1 from the one that started the sampling. In the sampling process, we are not interested in the chosen rings themselves but rather in their current heights in the virtual tree.

**Lemma 7.** *Let d be the height in the virtual tree which has the highest probability of being chosen in the sampling process. If we choose a random place, then the chosen ring has dimension d with probability $\Omega(\frac{1}{\log m})$.*

Using the above lemma, we prove that our network is balanced with high probability. This time we allow the global difference in levels of leaves to be 2, as we only prove that rings lie on levels differing from the most popular one by at most 1. In the time a ring has to change its dimension twice, $\Omega(\log m)$ v-peers are inserted into it (or removed from it) and it checks $\Omega(\log^2 m)$ random places, so the probability that it never hits the most popular level is bounded by $(1 - 1/(\log m))^{c \cdot \log^2 m} \le e^{-c \log m}$. If the scheme is run for polynomial number of steps, we get the following corollary.

**Corollary 1.** *If the lifetime of the network is bounded by a polynomial in m and the global checking scheme is used to detect imbalance, then with high probability the maximum difference in levels in the network is 2 during the whole runtime. Thus, the imbalance is at most $\frac{4 \cdot \beta}{\alpha}$ at all times.*

## 4.6  The Cost of Routing

Finally, we prove that from the point of view of an average peer the cost of routing a single message depends logarithmically on the number of peers rather than v-peers. This is important for at least two reasons: (i) we do not want strong peers to be slowed down by weak peers, (ii) if we make a mistake in setting the granularity of the network and the dimension of all peers is lower-bounded by some $d$, we do not want $d$ to be added to the routing time.

**Lemma 8.** *For a d-dimensional peer the cost of routing a message from this peer to any destination in a balanced network is at most $2 \cdot (\log m - (d - 2))$. For an average (of strength m/n) or stronger peer, this cost is at most $\log n$.*

# 5   Storing Data

In this section, we describe how data is stored in SkewCCC+ and extend our construction to storing data items of different sizes, and thus respecting different needs for storage space and possibly network capacity.

If the set of data items is static (not ruled by the adversary) and only standard items of unit size are to be used, i.e., the network is to be homogeneous with respect to item sizes, we use a fixed and global hash function $g$ which takes the name of an item and a positive integer and returns one bit. Using $g$ it is possible to generate an infinite pseudo-random sequence of bits for any item; we will use it to fix its place in the network, or actually on the virtual tree.

For an item $u$, the function $g$ taken on bit $0, 1, 2, \ldots$ fixes an infinite path in a binary tree. Using this path we fix a leaf in the virtual tree and store $u$ in the ring representing this leaf in the current SkewCCC+. Using standard routing described in Section 3.1, it is possible to find $u$ in $O(\log m - d)$ steps starting from any $d$-dimensional peer.

We do not fix any concrete strategy to choose on which v-peer in its ring the item should be placed. It can be stored on any v-peer in the ring with employment of any in-ring load balancing procedure or (to support redundancy) it can be stored on all v-peers in its ring. In the latter case we get the following redundancy and robustness properties: each item is stored on $\Theta(\log m)$ v-peers belonging to $\Theta(\log m)$ distinct peers.

When also the set of data items is ruled by an adaptive adversary or items are of different sizes, we propose to employ the same techniques for balancing the data set as we did for balancing the peer set. Two virtual trees spanning the same virtual space are used: one for v-peers and one for data, both serving heterogeneous items, where a larger (more dimensional) data item denotes a data item demanding larger storage space. In this case, instead of using the above hash function $g$, we use a family of changing hash functions similar to the hash functions used in the network construction. Large data items are partitioned into chunks similarly as peers were partitioned into sets of v-peers. Concerning peers, we assume that each of them has storage space proportional to its network bandwidth, i.e., to what we have defined as the strength of a peer.

We assume that both virtual trees are balanced at each moment and that the nodes-tree is lower than the data-tree. Otherwise, it would mean that there are more v-peers than data and not much optimization would be needed. A ring being a leaf $s$ in the nodes-tree serves all data in the subtree rooted in $s$ in the data-tree. We achieve the following properties: (i) each ring in the SkewCCC+ network takes care of the same total size of data up to a constant factor (it comes from both trees being balanced), (ii) inserting and removing of peers and data takes square logarithmic time and square-logarithmic amortized costs, (iii) if each chunk is copied to all v-peers in its ring, it is stored on $\Theta(\log m)$ distinct peers.

# 6    Conclusion

We have presented a Peer-to-Peer network construction which has two main qualities: (i) it allows for smooth serving of heterogeneous peers, and (ii) it maintains balance in the presence of an adaptive adversary. Costs of constructing and maintaining the network are polylogarithmic per operation (in the total capacity of the network). The costs for routing are logarithmic in the number of peers for peers of at least average strength.

For the balance feature, if we opt for the local balance checking algorithm, it achieves polynomial imbalance of $O(\sqrt{m})$ (we may reduce it to $O(\sqrt[k]{m})$ for any constant $k$ if we look further than direct neighbors) and if we allow the global sampling routine, we achieve constant imbalance, with high probability. We stress that the analysis assumes to compete against an adaptive adversary and that in previous constructions the adversary was always an oblivious one, which made the analysis much simpler.

Last but not least, our network construction has nice properties concerning data management. First, essentially the same strategies can be used to distribute heterogeneous data. Second, if we require robustness (i.e., we want to store data in multiple places), we may just replicate a data item within the whole ring to which it is mapped. Since any peer has at most four of its v-peers in a single ring, through such replication the data item is stored on $\Omega(\log m)$ different physical peers.

# References

1. Aberer, K., Cudré-Mauroux, P., Datta, A., Despotovic, Z., Hauswirth, M., Punceva, M., Schmidt, R.: P-grid: a self-organizing structured p2p system. SIGMOD Record 32(3), 29–33 (2003)
2. Adler, M., Halperin, E., Karp, R., Vazirani, V.: A stochastic process on the hypercube with applications to peer-to-peer networks. In: Proc. of the 35th ACM Symp. on Theory of Computing (STOC), pp. 575–584 (2003)
3. Aspnes, J., Shah, G.: Skip graphs. ACM Transactions on Algorithms 3(4) (2007)
4. Awerbuch, B., Scheideler, C.: Group spreading: A protocol for provably secure distributed name service. In: Díaz, J., Karhumäki, J., Lepistö, A., Sannella, D. (eds.) ICALP 2004. LNCS, vol. 3142, pp. 183–195. Springer, Heidelberg (2004)
5. Awerbuch, B., Scheideler, C.: The hyperring: a low-congestion deterministic data structure for distributed environments. In: Proc. of the 15th ACM-SIAM Symp. on Discrete Algorithms (SODA), pp. 318–327 (2004)
6. Bharambe, A.R., Agrawal, M., Seshan, S.: Mercury: supporting scalable multi-attribute range queries. In: Proc. of the ACM SIGCOMM, pp. 353–366 (2004)
7. Bhargava, A., Kothapalli, K., Riley, C., Scheideler, C., Thober, M.: Pagoda: a dynamic overlay network for routing, data management, and multicasting. In: Proc. of the 16th ACM Symp. on Parallel Algorithms and Architectures (SPAA), pp. 170–179 (2004)
8. Bienkowski, M., Brinkmann, A., Korzeniowski, M.: Degree 3 suffices: A large-scale overlay for P2P networks. In: Baker, T.P., Bui, A., Tixeuil, S. (eds.) OPODIS 2008. LNCS, vol. 5401, pp. 184–196. Springer, Heidelberg (2008)

9. Brinkmann, A., Salzwedel, K., Scheideler, C.: Compact, adaptive placement schemes for non-uniform requirements. In: Proc. of the 14th ACM Symp. on Parallel Algorithms and Architectures, SPAA (2002)
10. Durand, M., Flajolet, P.: Loglog counting of large cardinalities. In: Di Battista, G., Zwick, U. (eds.) ESA 2003. LNCS, vol. 2832, pp. 605–617. Springer, Heidelberg (2003)
11. Girdzijauskas, S., Datta, A., Aberer, K.: Structured overlay for heterogeneous environments: Design and evaluation of oscar. ACM Transactions on Autonomous and Adaptive Systems 5(1) (2010)
12. Godfrey, P., Stoica, I.: Heterogeneity and load balance in distributed hash tables. In: Proc. of the 24th IEEE INFOCOM (2005)
13. Guerraoui, R., Handurukande, S.B., Huguenin, K., Kermarrec, A.-M., Fessant, F.L., Riviere, E.: Gosskip, an efficient, fault-tolerant and self organizing overlay using gossip-based construction and skip-lists principles. In: Peer-to-Peer Computing, pp. 12–22 (2006)
14. Harvey, N.J.A., Jones, M.B., Saroiu, S., Theimer, M., Wolman, A.: Skipnet: a scalable overlay network with practical locality properties. In: Proc. of the 4th USENIX Symposium on Internet Technologies and Systems (2003)
15. Karger, D., Lehman, E., et al.: Consistent hashing and random trees: Distributed caching protocols for relieving hot spots on the world wide web. In: Proc. of the 29th ACM Symp. on Theory of Computing, STOC (1997)
16. Kuhn, F., Schmid, S., Wattenhofer, R.: A self-repairing peer-to-peer system resilient to dynamic adversarial churn. In: van Renesse, R. (ed.) IPTPS 2005. LNCS, vol. 3640, pp. 13–23. Springer, Heidelberg (2005)
17. Leighton, F.: Introduction to parallel algorithms and architectures: arrays, trees, hypercubes. Morgan Kaufmann Publishers, San Francisco (1992)
18. Liang, J., Kumar, R., Ross, K.: The kazaa overlay: A measurement study. In: Proceedings of the 19th IEEE Annual Computer Communications Workshop (2004)
19. Malkhi, D., Naor, M., Ratajczak, D.: Viceroy: A scalable and dynamic emulation of the butterfly. In: Proc. of the 21st ACM Symp. on Principles of Distributed Computing (PODC), pp. 183–192 (2002)
20. Naor, M., Wieder, U.: Novel architectures for P2P applications: The continuous-discrete approach. ACM Transactions on Algorithms 3(3) (2007); Also appeared in Proc. of the 15th SPAA, pp. 50–59 (2003)
21. Ratnasamy, S., Francis, P., Handley, M., Karp, R.M., Shenker, S.: A scalable content-addressable network. In: Proc. of the ACM SIGCOMM, pp. 161–172 (2001)
22. Stoica, I., Morris, R., Liben-Nowell, D., Karger, D.R., Kaashoek, M.F., Dabek, F., Balakrishnan, H.: Chord: a scalable peer-to-peer lookup protocol for internet applications. IEEE/ACM Transactions on Networking 11(1), 17–32 (2003); Also appeared in Proc.of the ACM SIGCOMM, pp. 149–160 (2001)
23. Xiao, L., Zhuang, Z., Liu, Y.: Dynamic layer management in superpeer architectures. IEEE Transactions on Parallel and Distributed Systems 16(11), 1078–1091 (2005)
24. Yang, B., Garcia-Molina, H.: Designing a super-peer network. In: Proc. of the 19th International Conference on Data Engineering, ICDE (2003)

# On the Automated Implementation of Time-Based Paxos Using the IOA Compiler

Chryssis Georgiou[1], Procopis Hadjiprocopiou[1], and Peter M. Musial[2]

[1] Department of Computer Science, University of Cyprus, 1678 Nicosia, Cyprus
{chryssis,cs02cp2}@cs.ucy.ac.cy
[2] Department of Computer Science, University of Puerto Rico Rio Piedras, USA
peter.musial@uprrp.edu

**Abstract.** Paxos is a well known algorithm for achieving consensus in distributed environments with uncertain processing and communication timing. Implementations of its variants have been successfully used in the industry (eg., *Chubby* by Google, *Autopilot cluster management* in Bing by Microsoft, and many others). This paper addresses the challenge of the manual coding of complex distributed algorithms, such as Paxos, where this is an error prone process. Our approach in ensuring correct implementation is to use a verified automated translator to compile a source specification that has been proven to be itself correct. We use specification of the Paxos algorithm in the *General Timed Automata* (GTA) model, an extension of I/O Automata, as input to an augmented compiler for the *Input/Output Automata notation* (a.k.a., the IOA compiler) in order to generate executable Java code. The resulting code is interfaced with MPI for communication needs. We have extended the IOA compiler to support a version of the GTA model, which uses time-passage actions such as $\nu(t)$, to model the passage of time by $t$ time units. A time-based version of Paxos is used to demonstrate the capabilities of our extension. In this paper we describe the process to be followed in order to compile time-based Paxos, or similar algorithms. The utility of our approach is supported by an experimental evaluation of our Paxos implementation on a collection of workstations. To the best of our knowledge, our case study constitutes the first example of a time-dependent distributed algorithm that has been specified, verified and implemented in an automated way, using a common formal methodology.

## 1 Introduction

Reasoning about the behavior of complex distributed systems and algorithms is a challenging task. Over the years, several formal methodologies for specifying distributed systems have been proposed and associated techniques and tools have been developed for verifying such systems (e.g., [3,17,20,27,22,28]). However, the benefits of using formal methods has not reached its full potential due to the remaining challenge of implementing such systems; usually the programmer has to manually map the functionality of the abstract specification to detailed programs in order to be executed on target distributed platforms. This raises the question whether the correctness of the abstract specification is maintained during the coding process. To this respect, some tools have been developed in an attempt to provide automated simulation or implementation

C. Lu, T. Masuzawa, and M. Mosbah (Eds.): OPODIS 2010, LNCS 6490, pp. 235–252, 2010.

of formally specified code (e.g., [2,6,7,1,24]). To the best of our knowledge, the IOA Toolkit [1] is the only system to date that combines a language with formally specified semantics (IOA language and checker), automated proof assistants (IOAtoLR theorem prover), simulator (IOA simulator) *and* compiler (IOAtoJava code generator). A number of *asynchronous* algorithms, specified and proved correct using the IOA framework, have been successfully implemented in an automated way using the IOA code generator (see [14,30,15]); the generator translates the IOA specification of a given algorithm to Java code which then can be executed on a network of workstations, where communication is established using MPI [10]. However, before our work, the IOA code generator did not support timing issues.

Existing distributed systems can be viewed as *partially synchronous systems* in the sense that some bounds on processes computation time and messages delays can be estimated and be assumed, but cannot be guaranteed to hold at all times; that is, these bounds might be violated, leading to *timing failures*. Moreover, implementations of algorithms and programs on such systems usually make use of timeouts in order to render some progress of the computation (that is, to provide some liveness guarantees) and to detect component failures. Being able to specify, prove correct and automatically implement such algorithms on a real distributed system using a common formal methodology is the focus of this work.

We have extended the IOAtoJava code generator (or simply IOA compiler) to handle actions modeling passage of time. More specifically, we have extended the IOA compiler (including the IOA syntax checker and IOA composer) to support a version of the General Timed Automata (GTA) model, a timed I/O Automaton model introduced by Lynch and Vaandrager [26]. To demonstrate the functionality of this extension, we used a timed specification of Paxos algorithm [21] as an input to the augmented compiler. The GTA model provides a systematic way of describing the timing behaviors of partially synchronous distributed systems subject to timing failures. The model (and variations of it) can be used for the study of the performance and fault-tolerance analysis (i.e., the liveness) of practical distributed systems under stabilization conditions (see for example the work in [8]). Lynch and Shvartsman [25] produced a GTA-based specification of a timed version of the Paxos algorithm, they proved its correctness (safety) and performed a latency analysis conditioned on certain timing and failure assumptions. The proof presented in [25] was checked to be correct using the interactive IOAtoLR theorem prover by Win and Immorlica in [18] (see also [31]). The specification we used to produce an automated implementation of Paxos using our extended version of the IOA compiler was based on the one in [25].

The rest of the paper is organized as follows. Section 2 overviews the I/O Automata and GTA models as well as the IOA notation and compiler. Also the Paxos algorithm is discussed. In Section 3 we present an in-depth analysis of the procedure for compiling and executing Paxos. Experimental results obtained by implementing Paxos on a network of workstations are presented in Section 4. We conclude in Section 5.

## 2   Background

In this section we provide the necessary background required in the remainder sections.

## 2.1   I/O Automata and the GTA Models

The I/O Automata framework was introduced by Lynch and Tuttle in [27]. A detailed description of this model can be obtained there and in [23, Chapter 8]. An I/O Automaton is a labeled state machine in which a set of transitions connects the actions with the states. It entails a set of states (not necessarily finite) with a nonempty subset of start states, a transition relation, and a set of *actions*. These actions are classified as *input*, *output* and *internal*. The utilization of input and output actions enables the communication of an automaton with its environment. Input actions are controlled by the external environment, whereas internal and output transitions are controlled by the automaton. Actions are given in a precondition-effect style. An action is said to be *enabled* if its preconditions are satisfied. Input actions are always enabled. A transition (also called a step) is given in the form $(s,\pi,s')$ where $s$, $s'$ are states and $\pi$ an action. I/O Automata support the operation of (parallel) *composition* where automata can be combined to form a larger, multifunctional automaton representing a complicated distributed system. The I/O Automata model is *nondeterministic* since in any given state any number of actions may be enabled and there are no restrictions on when an enabled action should be performed.

The GTA model of Lynch and Vaandrager [26] (see also [23, Chapter 23]) is a variant of the I/O Automata model that enables the modeling of timing restrictions. These restrictions can be encoded directly into the states and transitions of the automaton. In addition to input, output and internal actions, a GTA uses *time-passage* actions to model the passage of time. In particular, an action $\nu(t)$ of type time-passage specifies the passage of time by $t$ time-units, $t \in \Re^{+}$. Like internal and output actions, time-passage actions are also controlled by the automaton. Unlike I/O automata, GTAs do not have *tasks* components. Hence, a GTA is composed of four components: (*i*) the signature which contains the input, output, internal and time-passage actions, (*ii*) a set of states, (*iii*) a set of initial states, and (*iv*) the state-transition relation (steps). The GTA model supports the composition of automata similarly to the I/O automaton model. Particularly, a composition of (compatible) GTA automata yields a GTA automaton.

For the purposes of this work, we consider a *free version* of GTA [25] which is similar to the concept of Clock GTA as introduced by De Prisco [8]. In particular, if an automaton $A$ is a GTA, then the free version of $A$ (denoted by $free(A)$ in [25]) is a variant of $A$ that behaves like $A$, except that it relaxes time constraints by allowing any amount of time to pass in situations where $A$ specifies that a particular amount of time should pass. This enables our extended version of the IOA compiler to handle situations in which the exact time constraints are not met by the program (e.g., due to unexpected processing and communication delays).

## 2.2   The IOA Language and the IOA Compiler

The IOA notation is a language used to describe I/O Automata, and can be used both as a formal specification language and a programming language [11]. States are described by the means of the values of variables and transitions in precondition-effect style, instead of state-action-state triples. Preconditions and parameters of the transition must hold whenever this action is executed. The IOA language supports axiomatic and operational descriptions of programing implementations. The language inherits the non-deterministic nature of the I/O Automata model. The IOA notation is supported by the IOA Toolkit [1] via a sequence of tools, such as the checker, the simulator, the theorem prover, and also the compiler. The compiler translates IOA code into Java code.

It was proven that a restricted set of source IOA specifications [30] can be compiled to executable Java code while preserving the *safety* properties of the source specification. To name few such restrictions, specifications must be presented in a node-channel form (discussed next), and specifications must be input delay insensitive. As noted in [14] and [30], a challenging problem (which remains open) is to enable the code generator to also provide some kind of *liveness* guarantees.

Let us now turn our attention to Paxos. To be suitable for compilation, the Paxos specification must be in the node-channel form. Meaning, the algorithm will have two components: First, modeling algorithm code being executed on each network location (or *algorithm automaton*), such as ballot preparation, voting, and reaching the consensus decision. Second, modeling communication channels (or *channel automata*) between different network locations. During the specification phase such channels will be abstract, but with specific safety properties (ex., lossy, reliable, secure, etc.). For the moment let us assume that a node-channel representation of Paxos exists.

Unfolded below is a high level description of the procedure required for the compilation and execution of Paxos. A detailed, algorithm-independent, step-by-step description of the compilation procedure can be found in [14]. We start with the syntactically correct IOA specification of Paxos (described in detail in the next section) in the node-channel form, which can be verified using the IOA *checker*.

Next step is to replace the abstract communication channel with a specific implementation. In our case communication is implemented using the Message Passing Interface (MPI) [10], which is supported by the IOA compiler. The MPI channel is modeled as a *channel automaton* that is a composition of *SendMediator* and *ReceiveMediator* automata. These automata provide the linking to the MPI native libraries and an appearance of interfacing with the abstract channel. All communication between nodes in Paxos is modeled as point-to-point connections. Note that the use of MPI with the Paxos specification does not affect the safety properties of the specification. Preserving the liveness properties, as mentioned above, remains an open challenge. However, our experiments do suggest that under the scenarios considered, the use of MPI does not fault executions of Paxos.

Before the specification is fed to the compiler, additional annotations must be given to resolve nondeterminism. The nondeterminism, inherent from the IOA model, is resolved by requiring the programmer to write a *schedule*. A schedule is a function of the state of the local node that picks the next action to execute at the node. That is, the schedule function selects the next enabled transition as well as the values of its parameters and operates the effects of that transition. In format, a schedule is written at the IOA level in an auxiliary non-determinism resolution language (NDR) consisting of imperative programming constructs similar to those used in IOA effects clauses. Therefore, we developed a (non-trivial) schedule appropriate for Paxos which is contained in Figure 9.

The following steps are independent of input specification. The *composite node automaton* is described as the composition of the algorithm automaton with the channel mediator automata. A *composer* expands this composition into a new, equivalent IOA program in primitive where each piece of the automaton is explicitly instantiated. The resulting *automaton* is annotated with the schedule that describes sequence of computations per each node. The automaton along with its schedule is the final input program to the compiler. The *composite node automaton* augmented with a schedule is now ready for compilation. All the nodes in the system differ in parameterization and input. A common information can be provided to the nodes through the automaton parameters just before the execution of the system. The rank of each node *MPIrank*, described as a unique non-negative integer, is provided by MPI. Another operator supported by MPI is the *MPIsize* which records the number of nodes in the system. The compiler translates each scheduled node automaton into its own Java program suitable to run on the target host.

### 2.3   The Paxos Algorithm

Reaching *consensus* is a fundamental problem in distributed systems. The consensus problem addresses the situation in which there is a set of $n$ processes; each process can propose a value, but in order for the system to reach a consensus state, every process must decide on the same value. In particular three conditions must hold: (a) *Agreement*, all (correct) processes agree on the same value. (b) *Validity*, the agreed value was among the ones proposed by the processes. (c) *Termination*, eventually each (correct) process decides. The first two conditions are *safety* conditions, that is, they must hold at all times. The third one is a *liveness* condition and it can only be met under certain constraints (e.g., it is well known that consensus cannot be solved in a purely asynchronous systems in the presence of a single process crash failure [9]). Distributed consensus has been extensively studied under various system and failure models, see e.g., [23,4].

Paxos is an algorithm designed to solve the consensus problem. It was presented by Lamport in 1990 and was published in 1998 [21]. A considerable advantage of this popular algorithm is that it tolerates processes crashes (and recoveries), message loss, duplication and reordering as well as timing failures. Paxos is guaranteed to work safely (that is, it satisfies agreement and validity) regardless of process, channel and timing failures. When the distributed system stabilizes (that is, there are no failures and a majority of the processes are not crashed, for a long period of time), termination is also achieved [8].

*Description of Paxos.* In brief, Paxos works as follows: a leader starts ballots, tries to associate a value to each ballot, and tries to collect enough approval for each ballot to use the value of that ballot as the decision value. The leader bases its choice of a value to associate with a ballot on the information returned by a quorum of processes[1]. Once the value is associated with the ballot, the leader tries to collect approval from a quorum of processes: if it succeeds, the ballot's value becomes the final consensus decision value. In general, several leaders may operate at the same time and may interfere with each

---

[1] Quorums are sets of processes such that each quorum has a pairwise intersection with any other quorum. Majorities are special cases of quorums.

other's work. However, under a stable state only one leader operates and ensures that a ballot completes. We now outline the main phases of Paxos.

(1) The leader starts a new ballot and informs the others about it.

(2) A process that learns about the new ballot abstains from any earlier ballot for which it has not voted for. In response, a process replies to the leader with the value of the ballot for which it last voted for.

(3) Once the leader receives responses from a quorum, it chooses a value for the ballot that is based on the received values and announces that value to others.

(4) A process that learns about a new value may vote for the ballot, if it has not already abstained. If the process votes, then it informs the leader and others about its vote.

(5) The leader decides on the ballot's value once it receives messages from a quorum with a vote for that value. In case that the leader has failed, a separate leader election service is used to elect a new one. Timeouts are used to determine which processes are operational, and among these, the one with the highest id is elected as the leader. After the election, the new leader starts a new ballot.

(6) Timeouts are also used for the leader to decide when it should start new ballots (that is, there is a limit on how long it takes for a given ballot to be accepted by a quorum of processes).

Based on the above description, there are two timing-dependent components: the leader-election service that determines when a new election should be triggered, and the mechanism that determines when a leader should trigger a new ballot.

*Specification and Correctness of Paxos.* A manuscript by Lynch and Shvartsman [25] provides a formal presentation of the Paxos algorithm. The presentation includes a General Timed Automata specification of the algorithm, a correctness proof (safety) and a performance analysis. The correctness proof, which ensures the agreement and validity properties, was done by hand and it is based on a mapping to an abstract state machine representing a non-distributed version of the algorithm. The performance analysis proves latency bounds, conditioned on certain timing and failure assumptions.

In [18,31] using a time-free version of the Paxos specification of [25] (essentially the last two timing-dependent phases were not considered), and using the IOA2LSL translation tool of the IOA toolkit, the safety of Paxos was mechanically checked. More precisely, it has been shown that every possible externally observable outcome of the Paxos algorithm is also an externally outcome of a general consensus specification. That is, a forward simulation relation from the Paxos automaton to the consensus automaton was defined. Furthermore, the automata and forward simulation conjecture were translated into a readable form by the Larch Prover [12] using an automated translation by the IOA2LSL Tool of the IOA toolkit.

It is worth mentioning that Musial [29] has also translated a version of the Paxos specification of [25] to Java code. The communication medium used was Java Sockets with TCP (instead of MPI) but the translation was done in a manual manner (as opposed to the automated translation offered by the IOA Toolkit). It is also worth mentioning that work is underway in enabling the IOA compiler to also use Java Sockets and TCP [16].

In [5] a deconstruction of (untimed) Paxos into two main abstractions, register and leader, is presented. The eventual register abstraction encapsulates the safety properties of Paxos whereas the eventual leader election abstraction encapsulates its liveness.

The IOA Paxos specification presented in this paper (see next section) makes a similar deconstruction of (timed) Paxos: safety is encapsulated via a Paxos Process automaton and liveness via a Ballot Trigger automaton (which includes a leader election module).

## 3    Implementation of the Paxos Algorithm

### 3.1    Extending the IOA Compiler

In order to implement Paxos, we had to extend the IOA toolkit to support timing issues. In particular, we had to enable the checker, composer and compiler to support the free variant of the GTA model (discussed in Section 2.1). Recall that GTA, besides the action types input, output and internal of IOA, also requires a fourth action type, that of time-passage (that specifies the passage of time). Introducing this new action type was a non-trivial task which involved making several changes and adjustments to various parts of the checker, composer and compiler code.

In addition, for the successful implementation of the time-based Paxos we implemented a set of operators and data types. Each IOA data type is implemented by a hand-coded Java class. A library of such classes for the standard IOA data types is included in the compiler. Each IOA data type (e.g., Set [ ]) and operator (e.g., Set [ ] → Nat) is matched with its Java implementation class using a data type registry [30], which we extended in this work. Examples of operators that we have developed and included in the compiler to support the implementation of Paxos are (their usage is shown in later sections): *maxElement, maxBallot, getprocid, getseqno, setBallot, allessdead, ifProposed,existVal, valProposed, notnil, internalDecideOp, timePsg, ifmajv* and *votedBallot*. Examples of developed data types are *Ballot* and *Last* (their Java code can be found in the full paper [13]).

Recall that in [30] it was shown that the IOA compiler preserves the safety properties of the source IOA code (the specification of the algorithm to be implemented). As the safety properties are not affected by timing issues, it follows that the Java code generated by our extended version of the compiler for Paxos preserves the safety properties of the source GTA specification. As already mentioned, preserving some liveness guarantees in an automated manner is an open research question [14,30].

Although in this work we have focused on Paxos, we believe that our extended version of the IOA compiler (including checker and composer) can be used for the automated implementation of other timing-dependent distributed algorithms where their computational progress relies on timeouts, and which adhere to the aforementioned restrictions imposed by the IOA compiler.

### 3.2    Procedure

The compilation steps of the time-based Paxos specification are as outlined in Section 2.2, where instead of using the IOA compiler we used our developed extended version (that supports the free variant of the GTA model).

**Paxos Specification.** Our Paxos specification is based on the one given in [25], but it had to be expressed in the IOA notation suitable for compilation. In addition we had to

```
Signature                                          Internal:
Input:                                               abstain (const MPIrank:Int, BAbstain:Set[Ballot])
 init (const MPIrank:Int, vInit:Int)                 vote (const MPIrank:Int, bVote:Ballot)
 fail (const MPIrank:Int)                            internalDecide(const MPIrank:Int,
 newBallot(const MPIrank:Int)                               bInternDecide:Ballot)
 RECEIVE(m:Message, const MPIrank:Int,u:Int)         valueDecision(const MPIrank:Int, You:Int,
Output:                                                     LatestVal:Int, ballot:Ballot)
 decide(const MPIrank: Int, vDecide:Int)             gossip(const MPIrank:Int)
 SEND(m:Message , const MPIrank:Int ,u:Int)        TimePassage:
 assignVal(const MPIrank:Int,                        v(T)
       bAssignVal:Ballot, vAssignVal:Int)
 makeBallot(const MPIrank: Int,bMakeBallot:Ballot)

States:                                            seqNo:Int := 0
mode:ModeType := idle                              lastProposedBallot:Ballot := setBallot(-1,-1)
proposed:Array[Int,Set[Int]]:=constant({})         lastvotedvalue:Int := -1
failed:Bool := false                               lastValue:Array[Ballot,Set[Last]]
ballots:Set[Ballot]:={}                            leader:Int := -1
val:Array[Int,Array[Ballot,Null[Int]]] :=          assignvalue:Int := -1
      constant(constant(nil))                      tempLast:Last
voted:Array[Int,Array[Int,Set[Ballot]]] :=         tempVal:Int
      constant(constant({}))                       tempballot:Ballot
abstained:Array[Int,Array[Int,Set[Ballot]]] :=     tempbalDecide:Ballot
         constant(constant({}))                    nodes:Set[Int]
doMakeBallot:Array[Int,Bool]:=constant(false)      countVote:Int := 0
succeeded:Array[Int,Set[Ballot]]:=constant({})     balvalsucc:Int := -1
done:Array[Int,Bool] := constant(false)            Clock:Real := 0
neighbours:Set[Int] := {}                          nextGossipTime:Real := 0
tempnghbrs:Array[Ballot,Set[Int]] := constant({})  period:Real
rcvBallots:Set[Ballot] := {}                       T:Real
sendVote:Bool := false                             mProposed:Int := -1
readyAssign:Bool := false                          mBallots:Ballot := setBallot(-1,-1)
ballotsucceeded:Ballot := setBallot(-1,-1)         mVal:Int := -1
queueOut:Map[Link,Seq[Message]]                    mVoted:Ballot := setBallot(-1,-1)
queueIn:Map[Link,Seq[Message]]                     mAbstained:Set[Ballot] := {}
lnks:Set[Link] := {}
```

**Fig. 1.** *PaxosProcess(i):* Signature and State variables

develop several auxiliary operators and data structures. The specification includes two automata: the *PaxosProcess* and *BallotTrigger*. The former implements the first four main phases of Paxos as outlined in Section 2.3 whilst the two last (timing-dependent) phases are implemented by the latter. Note that for simplicity of presentation we used majorities instead of quorums. We present the specification of each automaton along with the new operators and data structures we have developed. Each automaton specification was syntactically checked using our updated version of the IOA checker.

*PaxosProcess Automaton.* Figure 1 shows the signature and the state variables of the *PaxosProcess(i)* automaton. The analysis of the new data types and operators follows in this section. Figure 2 shows the transitions of actions *init, newBallot* and *makeBallot*.

The *init* action proposes and records the submitted value. It also changes the mode to active and sends the value *vProposed* to the other processes. The *newBallot* input action notifies the *PaxosProcess(i)* to originate a new ballot. The *makeBallot* action is triggered once a request for a new ballot has arrived. In this action a new sequence number that is bigger than any previously known sequence number is selected, and then the leader sends the new ballot $b$. The new ballot identifier is a two field record of the sequence number and the identifier of the new ballot's originator. At this point no value is associated with the ballot. The *maxBallot* operator that is imported in the *makeBallot* action, identifies and returns the largest ballot that has been witnessed so far. In case a process has crashed, the *fail* action is executed (variable *failed* is set to true). It is important to highlight that only a leader process can start a new ballot.

Figure 3 contains the transitions of actions *abstain, valueDecision* and *assignVal*. The *PaxosProcess(i)* automaton uses the *abstain* action to abstain from all the ballots of a set $B$. This is allowed when the known identifier of a ballot is larger than any other

```
input init(i,vInit)                        output makeBallot(i,MakeBallot)
eff                                        pre ¬failed;
 if ¬failed then                            mode=active;
  if (mode=idle) then                       doMakeBallot[i];
   mode := active;                          (getprocid(maxBallot(ballots))<
   Clock := clock;                             getprocid(MakeBallot))∨
   proposed[i] := proposed[i] ∪ {vInit};    ((getseqno(maxBallot(ballots))<
   for k:Int in nodes-{i} do                   getseqno(MakeBallot)));
    queueOut[[i,k]] := queueOut[[i,k]] ⊢    getprocid(MakeBallot) = i;
     sProposed([PROPOSED,[i,k],vInit]);     eff
   od;                                       seqNo := seqNo + 1;
   mProposed:=vInit;                         ballots := insert(MakeBallot, ballots);
  fi; fi;                                    lastProposedBallot := MakeBallot;
                                             doMakeBallot[i] := false;
input newBallot(i)                           for k:Int in nodes-{i} do
eff                                           queueOut[[i,k]] := queueOut[[i,k]] ⊢
 if ¬failed then                               sBallot([BALLOT,[i,k],MakeBallot]);
  if mode= active then                       od;
   doMakeBallot[i] := true;                  rcvBallots := {};
  fi; fi;                                    rcvBallots := insert(MakeBallot, rcvBallots);
                                             mBallots:=MakeBallot;
```

**Fig. 2.** *PaxosProcess(i):* Transitions of actions *init, newBallot, makeBallot*

```
internal abstain(i,BalAbstain)              od;
pre                                         else
 mode=active;                                tempLast:=setLast(1,latestVal);
 ¬failed;                                     lastValue[ballot] := insert(tempLast,
 getseqno(maxBallot(BalAbstain))<              lastValue[ballot]);
 getseqno(maxBallot(ballots))∨              fi
 getprocid(maxBallot(BalAbstain))<         else
 getprocid(maxBallot(ballots));             tempVal:=0;
 (voted[i][i] ∪ abstained[i][i]) ∩ BalAbstain={};  for k:Last in lastValue[ballot] do
eff                                          if getnodeNum(k)>tempVal then
 abstained[i][i] := abstained[i][i] ∪ BalAbstain;   tempVal := getnodeNum(k);
 for k:Int in nodes-{i} do                    assignvalue := getvalue(k);
  queueOut[[i,k]] := queueOut[[i,k]] ⊢       fi
   sAbstain([ABSTAIN,[i,k],BalAbstain,      od;
   getprocid(maxBallot(BalAbstain))]);      if assignvalue=(-1) then
 od;                                         assignvalue := chooseRandom(proposed[i]);
 for j:Ballot in BalAbstain do              fi
  rcvBallots := delete(j,rcvBallots);       readyAssign := true;
 od;                                       fi; fi;
 mAbstained := BalAbstain;
                                           output assignVal (i,balAssignVal,valAssignVal)
internal valueDecision(i, u,latestVal,ballot)  pre
pre mode≠idle;                              ¬failed;
 head(queueIn[[i,u]])=sLatestValue([LATESTVAL,  mode=active;
 [u,i],latestVal,ballot]);                  readyAssign;
eff                                         balAssignVal ∈ ballots;
 queueIn[[i,u]] := tail(queueIn[[i,u]]);    getprocid(balAssignVal)=i;
 tempnghbrs[ballot]:=insert(u,tempnghbrs[ballot]);  val[i][balAssignVal]=nil;
 if ¬(ballot ∈ abstained[i][i]) then        ifProposed(proposed,valAssignVal);
  if ((size(tempnghbrs[ballot]))<           (allessdead(ballots,balAssignVal,abstained[i],nodes)∨
   (div(size(neighbours),2)))) then         existval(val,valAssignVal,abstained[i],ballots,nodes))
   if (latestValue(lastValue[ballot],latestVal))then  eff
    for k:Last in lastValue[ballot] do      val[i][balAssignVal]:=embed(valAssignVal);
     if getvalue(k)=latestVal then          for k:Int in nodes-{i} do
     lastValue[ballot] := delete(k,          queueOut[[i,k]] := queueOut[[i,k]] ⊢
      lastValue[ballot]);                      sValue([VALUE,[i,k],balAssignVal,valAssignVal]);
     tempLast := setLast(getnodeNum(k)+1,   od;
      getvalue(k));                         readyAssign:=false;
     lastValue[ballot] := insert(tempLast,  mVal:=valAssignVal;
      lastValue[ballot]);
    fi
```

**Fig. 3.** *PaxosProcess(i):* Transitions of actions *abstain, valueDecision* and *assignVal*

ballot in $B$, and provided that it has not already voted for any of the ballots of the set $B$ in an earlier state. After the initiation of a ballot process, a value for the ballot has to be chosen. The internal action *ValueDecision* is used to choose a value for the ballot $b$. The specified transition is being executed only by the leader. All processes have to send the value of the latest ballot that they have voted for (if voted) to the leader. When the leader receives the values from a majority of the processes it chooses a the value for ballot $b$. The leader ignores all values equal to $-1$ (indicating that the sender has not voted for any ballot yet). The prevailed value will be assigned to ballot $b$.

```
internal vote(i,balVote)                          internal internalDecide(i,balInternDecide)
pre                                               pre
  mode=active;                                      ¬failed;
  ¬failed;                                          mode=active;
  valproposed( ballots,balVote);                    internalDecideOp(nodes,balInternDecide,voted[i])
  notnil(val,balVote);                            eff
  ¬(balVote ∈ abstained[i][i]);                     succeeded[i] := succeeded[i] ∪ {balInternDecide};
  ¬(balVote ∈ voted[i][i]);                         if (val[getprocid(balInternDecide)]
eff                                                     [balInternDecide]≠nil) then
  voted[i][i]:=voted[i][i] ∪ {balVote};               balvalsucc:=val[getprocid(balInternDecide)]
  for k:Int in nodes-{i} do                             [balInternDecide].val fi;
  queueOut[[i,k]] := queueOut[[i,k]] ⊢ sVote([       ballotsucceeded:=balInternDecide;
  VOTE,[i,k],balVote,getprocid(balVote)]);
  countVote:=countVote+1;                         output decide(i, valDecide)
  od;                                             pre
  sendVote:=false;                                  ¬failed;
  rcvBallots:=delete(balVote,rcvBallots);           ¬done[i];
  lastvotedvalue:=val[getprocid(balVote)]           mode = active;
    [balVote].val;                                  ballotsucceeded ∈ succeeded[i];
  mVoted:=balVote                                   embed(valDecide) = val[getprocid(ballotsucceeded)]
                                                      [ballotsucceeded];
                                                  eff
                                                    done[i] := true;
```

**Fig. 4.** *PaxosProcess(i):* Transitions of actions *vote, internalDecide* and *decide*

*PaxosProcess(i)* uses the internal action *assignValue* to assign the value $v$ to ballot $b$. The possibility to assign a value $v$ to a ballot is based on an important consistency check with smaller ballots. Specifically, *PaxosProcess(i)* checks whether $b$ is a known ballot and that $i$ is the originator of ballot $b$. So far, no value has yet been assigned to $b$, as far as $i$ knows. But since $i$ is the process that originally started ballot $b$, $i$ is the one that has the ability to assign the value $v$ to $b$. Value $v$ must be known to be the initial value of a process. Besides, all smaller ballots either must have the value $v$, or are known as "dead". The specified transition uses the operators *ifProposed, allessdead, existval* and *dead*. The *ifProposed* operator examines whether value $v$ is one of the values that had been proposed by processes. The *allessdead* operator checks if all the ballots that are smaller than $b$ are dead. Also, *existval* checks if $v$ has been assigned to all the smaller non-dead ballots. Once the value has been assigned to the ballot, the leader notifies the other processes about the new value.

Figure 4 depicts the transitions of actions *vote, internalDecide* and *decide*. For the system to reach a consensus state, processes have to accept the value of the ballot by voting the ballot. *PaxossProcess(i)* may vote for a ballot $b$ if it is known that a value has been assigned to $b$, and if $i$ has not yet abstained from $b$. The responsibility of action *vote(i, b)* is for process $i$ to vote for ballot $b$ and to inform the environment about its participation, by sending a Vote message. This action consists of the operators *valproposed* and *notnil; valproposed* checks whether ballot $b$ has been proposed by a process, whereas the *notnil* operator examines if a value has been given to $b$.

Once it is known that a majority of processes have approved the ballot $b$ with value $v$, *PaxosProcess(i)* may decide that the system has reached consensus by executing the internal action *internalDecide(i,b)*. This action, using the *internalDecideOp* operator, checks whether a majority of processes have accepted $b$.

Finally, *PaxosProcess(i)* announces the decision to the external environment with the *decide(i)* action. The SEND and RECEIVE actions are used to propagate information among processes reaching consensus. The information includes the proposed and sets of ballots, and the value, voted and abstained maps. The transitions of actions SEND and RECEIVE can be found in the full paper [13].

For the best manipulation of messages, we created two queue-type data structures *queueOut* and *queueIn*, in which we record the out and in bound messages respectively. In action *SEND(m,i,u)*, process $i$ sends the message that is at the top of queueOut to receiver $u$. Once the message is sent, it is removed from the queue. A process can decide and terminate when it sends all the voted messages that exist in *queueOut*.

In *RECEIVE(m,i,u)*, the received messages are stored in the queue named queueIn for further utilization. Paxos restricts the communications among processes so as only important information to be sent, thus sending periodically gossip messages at interval of *period*. This message restriction is achieved through the *v(T)* and *gossip(i)* actions of the *PaxosProcess(i)* automaton described in Figure 5.

```
timePassage v(T)                                eff
pre ¬failed;                                        for k:Int in nodes-{i} do
    isEmptyQue(queueOut);                               queueOut[[i,k]] := queueOut[[i,k]] ⊢
eff Clock := Clock + T;                                     sState([[i,k],mProposed,mBallots,
                                                            mVal,mVoted,mAbstained]));
internal gossip(i)                              od;
pre ¬failed;                                        nextGossipTime := nextGossipTime + period;
    Clock ≥ nextGossipTime;
```

**Fig. 5.** *PaxosProcess(i): v(T) and gossip*

In particular, the *v(T)* action models the passage of time. The *Clock* variable (initialized to zero) is increased by $T$ units, $T$ being a predefined quantity and specifies the (worst-case) time needed for all the abovementioned transitions to take place; as we explain later, both $T$ and *period* are system-dependent and therefore these parameters must be computed based on timing properties of the target deployment platform.

*BallotTrigger Automaton.* The *BallotTrigger* automaton is the one to specify how a new leader is elected and when a leader generates a new ballot. That is, this automaton is the one to specify the main timing issues of time-based Paxos. The *BallotTrigger(i)* signature and state variables are presented in Figure 6. Figure 7 presents the transitions of *BallotTrigger(i)*.

```
Signature                                       Output:
  Input:                                            newBallot(const MPIrank: Int)
    init(const MPIrank: Int,vInit: Int)             sendAlive(const MPIrank:Int,u:Int)
    fail(const MPIrank: Int)                         SEND(m:Message,const MPIrank: Int,u:Int)
    decide(const MPIrank: Int,vDecide:Int)      Internal:
    assignVal(const MPIrank: Int,                   nodeTimeout(const MPIrank:Int,u:Int)
        bAssignVal:Ballot,vAssignVal:Int)       TimePassage:
    RECEIVE(m:Message, const MPIrank: Int, u:Int)   v(T)

States                                              delay:Real
  mode:Mode := idle                                 period:Real
  suspected:Set[Int] := {}                          T:Real
  timeout:Array[Int,Real]                           nodes:Set[Int]
  nextBallotTime:Real := -1                         done:Bool := false
  nextSendTime:Array[Int,Real]                      queueOut:Map[Link, Seq[Message]]
  leader:Int := -1                                  queueIn:Map[Link, Seq[Message]]
  Clock:Real := 0
  failed: Bool := false
```

**Fig. 6.** *BallotTrigger(i): Signature and State variables*

The *BallotTrigger(i)* automaton handles the event of the ballot voting timeout as follows. If a ballot voting does not complete within a predefined time interval, it is terminated by having the leader initiate a new ballot voting. (Assume that $i$ is the current leader.) Particularly, the leader measures the time starting from the execution of action

```
Transitions

input RECEIVE(m,i,u)                          if ¬failed then
eff                                             if mode=active then
 if ¬failed then                                 nextBallotTime:=Clock+delay;
  if mode= active ∧ tag(m)=sAlive then          fi; fi;
   queueIn[[i,u]] := queueIn[[i,u]] ⊢ m;
   if head(queueIn[[i,u]]) = sAlive([ALIVE,[u,i]]) then   input decide(i,v)
    queueIn[[i,u]] := tail(queueIn[[i,u]]);     eff if ¬failed then
    timeout[u] := Clock+delay;                   if mode=active then
    if u ∈ suspected then                          done:=true;
     nextSendTime[u] := Clock;                     nextBallotTime:=-1;
     suspected := suspected - {u};                fi
     if u⊳leader then                            fi;
      leader:=u;
     fi; fi;                                     output newBallot(i)
     if u≠leader then                            pre ¬failed;
      nextBallotTime := -1;                       mode=active;
 fi; fi; fi; fi;                                  Clock ≥nextBallotTime∧ ¬(nextBallotTime=(-1));
                                                  ¬done;
input init(i,v)                                 eff nextBallotTime:=Clock+delay;
eff
 if ¬failed then                                timePassage v(T)
  if(mode=idle) then                            pre ¬failed;
   mode := active;                               (Clock+T)≤(nextBallotTime)∨ nextBallotTime=(-1);
   Clock := clock;                               timePsg(Clock,T, timeout);
   leader := maxElement(nodes);                  timePsg(Clock,T,nextSendTime);
   for k:Int in (nodes - {i}) do
    nextSendTime[k] := Clock;                   eff Clock:=Clock+T;
    timeout[k] := Clock+delay;
   od;                                          input fail(i)
   if i=leader then                             eff mode:=failed;
    nextBallotTime := Clock;
 fi; fi; fi;
                                                output sendAlive(i,u)
internal nodeTimeout(i,u)                       pre ¬failed;
pre ¬failed;                                     mode= active;
 mode=active;                                    ¬ (u ∈ suspected);
 Clock ≥ timeout[u]∧ ¬(timeout[u]=(-1));         nextSendTime[u] ≤ (Clock +T);
eff                                             eff
 suspected:=suspected ∪ {u};                     queueOut[[i,u]]:=queueOut[[i,u]] ⊢ sAlive([ALIVE,[i,u]]);
 timeout[u] :=-1;                                nextSendTime[u] := Clock + delay;
 nextSendTime[u] :=-1;
 if leader=u then                               output SEND (m,i,u)
  leader := maxElement((nodes-suspected));      pre ¬failed;
 fi                                              mode= active;
 if i=leader ∧ i<u ∧ ¬done then                 queueOut[[i, u]] ≠ ({});
  nextBallotTime := Clock;                       m =head(queueOut[[i,u]]);
 fi                                             eff
                                                if m =head(queueOut[[i,u]]) then
input assignVal(i,b, v)                          queueOut[[i,u]] := tail(queueOut[[i,u]]);
eff                                             fi
```

**Fig. 7.** *BallotTrigger(i):*Transitions

*newBallot(i)* and checks whether the *decide( *,i)* action is executed within the predefined time period. If the execution has not been completed and $i$ is still the leader, then the *newBallot(i)* action is triggered for the initiation of a new ballot voting. The next *BallotTime* variable determines the time when the leader should create a new ballot, whilst the *nextSendTime* defines the time that the acknowledgment message will be sent.

Another responsibility of the *BallotTrigger* automaton is to execute a failure detection mechanism in order for a new leader to be elected, when the current one seems to have crashed. In particular, the automaton implements process crash detection by having the processes interchanging "alive" messages at regular time intervals. When a process $i$ does not receive the alive message of process $u$ within a predetermined time interval, then $i$ inserts $u$ into a set of "suspected" processes (this is implemented by the *nodeTimeout(i,u)* action). The *sendAlive(i,u)* action allows process $i$ to send an alive message to process $u$ after the passage of time and when $u$ is not a suspected process. The receipt of alive messages is implemented using the *recvAlive(i,u)* action. So, when process $i$ receives a message from process $u$, the timeout variable ($Clock + delay$) is renewed for process $u$. The variable *delay* is system-dependent and hence, as with $T$, its value was computed based on empirical performance measurements of our deployment platform (more details are provided in Section 4). Due to the fact that the system is partially synchronized (and hence, it exhibits timing failures) it is possible that $i$ might not receive $u$'s alive message within the predetermined period and place $u$ in the suspected set, although $u$ is in fact still operational. However, when $i$ receives the delayed message, it removes $u$ from the suspected set. When the leader is included in the set of suspected processes of some process, a new leader election operation is triggered.

The *BallotTrigger(i)* automaton contains the input actions *init* and *decide* for the processes to reach consensus. As an effect of the *init* action, the automaton's state toggles from idle to active, and the current timing value is assigned to the Clock value of the automaton. Initially, each process is assigned as a leader. However, when process $i$ receives an alive message from process $u$ that has greater $id$, then $i$ grants its leadership to $u$. In the end, after correct processes exchange alive messages, the leader is the one with the highest $id$. The input action *decide(i,u)* is activated when consensus is achieved. Consequently, the *decide* variable is toggled to true, and the value $-1$ (coding infinity) is assigned to *nextBallotTime*. It is important to mention that the action *decide(i,u)* of the *PaxosProcess(i)* automaton activates the corresponding action of the *BallotTrigger* automaton when the two automata are composed (the two automata have been specified in such a way that are *composition compatible* [23]).

The passage of time is specified via the *v(T)* action. The *Clock* variable is increased by $T$ units, $T$ being a predefined quantity and specifies the (worst-case) time needed for all the above mentioned transitions to take place. Finally, the action *fail(i)* specifies the crash of process $i$ (the process state changes from active to failed, and hence no further actions can be triggered from $i$).

**Obtaining the PaxosNode Automaton and Resolving Nondeterminism.** As mentioned in Section 2.2 after the description of the system into IOA language the programmer must combine the algorithm automaton with auxiliary, channel automata. The developed automaton named PaxosNodeCom (Figure 8), composes the algorithm automata (PaxosProcess and BallotTrigger) with the mediator automata responsible for the establishment of the communication (via MPI) among processes. The *SendMediator* automaton consists of the actions *Isend, resp_Isend* and *resp_test*, while the *ReceiveMediator* consists of the actions *Iprobe, resp_Iprobe, receive* and *resp_receive* (which specify standard MPI constructs). More on these mediator automata can be found in [14,30]. The PaxosNodeCom automaton is fed to the composer which generates the PaxosNode automaton (it includes all states and transitions of the composed automata).

```
automaton PaxosNode(MPIrank:Int,MPIsize:Int)       MPIrank,j);
  components                                         SM[j: Int]: SendMediator(Message, Int,
  P: PaxosProcess(MPIrank,MPIsize);                    MPIrank, j)
  B:BallotTrigger(MPIrank,MPIsize);
  RM[j: Int]: ReceiveMediator(Message, Int,
```

**Fig. 8.** *PaxosNodeCom:* Composition Automaton

After the composition, and before compilation, we included a schedule, presented in Figure 9, to resolve nondeterminism. The schedule consists of the operators *ifmajv* and *votedBallot*. The first checks whether a majority of processes have approved a proposed ballot, whereas the latter operator returns the approved ballot.

```
schedule                                          lnk.u);
states                                          fi
                                                if P.queueIn[[lnk.i,lnk.u]]≠{} ∧ P.mode=active
links:Set[Link],                                  ∧ B.mode=active
lnk:Link,                                         ∧ tag(head(P.queueIn[[lnk.i,lnk.u]]))=sLatestValue
newBallot:Ballot,                               then
tmpBallots:Set[Ballot]:={},                       fire internal valueDecision(lnk.i,lnk.u,(head(
tempVal:Int:=-1,                                    P.queueIn[[lnk.i,lnk.u]])).sLatestvalue,
temprcvBallot:Ballot,                              (head(P.queueIn[[lnk.i,lnk.u]])).sLatestValue.ballot);
setTemp:Set[Ballot]:={},                        fi
flag:Int:=0                                     if P.mode=active ∧ B.mode=active ∧ P.readyAssign=true
                                                  ∧ P.val[MPIrank][P.lastProposedBallot]=nil
do                                                ∧ ifProposed(P.proposed,P.assignvalue)
  fire input init(MPIrank,valInit);               ∧ (allessdead(P.ballots,P.lastProposedBallot,
  while(P.done[MPIrank]≠true ∧ P.failed=false ∧    P.abstained[MPIrank],P.quorum) ∨ existval(P.val,
    ¬(B.failed) ∧ B.mode=active) do               P.assignvalue,P.abstained[MPIrank],P.ballots,P.quorum))
    links:=P.lnks;                              then
    while(¬isEmpty(links) ∧ P.done[MPIrank]≠true ∧  tempVal:=P.assignvalue;
      P.failed=false) do                          fire output assignVal(MPIrank,P.lastProposedBallot,
      lnk := chooseRandom(links);                   tempVal);
      links := delete(lnk, links);              fi
      if ¬P.failed ∧ P.mode=active ∧ B.mode=active ∧  if P.rcvBallots≠{} then
        B.Clock≥B.nextBallotTime ∧ ¬(B.nextBallotTime=(-1))  temprcvBallot:=chooseRandom(P.rcvBallots);
        ∧ ¬B.done then                            if notnil(P.val,temprcvBallot) ∧ B.mode=active ∧
        fire output newBallot(MPIrank);           P.mode=active ∧¬(P.failed) ∧ valproposed(
        if P.doMakeBallot[MPIrank] then           P.ballots,temprcvBallot) ∧ ¬(temprcvBallot ∈
        newBallot:=setBallot(P.seqNo+1, MPIrank);   (P.abstained[MPIrank])[MPIrank]) ∧ ¬(temprcvBallot
        fire output makeBallot(MPIrank, newBallot);  ∈(P.voted[MPIrank])[MPIrank]) then
        setTemp:=((P.ballots-(maxBallot(P.ballots)))  fire internal vote(MPIrank, temprcvBallot);
          P.abstained[MPIrank][MPIrank]) -        fi; fi;
          P.voted[MPIrank][MPIrank];              if ifquorumv(P.voted[MPIrank],P.quorum) ∧
        if setTemp ≠ {} then                      P.mode=active∧B.mode=active∧P.balvalsucc=(-1) then
          fire internal abstain(lnk.i,setTemp);   fire internal internalDecide(MPIrank,
        fi; fi; fi;                                 votedBallot(P.voted[MPIrank], P.quorum));
      setTemp := ((P.ballots-(maxBallot(P.ballots)))  fi
        P.abstained[MPIrank][MPIrank]) -          if P.ballotsucceeded ∈ P.succeeded[MPIrank]∧
        P.voted[MPIrank][MPIrank];                P.balvalsucc≠(-1) ∧ P.countVote=0 ∧ P.mode=active
      if(setTemp≠{}) then                          ∧ B.mode=active ∧ embed(P.balvalsucc)=
        fire internal abstain(lnk.i,setTemp); fi;  P.val[MPIrank][P.ballotsucceeded] then
      if flag=0 then                              fire output decide(MPIrank, P.balvalsucc);
        if P.queueOut[lnk]≠{} then                fi
          fire output SEND(head(P.queueOut[lnk]),MPIrank,lnk.u);  if B.mode= active ∧ P.mode=active
          elseif B.queueOut[lnk]≠{} then           ∧ ¬lnk.u ∈ B.suspected
          fire output SEND(head(B.queueOut[lnk]),MPIrank,lnk.u);  ∧ B.nextSendTime[lnk.u]≤B.Clock then
          flag:=1;                                 fire output sendAlive(lnk.i,lnk.u);
        fi                                        fi
      else if flag=1 then                         if B.mode=active ∧ P.mode=active
        if B.queueOut[lnk]≠{} then                 ∧ B.Clock≥B.timeout[lnk.u]
          fire output SEND(head(B.queueOut[lnk]),MPIrank,lnk.u);  ∧ ¬(B.timeout[lnk.u]=(-1)) then
        elseif P.queueOut[lnk] ≠ {} then           fire internal nodeTimeout(lnk.i,lnk.u);
          fire output SEND(head(P.queueOut[lnk]),MPIrank,lnk.u);  fi
          flag:=0;                                if (¬P.failed ∧ isEmptyQue(P.queueOut))
        fi; fi;                                     ∨ (¬B.failed ∧ (B.Clock+B.T≤B.nextBallotTime
      if SM[lnk.u].status=idle ∧ SM[lnk.u].toSend≠{} then  ∨ B.nextBallotTime=-1)
        fire output Isend(head(SM[lnk.u].toSend),MPIrank,lnk.u);  ∧ timePsg(B.Clock,B.T,B.timeout)
      fi                                           ∧ timePsg(B.Clock,B.T,B.nextSendTime))
      if SM[lnk.u].status=idle ∧ SM[lnk.u].handles≠{} then  ∧ P.mode=active ∧ B.mode=active then
        fire output test(head(SM[lnk.u].handles),MPIrank,lnk.u);  fire timePassage v(P.T);
      fi                                          fi
      if RM[lnk.u].status=idle ∧ RM[lnk.u].ready=false then  if (¬(P.failed) ∧ P.Clock ≥ P.nextGossipTime) then
        fire output Iprobe(MPIrank, lnk.u); fi;    fire internal gossip(MPIrank);
      if RM[lnk.u].status=idle ∧ RM[lnk.u].ready=true then  fi
        fire output receive(MPIrank, lnk.u); fi;  od; od; od;
      if RM[lnk.u].toRecv ≠ {} then
        fire output RECEIVE(head(RM[lnk.u].toRecv), MPIrank,
```

**Fig. 9.** *Paxos* Schedule

Finally, the scheduled *PaxosNode* automaton (which includes the schedule) was fed to our updated version of the IOA compiler (which can handle time-passage action types and includes the developed operators and data structures) and we obtained the Paxos.java file which was then compiled into a *class* file (a JVM executable).

## 4   Experimentation

To demonstrate the functionality of the augmented compiler, we have run the derived Java code (which implements Paxos) on a network of workstations and obtained some experimental data.

*Platform and Preparation.* Our experimentation platform consists of a cluster of 17 local machines. Each machine is powered by an Intel Pentium V 1.5 GHz CPU and is running Linux (Fedora Core v5 OS).

As aforementioned in the previous section, the time-related parameters $T$, *delay*, and *period* used abstractly in the specification are system-dependent. In particular, $T$ is the (worst-case) time needed for a node to perform a certain sequence of actions (as specified in the previous section). This time may vary on different platforms. Parameter *delay* includes the (worst-case) time for a message round-trip, local computation and other Java-related delays. Clearly, this depends on the implementation platform. Parameter *period* may be thought as programmer-defined, in the sense that it is up to the programmer to decide on how often the nodes should gossip. Of course, this decision also depends on the deployment platform, as the period should depend on the message round-trip time, the network topology, and the system load.

Therefore, in order to identify sensible values for these parameters for our deployment platform, we performed some initial experiments with simple executions of the code, taking into consideration the system's ping times and the performance analysis presented in [25]. From this preprocessing phase the following values (in *msecs*) were finally chosen: $T = 22$, $delay = 8822$, and $period = 24$.

*Scenarios and Results.* It is noteworthy that Paxos is capable of dealing with small transient failures which are concealed by the use of majority voting. MPI is not fault tolerant and when nodes fail the system can suffer a failure as a whole – due to resource depletion. Removing this limitation is subject of future work where the MPI mediator automata can be replaced with Java mediator automata (as proposed in [16]) that allow more dynamic behaviors.

Despite the above note, a practical evaluation of our automatically generated code is still meaningful: first, it demonstrates that indeed the generated code is executable, second that the resulting program behaves as expected, and finally that we obtain a reasonable performance. To this end we present three scenarios.

The first scenario aims to identify the average execution time and number of sent messages for achieving consensus on a single ballot voting, while the number of participants increases gradually from 2 to 17. The objective of the second scenario is to test the resilience of Paxos to message loss. Since MPI is not fault-tolerant, we introduce code on the sender side that randomly chooses messages to be dropped before the MPI send primitive is invoked. Scenarios 2A and 2B drop 10% and 20% of messages respectively. The third scenario seeks to measure the performance of our Paxos implementation in the presence of leader crashes. In particular, in this scenario we simulate the leader crash (by setting the status of the leader as failed in the schedule block) and hence we force the algorithm to initiate the leader election and new ballot mechanisms.

Each scenario was run 10 times and thus each plot point depicts the average of the runs. Figures 10(a) to 10(d) illustrate the average execution time and sent messages respectively for each scenario. The first scenario is used as baseline against the other two scenarios.

The experimental data in Figures 10(a) and 10(b) demonstrate, as expected, that Paxos is able to cope well with message omission. The difference in message count in Figure 10(b) is negligible between the scenarios, which is to be expected, since

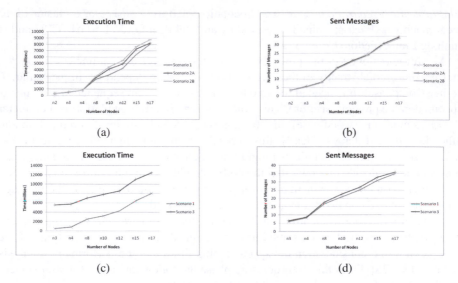

**Fig. 10.** Experimental results: (a) Avg. execution time for scenarios 1, 2A, & 2B, (b) Avg. number of sent messages for scenarios 1, 2A, & 2B, (c) Avg. execution time for scenarios 1 & 3, and (d) Ave. of sent messages for scenarios 1 & 3.

non-leader node message loss is amortized by the use of majority voting, whereas, leader message loss contributes only to a few additional messages; it does, however, contribute to timeouts and hence the increased operation latency as depicted in Figure 10(a).

The timing in Figure 10(c) predictably favors the case when the leader is stable. It is important to point out that the performance in scenarios 1 and 3 is parallel where the difference reflects the timeout until a new leader election is triggered. We also observe a linear decay in performance as the number of nodes increases, which is to be expected. However, we do not expect this behavior to last indefinitely, especially when the network becomes saturated.

## 5    Conclusions

In this paper we have described our experience in specifying, compiling and running a time-based version of the popular Paxos consensus algorithm. In particular, by using a GTA specification of Paxos (which was proved to be correct in [25] and machine-checked in [18,31]) and by extending the IOA checker, composer and compiler (of the IOA Toolkit) in supporting a variant of the GTA framework, we have managed to develop an automated implementation of time-based Paxos. To the best of our knowledge, our work constitutes the first example of a time-dependent complex distributed algorithm that has been specified, verified and implemented in an automated way, using a common formal methodology (IOA in our case).

Several future research directions emanate from our work. First, it would be interesting to assess the efficiency of the automated implementation produced by the compiler. One way is to compare our implementation of Paxos with the one of Musial [29] which was done in a manual way. However, at this point such a comparison would not be fair, as the implementation of Musial uses Java Sockets and TCP, and not MPI for communication. This brings us to a second future objective. Currently the compiler is limited to static participation and use in LANs due to the use of MPI. The compiler design is general enough to enable the use of other communication paradigms. In [16] an alternative communication paradigm is suggested (Java Sockets with TCP) that enables the automated implementation of algorithms that have dynamic participation (nodes may join and leave the computation at any time). Ongoing work is attempting to incorporate this alternative paradigm into (our version of) the IOA compiler.

The TIOA framework (an extension of the IOA framework) models distributed systems with timing constraints as collections of interacting state machines, called Timed Input/Output Automata (an extension of Input/Output Automata) [19]. This framework can be considered more general than GTA, since a state in TIOA not only can be changed by discrete transitions but also by trajectories. A trajectory is a (continuous or discontinuous) function that describes the evolution of the state variables over intervals of time. Therefore, it seems that TIOA can be used to specify a wider family of time-based algorithms (and not just the ones that their computational progress depends on timeouts – like Paxos). A TIOA toolkit is underway [24] which currently includes a TIOA checker, a theorem prover and a TIOA simulator with limited functionality. A very challenging research direction is to develop a TIOA code generator. Our work can be considered an important step towards that direction.

# References

1. IOA language and toolset, http://theory.lcs.mit.edu/tds/ioa/
2. INMOS Ltd: occam Programming Manual (1984)
3. Alur, R., Dill, D.: A theory of timed automata. Theoretical Computer Science 126, 183–235 (1994)
4. Attiya, H., Welch, J.: Distributed Computing: Fundamentals, Simulations, and Advanced Topics, 2nd edn. Wiley Interscience, Hoboken (2004)
5. Boichat, R., Dutta, P., Frolund, S., Guerraoui, R.: Deconstructing Paxos. SIGACT News 34(1), 47–67 (2003)
6. Cleaveland, R., Gada, J.N., Lewis, P.M., Smolka, S.A., Sokolsky, O., Zhang, S.: The Concurrency Factory: Practical tools for specification, simulation, verification and implementation of concurrent systems. In: DIMACS Workshop, pp. 75–89 (1994)
7. Cleaveland, R., Parrow, J., Steffen, B.U.: The concurrency workbench: A semantics-based tool for the verification of concurrent systems. ACM TOPLAS 15(1) (1993)
8. De Prisco, R.: Revisiting the Paxos Algorithm. Master's thesis, Laboratory for Computer Science, Massachusetts Institute of Technology (1997); Also as TR: MIT-LCS-TR-717
9. Fisher, M., Lynch, N., Paterson, M.: Impossibility of distributed consensus with one faulty process. Journal of the ACM 32, 374–382 (1985)
10. M. P. I. Forum. MPI: A message-passing interface standard. International Journal of Supercomputer Applications 8(3/4) (1994)
11. Garland, S., Lynch, N., Tauber, J., Vaziri, M.: IOA user guide and reference manual. Technical Report MIT/LCS/TR-961 (July 2004), http://theory.lcs.mit.edu/tds/ioa/manual.ps
12. Garland, S.J., Guttag, J.V.: A guide to LP, the Larch Prover. Research Report 82, Digital Systems Research Center (1991)

13. Georgiou, C., Hadjiprocopiou, P., Musial, P.M.: On the automated implementation of time-based Paxos using the IOA compiler. Technical Report (2010), http://www.cs.ucy.ac.cy/~chryssis/pubs/tpaxos.pdf

14. Georgiou, C., Lynch, N., Mavrommatis, P., Tauber, J.A.: Automated implementation of complex distributed algorithms specified in the IOA language. Journal of Software Tools for Technology Transfer 11(2), 153–171 (2009)

15. Georgiou, C., Mavrommatis, P., Tauber, J.A.: Implementing asynchronous distributed systems using the IOA toolkit. Technical Report MIT/LCS/TR-966 (2004)

16. Georgiou, C., Musial, P.M., Shvartsman, A.A., Sonderegger, E.L.: An abstract channel specification and an algorithm implementing it using Java sockets. In: Proceedings of the 7th IEEE International Symposium on Network Computing and Applications (NCA 2008), pp. 211–219 (2008)

17. Hoare, C.: Communicating Sequential Processes. Prentice-Hall International, UK (1985)

18. Immorlica, N., Win, T.: A Case Study: Proving Paxos with the IOA Toolkit (2002) (manuscript)

19. Kaynar, D., Lynch, N., Segala, R., Vaandrager, F.: The Theory of Timed I/O Automata (Synthesis Lectures in Computer Science). Morgan & Claypool Publishers, San Francisco (2006)

20. Lamport, L.: The temporal logic of actions. ACM Transactions on Programming Languages and Systems 16(3), 872–923 (1994)

21. Lamport, L.: The part-time parliament. ACM Transactions on Computer Systems 16(2), 133–169 (1998)

22. Larsen, K., Pettersson, P.: Uppaal in a nutshell. Journal of Software Tools for Technology Transfer 1(1/2), 134–152 (1997)

23. Lynch, N.: Distributed Algorithms. Morgan Kaufmann, San Francisco (1996)

24. Lynch, N., Michel, L., Shvartsman, A.: Tempo: A toolkit for the timed Input/Output automata formalism. In: Proceedings of the 1st International Conference on Simulation Tools and Techniques for Communications, Networks, and Systems, SIMUTools 2008 (2008)

25. Lynch, N., Shvartsman, A.: Paxos made even simpler, and formal (2002) (manuscript)

26. Lynch, N., Vaandrager, F.: Forward and backward simulations – Part II: Timing-based systems. Information and Computation 128(1), 1–25 (1996)

27. Lynch, N., Tuttle, M.: An introduction to Input/Output Automata. CWI-Quarterly 2(3), 219–246 (1989)

28. Milner, R.: Communication and Concurrency. Prentice-Hall International, UK (1989)

29. Musial, P.M.: From High Level Specification to Executable Code: Specification, Refinement, and Implementation of a Survivable and Consistent Data Service for Dynamic Networks. PhD thesis, Dept. of Computer Science and Engineering, University of Connecticut (2007)

30. Tauber, J.A.: Verifiable Compilation of I/O Automata without Global Synchronization. PhD thesis, Dept. of Electrical Engineering and Computer Science, M.I.T. (2005)

31. Win, T.N.: Theorem-proving distributed algorithms with dynamic analysis. Master's thesis, Dept. of Electrical Engineering and Computer Science, M.I.T. (2003)

# Partitioning Real-Time Systems on Multiprocessors with Shared Resources*

Farhang Nemati, Thomas Nolte, and Moris Behnam

Mälardalen Real-Time Research Centre, Västerås, Sweden
{farhang.nemati,thomas.nolte,moris.behnam}@mdh.se

**Abstract.** In this paper we propose a blocking-aware partitioning algorithm which allocates a task set on a multiprocessor (multi-core) platform in a way that the overall amount of blocking times of tasks are decreased. The algorithm reduces the total utilization which, in turn, has the potential to decrease the total number of required processors (cores). In this paper we evaluate our algorithm and compare it with an existing similar algorithm. The comparison criteria includes both number of schedulable systems as well as processor reduction performance.

## 1 Introduction

Two main approaches for scheduling real-time systems on multiprocessors exist; global and partitioned scheduling [1–4]. Under global scheduling, e.g., Global Earliest Deadline First (G-EDF), tasks are scheduled by a single scheduler and each task can be executed on any processor. A single global queue is used for storing jobs. A job can be preempted on a processor and resumed on another processor, i.e., migration of tasks among processors is permitted. Under a partitioned scheduling, tasks are statically assigned to processors and tasks within each processor are scheduled by a uniprocessor scheduling protocol, e.g., Rate Monotonic (RM) and EDF. Each processor is associated with a separate ready queue for scheduling task jobs.

Partitioned scheduling protocols have been used more often and are supported (with fixed priority scheduling) widely by commercial real-time operating systems [5], inherent in their simplicity, efficiency and predictability. Besides, the well studied uniprocessor scheduling and synchronization methods can be reused for multiprocessors with fewer changes (or no changes). However, partitioning (allocating tasks to processors) is known to be a bin-packing problem which is a NP-hard problem in the strong sense; hence finding an optimal solution in polynomial time is not realistic in the general case. Thus, to take advantage of the performance offered by multi-cores, scheduling protocols should be coordinated with appropriate partitioning algorithms. Heuristic approaches and sufficient feasibility tests for bin-packing algorithms have been developed to find a near-optimal partitioning [1, 3]. However, the scheduling protocols and existing partitioning algorithms for multiprocessors (multi-cores) mostly assume independent tasks while in real applications, tasks often share resources.

---

* This work was partially supported by the Swedish Foundation for Strategic Research (SSF) via Mälardalen Real-Time Research Centre (MRTC) at Mälardalen University.

C. Lu, T. Masuzawa, and M. Mosbah (Eds.): OPODIS 2010, LNCS 6490, pp. 253–269, 2010.

We have developed a heuristic partitioning algorithm [6], under which our system assumptions include presence of mutually exclusive shared resources. The heuristic partitions a system (task set) on an identical shared memory single-chip multiprocessor platform. The objective of the algorithm is to decrease blocking overheads by assigning tasks to appropriate processors (partitions). This consequently increases the schedulability of the system and may reduce the number of processors. Our heuristic identifies task constraints, e.g., dependencies between tasks, timing attributes, and resource sharing, and extends the best-fit decreasing (BFD) bin-packing algorithm with blocking time parameters. In practice, industrial systems mostly use Fixed Priority Scheduling (FPS) protocols. The Multiprocessor Priority Ceiling Protocol (MPCP) which was proposed by Rajkumar in [7], for many years, has been a standard multiprocessor synchronization protocol under fixed priority partitioned scheduling. Thus, both our algorithm and an existing similar algorithm proposed in [5] assume that MPCP is used for lock-based synchronization. We have investigated MPCP in more details in [6]. Our algorith, however, can be easily extended to other synchronization protocols under partitioned scheduling policies. The algorithm proposed in [5] is named the Synchronization-Aware Partitioning Algorithm (SPA), and our algorithm is named the Blocking-Aware Partitioning Algorithm (BPA). From now on we refer them as SPA and BPA respectively.

## 1.1  Contributions

The contributions of this paper are threefold:
(1) We propose a blocking-aware heuristic algorithm to allocate tasks onto the processors of a single chip multiprocessor (multi-core) platform. The algorithm extends a bin-packing algorithm with synchronization parameters.
(2) We implement our algorithm together with the best known existing similar heuristic [5]. The implementation is modular in which any new partitioned scheduling and synchronization protocol as well as any new partitioning heuristic can easily be inserted.
(3) We evaluate our algorithm together with the existing heuristic and compare the two approaches to each other as well as to an blocking-agnostic bin-packing partitioning algorithm, used as reference. The blocking-agnostic algorithm, in the context of this paper, refers to a bin-packing algorithm that does not consider blocking parameters to increase the performance of partitioning, although blocking times are included in the schedulability test.

The rest of the paper is as follows: we present the task and platform model in Section 2. We explain the existing algorithm (SPA) and present our partitioning algorithms (BPA) in Section 3. In Section 4 the experimental results of both algorithms are presented and the results are compared to each other as well as to the blocking-agnostic algorithm.

## 1.2  Related Work

A significant amount of work has been done in the domain of task allocation on multiprocessors and distributed systems. The emerging of multi-core architectures has increased the interest in the multiprocessor methods. However, in this paper we present the most related works to our approach.

Tindell et al. [8] describe a method called *simulated annealing* for partitioning a task set on a distributed system. The simulated annealing technique is not a heuristic solution but a global optimization method which is used to find a near-optimal solution. The important factor in simulated annealing is that it includes jumps to new solutions to be able to get a better one. The simulated annealing techniques do not include heuristics and it is usually difficult to find a good or even any feasible partitioning [9].

The *Slack Method* presented in [9] is a partitioning heuristic in which the first step is to divide the tasks into sets of communicating tasks (precedence constraint). The size of each set then is reduced based on the concept of *task slack* which is the delay a task can tolerate without missing its deadline. The second step is to map the sets of tasks onto the processors in a way to reduce the communication among processors.

A study of bin-packing algorithms for designing distributed real-time systems is presented in [10]. The method partitions software into modules to be allocated on hardware nodes. In their approach they use two graphs; a graph which models software modules and a graph that represents the hardware architecture. The authors extend the bin-packing algorithm with heuristics to minimize the number of bins (processors) needed and the bandwidth required for the communication between nodes. However, their partitioning method assumes independent tasks.

Baruah and Fisher have presented a bin-packing partitioning algorithm (first-fit decreasing (FFD) algorithm) in [11] for a set of sporadic tasks on multiprocessors. The tasks are indexed in non-decreasing order based on their relative deadlines and the algorithm assigns the tasks to the processors in first-fit order. The algorithm, however, assumes independent tasks. On the other hand their algorithm has been developed under the EDF scheduling protocol while most existing real-time systems use fixed priority scheduling policies. The focus of our proposed heuristic, in this paper, is fixed priority scheduling protocols, although it can easily be extended to other policies.

Of great relevance to our work presented in this paper is the work presented by Lakshmanan et al. in [5]. In the paper they investigate and analyze two alternatives of execution control policies (suspend-based and spin-based remote blocking) under MPCP. They have developed a blocking-aware task allocation algorithm (an extension to BFD) and evaluated it under both execution control policies.

In their partitioning algorithm, the tasks that directly or indirectly share resources are put into what they call bundles (in this paper we call them macrotasks) and each bundle is tried to be allocated onto a processor. The bundles that cannot fit into any existing processors are ordered by their cost, which is the blocking overhead that they introduce into the system. Then the bundle with minimum cost is broken and the algorithm is run from the beginning. However, their algorithm does not consider blocking parameters when it allocates the current task to a processor, but only its size (utilization). Furthermore, no relationship (e.g., as a cost based on blocking parameters) among individual tasks within a bundle is considered which could help to allocate tasks from a broken bundle to appropriate processors to decrease the blocking times. However, their experimental results show that a blocking-aware bin-packing algorithm for suspend-based execution control policy does not have significant benefits compared to a blocking-agnostic bin-packing algorithm. Firstly, for the comparison, they have only focused on the processor reduction issue; they suppose that the algorithm is better if it reduces the number of processors. They have not considered the worst case as it could be the case that an algorithm fails to schedule a task set. In our experimental evaluation, besides processor reduction, we have considered this issue as well. If an algorithm can schedule some task sets while others fail, we consider it as a benefit. Secondly, in their experiments they have not investigated the effect of some parameters such as the different number of resources, variation in the number and length of critical sections of tasks. By considering these parameters, our experimental results show that in most cases our blocking-aware algorithm has significantly better results than blocking-agnostic algorithms. However, according to our experimental results, their heuristic performs slightly better than the blocking-agnostic algorithm, and our algorithm performs significantly better than both.

In the context of multiprocessor synchronization, Rajkumar et al. for the first time proposed a synchronization protocol in [12] which later [7] was called Distributed Priority Ceiling Protocol (DPCP). DPCP extends PCP to distributed systems and it can be used with shared memory multiprocessors. However, a major motivation of increasing interest in the multiprocessor methods is the emerging of multi-core platforms for which DPCP is not an appropriate synchronization protocol. Rajkumar in [7] presented MPCP, which extends PCP to multiprocessors hence allowing for synchronization of tasks sharing mutually exclusive resources using partitioned FPS. Considering that MPCP has been a standard multiprocessor synchronization protocol, our partitioning algorithm attempts to decrease blocking times under MPCP and consequently decrease worst case response times which in turn may reduce the number of needed processors. Gai et al. [13, 14] present MSRP (Multiprocessor SRP), which is a P-EDF (Partitioned EDF) based synchronization protocol for multiprocessors. The shared resources are classified as either (i) local resources that are shared among tasks assigned to the same processor, or (ii) global resources that are shared by tasks assigned to different processors. In MSRP, tasks synchronize local resources using SRP [2], and access to global resources is guaranteed a bounded blocking time. Lopez et al. [15] present an implementation of SRP under P-EDF. Devi et al. [16] present a synchronization technique under G-EDF. The work is restricted to synchronization of non-nested accesses to short and simple objects, e.g., stacks, linked lists, and queues. In addition, the main focus of the method is soft real-time systems.

Block et al. [17] present Flexible Multiprocessor Locking Protocol (FMLP), which is the first synchronization protocol for multiprocessors that can be applied to both partitioned and global scheduling algorithms, i.e., P-EDF and G-EDF. An implementation of FMLP has been described in [18]. However, although in a longer version of [17][1], the blocking times have been calculated, but to our knowledge there is no concrete schedulability test for FMLP under global scheduling protocols. However, Brandenburg and Anderson in [19] have extended partitioned FMLP to fixed priority scheduling policy and derived a schedulability test for it. In a later work [20], the same authors have compared DPCP, MPCP and FMLP. However, as the partitioned scheduling approaches suffer from bin-packing problem, we believe to achieve a better and fair comparison of the approaches, they should be coordinated with task allocation algorithms.

Recently, Easwaran and Andersson have proposed a synchronization protocol [21] under global fixed priority scheduling protocol. In this paper, for the first time, the

---

[1] Available at http://www.cs.unc.edu/~anderson/papers/rtcsa07along.pdf

authors have derived schedulability analysis of the priority inheritance protocol under global scheduling algorithms.

## 2    Task and Platform Model

In this paper we assume a task set that consists of $n$ sporadic tasks, $\tau_i(T_i, C_i, \rho_i, \{c_{i,p,q}\})$ where $T_i$ denotes the minimum inter-arrival time between two successive jobs of task $\tau_i$ with worst-case execution time $C_i$ and $\rho_i$ as its priority. The tasks share a set of resources, $R$, which are protected using semaphores. The set of critical sections, in which task $\tau_i$ requests resources in $R$ is denoted by $\{c_{i,p,q}\}$, where $c_{i,p,q}$ indicates the maximum execution time of the $p^{th}$ critical section of task $\tau_i$ in which the task locks resource $R_q \in R$. Critical sections of tasks should be sequential or properly nested. The deadline of each job is equal to $T_i$. A job of task $\tau_i$, is specified by $J_i$. The utilization factor of task $\tau_i$ is denoted by $u_i$ where $u_i = C_i/T_i$.

We also assume that the multiprocessor (multi-core) platform is composed of identical, unit-capacity processors (cores) with shared memory. The task set is partitioned into partitions $\{P_1, \ldots, P_m\}$, and each partition is allocated onto one processor (core), thus $m$ represent the minimum number of processors needed.

## 3    The Blocking Aware Partitioning Algorithms

### 3.1    Blocking-Aware Partitioning Algorithm (BPA)

In this section we propose a partitioning algorithm that groups tasks into partitions so that each partition can be allocated and scheduled on one processor. The objective of the algorithm is to decrease the overall blocking times of tasks. This generally increases the schedulability of a task set which may reduce the number of required partitions (processors).

Considering the blocking factors of tasks under MPCP, tasks with more and longer global critical sections lead to more blocking times. This is also shown by experiments presented in [14]. Our goal is to (i) decrease the number of global critical sections by assigning the tasks sharing resources to the same partition as far as possible, (ii) decrease the ratio and time of holding global resources by assigning the tasks that request the resources more often and hold them longer to the same partition as long as possible.

In our previous work [22] we have presented a partitioning framework in which tasks are grouped together based on task preferences and constraints. The framework partitions tasks based on a cost function which is derived from task preferences and constraints. The framework attempts to allocate the tasks that directly or indirectly share resources onto the same processor. Tasks that directly or indirectly share resources are called *macrotasks*, e.g., if tasks $\tau_i$ and $\tau_j$ share resource $R_p$ and tasks $\tau_j$ and $\tau_k$ share resource $R_q$, all three tasks belong to the same macrotask. However, there are cases that a macrotask cannot fit in one processor (i.e., assuming that the tasks in the macrotask are the only tasks allocated on a processor, still it can not be scheduled by the processor). In this case tasks belonging to the same macrotask can be allocated to different partitions (processors).

The goal of the framework presented in [22] is to put the tasks into appropriate partitions so that the costs are minimized. The framework may have different partitioning strategies, e.g., increasing cache hits, decreasing blocking times, etc. The strategy of partitioning may differ, depending on the nature of a system, and result in different partitions. The framework is a general partitioning approach without deeply focusing on any specific strategy and thus we have not presented any evaluation except one example. Obviously, for different partitioning strategies (e.g., increasing cache hits) the guiding heuristics as well as the implementation of the algorithm will be completely different. In current work, however, we specifically focus on a partitioning strategy for decreasing remote blocking overheads of tasks which leads to increasing the schedulability of a task set and possibly will reduce the number of processors required for scheduling the task set. We derive heuristics to specifically guide the partitioning algorithm to reduce the remote blocking times. We have also performed detailed experimental evaluation according to different resource sharing parameters.

We have developed a blocking-aware algorithm that is an extension to the BFD algorithm. In a blocking-agnostic BFD algorithm, bins (processors) are ordered in non-increasing order of their utilization and tasks are ordered in non-increasing order of their size (utilization). The algorithm attempts to allocate the task from the top of the ordered task set onto the first processor that fits it (i.e., the first processor on which the task can be allocated while all processors are schedulable), beginning from the top of the ordered processor list. If none of the processors can fit the task, a new processor is added to the processor list. At each step the schedulability of all processors should be tested, because allocating a task to a processor can increase the remote blocking time of tasks previously allocated to other processors and may make the other processors unschedulable. This means, it is possible that some of the previous processors become unschedulable even if a task is allocated to a new processor, which makes the algorithm fail.

**The Algorithm:** The algorithm performs partitioning of a task set in two rounds and the result will be the output of the round with better partitioning results. However, the algorithm performs a few common steps before starting to perform the rounds. Each round allocates tasks to the processors (partitions) in a different strategy. When a BFD algorithm allocates an object (task) to a bin (processor), it usually puts the object in a bin that fits it better, and it does not consider the unallocated objects that will be allocated after the current object. The rationale behind the two rounds is that the heuristic tries to consider both past and future by looking at tasks allocated in the past and those that are not allocated yet. In the first round the algorithm considers the tasks that are not allocated to any processor yet; and tries to take as many as possible of the best related tasks (based on remote blocking parameters) with the current task. On the other hand, in the second round it considers the already allocated tasks and tries to allocate the current task onto the processor that contains best related tasks to the current task. In the second round, the algorithm performs more like the usual bin packing algorithms (i.e., tries to find the best bin for the current object), although it considers the remote blocking parameters while allocating a task to a processor. Any time the algorithm performs schedulability test, for more precise schedulability analysis, it always performs response time analysis [23].

The common steps of the algorithm before the two rounds are performed are as follow:

**1.** Each task is assigned a weight. The weight of each task, besides its utilization, should depend on parameters that lead to potential remote blocking time caused by other tasks:

$$w_i = u_i +$$
$$\lceil (\sum_{\rho_i < \rho_k} \mathrm{NC}_{i,k}\beta_{i,k}\lceil \frac{T_i}{T_k}\rceil + \mathrm{NC}_i \max_{\rho_i \geq \rho_k} \beta_{i,k})/T_i \rceil \tag{1}$$

where, $\mathrm{NC}_{i,k}$ is the number of critical sections of task $\tau_k$ in which it shares a resource with $\tau_i$, among these critical sections $\beta_{i,k}$ is the longest one, and $\mathrm{NC}_i$ is the total number of critical sections of $\tau_i$.

Considering the remote blocking terms of MPCP [6], the rationale behind the definition of weight is that the tasks that can be punished more by remote blocking become heavier. Thus, they can be allocated earlier and attract as many as possible of the tasks with which they share resources.

**2.** Macrotasks are generated, i.e., the tasks that directly or indirectly share resources are put into the same macrotask. A macrotask has two alternatives; it can either be broken or unbroken. If a macrotask cannot fit in one processor, (i.e., it is not possible to schedule the macrotask on a single processor even if there is no any other tasks), it is set as broken, otherwise it is denoted as unbroken. Please observe that the test of fitting a macrotask in a single processor (to set it as broken or unbroken) is only done at the beginning. Later on at any time the algorithm tests fitting an unbroken macrotask in a processor, the macrotask may co-exist with other tasks and/or macrotasks on the same processor.

If a macrotask is unbroken, the partitioning algorithm always allocates all tasks in the macrotask to the same partition (processor). This means that all tasks in the macro-task will share resources locally relieving tasks from remote blocking. However, tasks within a broken macrotask will be distributed into more than one partition. Similar to tasks, a weight is assigned to each unbroken macrotask, which equals to the sum of the utilizations (not weights) of its tasks . This is because all the tasks within an unbroken macrotask will always be allocated on the same processor and the tasks will not suffer from any remote blocking, hence there is no need to consider blocking parameters in the weight of an unbroken macrotask.

**3.** The unbroken macrotasks together with the tasks that do not belong to any unbroken macrotasks are ordered in a single list in non-increasing order of their weights. We denote this list the *mixed list*.

The strategy of allocation of tasks in both rounds depends on attraction between tasks. The attraction function of task $\tau_k$ to a task $\tau_i$ is defined based on the potential remote blocking overhead that task $\tau_k$ can introduce to task $\tau_i$ if they are allocated onto different processors. We represent the attraction of task $\tau_k$ to task $\tau_i$ as $v_{i,k}$ which is defined as follows:

$$v_{i,k} = \begin{cases} \mathrm{NC}_{i,k}\beta_{i,k}\lceil \frac{T_i}{T_k}\rceil & \rho_i < \rho_k; \\ \mathrm{NC}_i\beta_{i,k} & \rho_i \geq \rho_k \end{cases} \tag{2}$$

The rationale of the attraction function is to allocate the tasks that may remotely block a task, $\tau_i$, to the same processor as of $\tau_i$ (in order of the amount of remote blocking overhead) as far as possible. Please notice, the definition of weight (Equation 1) and attraction function (Equation 2) are heuristics that guide the algorithm under MPCP. However, these functions may differ under other synchronization protocols, e.g., MSRP and partitioned FMLP, which have different remote blocking terms.

There can be the case in which all tasks sharing resources end up in one macrotask. In this case if the macrotask can fit in one processor, there is no need to use MPCP or any other multiprocessor synchronization protocol, because there will not be any global resources in the system. On the other hand, if the macrotask does not fit in one processor (i.e., should be broken) the algorithm attempts, by using weight (Equation 1) and attraction (Equation 2) functions to put attracted tasks on the same processor as far as possible which leads to reducing the remote blocking overhead.

Now we present the continuation of the algorithm in two rounds:

**First Round:**  After the common steps the following steps are repeated within the first round until all tasks are allocated to processors (partitions):
1. All processors are ordered in their non-increasing order of utilization.
2. The object at the top of the mixed list is picked. **(i)** If the object is a task, $\tau_i$, and it does not belong to a broken macrotask ($\tau_i$ does not share any resource) $\tau_i$ will be allocated onto the first processor that fits it (all tasks on the processor are still schedulable), beginning from the top of the ordered processor list (similar to blocking-agnostic BFD). If none of the processors can fit $\tau_i$ a new processor is added to the list and $\tau_i$ is allocated onto it. **(ii)** If the object is an unbroken macrotask, all its tasks will be allocated onto the first processor that fits all of them. If none of the processors can fit the macrotask, it (all its tasks) will be allocated onto a new processor.**(iii)** If the object is a task, $\tau_i$, that belongs to a broken macrotask, the algorithm orders the tasks (those that are not allocated yet) within the macrotask in non-increasing order of attraction to $\tau_i$ based on equation 2. We call this list the *attraction list* of $\tau_i$. Task $\tau_i$ itself will be on the top of its attraction list. The best processor for allocation is selected, which is the processor that fits the most tasks from the attraction list, beginning from the top of the list. As many as possible of the tasks from the attraction list are then allocated to the processor. If none of the existing processors can fit any of the tasks, a new processor is added and as many tasks as possible from the attraction list are allocated to the processor. However, if the new processor cannot fit any task from the attraction list, i.e., at least one of the processors become unschedulable, the first round fails and the algorithm moves to the second round and restarts.

**Second Round:**  The following steps are repeated until all tasks are allocated to processors:
1. The object at the top of the mixed list is picked. **(i)** If the object is a task and it does not belong to a broken macrotask, this step is performed the same way as in the first round. **(ii)** If the object is an unbroken macrotask, in this the algorithm performs the same way as in the first round. **(iii)** If the object is a task, $\tau_i$, that belongs to a broken macrotask, the processors are put in a ordered list, denoted as *Plist*. However the processors are put in *Plist* in two steps. First, the processors that include some tasks from $\tau_i$'s macrotask

are added to *Plist* in non-increasing order of processors' attraction to $\tau_i$ (according to equation 2), i.e., the processor which has the greatest sum of attractions of its tasks to the picked task ($\tau_i$) is the most attracted processor to $\tau_i$ and is added to *Plist* first. Second, the processors that do not contain any task from $\tau_i$'s macrotask are added to *Plist* in non-increasing order of their utilization. After the two steps, the processors which contain at least one task from $\tau_i$'s macrotask will be located at the top of the ordered list, *Plist*, followed by the processors not containing any task from $\tau_i$'s macro task. The rationale behind this is that the algorithm first attempts to allocate $\tau_i$ on a processor containing some tasks from $\tau_i$'s macro task and if not succeeded then it tries other processors. The picked task ($\tau_i$) will be allocated onto the first processor from the processor list (*Plist*) that will fit $\tau_i$. Task $\tau_i$ will be allocated to a new processor if none of the existing ones can fit it. And the second round of the algorithm fails if allocating the task to the new processor makes some of the processors unschedulable.

If both rounds fail to schedule a task set the algorithm fails. If one of the rounds fails the result will be the output of the other one. If both rounds succeed to schedule the task set, the one with fewer partitions (processors) will be the output of the algorithm.

## 3.2   Synchronization-Aware Partitioning Algorithm (SPA)

We have implemented the best known existing partitioning algorithm proposed in [5] in our experimental evaluation framework. The implementation of the algorithm required details of the algorithm which were not presented in [5], hence, in this section we present the algorithm in more details.

**1.** First, the macrotasks are generated. In [5], macrotasks are denoted as bundles. A number of processors (enough processors that fit the total utilization of the task set) are added.

**2.** The macrotasks together with other tasks are ordered in a list in non-increasing order of their utilization. The algorithm attempts to allocate each macrotask (i.e., allocate all tasks within the macrotask) onto a processor. Without adding any new processor, all macrotasks and tasks that fit are allocated onto the processors. The macrotasks that can not fit are put aside. After any allocation, the processors are ordered in their non-increasing order of utilization.

**3.** The remaining macrotasks are ordered in the order of the cost of breaking them. The cost of breaking a macrotask is defined based on the estimated cost (blocking overhead) introduced into the tasks by transforming a local resource into a global resource (i.e., the tasks sharing the resource are allocated to different processors). The estimated cost of transforming a local resource $R_q$ into a global resource is calculated as follows:

$$\mathrm{Cost}(R_q) = \text{Global Overhead} - \text{Local Discount} \tag{3}$$

The Global Overhead is calculated as follows:

$$\text{Global Overhead} = \max(|Cs_q|)/\min_{\forall \tau_i}\{\rho_i\} \tag{4}$$

where $\max(|Cs_q|)$ is the length of longest critical section accessing $R_q$.

The Local Discount is defined as follows:

$$\text{Local Discount} = \max_{\forall \tau_i \text{ accessing } R_q} (\max(|Cs_{i,q}|)/\rho_i) \quad (5)$$

where $\max(|Cs_{i,q}|)$ is the length of longest critical section of $\tau_i$ accessing $R_q$.

The cost of breaking any macrotask, $\text{mTask}_k$, is calculated as the summation of blocking overhead caused by transforming its accessed resources into global resources.

$$\text{Cost}(\text{mTask}_k) = \sum_{\forall R_q \text{ accessed by } \text{mTask}_k} \text{Cost}(R_q) \quad (6)$$

**4.** The macrotask with minimum breaking cost is picked and is broken in two pieces such that the size of one piece is as close as the largest utilization available among processors. This means, tasks within the selected macrotask are ordered in decreasing order of their size (utilization) and the tasks from the ordered list are added to the processor with the largest available utilization as far as possible. In this way, the macrotask has been broken in two pieces; (i) the one including the tasks allocated to the processor and (ii) the tasks that could not fit in the processor. If the fitting is not possible a new processor is added and the whole algorithm is repeated again.

Firstly, as one can see, the SPA algorithm does not consider blocking parameters when it allocates the current task to a processor, but only its utilization, i.e. the tasks are ordered in order of their utilization only. However, our algorithm assigns a weight (Equation 1) which besides the utilization includes the blocking terms as well. Secondly, no relationship (e.g., as a cost based on blocking parameters) among individual tasks within a bundle (macrotask) is considered which could help to allocate tasks from a broken bundle to appropriate processors to decreases the blocking times. In our heuristic, we have defined an attraction function (Equation 2), which attempts to allocate the most attracted tasks from the current task's broken macrotask, on a processor. As the experimental evaluation in Section 4 shows, considering these issues can improve the partitioning significantly.

## 4 Experimental Evaluation and Comparison of Algorithms

In this section we present our experimental results of our blocking-aware bin-packing algorithm (BPA) together with the blocking-aware algorithm recently proposed in [5] (SPA), as well as the reference blocking-agnostic algorithm. For a number of systems (task sets), we have compared the performance of the algorithms in two different aspects; (1) Given a number of systems, the total number of systems that each of the algorithms can schedule, (2) The processor reduction aspect of algorithms.

### 4.1 Experiment Setup

We generated systems (task sets) for different workloads; we denote workload as a defined number of fully utilized processors, e.g., the workload equal to 3 fully utilized

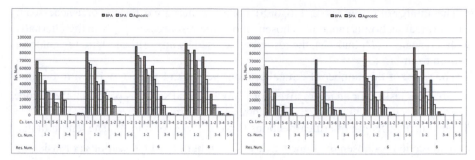

(a) Workload: 3 processors, 3 tasks per processor

(b) Workload: 3 processors, 6 tasks per processor

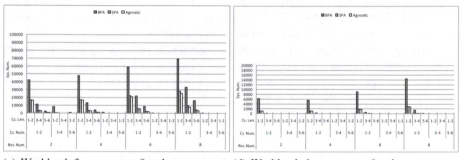

(c) Workload: 3 processors, 9 tasks per processor

(d) Workload: 6 processors, 6 tasks per processor

**Fig. 1.** Total number of task sets each algorithm schedules

processors means the summation of utilizations of all tasks in the system equals to 3. Please notice that the definition of the workload as a number of processors is only to show the total utilization of the task set and it is not the same as the number of required processors (which may be more than the workload) to schedule the task set. Given a workload, the full capacity of each processor (utilization of 1) is randomly divided among a defined number of tasks. Usually for generating systems, utilization and periods are randomly assigned to tasks, and worst case execution times of tasks are calculated based on them. However, in our system generation, the worst case execution times (WCET) of tasks are randomly assigned and the period of each task is calculated based on its utilization and WCET. The reason is that we had to restrict that the WCET of a task not to be less than the total length of its critical sections. Since we have limited the maximum number of critical sections to 6 and the maximum length of any critical section to 6 time units, hence the WCET of each task is greater than 36 ($6 \times 6$) time units. The WCET of each task was randomly chosen between 36 and 150 time units. The system generation was based on different settings; the input parameters for settings are as follows:

**1.** Workload (3, 4, 6, or 8 fully utilized processors).

**2.** The number of tasks per processor (3, 6 or 9 tasks per processor), e.g., 3 tasks per processor means that the utilization of one processor (utilization = 1) is randomly distributed among 3 tasks.

**3.** The number of resources (2, 4, 6, or 8). For each alternative, the resource accessed by each critical section is randomly chosen among the resources, e.g, given the alternative with 2 resources ($R_1$ and $R_2$), the resource accessed by any critical section is randomly chosen from $\{R_1, R_2\}$.

**4.** The range of the number of critical sections per task (1 to 2, 3 to 4 or 5 to 6 critical sections per task). For an alternative (e.g., 1 to 2 critical sections per task), the number of critical sections of any task $\tau_i$ is randomly chosen from $\{1, 2\}$.

**5.** The range of length of critical sections (1 to 2, 3 to 4, or 5 to 6). The length of each critical section is chosen the same way as the number of critical sections per task.

For each setting, we generated 100.000 systems, and combining the parameters of settings, i.e., (workloads)×(tasks per processor)×(resources)×(critical sections per task)×(critical section lengths)= $4 \times 3 \times 4 \times 3 \times 3 = 432$ different settings, total number of systems generated for the experiment were 43.200.000.

With the generated systems we were able to evaluate the partitioning algorithms with respect to different factors, i.e., various workloads (number of fully utilized processors), number of tasks per processor, number of shared resources, number of critical sections per task, and length of critical sections.

## 4.2 Results

In this section we present the evaluation results of our proposed blocking-aware algorithm (BPA), an existing blocking-aware algorithm [5] (SPA) and the blocking-agnostic algorithm.

The first aspect of comparison of the results from the algorithms is, given a number of systems, the total number of systems each algorithm successfully schedules (Figure 1). Figures 1(a), 1(b) and 1(c) represent the results for 3, 6 and 9 tasks per processor respectively. The vertical axis shows the total number of systems that the algorithms could schedule successfully. The horizontal axis shows three factors in three different lines; the bottom line shows the number of shared resources within systems (Res. Num.), the second line shows the number of critical sections per task (Cs. Num.), and the top line represents the length of critical sections within each task (Cs. Len.), e.g., Res. Num.=4, Cs. Num.=1-2, and Cs. Len.=1-2 represents the systems that share 4 resources, the number of critical sections per each task are between 1 and 2, and the length of these critical sections are between 1 and 2 time units. For some settings the number of schedulable systems were too few to be shown on the graphs, thus we omitted these settings from the graphs, e.g., The results for the combination of the number of critical sections = 3-4 and the length of critical sections = 5-6 are not shown in Figure 1.

As depicted in Figure 1, considering the total number of systems that each algorithm succeeds to schedule, our blocking-aware algorithm (BPA) performs better (can schedule more systems) compared to the SPA and the blocking-agnostic algorithm. However the SPA performs better than the blocking-agnostic algorithm. As shown in the figure, by increasing the number of resources, the number of successfully scheduled systems in all algorithms is increased. The reason for this behavior is that with fewer resources, more tasks share the same resource introducing more blocking overheads which leads to fewer schedulable systems. However, it is illustrated that the blocking-aware algorithms perform better as the number of resources is increased. It is also shown that increasing the number and/or the length of critical sections generally reduces the number of schedulable systems significantly. The reason is that more and longer critical sections introduce greater blocking overhead into the tasks making fewer systems schedulable.

As the number of tasks per processor is increased from 3 (Figures 1(a)) to 6 (Figures 1(b)) and to 9 (Figures 1(c)), the BPA performs significantly better (i.e., schedules significantly more systems) than the SPA and blocking-agnostic bin-packing. However, as one can see, the SPA does not perform significantly better than the blocking-agnostic algorithm as the number of tasks per processor are increased. Increasing the number of tasks per processor lead to smaller tasks (tasks with smaller $u_i$). The BPA allocates tasks from a broken macrotask based on Equations 1 and 2, which are functions of the blocking parameters (the number and length of critical sections) as well as the size of the tasks. On the other hand, with the smaller size of tasks, the blocking parameters have a bigger role in these functions, hence more dependent tasks are allocated to the same processor. This leads to less blocking overhead and increased schedulability, hence more systems are scheduled by BPA as the tasks per processor are increased. On the other hand, in SPA, allocation of tasks from a broken macrotask is only based on their utilization, and this does not necessarily allocates highly dependent tasks to the same processor.

As the workload (the number of fully utilized processors) is increased, although the BPA still performs better than the SPA and the blocking-agnostic algorithm, generally the number of schedulable systems by all algorithms is significantly reduced (Figure 1(d)). The reason for this behavior is that the number of tasks within systems are relatively many (36 tasks per each system in Figure 1(d)) and the workload is high (6 fully utilized processors), and all the tasks within systems share resources. On the other hand, the MPCP is pessimistic. This introduces a lot of interdependencies among tasks and consequently a huge amount of blocking overheads, making fewer systems schedulable. In practice in big systems with many tasks, not all of the tasks share resources, which leads to fewer interdependencies among tasks and less blocking times. However, we continued the experiment with higher workload in the same way as the other experiments (that all tasks share resources) to be able to compare the results with the previous results. We believe that realistic systems, even with high workload and many tasks can benefit from our partitioning algorithm to increase the performance.

The second aspect for comparison of performance of the algorithms is the processor reduction aspect. To show this, for each algorithm, we ordered the total schedulable systems in order of the number of required processors. Figure 2 illustrates the results for the workload of 3 fully packed processors and different number of tasks (3, 6 and 9) per processor. For each algorithm, the schedulable systems by each number of processors are shown as percentage of the total scheduled systems by that algorithm. As the results show, for 3 tasks per processor all three algorithms perform almost the same (Figure 2(a)), i.e., each algorithm schedules around 80% of its schedulable systems by 4 processors, 15% to 18% by 5 processors and less than 3% by 6 processors, etc. The reason is that the tasks are large (the utilization of a processor is distributed among 3 task), thus the blocking-aware algorithms do not have much possibility to increase the

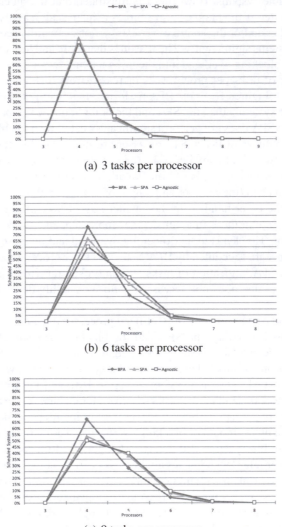

(a) 3 tasks per processor

(b) 6 tasks per processor

(c) 9 tasks per processor

**Fig. 2.** Percentage of systems each algorithm schedules, ordered by required number of processors

performance. However as the number of tasks per processor is increased (Figures 2(b) and 2(c) for 6 and 9 tasks per processor respectively), the blocking-aware algorithms, generally, perform better in processor reduction aspect. Especially the BPA, performs significantly better than the the SPA and the blocking-agnostic algorithm. This means that BPA reduces the required number of processors compared to SPA and the blocking-agnostic algorithm, e.g., as shown in Figure 2(c), 68% and 28% of the systems scheduled by BPA require 4 and 5 processors respectively, while 54% and 37% of systems scheduled by SPA can be scheduled by 4 and 5 processors respectively. This means a bigger part (68%) of systems scheduled by BPA require only 4 processors while with SPA this number is smaller (54%).

# 5  Conclusion

In this paper we have proposed a heuristic blocking-aware algorithm, for identical unit-capacity multiprocessor systems, which extends a bin-packing algorithm with synchronization parameters. The algorithm allocates a task set onto the processors of a single-chip multiprocessor (multi-core) with shared memory. The objective of the algorithm is to decrease blocking times of tasks by means of allocating the tasks that directly or indirectly share resources onto appropriate processors. This generally increases schedulability of a task set and may lead to fewer required processors compared to blocking-agnostic bin-packing algorithms. We have also presented and implemented an existing similar blocking-aware algorithm originally proposed in [5].

Since in practice most systems use fixed priority scheduling protocols, we have developed our algorithm under MPCP, a standard synchronization protocol for multiprocessors (multi-cores) which works under fixed priority scheduling. Another reason to implement our algorithm under MPCP was to be able to compare our approach to the existing similar approach [5] which has also been developed under MPCP. However, our approach is not limited to MPCP and it can easily be extended to other synchronization protocols such as MSRP and partitioned FMLP.

Our experimental results confirm that our algorithm mostly performs significantly better than the blocking-agnostic as well as the existing heuristic with respect to the number of schedulable systems and the number of required processors. However, given a NP-hard problem, a bin-packing algorithm may not achieve the optimal solution, i.e, there can exist systems that only one of the algorithms can schedule. Thus using a combination of heuristics improves the results with respect to the total number of schedulable systems and processor reduction.

A future work will be extending our partitioning algorithm to other synchronization protocols, e.g., MSRP and FMLP for partitioned scheduling. A very interesting future work is to apply our approach to different synchronization protocols and investigate the effect of bin-packing on those protocols and compare the improvement in their performance. Another interesting future work is to apply our approach to real systems and study the performance gained by the algorithm on these systems. In the domain of multiprocessor scheduling and synchronization our future work also includes investigating global and hierarchical scheduling protocols and appropriate synchronization protocols.

# Acknowledgments

The authors wish to thank Karthik Lakshmanan for fruitful discussions, helping out in improving the quality of this paper.

# References

1. Baker, T.: A comparison of global and partitioned EDF schedulability test for multiprocessors. Technical report (2005)
2. Baker, T.: Stack-based scheduling of real-time processes. Journal of Real-Time Systems 3(1), 67–99 (1991)
3. Carpenter, J., Funk, S., Holman, P., Srinivasan, A., Anderson, J., Baruah, S.: A categorization of real-time multiprocessor scheduling problems and algorithms. In: Handbook on Scheduling Algorithms, Methods, and Models. Chapman Hall/CRC, Boca (2004)
4. Devi, U.: Soft real-time scheduling on multiprocessors. In: PhD thesis (2006), http://www.cs.unc.edu/~anderson/diss/devidiss.pdf
5. Lakshmanan, K., de Niz, D., Rajkumar, R.: Coordinated task scheduling, allocation and synchronization on multiprocessors. In: Proceedings of 30th IEEE Real-Time Systems Symposium (RTSS 2009), pp. 469–478 (2009)
6. Nemati, F., Nolte, T., Behnam, M.: Blocking-aware partitioning for multiprocessors. Technical report, Mälardalen Real-Time research Centre (MRTC), Mälardalen University (March 2010), http://www.mrtc.mdh.se/publications/2137.pdf
7. Rajkumar, R.: Synchronization in Real-Time Systems: A Priority Inheritance Approach. Kluwer Academic Publishers, Dordrecht (1991)
8. Tindell, K.W., Burns, A., Wellings, A.J.: Allocating hard real-time tasks: An NP-hard problem made easy. Journal of Real-Time Systems 4(2), 145–165 (1992)
9. Altenbernd, P., Hansson, H.: The slack method: A new method for static allocation of hard real-time tasks. Journal of Real-Time Systems 15(2), 103–130 (1998)
10. de Niz, D., Rajkumar, R.: Partitioning bin-packing algorithms for distributed real-time systems. Journal of Embedded Systems 2(3-4), 196–208 (2006)
11. Baruah, S., Fisher, N.: The partitioned multiprocessor scheduling of sporadic task systems. In: Proceedings of 26th IEEE Real-Time Systems Symposium (RTSS 2005), pp. 321–329 (2005)
12. Rajkumar, R., Sha, L., Lehoczky, J.P.: Real-time synchronization protocols for multiprocessors. In: Proceedings of the 9th Real-Time Systems Symposium, RTSS 1988 (1988)
13. Gai, P., Lipari, G., Natale, M.D.: Minimizing memory utilization of real-time task sets in single and multi-processor systems-on-a-chip. In: Proceedings of 22nd IEEE Real-Time Systems Symposium (RTSS 2001), pp. 73–83 (2001)
14. Gai, P., Di Natale, M., Lipari, G., Ferrari, A., Gabellini, C., Marceca, P.: A comparison of MPCP and MSRP when sharing resources in the janus multiple processor on a chip platform. In: Proceedings of 9th IEEE Real-Time And Embedded Technology Application Symposium (RTAS 2003), pp. 189–198 (2003)
15. López, J.M., Díaz, J.L., García, D.F.: Utilization bounds for EDF scheduling on real-time multiprocessor systems. Journal of Real-Time Systems 28(1), 39–68 (2004)
16. Devi, U., Leontyev, H., Anderson, J.: Efficient synchronization under global EDF scheduling on multiprocessors. In: Proceedings of 18th IEEE Euromicro Conference on Real-time Systems (ECRTS 2006), pp. 75–84 (2006)
17. Block, A., Leontyev, H., Brandenburg, B., Anderson, J.: A flexible real-time locking protocol for multiprocessors. In: Proceedings of 13th IEEE Conference on Embedded and Real-Time Computing Systems and Applications (RTCSA 2007), pp. 47–56 (2007)
18. Brandenburg, B., Calandrino, J., Block, A., Leontyev, H., Anderson, J.: Synchronization on multiprocessors: To block or not to block, to suspend or spin? In: Proceedings of 14th IEEE Real-Time and Embedded Technology and Applications Symposium (RTAS 2008), pp. 342–353 (2008)

19. Brandenburg, B., Anderson, J.: An implementation of the PCP, SRP, D-PCP, M-PCP, and FMLP real-time synchronization protocols in LITMUS. In: Proceedings of 14th IEEE International Conference on Embedded and Real-Time Computing Systems and Applications (RTCSA 2008), pp. 185–194 (2008)
20. Brandenburg, B.B., Anderson, J.H.: A comparison of the M-PCP, D-PCP, and FMLP on LITMUSRT. In: Baker, T.P., Bui, A., Tixeuil, S. (eds.) OPODIS 2008. LNCS, vol. 5401, pp. 105–124. Springer, Heidelberg (2008)
21. Easwaran, A., Andersson, B.: Resource sharing in global fixed-priority preemptive multiprocessor scheduling. In: Proceedings of 30th IEEE Real-Time Systems Symposium (RTSS 2009), pp. 377–386 (2009)
22. Nemati, F., Behnam, M., Nolte, T.: Efficiently migrating real-time systems to multi-cores. In: Proceedings of 14th IEEE Conference on Emerging Techonologies and Factory, ETFA 2009 (2009)
23. Burns, A.: Preemptive priority based scheduling: An appropriate engineering approach. In: Principles of Real-Time Systems, pp. 225–248. Prentice Hall, Englewood Cliffs (1994)

# On Best-Effort Utility Accrual Real-Time Scheduling on Multiprocessors

Piyush Garyali, Matthew Dellinger, and Binoy Ravindran

ECE Dept., Virgina Tech, Blacksburg, VA 24061, USA
{piyushg,mdelling,binoy}@vt.edu

**Abstract.** We consider the problem of scheduling dependent real-time tasks for overloads on a multiprocessor system, yielding best-effort timing assurance. The application/scheduling model includes tasks with time/utility function time constraints, mutual exclusion constraints, and arbitrary arrival, execution-time and resource access behaviors, with timeliness optimization objective of maximizing total accrued utility while ensuring mutual exclusion constraints and deadlock-freedom. Since this problem is NP-hard, we develop a class of polynomial-time heuristic algorithms, called the *Global Utility Accrual* (GUA), and present two algorithm instances, namely, *Non-Greedy Global Utility Accrual* (NG-GUA) and *Greedy Global Utility Accrual* (G-GUA). We establish several properties of the algorithms including conditions under which optimal total utility is accrued, mutual exclusion constraints are satisfied, and deadlock-freedom is achieved. We develop a Linux-based real-time kernel called ChronOS, extended from PREEMPT_RT real-time patch. ChronOS provides a framework for implementation of a variety of multiprocessor schedulers. Our experimental studies with ChronOS reveal the effectiveness of GUA algorithms under a broad range of workloads.

**Keywords:** real-time, multiprocessors, scheduling, time/utility functions.

## 1 Introduction

Recently, there has been a shift in the computer industry from increasing clock rates to designing multi-core and hyper-threading architectures in a quest to produce faster computers [29]. Motivated by heat/power issues, most chip manufacturers have chosen the route of increasing system- and chip-level parallelism, as opposed to increasing clock rates, to improve performance. Consequently, the design of multiprocessor real-time scheduling algorithms has become important so that real-time applications can take advantage of these emerging architectures.

One unique aspect of multiprocessor real-time scheduling is the degree of run-time migration allowed for job instances of a task across processors (at scheduling events). Example migration models include: (1) *full migration*, where jobs are allowed to arbitrarily migrate across processors during their execution. This usually implies a global scheduling strategy, where a single shared scheduling queue is maintained for all processors and a processor-wide scheduling decision is made by a single (global) scheduling algorithm; (2) *no migration*, where tasks

C. Lu, T. Masuzawa, and M. Mosbah (Eds.): OPODIS 2010, LNCS 6490, pp. 270–285, 2010.

are statically (off-line) partitioned and allocated to processors. At run-time, job instances of tasks are scheduled on their respective processors by processors' local scheduling algorithm, such as single processor scheduling; and (3) *restricted migration*, where some form of migration is allowed—e.g., at job boundaries.

The Pfair class of algorithms [8] that allow full migration and fully dynamic priorities have been shown to be theoretically optimal—i.e., they achieve a *schedulability utilization bound*, $U$, below which all tasks meet their deadlines, that equals the total capacity of all $m$ processors i.e., $U = m$. Under Pfair, tasks are decomposed into several small uniform segments, which are then quantum-scheduled and may cause frequent scheduling and migration. Thus, algorithms other than Pfair have also been intensively studied though their utilization bounds are lower. Examples of global algorithms include global-EDF [9], global-non-preemptive-EDF [7] with a bound at most $U \approx m/2$, LLREF [12], LRE-TL [16], PG/PCG [10], NVNLF [15] with a bound $U = m$, and global-RM-US with a bound $U \approx m/3$ [4]. In the partitioned space, examples include partitioned dynamic priority algorithms, such as partitioned-EDF [3] with a bound $U \approx m/2$, and fixed priority algorithms such as partitioned-DMS with a bound $U \approx m/2$ [5] and PDMS-HPTS-DS with a bound of 65% [23].

Majority of these scheduling efforts focus on application contexts where key aspects of application behavior—e.g., task arrivals, execution times, resource accesses—are deterministically bounded or known. Although this is an extremely important subspace of the real-time problem space, there also exist some real-time applications with behaviors outside this envelope—e.g., unpredictable task arrival and execution-time behaviors, caused due to data- and context-dependent executions, resulting in transient and permanent overloads (i.e., $U > m$) [14,2]. During overloads, applications such as [14] desire graceful timeliness degradation and "best-effort" timing assurance in the sense that as many processor cycles as needed are assured to be allocated to the most important task, less so are allocated to the least important task, and so on [26,20]. (Note that task importance may be orthogonal to task urgency.) An interesting feature of these applications is that their task execution-time magnitudes are relatively longer—e.g., milliseconds to minutes. This allows relatively time-expensive real-time scheduling.

Past works on overload-scheduling with best-effort timing assurances (e.g., LBESA [26], DASA [13], GUS [24,25], $D^{over}$ [21]) have focused on single processor systems, with a few exceptions.[1] The only efforts in this space that consider multiprocessors include MOCA [22] and gMUA [11]. Both these algorithms, however, exclude task dependencies that arise due to synchronization constraints.

In this paper, we focus on this multiprocessor problem space, directly motivated by applications such as [14,2]. We consider tasks with time/utility function (TUF) time constraints [19] that subsume deadlines and allow task urgency to be expressed independent of task importance. Tasks have unknown arrival behaviors and are subject to execution overruns, causing overloads. In addition, tasks

---

[1] LBESA's and DASA's design were directly motivated by the "best-effort" real-time notion, and have been transferred to the application in [14] due to its matching operational requirements.

have mutual exclusion constraints; they use lock-based concurrency control, with unknown lock-access and release behaviors. We consider the timeliness objective of maximizing the total accrued timeliness utility, while satisfying mutual exclusion constraints and freedom from deadlocks. This problem is NP-hard. We develop a class of polynomial-time heuristic algorithms called the GUA class of algorithms, and present two algorithm instances, namely, NG-GUA and G-GUA. We establish several properties of the algorithms including conditions under which optimal total utility is obtained, mutual exclusion constraints are satisfied, and deadlock-freedom is achieved.

We develop a Linux-based real-time OS kernel called ChronOS, extended from the PREEMPT_RT real-time Linux patch, which provides optimized interrupt service latencies and real-time locking primitives. ChronOS provides a scheduling framework for the implementation of a broad range of scheduling algorithms as scheduler plugins. We implement the GUA algorithms and their competitors (e.g., G-EDF, G-NP-EDF, gMUA, P-EDF, P-DASA) in ChronOS and conduct experimental studies. Our results reveal the effectiveness of the GUA algorithms under a broad range of workloads.

Thus, the paper's contribution is the GUA class of algorithms that allow tasks to be subject to run-time uncertainties, overloads and dependencies, and yield optimal total utility (when possible) and best-effort timeliness behavior otherwise — the first such multiprocessor real-time scheduling algorithms.

The rest of the paper is organized as follows: Section 2 describes our models and objective. Section 3 presents the GUA class of algorithms. The algorithms' rationale, design, and properties are described in this section. We report our experimental studies in Section 4. Finally, we conclude in Section 5.

## 2    Models and Objective

We consider Clark's phase abstraction [13] as the unit of scheduling. A phase describes a single flow of execution. Phases arrive arbitrarily and may be preempted arbitrarily.

Phases have time constraints. A time constraint has a "scope"—a segment of the phase control flow that is associated with the time constraint [28]. Such a scope is called a "scheduling segment". Each phase has a single scheduling segment. A phase $i$'s scheduling segment's time constraint is specified using a TUF. TUFs can only be downward step-shaped—i.e., a constant maximum utility $u_i$ is accrued if the segment completes before a deadline time $d_i$; zero utility otherwise.

A good-faith estimate of a phase $i$'s scheduling segment's execution time, $e_i$, is available (through off-line measurements). This time estimate is not the worst-case; it can be violated at run-time (e.g., due to context dependence) and can cause processor overloads.

A phase enters and exits a scheduling segment by invoking scheduler APIs—e.g., Real-Time CORBA's [28] begin_scheduling_segment and end_scheduling _segment APIs. When a scheduling segment is entered, a phase passes its scheduling parameters (e.g., $d_i, u_i, e_i$) to the API.

Phases may access non-CPU resources, which are serially reusable. Resources can be shared and be subject to mutual exclusion constraints. A phase may request multiple shared resources during its lifetime. The requested time intervals for holding resources may be nested, overlapped, or disjoint. Phases may request and release resources arbitrarily—i.e., which phase needs which resource and in what order is unknown. A phase can only be blocked on one resource.

An `abort_handler` is associated with each phase scheduling segment. We consider a termination model for all failures encountered during phase executions including time-constraint violations and logical errors, which raises an exception that is handled by the phase. The handler performs compensating actions that are necessary to avoid inconsistencies and ensure the safety of the external state.

*Scheduling Objective.* Our objective is to schedule the phases on an $m$-processor system such that the sum of the utility accrued by the completion of the phases is maximized, as much as possible, while satisfying phase mutual exclusion constraints and ensuring deadlock-freedom. Additionally, the number of phase deadlines missed must be minimized as much as possible.

This problem of selecting tasks to execute at a specific instant during overloads is NP-hard because its one-processor version is NP-hard [13]. Thus, the GUA algorithms presented here are polynomial-time heuristic algorithms.

## 3   GUA Class of Algorithms

### 3.1   Basic Rationale

Since the phase model is dynamic—i.e., when phases will arrive, how long they will execute, which set of resources will be needed by which phases, the length of time for which those resources will be needed, the order of accessing the resources are all statically unknown—future scheduling events such as new phase arrivals and new resource requests cannot be considered at a scheduling event. Thus, a schedule must be constructed solely exploiting the current system knowledge.

Since the primary scheduling objective is to maximize the total utility, a reasonable heuristic is a "greedy" strategy: Favor "high return" phases over low return ones, and complete as many of them as possible before phases' termination times and also as early as possible.

The potential utility that can be accrued by executing a phase is an indication of its "return on investment". We measure this using a metric called the *Potential Utility Density* (or PUD) pioneered in [13]. A phase's PUD measures the utility that can be accrued per unit time by immediately executing the phase and those phase(s) that it (directly or transitively) depends upon for locked resources.

**Ensuring Mutual Exclusion:** The presence of phase dependencies can result in many phase dependency chains, similar to single-processor case. But unlike single-processors, as there are $m > 1$ processors for multiprocessors, up to $m$ of these chains (or phases at the head of those chains) can be potentially executed. In [13], dependency chains are computed at per phase level which works well for single-processors as only one phase needs to be selected. However, this

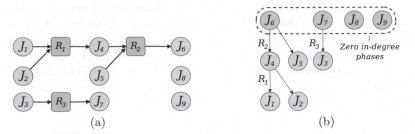

**Fig. 1.** (a) A phase and resource dependency chain (b) DAG representation showing the zero in-degree phases

method cannot be applied to multiprocessors. In order to ensure mutual exclusion, two phases that are dependent on each other should not be allowed to be executed on processors concurrently. Hence, there is a need to find the dependency relationship of all phases in an effective way. We solve this problem by constructing a directed acyclic graph (DAG) to represent dependency relationships between phases. Fig 1(a) shows the dependency relationship between phases and resources—e.g., phase $J_1$ requires a resource that is owned by phase $J_4$. Note that a phase cannot be blocked on multiple resources. In Fig 1(b), we represent the dependency relationship using DAG (the node represents the phase and the edge represents resource relationship). Thus, at the end of graph creation, we consider the zero in-degree (ZID) phases, which have zero input edges and hence are not dependent on other phases, as eligible for the final schedule.

**Maximizing Accrued Utility:** Once the ZID phases are found, we need to determine $m$ ZID phases that have highest execution eligibility. PUD metric [13] has been shown to be highly effective in determining phase execution eligibility for single-processors as a single phase needs to be selected at end of the schedule. On multiprocessors, unlike single-processors, many non-dependent phases can be concurrently dispatched for execution. However, PUD metric alone cannot be used to pick one phase over another, as there could be a ZID phase that currently owns a resource, blocking other phases in the system, but has a lower PUD. As a result, the phase could be pushed to the back of the queue preventing other eligible phases that are currently blocked on it, from executing. The challenge is to find a metric that provides a way to represent overall benefit the system can accrue if a particular phase is selected for execution. To solve this, we define two metrics—*Local Value Density* (LVD), which is equivalent to PUD of a phase, and *Global Value Density* (GVD), which is defined as the sum of the LVDs of individual phases that are in a dependency relation with a ZID phase. For example, in Fig 1(b), $\text{GVD}(J_6) = \text{LVD}(J_6) + \text{LVD}(J_4) + \text{LVD}(J_5) + \text{LVD}(J_1) + \text{LVD}(J_2)$. The GVD for a ZID phase $J$ represents the aggregate value density for entire dependency chain of $J$. This gives a fair representation of the dependency relationship for that phase, and provides the highest execution eligibility for a phase that is currently blocking other phases.

**Deadlock Detection and Resolution:** A deadlock can occur when a phase $J_a$, which owns resource $R_a$, makes a request for another resource $R_b$, owned by a phase $J_b$, wherein phase $J_b$ directly or through its dependency chain requests for the resource $R_a$ (owned by $J_a$). Thus, a deadlock represents a cycle in the dependency chain of a phase which can be detected using a cycle detection algorithm. In [13], deadlocks are detected at the phase level when the individual phase dependency chains are being computed. However, as we construct a DAG to represent the dependency relationship of all the phases, deadlock detection and resolution can be integrated with DAG construction. A DAG can be created for a phase and all its dependencies in a single pass. During each step, phases in the dependency chain can be maintained in a list such that if a dependent phase is again added to the list, a deadlock is detected. In order to resolve the deadlock, one of the phases needs to be rejected. In the design of GUA algorithms, we select the least LVD phase as it contributes the least utility to the total accrued utility.

## 3.2 Overview

GUA's scheduling events include the arrival of a phase, completion of a phase, a resource request and a resource release. To describe the algorithms we define the variables and auxiliary functions. For a phase J, J.RemExec is the estimated remaining execution cost of the phase and J.Utility denotes the TUF at the time of the scheduling event. The following auxiliary functions are used:

InsertEdge(J, DepJ) inserts an edge between phases J and DepJ.
RemoveEdge(J) removes all in-degree and out-degree edges of phase J.
InsertList(J, $\sigma$) inserts the phase J in the list $\sigma$.
InsDeadLnPos(J, $\sigma$) inserts phase J in the list $\sigma$ at its deadline position.
FindZIDPhases($\sigma$) returns the ZID phases from the list $\sigma$.
RemoveLeastLVD($\sigma$) removes the phase with the least LVD from the list $\sigma$.
RemoveLeastGVD($\sigma$) removes the phase with the least GVD from the list $\sigma$.
FindPIPDeadLn($\sigma$) finds the earliest deadline amongst the dependents of the
    ZID phases in list $\sigma$ to ensure Priority Inheritance Protocol (PIP) behavior.
ComputeGVD($\sigma$) computes the GVD for the ZID phases in the list $\sigma$.
SortByGVD($\sigma$) sorts the list $\sigma$ by the decreasing value of GVD.
IsPresent(J, $\sigma$) returns true if the phase J is present in the list $\sigma$.
IsFeasible($\sigma$) returns true if schedule in $\sigma$ is feasible, i.e., the predicted com-
    pletion time of each phase in $\sigma$ must never exceed its deadline.
Owner(R) returns the phase that holds resource R.
ResRequested(J) returns the resource requested by phase J.
FindProcessor() returns the ID of the processor on which the currently as-
    signed phases have the shortest sum of allocated execution times.
FindProcessor(cpu_mask) is an extended version of FindProcessor() that
    takes a cpu_mask (mask of processors that have been checked earlier and
    should be avoided). If all the processors are in the mask, it returns NULL.
AddCpuToMask(p, cpu_mask) adds processor p to the cpu_mask.

---

**Algorithm 1.** Creation of DAG with detection/resolution of deadlocks

---

```
 1: Procedure: CreateDAGwithDRD (σ_T)
 2: Input: σ_T              // List of released phases
 3: Vars: J, V, next        // Phase pointers
 4: Vars: σ_J               // Phase J's list of dependents
 5: for each phase J in σ_T do
 6:     σ_J = φ ;
 7:     J.Lvd = J.Utility/J.RemExec;
 8:     InsertList(J, σ_J);
 9:     next = Owner(ResRequested(J));
10:     while next ≠ φ do
11:         if IsPhaseAborted(next) then
12:             break;
13:         if IsPresent(next, σ_J) == false then
14:             InsertEdge(J, next) ;
15:             InsertList(next, σ_J);
16:             J = next;
17:             next = Owner(ResRequested(next));
18:         else
19:             V = FindLeastLVD(σ_J);
20:             AbortPhase(V);
21:             RemoveEdge(V);
22:             break;
```

---

FindLeastLVD(J) finds the phase with the least LVD in dependency chain of J.
UpdateCpuEC(p, J, b) adds J.RemExec to the sum of remaining execution
    times for phases allocated on processor p if b is true, subtracts otherwise.
AbortPhase(J) sends an aborting signal to the phase J.
IsPhaseAborted(J) returns true if phase J has been marked for abortion.
HeadOf(σ) returns the phase J which is at the head of the list σ.

Algorithm 1 describes the pseudo-code for CreateDAGwithDRD($\sigma_T$) that uses
the list of phases, $\sigma_T$, and creates a DAG representation along with deadlock
detection and resolution. We refer to the phase that has requested a resource as
a *child* while the phase that owns the resource being requested as a *parent*. In
lines 5-22, the algorithm iterates over the list, $\sigma_T$, and for each phase, $J$, checks
if a *parent* node exists and adds an edge from the *parent* to the *child* (line 14).
In lines 16-17, the algorithm sets the current *parent* node as the new *child* and
checks if it has requested a resource. The steps are repeated for all the phases
in the dependency chain of $J$.

In order to detect deadlocks we use list $\sigma_J$ (line 6) to which we add all the
dependencies for phase $J$ (line 15). Before adding an edge between a *child* and a
*parent*, we check if the phase exists in $\sigma_J$ (line 13). The existence of the phase in
$\sigma_J$ indicates that the phase has already been added to the graph, thus detecting
a deadlock. To resolve the deadlock, we find the least LVD phase in $\sigma_J$, abort
the phase and remove it from the graph (lines 19-21).

### 3.3    Non-Greedy Global Utility Accrual (NG-GUA)

Algorithm 2 describes the NG-GUA scheduling algorithm. For a given list of phases $\sigma_T$, the DAG is created (line 4) using `ComputeDAGwithDRD()`. To ensure mutual exclusion we find the ZID phases and compute their GVD (lines 5-6). In the presence of dependencies, NG-GUA defaults to G-EDF with PIP. We compute the PIP deadlines for each of the ZID phases (line 7). The PIP deadline of a ZID phase $J_z$ is the earliest deadline of a phase $J_i$ which is dependent on $J_z$. In line 8, we sort the ZID phases by their PIP deadlines. The key idea here is to sort the ZID phases by the deadlines of the phases which have an earlier deadline but are currently blocked on a resource that is being held by the ZID phases, thus ensuring a PIP behavior[2].

In lines 9-11, we use `FindProcessor()` to assign phases to individual processor lists $\sigma_p$. The processor that yields the shortest sum of allocated execution times of all jobs in its local schedule is selected for assignment. The rationale for this choice is that the shortest summed execution time processor results in the nearest scheduling event for completing a job after assigning each job.

In lines 12-14, we check each of $\sigma_p$ lists for schedule feasibility using the `IsFeasible()` method. During overloads, the schedule might not be feasible. Hence, NG-GUA attempts to maximize the total utility by allowing phases that have a higher value density to be executed. In line 14, we remove the phase in $\sigma_p$ that has the least GVD and check the schedule for feasibility. Lines 12-14 are repeated until a feasible schedule is found. Finally, the head of the final feasible schedule, $\sigma_p$, for each processor $p$ is dispatched (lines 15-17). In the absence of dependencies, all the phases are treated as ZID phases and the PIP deadlines for each phase is equivalent to the phase's deadline. Hence, after the sort (line 8), NG-GUA defaults to a G-EDF order.

Algorithm 2 is referred to as non-greedy because it defaults to a deadline order rather than a value density order along with support for priority inheritance protocol, thus following a G-EDF with PIP behavior during underloads and maximizing total accrued utility during overloads. A sample schedule for NG-GUA is presented in [17].

### 3.4    Greedy Global Utility Accrual (G-GUA)

Algorithm 3 describes the G-GUA scheduling algorithm. Lines 6-8 are similar to the NG-GUA algorithm, described in Section 3.3. We create the DAG, find the ZID phases and compute their GVD. G-GUA does not default to G-EDF with PIP. Hence, we do not need to find the PIP deadlines.

G-GUA differs from NG-GUA in two ways— (i) the ZID phases are sorted by GVD instead of the PIP deadlines (line 9); and (ii) instead of assigning phases to all the processors and then running the feasibility check, G-GUA follows a greedier approach to accrue total utility. For all individual GVD-sorted ZID phases in $\sigma_d$ (lines 10-24), G-GUA assigns the phase to a processor which has the smallest

---

[2] In the absence of dependencies, the PIP deadline of a ZID phase $J_z$ can be considered equal to the deadline of $J_z$.

---

**Algorithm 2.** NG-GUA: Non-greedy Global Utility Accrual

---

1: **Input:** $\sigma_T$           // List of released phases
2: **Vars:** $\sigma_1 \cdots \sigma_m$   // Per processor ready queues for $m$ processors
3: **Vars:** $\sigma_z$           // Zero in-degree phase list
4: `ComputeDAGwithDRD`($\sigma_T$);
5: $\sigma_z \leftarrow$ `FindZIDPhases`($\sigma_T$);
6: `ComputeGVD`($\sigma_z$);
7: $\sigma_z \leftarrow$ `FindPIPDeadLn`($\sigma_z$);
8: $\sigma_d \leftarrow$ `SortByPIPDeadLn`($\sigma_z$);
9: **for** *each phase J in* $\sigma_d$ **do**
10:   | $p \leftarrow$ `FindProcessor`();
11:   | `InsertList`(J, $\sigma_p$);

12: **for** *each processor p* **do**
13:   | **while** `IsFeasible`($\sigma_p$) == *false* **do**
14:   |   | `RemoveLeastGVD`($\sigma_p$);

15: **for** *each p processor's schedule* $\sigma_p$ *in m* **do**
16:   | $Job_p \leftarrow$ `HeadOf`($\sigma_p$) ;
17: **return** { $Job_1, \cdots, Job_m$ };

---

sum of total phase remaining execution cost and checks for feasibility of schedule on that processor. If the schedule is feasible, G-GUA moves to the next phase in $\sigma_d$. However, during overloads, if the schedule is not feasible (after the phase was added to the first processor it was assigned to), G-GUA removes it from that processor's list and tries the same phase on all the other available processors (using `cpu_mask`, lines 18-22). The key idea is to ensure that a high GVD phase is checked on all processors before being rejected. In lines 25-27, the head of the final feasible schedule ($\sigma_p$) for each processor $p$ is taken and dispatched to the individual processor for scheduling. In the absence of dependencies, all the phases are treated as ZID phases. Thus, GVD for each phase is equivalent to the phase's LVD.

G-GUA is greedier than NG-GUA for accrued utility during overloads. It does not default to any deadline-based scheduling algorithm and attempts to maximize accrued utility both during underloads and overloads. A sample schedule is presented in [17].

### 3.5   Algorithm Properties

The properties of NG-GUA and G-GUA are summarized in this section. For brevity, the proofs have been omitted and provided in [17].

**Theorem 1.** *During underloads NG-GUA without dependencies defaults G-EDF.*

**Theorem 2.** *During underloads NG-GUA with dependencies defaults G-EDF-PIP.*

**Theorem 3.** *gMUA is a special case of NG-GUA without dependencies.*

**Theorem 4.** *Both NG-GUA and G-GUA ensure mutual exclusion.*

**Theorem 5.** *For both algorithms, an application always makes progress if there is work offered and the application is not deadlocked.*

---

**Algorithm 3.** G-GUA: Greedy Global Utility Accrual

---

```
 1: Input: σ_T          // List of released phases
 2: Vars: σ_1 ··· σ_m  // Per processor ready queues for m processors
 3: Vars: σ_z          // Zero in-degree phase list
 4: Vars: cpu_mask
 5: Vars: not_fes
 6: ComputeDAGwithDRD(σ_T);
 7: σ_z ← FindZIDPhases(σ_T);
 8: ComputeGVD(σ_z);
 9: σ_d ← SortByGVD(σ_z);
10: for each phase J in σ_d do
11:     cpu_mask = 0; not_fes = true;
12:     while not_fes == true do
13:         p ← FindProcessor();
14:         if p == φ then
15:             break;
16:         InsDeadLnPos(J, σ_p);
17:         UpdateCpuEC(p, J, true);
18:         if IsFeasible(σ_p) == false then
19:             RemoveList(J, σ_p);
20:             UpdateCpuEC(p, J, false);
21:             AddCpuToMask(p, cpu_mask);
22:             not_fes = true;
23:         else
24:             not_fes = false;
25: for each p processor's schedule σ_p in m do
26:     Job_p ← HeadOf(σ_p) ;
27: return { Job_1, ··· , Job_m };
```

---

**Property 1.** *In [18], Theorem 16.3.1 shows that when the schedule length is used as a criteria, a greedy algorithm that schedules the ZID nodes in a DAG produces a schedule that is within a factor of two from being optimal. Further, for a multi-threaded application with $P$ threads, work $T_1$ and critical path length $T_\infty$, the length of the schedule is bounded by $\frac{T_1}{P_A} + \frac{T_\infty(P-1)}{P_A}$, where $P_A$ is defined as the average number of threads executed at each scheduling interval.*

**Theorem 6.** *Property 1 applies for both NG-GUA and G-GUA.*

**Theorem 7.** *For $m$ processors and $n$ phases, the asymptotic cost for both NG-GUA and G-GUA is $O(mn \log n)$.*

## 4    Experimental Evaluation

### 4.1    ChronOS Real-Time Linux

In order to implement and evaluate the performance of NG-GUA and G-GUA with other state-of-the-art scheduling algorithms, we created a real-time Linux kernel, called ChronOS [1], based on the PREEMPT_RT patch [27]. The key motivation was to take advantage of the PREEMPT_RT real-time patch which enables complete preemption in Linux and improves interrupt latencies. ChronOS provides a set of APIs and a scheduler plugin infrastructure using which various single-processor and multiprocessor (including utility accrual and non-utility accrual) scheduling algorithms can be implemented. ChronOS is the first academic real-time Linux kernel based on the PREEMPT_RT patch. The architectural details of ChronOS are discussed in [1,17].

## 4.2 Experimental Setup

Both NG-GUA and G-GUA do not assume any specific task arrival model (e.g., periodic, aperiodic, sporadic). Tasks can arrive at any time in the system and generate scheduling events. However, in order to evaluate NG-GUA/G-GUA against other state-of-the-art algorithms, we use a periodic model which helps quantify the schedulability criteria of the algorithms and allows us to compare the performance with other scheduling algorithms. We create a synthetic real-time test application in ChronOS which enables evaluation using a wide range of workloads. Tasks are represented as threads and the application periodically fires threads with specified time-constraints. For each task, we use a burn_cpu(exec_cost) method, which takes the execution cost of the task as an input and burns processor cycles for that amount of time.

We conduct the experiments on a quad-core platform based on AMD Phenom 9650 processor with 2.3 GHz frequency and 2 MB L3 cache, and measure the Deadline Satisfaction Ratio (DSR) and the Accrued Utility Ratio (AUR). At a given utilization load $U$, the DSR is measured as the ratio of the tasks that met their deadlines to the total number of tasks released in the system. In a similar fashion, the AUR is measured as the total accrued utility of the tasks that met their deadlines to the total possible accrued utility in the system.

We consider two types of task-sets in this paper—12 tasks (12T) with periods in the range $[300ms - 20000ms]$ and 27 tasks (27T) with periods in the range $[50ms - 7500ms]$. The utilization load per task are in the range $[0.01 - 0.5]$. Note that this utilization range favors G-EDF. We use a downward "step" TUF and consider three models: (i) *Increasing Utility* (IU), utilities assigned to the tasks are proportional to their deadlines. The task with the earliest deadline has the least utility and vice-versa; (ii) *Decreasing Utility* (DU), utilities assigned to a task are inversely proportional to their deadlines. The task with the earliest deadline has the highest utility and vice-versa; and (iii) *Random Utility* (RU), tasks are assigned random utilities with no two tasks having the same utility. These models are used to ascertain whether, irrespective of the TUF ordering, our algorithms perform comparable to the competitors and to ensure that we do not create a bias based on the TUF assignment against the deadline-based algorithms. The data points on all results are shown as an average of ten samples along with the standard deviation.

We have additional extensive results using a wide range of task-sets on two, four, and eight processor platforms and those have been excluded here due to space limitation. We present those results in [17].

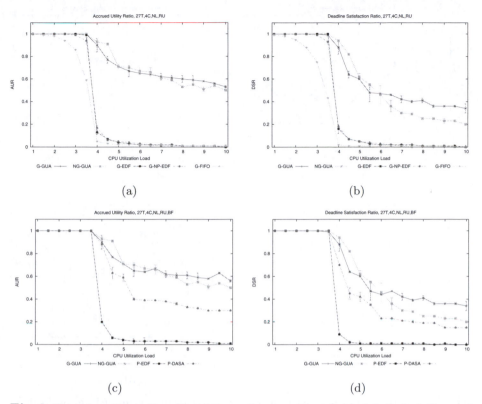

**Fig. 2.** Experimental results without dependencies against (a-b) global scheduling and (c-d) partitioned scheduling algorithms

### 4.3    Results without Dependencies

**Comparison with global scheduling algorithms**. We compare against G-EDF, G-NP-EDF and G-FIFO. These are selected as baseline competitors. Fig 2(a) and Fig 2(b) show the AUR and DSR results, respectively, for 27T using RU on a 4-core platform. No locks have been used. We observe that both NG-GUA and G-GUA are able to accrue higher utility during overload conditions as compared to the deadline-based scheduling algorithms. As a consequence, the algorithms are able to satisfy more task deadlines during overloads when compared to the deadline-based scheduling algorithms. On a 4-core platform, G-EDF is able to meet all deadlines upto $\approx 380\%$ CPU utilization load, after which it suffers from a domino effect. G-NP-EDF starts missing deadlines earlier than G-EDF. On the other hand, NG-GUA not only defaults to G-EDF during underloads, it is able to sustain higher DSR during overloads. As NG-GUA defaults to a deadline-based order, we observe that it is able to meet more deadlines than G-GUA in Fig 2(b). The performance improvement for AUR is manifold. We observe $\approx 900\%$ improvement in AUR during overloads for both G-GUA and NG-GUA over the deadline-based algorithms.

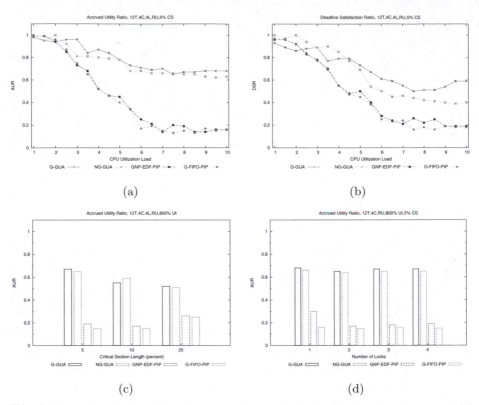

**Fig. 3.** Experimental results with dependencies by (a-b) varying utilization load (c) varying critical section length (d) varying number of locks

**Comparison with partitioned scheduling algorithms**. Fig 2(c) and Fig 2(d) show the AUR and DSR results, respectively, for 27T using RU on a 4-core platform with the task-set partitioned using Baruah's optimized first-fit (BF) heuristic [6]. The task-set was partitioned off-line and assigned to the individual processors using ChronOS APIs. We compare our algorithms against P-EDF and P-DASA. In Fig 2(d), we observe that P-EDF is able to meet all deadlines upto $\approx 390\%$ CPU utilization, after which it suffers from a domino effect. P-DASA uses the single-processor utility accrual scheduling algorithm, DASA, on individual processors. DASA defaults to EDF during underloads and maximizes accrued utility during overloads. We observe a similar behavior in Fig 2(c). However, both NG-GUA and G-GUA perform better than P-EDF and P-DASA during overloads by yielding a "best-effort" utility accrual behavior, with an improvement of $\approx 50\%$ in AUR over P-DASA.

### 4.4   Results with Dependencies

To compare the performance of G-GUA and NG-GUA in the presence of dependencies against global scheduling algorithms, we consider three models: (i) *varying utilization load*, keeping the number of locks and critical section length fixed;

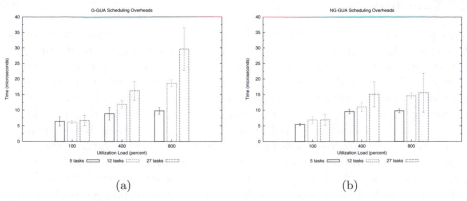

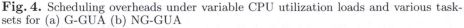

|     |     |
| --- | --- |
| (a) | (b) |

**Fig. 4.** Scheduling overheads under variable CPU utilization loads and various task-sets for (a) G-GUA (b) NG-GUA

(ii) *varying critical section length*, keeping the utilization load and the number of locks fixed; and (iii) *varying number of locks*, keeping the utilization load and the critical section length fixed. We implement locks using `futexes`, which allow us to share a context between kernel-space and user-space. We consider the critical section length as a percentage of the total execution cost of the task. The tasks request the locks sequentially.

Fig 3(a) and Fig 3(b) show the AUR and DSR results for 12T with 4 locks per task, using RU on a 4-core platform. The tasks use locks sequentially. Locks have a fixed critical section length of 5% of total task execution cost. In the presence of locks, none of the scheduling algorithms are able to meet all deadlines during underloads. We observe that both G-GUA and NG-GUA provide better accrued utility as well as deadline satisfaction during overloads. In Fig 3(a), G-GUA and NG-GUA provide a consistent 80% accrued utility benefit when compared with deadline-based algorithms. Fig 3(c) provides AUR results for a fixed utilization load of 800% and 4-locks by varying critical section length. With an increase in critical section length, the overall AUR decreases. G-GUA is able to provide an improvement of $\approx$ 5% over NG-GUA. Fig 3(d) shows AUR results for fixed utilization load of 800% and critical section length of 5% while varying number of locks. We observe that both NG-GUA and G-GUA consistently accrue higher utility when compared to deadline-based algorithms.

### 4.5   Scheduling Overheads

As mentioned earlier, both NG-GUA and G-GUA have a worst-case asymptotic cost of $O(mn \log n)$. Fig 4(a) and Fig 4(b) show the scheduling overheads for G-GUA and NG-GUA on ChronOS. We observe that with an increase in the number of tasks and also the task utilization load, the scheduling overhead of both algorithms increase. In particular, G-GUA is seen to have a higher overhead compared to NG-GUA. This is primarily because G-GUA is more greedy for accruing overall utility as compared to NG-GUA. For a 27T task-set, we observe $\approx 30\mu s$ overhead for G-GUA and $\approx 15\mu s$ overhead for NG-GUA.

## 5   Conclusions

This paper focuses on the dynamic, multiprocessor real-time scheduling problem space—i.e., those characterized by execution overruns, unpredictable task arrivals, causing transient and permanent overloads. The paper demonstrates that it is possible to design scheduling algorithms for this problem space, such that they yield an optimal timeliness behavior (e.g., meeting all deadlines; obtaining maximum total utility), when total utilization demand does not exceed the algorithms' utilization bound, and a best-effort timeliness behavior at all other times. This approach was pioneered in the Alpha OS kernel [20], which included two generations of TUF scheduling algorithms for scheduling single-processor systems [26,13]. At its core, the paper's algorithms demonstrate that a similar approach can also be successfully extended for multiprocessors. Additionally, the ChronOS real-time Linux kernel, provides a framework for implementing a broad range of multiprocessor real-time schedulers, while taking advantage of PREEMPT_RT patch's optimized interrupt service latencies and real-time locking primitives. Ongoing work is transitioning the GUA algorithms and ChronOS kernel to a US Department of Defense system.

There are several directions for future work. Immediate directions include improving the utilization bound from $\approx m/2$ and reducing time overheads. Other directions include developing scalable and approximate algorithms for GUA's problem space with lower bounds on accrued utility and satisfied deadlines.

## Acknowledgments

This work is supported by US National Science Foundation under Grant 0915895.

## References

1. ChronOS Real-time Linux, http://www.chronoslinux.org
2. Allen, R., Garlan, D.: A Case Study in Architectural Modelling: The AEGIS System. In: Proceedings of the 8th International Workshop on Software Specification and Design, IWSSD 1996, p. 6. IEEE Computer Society, Washington (1996)
3. Anderson, J.H., Bud, V., Devi, U.C.: An EDF-based Scheduling Algorithm for Multiprocessor Soft Real-Time Systems. In: Proceedings of the 17th Euromicro Conference on Real-Time Systems, ECRTS 2005, pp. 199–208 (2005)
4. Andersson, B., Baruah, S., Jonsson, J.: Static-priority Scheduling on Multiprocessors. In: RTSS 2001, pp. 193–202 (December 2001)
5. Andersson, B., Jonsson, J.: The Utilization Bounds of Partitioned and Pfair Static-priority Scheduling on Multiprocessors are 50%. In: ECRTS 2003, pp. 33–40 (July 2003)
6. Baruah, S., Fisher, N.: The Partitioned Multiprocessor Scheduling of Deadline-Constrained Sporadic Task Systems. IEEE Trans. Comput. 55(7), 918–923 (2006)
7. Baruah, S.K.: The Non-preemptive Scheduling of Periodic Tasks upon Multiprocessors. Real-Time Syst. 32(1-2), 9–20 (2006)
8. Baruah, S.K., Cohen, N.K., Plaxton, C.G., Varvel, D.A.: Proportionate Progress: A Notion of Fairness in Resource Allocation. Algorithmica 15(6), 600–625 (1996)
9. Bertogna, M., Cirinei, M., Lipari, G.: Improved Schedulability Analysis of EDF on Multiprocessor Platforms. In: ECRTS 2005, pp. 209–218 (2005)

10. Chen, S.Y., Hsueh, C.W.: Optimal Dynamic-Priority Real-Time Scheduling Algorithms for Uniform Multiprocessors. In: Proceedings of the 2008 Real-Time Systems Symposium, RTSS 2008, pp. 147–156. IEEE Computer Society, Washington (2008)
11. Cho, H.: Utility Accrual Real-Time Scheduling and Synchronization on Single and Multiprocessors. Ph.D. thesis, Virginia Tech. (2006)
12. Cho, H., Ravindran, B., Jensen, E.D.: An Optimal Real-Time Scheduling Algorithm for Multiprocessors. In: Proceedings of the 27th IEEE International Real-Time Systems Symposium, RTSS 2006, pp. 101–110. IEEE Computer Society, Washington (2006)
13. Clark, R.K.: Scheduling Dependent Real-Time Activities. Ph.D. thesis, CMU, CMU-CS-90-155 (1990)
14. Clark, R., Jensen, E.D., Kanevsky, A., Maurer, J., Wallace, P., Wheeler, T., Zhang, Y., Wells, D., Lawrence, T., Hurley, P.: An Adaptive, Distributed Airborne Tracking System ("Process the Right Tracks at the Right Time"). In: Rolim, J.D.P. (ed.) IPPS-WS 1999 and SPDP-WS 1999. LNCS, vol. 1586, pp. 353–362. Springer, Heidelberg (1999)
15. Funaoka, K., Kato, S., Yamasaki, N.: Work-Conserving Optimal Real-Time Scheduling on Multiprocessors. In: Euromicro Conference on Real-Time Systems, ECRTS 2008, pp. 13–22 (2-4, 2008)
16. Funk, S., Nanadur, V.: LRE-TL: An Optimal Multiprocessor Scheduling Algorithm for Sporadic Task Sets. In: 17th International Conference on Real-Time and Network Systems, pp. 159–168 (2009)
17. Garyali, P.: On Best-Effort Utility Accrual Real-Time Scheduling on Multiprocessors. Master's thesis, Virginia Tech. (2010)
18. Herlihy, M., Shavit, N.: The Art of Multiprocessor Programming. Morgan Kaufmann, San Francisco (2008)
19. Jensen, E., Locke, C., Tokuda, H.: A Time Driven Scheduling Model for Real-Time Operating Systems. In: IEEE RTSS, pp. 112–122 (1985)
20. Jensen, E., Northcutt, J.: Alpha: A Non-proprietary OS for Large, Complex, Distributed Real-Time Systems. In: Proceedings of IEEE Workshop on Experimental Distributed Systems, pp. 35–41 (11-12, 1990)
21. Koren, G., Shasha, D.: D-OVER; An Optimal On-line Scheduling Algorithm for Overloaded Real-Time Systems. In: Real-Time Systems Symposium 1992, pp. 290–299 (2-4, 1992)
22. Koren, G., Shasha, D.: MOCA: A Multiprocessor On-line Competitive Algorithm for Real-Time System Scheduling. Theor. Comput. Sci. 128(1-2), 75–97 (1994)
23. Lakshmanan, K., Rajkumar, R., Lehoczky, J.: Partitioned Fixed-Priority Preemptive Scheduling for Multi-core Processors. In: 21st Euromicro Conference on Real-Time Systems, ECRTS 2009, pp. 239–248 (1-3, 2009)
24. Li, P.: Utility Accrual Real-Time Scheduling: Models and Algorithms. Ph.D. thesis, Virginia Tech. (July 2004)
25. Li, P., Wu, H., Ravindran, B., Jensen, E.D.: A Utility Accrual Scheduling Algorithm for Real-Time Activities with Mutual Exclusion Resource Constraints. IEEE Trans. Comput. 55(4), 454–469 (2006)
26. Locke, C.D.: Best-Effort Decision Making for Real-Time Scheduling. Ph.D. thesis, CMU, CMU-CS-86-134 (1986)
27. Molnar, I.: CONFIG PREEMPT REALTIME, "Fully Preemptible Kernel", vp-2.6.9-rc4-mm1-t4, http://lwn.net/Articles/105948/
28. OMG: Real-time CORBA 2.0: Dynamic Scheduling Specification. Tech. rep., Object Management Group (September 2001)
29. Patterson, D.: The Trouble With Multicore. IEEE Spectrum 47(7), 28–32 (2010)

# Tardiness Bounds for Global EDF with Deadlines Different from Periods

Jeremy Erickson[1], Nan Guan[2], and Sanjoy Baruah[1]

[1] The University of North Carolina at Chapel Hill, Chapel Hill, NC
{jerickso,baruah}@cs.unc.edu
[2] Uppsala University, Uppsala, Sweden
Nan.Guan@it.uu.se

**Abstract.** The Earliest Deadline First (EDF) scheduling algorithm is known to be suboptimal for meeting all deadlines under global scheduling on multiprocessor platforms. However, EDF is an attractive choice for scheduling soft-real-time systems on multiprocessors. Previous work has demonstrated that the maximum tardiness is bounded, and has derived formulas for computing tardiness bounds, in EDF-scheduled real-time systems that can be modeled as collections of recurrent tasks modeled using the well-known *implicit-deadline* (Liu and Layland) task model. This research extends the applicability of previous techniques to systems that are modeled using the more general *arbitrary* sporadic task model. It also improves on prior work even for implicit-deadline systems. An algorithm is derived here that computes tardiness bounds in polynomial time. Previously, these bounds could only have been approximated in sub-exponential time.

## 1 Introduction

In recent years, multiprocessors have become increasingly used for diverse types of applications. For example, even most consumer laptops and desktops are sold today with multicore processors. Furthermore, soft timing guarantees are sufficient for many real-time applications, such as network servers, virtual-reality systems, and multimedia systems. Due to these trends, the ability to run soft real-time workloads on multiprocessor systems is becoming increasingly important.

*Motivation and Previous Work.* Much previous work on soft real-time systems (e.g. [1]) considers **implicit-deadline** sporadic task systems on identical multiprocessors. Under this model, each task $\tau_i$ is characterized by a worst-case execution time (WCET) $C_i$ and a period or inter-arrival separation parameter $T_i$. The relative deadline of each task —the duration between the arrival of a job generated by the task and its deadline— is assumed to also be $T_i$ (i.e., relative deadlines are equal to periods). The ratio $C_i/T_i$ is denoted as a task's *utilization*, and the sum of utilizations over all tasks is denoted as the system's utilization.

It is known that global Earliest Deadline First (EDF) scheduling, to meet all deadlines, generally requires the system utilization to be less than the platform

C. Lu, T. Masuzawa, and M. Mosbah (Eds.): OPODIS 2010, LNCS 6490, pp. 286–301, 2010.

computing capacity. Often, significant capacity is lost. In the case of soft real-time systems where it is acceptable to not always meet all deadlines, we may be able to achieve greater platform utilization by tolerating some failure to meet deadlines.

To our knowledge, the first work analyzing this aspect of the behavior of global EDF for soft real-time scheduling was that of Devi and Anderson [1]. Their analysis was restricted to implicit-deadline systems. They derived a technique for computing a value $x$ such that no job of task $\tau_i$ misses its deadline by more than $x + C_i$ time units, where $C_i$ is the WCET for $\tau_i$. In the terminology that they introduced, $x + C_i$ is a **tardiness bound** on $\tau_i$. They demonstrated that finite tardiness bounds can be derived as long as no task's utilization exceeds 1, and the total utilization of the system does not exceed the number of processors. These conditions are both necessary and sufficient for tardiness to be bounded.

This result was improved in [2], by exploiting the observation that rather than computing a single value $x$ for all tasks, we can compute a separate $x_i$ for each task $\tau_i$. Thus, the tardiness bounds are of the form $x_i + C_i$. While these bounds are tighter than those provided in [1], this technique is still restricted to the case of implicit-deadline systems.

It is possible to trivially extend the results in [1,2] to **arbitrary** sporadic task systems (i.e., task systems in which the relative deadline of a task may be smaller than, equal to, or greater than its period parameter) by analyzing the ratio of execution to the minimum of deadline and period (or *density*) rather than utilization. However, doing so introduces excessive pessimism in both the condition for bounded tardiness and the actual tardiness bound.

A more general analysis of tardiness in multiprocessor scheduling of soft-real-time systems is provided in [3,4]. Rather than analyzing any particular scheduling algorithm, the analysis is expanded to a broader class of global scheduling algorithms called *window-constrained*. It can be shown that global EDF of arbitrary sporadic task systems is window-constrained; hence [3,4] demonstrates that the conditions for bounded tardiness discussed above continue to hold when deadlines differ from periods. However, due to the generality of the analysis in [3,4], using these results to provide a specific bound for global EDF which is less pessimistic than the general window-constrained case is difficult, except in the implicit deadline case where the bound is equivalent to that provided in [1] (and hence inferior to the one in [2]).

*Contributions.* There are two major contributions of this paper. *First*, we extend the technique introduced in [2] from implicit-deadline to arbitrary sporadic task systems. We characterize the tardiness bound for each task in any arbitrary sporadic task system satisfying the properties that (i) no task has utilization exceeding 1 and (ii) system utilization does not exceed the number of processors (these are precisely the conditions necessary and sufficient for bounded tardiness provided in [3,4]). Our bounds are far tighter than the general bound provided in [3,4]. *Second*, we derive a polynomial-time algorithm for computing these tardiness bounds exactly. This represents an improvement over the iterative algorithm, of unknown computational complexity that was presented in [2] for implicit-deadline task systems.

In addition, we have produced a technical report [5] which demonstrates a method for altering the analysis to allow yet tighter bounds in some cases when constrained-deadline tasks are present. This method was not included in this paper for space reasons.

*Significance of this research.* The seminal works of Anderson, Devi, and Leontyev [1,3,4] have established that global EDF exhibits bounded tardiness, and is therefore suitable for the scheduling of a certain class of soft-real-time systems. This having been established, it is important that we be able to determine, during system design time itself, guarantees on the maximum tardiness that the system may experience during run-time. The tighter the guarantees, the more useful they are for the system design process. In prior work [2], we have improved on the tightest previously-known tardiness bounds for implicit-deadline sporadic system, and have designed an algorithm for computing these bounds. The significance of the current research lies both in the extension of our bounds to arbitrary sporadic systems, and in the fact that we are now able to compute these bounds more efficiently (in time polynomial in the number of tasks in the system). Hence, tardiness bounds can now be used during the system design process for a wider class of systems, and they may be computed with less computational effort.

*Organization.* In Section 2 we derive a tardiness bound for arbitrary sporadic task systems that are scheduled on a multiprocessor platform using global EDF. In Section 3 we examine the problem of computing the best bounds which can be demonstrated with our method: in Subsection 3.1 we demonstrate that a unique optimal (relative to our technique) bound exists for any system with bounded tardiness, and in Subsection 3.2 we present a polynomial-time algorithm for computing these bounds exactly.

## 2    A Characterization of Tardiness

We are concerned with the scheduling of a system $\tau$ of $n$ sporadic tasks $\tau_1, \tau_2, \ldots, \tau_n$ upon a platform comprised of $m$ identical unit-capacity processors, using the preemptive global EDF scheduling algorithm. Each sporadic task $\tau_i = (T_i, C_i, D_i)$ releases *jobs* and is characterized by its minimum inter-arrival separation $T_i$, WCET $C_i$, and relative deadline $D_i$. We make no assumptions about $D_i$ other than non-negativity; it may be greater than, equal to, or less than $T_i$ for any particular task. We also refer to a task's *utilization* $\frac{C_i}{T_i}$ as $U_i$. Let $U(\tau)$ denote $\sum_{i=1}^{n} U_i$, and $m_0 \stackrel{\text{def}}{=} \lceil U(\tau) \rceil$. It is evident that $U(\tau) \leq m$ (and thus $m_0 \leq m$), combined with $\forall i, U_i \leq 1$, is a necessary condition for bounded tardiness [3]. As discussed above, [3,4] demonstrates that this is *sufficient* for bounded tardiness as well. We provide an alternate proof for bounded tardiness under these same conditions, which allows us to derive tighter tardiness bounds.

We exploit the property that over any interval of length $t$, the total amount of work that may be required by an arbitrary sporadic task $\tau_i$ is bounded. This bound is called the *demand bound function*, or $\text{DBF}(\tau_i, t)$. As shown in [6],

$$\text{DBF}(\tau_i, t) = C_i \times \max\left\{0, \left\lfloor \frac{t - D_i}{T_i} \right\rfloor + 1\right\} \tag{1}$$

In [2], $\text{DBF}(\tau_i, t)$ for implicit-deadline sporadic task $\tau_i$ is upper-bounded by $U_i t$. This bound does not hold for arbitrary sporadic tasks; in Lemma 1 below, we derive a constant $S_i$ such that demand is upper bounded by $U_i t + S_i$. In essence, $S_i$ can be viewed as a constant showing the result of a task having higher demand over a short interval than expected by utilization alone.

**Lemma 1.** *For all $\tau_i \in \tau$ and $t \geq 0$,*

$$\text{DBF}(\tau_i, t) \leq U_i t + S_i$$

*where $S_i$ is defined as follows:*

$$S_i \stackrel{\text{def}}{=} C_i \times \max\left\{0, 1 - D_i/T_i\right\} \tag{2}$$

*Proof.* By (1) DBF increases only at each point where $t - D_i$ is an integer multiple of $T_i$, i.e., where there exists an integer $k \geq 0$ such that $t - D_i = T_i k$ or $t = T_i k + D_i$.

We first consider $t \in [0, D_i)$. $t - D_i < 0$, so $\left\lfloor \frac{t - D_i}{T_i} \right\rfloor + 1 \leq 0$. Therefore, by (1), $\text{DBF}(\tau_i, t) = 0 \leq U_i t + S_i$.

Alternatively, suppose $t \geq D_i$. Let $k = \left\lfloor \frac{t - D_i}{T_i} \right\rfloor$. Then $t \in [T_i k + D_i, T_i(k + 1) + D_i)$. Therefore:

$$U_i t + S_i$$
$$\geq U_i t + C_i \left(1 - \frac{D_i}{T_i}\right)$$
$$\geq U_i(T_i k + D_i) + C_i \left(1 - \frac{D_i}{T_i}\right)$$
$$= C_i k + \frac{C_i D_i}{T_i} + C_i - \frac{C_i D_i}{T_i}$$
$$= C_i(k + 1)$$
$$= C_i \left(\left\lfloor \frac{t - D_i}{T_i} \right\rfloor + 1\right)$$
$$= \text{DBF}(\tau_i, t)$$

Thus, the lemma is true for all $t > 0$.

For the $n$-task system $\tau$, we wish to compute non-negative real numbers $x_1, x_2, \ldots, x_n$ such that it can be guaranteed that no job of $\tau_i$ has a tardiness greater than $C_i + x_i$, $1 \leq i \leq n$ when $\tau$ is EDF-scheduled on $m$ unit-speed processors. In order to derive such $x_i$'s we use the notion of *compliant vectors*, first defined in [2]. Our definition here generalizes the one in [2] by using the $S_i$ terms to account for deadlines which may differ from periods. As is the case in [2], these definitions are each with respect to a particular task system $\tau$ and a specified number of processors $m$.

**Definition 1 (compliant vectors).** *For any* $x = \langle x_1, x_2, \ldots, x_n \rangle$ *an ordered list of* $n$ *non-negative real numbers, let* $\mathbf{L}(x)$ *denote the sum of the* $(m_0 - 1)$ *largest* $(x_i U_i + C_i)$'s:

$$\mathbf{L}(x) \stackrel{def}{=} \sum_{(m_0-1) \text{ largest}} \left( x_i U_i + C_i \right) \tag{3}$$

*Also denote (using (2))*

$$\mathbf{S}(\tau) \stackrel{def}{=} \sum_{\tau_i \in \tau} S_i \tag{4}$$

*We say that* $x$ *is a **compliant vector** if*

$$\frac{\mathbf{L}(x) + \mathbf{S}(\tau) - C_i}{m} \leq x_i \tag{5}$$

*is satisfied for all* $i$, $1 \leq i \leq n$, *and a **non-compliant vector** otherwise.*

We consider a compliant vector $x = \langle x_1, x_2, \ldots, x_n \rangle$ and an arbitrary collection $I_o$ of jobs generated by $\tau$. Without loss of generality, we assume that the earliest arrival of a job in $I_o$ occurs at time 0. We analyze an arbitrary job $j_k$ with deadline $d_k$, under the assumption that no job with deadline before $d_k$ completes later than $(C_i + x_i)$ time units after its deadline. This assumption will be justified in the proof of Theorem 1. We denote by $I$ the set of jobs in $I_o$ with deadlines $\leq d_k$. By the definition of EDF priority, only jobs in $I$ can affect the scheduling of $j_k$.

For each $\tau_j$ and all $0 \leq t \leq d_k$, let $W_j(t)$ denote the amount of remaining execution of jobs of $\tau_j$ in $I$, at time instant $t$. Denote $W(t) = \sum_{\tau_j \in \tau} W_j(t)$. Alternatively, $W(t)$ can be viewed as the total amount of execution required by jobs in $I$ minus the amount of work completed in $[0, t)$. Lemma 2 below provides an upper bound on $W(t)$.

**Lemma 2.** *If* $\forall i$, *each job of* $\tau_i$ *with a deadline before* $d_k$ *experiences tardiness no greater than* $x_i + C_i$,
$$W(d_k) \leq \mathbf{L}(x) + \mathbf{S}(\tau)$$

*Proof.* We define a time interval as being *busy* if throughout the interval at least $m_0$ processors are executing, and *non-busy* otherwise. Let $t_0 = 0, t_1, t_2, \ldots, d_k$ be an arbitrary set of time instants over $[0, d_k)$ such that each $[t_i, t_{i+1})$ is either all busy or all non-busy. We prove the lemma by demonstrating that the following more general statement is true for $t \in \{t_0, t_1, \ldots, d_k\}$:

$$W(t) \leq U(\tau) \times (d_k - t) + \mathbf{L}(x) + \mathbf{S}(\tau). \tag{6}$$

We proceed by induction on the time instants $t_0, t_1, \ldots, d_k$.

*Base case (t = 0)* $W(t_0)$ denotes the cumulative execution requirement of all jobs arriving in, and having deadlines in, the interval $[0, d_k]$. Thus,

$$W(t_0) \leq \sum_{\tau_i \in \tau} \mathrm{DBF}(\tau_i, d_k)$$

$$\leq \sum_{\tau_i \in \tau} U_i d_k + S_i \ \text{(By Lemma 1)}$$

$$= U(\tau) d_k + \mathbf{S}(\tau)$$

*Induction step.* We begin by assuming (6) is true for $t = t_i$. We will show that it is true at $t_{i+1}$ as well. We must consider two cases:

*Case A.* Suppose $[t_i, t_{i+1})$ is a busy interval. In this case, the amount of work performed in the interval is at least $m_0(t_{i+1} - t_i)$. Thus, by the inductive hypothesis,

$$W(t_{i+1}) = W(t_i) - m_0(t_{i+1} - t_i)$$
$$\leq U(\tau)(d_k - t_i) + \mathbf{L}(\boldsymbol{x}) + \mathbf{S}(\tau) - m_0(t_{i+1} - t_i)$$
$$\leq U(\tau)(d_k - t_i) + \mathbf{L}(\boldsymbol{x}) + \mathbf{S}(\tau) - U(\tau)(t_{i+1} - t_i)$$
$$\text{(Because } U(\tau) \leq m_0)$$
$$= U(\tau)(d_k - t_{i+1}) + \mathbf{L}(\boldsymbol{x}) + \mathbf{S}(\tau),$$

and thus (6) also holds at $t_{i+1}$.

*Case B.* Alternatively, suppose $[t_i, t_{i+1})$ is non-busy. We say that a task $\tau_j$ is "executing at time instant $t_{i+1}^-$" if and only if $\exists \epsilon > 0$ such that $\tau_j$ is executing throughout $[t_{i+1} - \epsilon, t_{i+1})$. At most $m_0 - 1$ processors may be executing at time instant $t_{i+1}^-$ or the CPU would be busy before $t_{i+1}$. We consider tasks that:

Case B.1: Do not execute at time instant $t_{i+1}^-$;
Case B.2: Execute at $t_{i+1}^-$, but not throughout the interval $[t_i, t_{i+1})$; and
Case B.3: Execute throughout the interval $[t_i, t_{i+1})$.

*Case B.1.* Suppose task $\tau_j$ is not executing at $t_{i+1}^-$. In this case, it must have no active work remaining, or it would have executed. Thus, all remaining work comes from jobs that are released in, and have deadlines within, $[t_{i+1}, d_k]$. Thus,

$$W_j(t_{i+1}) \leq \mathrm{DBF}(\tau_j, d_k - t_{i+1})$$
$$\leq U_j(d_k - t_{i+1}) + S_j \ \text{(By Lemma 1)}$$

*Case B.2.* Suppose task $\tau_j$ is executing at $t_{i+1}^-$ but at some point in $[t_i, t_{i+1})$ is not executing. Let $t'$ denote the latest time during this interval in which it is idle. Because it is idle but at least one processor is available, no jobs of $t_j$ are active at this instant. Thus, all remaining work comes from jobs that are released in, and have deadlines within, $[t', d_k]$. Of this remaining work, $t_{i+1} - t'$ units of execution are completed within $(t', t_{i+1})$. Thus,

$$
\begin{aligned}
W_j(t_{i+1}) &\leq \mathrm{DBF}(d_k - t') - (t_{i+1} - t') \\
&\leq U_j(d_k - t') + S_j - (t_{i+1} - t') \text{ (By Lemma 1)} \\
&\leq U_j(d_k - t') + S_j - U_j(t_{i+1} - t') \\
&\qquad \text{(Because } U_j \leq 1) \\
&= U_j(d_k - t_{i+1}) + S_j
\end{aligned}
$$

*Case B.3.* Suppose task $\tau_j$ executes throughout $[t_i, t_{i+1})$. Let $J$ denote the job of $\tau_j$ that is executing at time instant $t_i$, and let $\delta$ denote the sum of the execution of $J$ completed before $t_i$ and the amount by which it finishes early. This is so that $C_j - \delta$ is the exact remaining amount of execution of $J$. Let $d$ denote the deadline of $J$. We will consider two subcases, depending on whether $d \geq t_i$ (i.e., $J$ is not tardy at $t_i$) or $d < t_i$ (i.e., $J$ is tardy at $t_i$).

*Case B.3.1.* Suppose $J$ is not tardy at time instant $t_i$. Then, all jobs of $\tau_j$ that contribute to $W_j(t_i)$ arrive at or after $t_i$, because they must arrive after $d$. Therefore,

$$
\begin{aligned}
W_j(t_i) &\leq \mathrm{DBF}(\tau_j, d_k - t_i) + C_j - \delta \\
&\leq U_j(d_k - t_i) + S_j + C_j - \delta \text{ (Lemma 1)} \\
&\leq U_j(d_k - t_i) + S_j + C_j \text{ (Because } \delta \geq 0) \tag{7}
\end{aligned}
$$

Because $\tau_j$ executes throughout $[t_i, t_{i+1})$,

$$
\begin{aligned}
W_j(t_{i+1}) &= W_j(t_i) - (t_{i+1} - t_i) \\
&\leq U_j(d_k - t_i) + S_j + C_j - (t_{i+1} - t_i) \text{ (By (7))} \\
&\leq U_j(d_k - t_i) + S_j + C_j - U_j(t_{i+1} - t_i) \\
&\qquad \text{(Because } U_j \leq 1) \\
&= U_j(d_k - t_{i+1}) + S_j + C_j
\end{aligned}
$$

*Case B.3.2.* Suppose $j$ is tardy at $t_i$. Then, it will complete execution at $t_i + C_j - \delta$, because it cannot be preempted (it is already tardy). We assumed that no job of $\tau_j$ with deadline $< d_k$ experiences a tardiness greater than $x_j + C_j$. Therefore,

$$
t_i + C_j - \delta \leq d + C_j + x_j
$$

and thus

$$
d \geq t_i - (\delta + x_j) \tag{8}
$$

$W_j(t_i)$ is composed of all jobs released after $d$, in addition to the remaining execution $C_j - \delta$. Thus:

$$
\begin{aligned}
W_j(t_i) &\leq \mathrm{DBF}(\tau_j, d_k - d) + C_j - \delta \\
&\leq U_j(d_k - d) + S_j + C_j - \delta \text{ (By Lemma 1)} \\
&\leq U_j(d_k - t_i + (\delta + x_j)) + S_j + C_j - \delta \tag{9} \\
&\qquad \text{(By (8))}
\end{aligned}
$$

$\tau_j$ executes over the entire interval $[t_i, t_{i+1})$, so:

$$W_j(t_{i+1}) = W_j(t_i) - (t_{i+1} - t_i)$$
$$\leq U_j(d_k - t_i + (\delta + x_j)) + S_j + C_j - \delta - (t_{i+1} - t_i)$$
$$\text{(By (9))}$$
$$\leq U_j(d_k - t_i + (\delta + x_j)) + S_j + C_j - U_i(\delta + (t_{i+1} - t_i))$$
$$\text{(Because } U_j \leq 1)$$
$$= U_j(d_k - t_i + (\delta + x_j) - \delta - t_{i+1} + t_i) + S_j + C_j$$
$$= U_j(d_k - t_{i+1}) + U_j x_j + S_j + C_j$$

Summing the $W_j(t_{i+1})$ upper-bounding values for all tasks over all cases, we see the following contributions to the upper bound on $W(t_{i+1})$:

- $U_j(d_k - t_{i+1})$ from each task, summing to $U(\tau)(d_k - t_{i+1})$.
- $S_j$ from each task, summing to $\mathbf{S}(\tau)$.
- At most $m_0 - 1$ values of $C_j$ or $C_j + U_j x_j$, leading to a sum of $\mathbf{L}(\boldsymbol{x})$.

Thus, we see that (6) applies for $t_{i+1}$. By mathematical induction, (6) is true for all $t \in \{t_0, t_1, t_2, \ldots d_k\}$, which demonstrates that the lemma is true.

Our next lemma uses this result to bound the completion time of a job under the same assumptions.

**Lemma 3.** *If $\forall i$, each job of $\tau_i$ with a deadline before $d_k$ experiences tardiness no greater than $x_i + C_i$, the job of $\tau_k$ with a deadline of $d_k$ will complete by $d_k + x_k + C_k$.*

*Proof.* By Lemma 2, we see that the total amount of work remaining at the deadline $d_k$ of the last task in $I$ is at most $\mathbf{L}(\boldsymbol{x}) + \mathbf{S}(\tau)$.

Let $J$ denote the job of $\tau_k$ with a deadline at $d_k$. If $J$ has completed execution by $d_k$ it is not tardy, so we assume it has not. Denote $t_f \geq d_k$ as the first time instant after $d_k$ at which some processor is idle. We consider two cases, depending on whether $J$ has completed by $t_f$ or not.

*Case 1.* Suppose $J$ has completed by $t_f$. In this case, tardiness is bounded from above by $t_f - d_k$. Since the amount of work remaining is upper-bounded by $\mathbf{L}(\boldsymbol{x}) + \mathbf{S}(\tau)$, we can conclude:

$$t_f \leq d_k + \frac{\mathbf{L}(\boldsymbol{x}) + \mathbf{S}(\tau)}{m}$$
$$= d_k + \frac{\mathbf{L}(\boldsymbol{x}) + \mathbf{S}(\tau) - C_k}{m} + \frac{C_k}{m}$$
$$\leq d_k + x_k + \frac{C_k}{m} \quad \text{(Inequality (5))}$$
$$\leq d_k + x_k + C_k$$

*Case 2.* Suppose $J$ has not completed by $t_f$. Let $\gamma$ denote the amount of execution remaining for $J$ at $t_f$. Since the total amount of remaining work at $d_k$ is at most $\mathbf{L}(\boldsymbol{x}) + \mathbf{S}(\tau)$ and at least $\gamma$ units of this remain to be completed, we conclude that $t_f \leq d_k + \frac{\mathbf{L}(\boldsymbol{x}) + \mathbf{S}(\tau) - \gamma}{m}$. The presence of an idled processor implies that either $J$ or a predecessor job also of task $\tau_j$ is executing at $t_f$, and $\tau_j$ will continue to execute until $J$ completes.

If $J$ is running at $t_f$, then its tardiness guarantees that under EDF it will not be preempted. Thus it completes its execution by time instant

$$
d_k + \left( \frac{\mathbf{L}(\boldsymbol{x}) + \mathbf{S}(\tau) - \gamma}{m} + \gamma \right)
$$

$$
= d_k + \left( \frac{\mathbf{L}(\boldsymbol{x}) + \mathbf{S}(\tau) - C_k}{m} \right) + C_k - (C_k - \gamma)\left( 1 - \frac{1}{m} \right)
$$

$$
\leq d_k + \left( \frac{\mathbf{L}(\boldsymbol{x}) + \mathbf{S}(\tau) - C_k}{m} \right) + C_k
$$

$$
\leq d_k + x_k + C_k \ \text{(Inequality (5))}
$$

Alternatively, if $J$ is not running at $t_f$, then one of its predecessors is running, so $\gamma = C_k$. Therefore, $t_f \leq d_k + (\mathbf{L}(\boldsymbol{x}) + \mathbf{S}(\tau) - C_k)/m \leq d_k + x_k$ holds. Let $J'$ denote the immediate predecessor of $J$. The deadline of $J'$ must be at or before $d_k - T_k$, so by the inductive hypothesis, tardiness for $J'$ is at most $x_k + C_k$. Thus, $J'$ completes by time $d_k - T_k + x_k + C_k \leq d_k + x_k$. Therefore, $J$ begins executing by $d_k + x_k$ and cannot be preempted, so is completed by $d_k + x_k + C_k$.

Thus, in any case, $J$ must complete by $d_k + x_k + C_k$ and the lemma stands proved.

**Theorem 1.** *Let $\boldsymbol{x} = \langle x_1, x_2, \ldots, x_n \rangle$ denote any compliant vector. For each $\tau_i \in \tau$, each job generated by task $\tau_i$ completes no later than $(C_i + x_i)$ time units after its deadline.*

*Proof.* We will use induction to prove the desired result. We consider the jobs in $I_o$ in order of non-decreasing deadlines; when considering a particular job, we assume that no previous job experienced tardiness greater than claimed by the theorem. The earliest deadline will serve as the base case, because the assumption is vacuous.

The assumption used in Lemma 2 and thus in Lemma 3 as well is implied by the inductive hypothesis[1]. Therefore, by mathematical induction, Theorem 1 is true.

## 3    The Minimum Compliant Vector

Theorem 1 expresses tardiness bounds in terms of compliant vectors: given a compliant vector $\boldsymbol{x}$, we are guaranteed that no job of task $\tau_i$ experiences a

---

[1] It is not precisely the same only in that our inductive hypothesis may apply to some jobs which share a deadline with the job under consideration. Only the tardiness of jobs with earlier deadlines must be bounded by $x_i + C_i$ in order to prove the lemmas.

tardiness greater than $C_i + x_i$, for all $i$, $1 \leq i \leq n$. Since our objective is to compute small tardiness bounds — the smaller the better — we seek to compute a compliant vector in which the components (the $x_i$'s) are as small as possible. The notion of a *minimal* compliant vector was introduced in [2] to formalize this desideratum: a minimal compliant vector is a compliant vector satisfying the additional property that decreasing the value of any component of the vector by an arbitrarily small amount will result in a non-compliant vector.

It was pointed out in [2] that it is not clear how we would compare two different minimal compliant vectors if the $i$'th component is smaller in one, and the $j$'th component is smaller in the other — presumably, semantic considerations would need to be taken into account in determining which task is more resilient to tardiness. We now prove that such a situation cannot in fact arise: in Section 3.1 below we show that for any arbitrary-sporadic task system $\tau$ there is a *unique* minimal compliant vector – a **minimum** compliant vector. In Section 3.2 we will demonstrate an improved algorithm for determining this minimum compliant vector (and thereby computing minimum tardiness bounds) which is much more efficient than that published in [2]. In fact, our algorithm runs in polynomial time and determines the exact minimum compliant vector (whereas the algorithm in [2] is an iterative algorithm that generates an approximation to the minimal complaint vector: the greater the number of iterations of the algorithm, the more accurate the approximation).

## 3.1 Existence and Uniqueness

We first demonstrate a lemma that characterizes the behavior of $\mathbf{L}(\boldsymbol{x})$

**Lemma 4.** *Suppose length $n$ vectors $\boldsymbol{x}$ and $\boldsymbol{y}$ differ at exactly $k$ values, and for these values, $y_i = x_i + \delta$, where $\delta$ is a positive constant. Denote $u = \min\{k, m_0 - 1\}$. Then the following inequality holds:*

$$\mathbf{L}(\boldsymbol{x}) \leq \mathbf{L}(\boldsymbol{y}) \leq \mathbf{L}(\boldsymbol{x}) + \delta \cdot u \tag{10}$$

*Furthermore, if $k = n$, the first inequality is strict.*

*Proof.* We will define a *candidate sum* for $\boldsymbol{x}$, as any sum of $m_0 - 1$ values of distinct $x_i U_i + C_i$. $\mathbf{L}(\boldsymbol{x})$ can trivially be seen to be the largest candidate sum for $\boldsymbol{x}$.

First, we prove $\mathbf{L}(\boldsymbol{x}) \leq \mathbf{L}(\boldsymbol{y})$. Consider the candidate sum $S$ for $\boldsymbol{y}$ computed by selecting the same indices for the terms as in $\mathbf{L}(\boldsymbol{x})$. Because $\forall i, x_i \leq y_i$, $\mathbf{L}(\boldsymbol{x}) \leq S$. Because $\mathbf{L}(\boldsymbol{y})$ is the largest candidate sum for $\boldsymbol{y}$, $S \leq \mathbf{L}(\boldsymbol{y})$. Thus, $\mathbf{L}(\boldsymbol{x}) \leq \mathbf{L}(\boldsymbol{y})$. If $k = n$, then all terms in $S$ must actually be greater than the corresponding terms in $\mathbf{L}(\boldsymbol{y})$, so the inequality is strict in this case.

Next, we prove $\mathbf{L}(\boldsymbol{y}) \leq \mathbf{L}(\boldsymbol{x}) + \delta \cdot u$. We will use proof by contradiction. Suppose $\mathbf{L}(\boldsymbol{y}) > \mathbf{L}(\boldsymbol{x}) + \delta \cdot u$. Consider the candidate sum $T$ for $\boldsymbol{x}$ computed by selecting the same indices for the terms as in $\mathbf{L}(\boldsymbol{y})$. Observe that at most $u$ terms contribute to the difference between $\mathbf{L}(\boldsymbol{y})$ and $T$. This holds because we assumed only $k$ terms differ between the two vectors, and at most $m_0 - 1$ terms of the sum utilize $x_i$ values. Because $\forall i, x_i \geq y_i - \delta$, $T \geq \mathbf{L}(\boldsymbol{y}) - \delta \cdot u$. Thus, $T > \mathbf{L}(\boldsymbol{x})$, which is a contradiction because $\mathbf{L}(\boldsymbol{x})$ must be a *maximal* candidate sum for $\boldsymbol{x}$.

Thus, the lemma stands proved.

We next use Lemma 4 to characterize any minimal compliant vector which could possibly exist.

**Lemma 5.** *If $\boldsymbol{y}$ is a minimal compliant vector, then $\forall i, y_i = \frac{\mathbf{L}(\boldsymbol{y})+\mathbf{S}(\tau)-C_i}{m}$.*

*Proof.* We will use proof by contradiction. Observe that by definition, $\forall i$,

$$y_i \geq \frac{\mathbf{L}(\boldsymbol{y}) + \mathbf{S}(\tau) - C_i}{m} \tag{11}$$

Thus we need only consider the case in which $\boldsymbol{y}$ is a minimal compliant vector, but $\exists j$ such that

$$y_j > \frac{\mathbf{L}(\boldsymbol{y}) + \mathbf{S}(\tau) - C_i}{m} \tag{12}$$

Consider a $\boldsymbol{x}$ such that $x_i = y_i$ whenever $i \neq j$ and

$$x_j = \frac{\mathbf{L}(\boldsymbol{y}) + \mathbf{S}\tau - C_i}{m} \tag{13}$$

Note that $\boldsymbol{x}$ and $\boldsymbol{y}$ are of the form of Lemma 4 with $k = 1$. (The value of $\delta$ is irrelevant in this case.) Thus, $\mathbf{L}(\boldsymbol{x}) \leq \mathbf{L}(\boldsymbol{y})$.

Because $\boldsymbol{y}$ is compliant, $\forall i \neq j$,

$$\frac{\mathbf{L}(\boldsymbol{x}) + \mathbf{S}(\tau) - C_i}{m}$$
$$\leq \frac{\mathbf{L}(\boldsymbol{y}) + \mathbf{S}(\tau) - C_i}{m}$$
$$\leq y_i$$
$$= x_i$$

Also, by construction,

$$\frac{\mathbf{L}(\boldsymbol{x}) + \mathbf{S}(\tau) - C_j}{m}$$
$$\leq \frac{\mathbf{L}(\boldsymbol{y}) + \mathbf{S}(\tau) - C_j}{m}$$
$$= x_j$$

Thus, $\boldsymbol{x}$ is compliant, which contradicts the minimality of $\boldsymbol{y}$.

With these lemmas in place, we are now ready to prove the main result of this section:

**Theorem 2.** *For any given task set $\tau$, there exists a unique minimal compliant vector (which we refer to as the **minimum compliant vector**).*

*Proof.* We begin the proof by defining some notation we will use throughout the proof.

- $v(s)$ is defined as $\boldsymbol{v}$ where $v_i = s + \frac{\mathbf{S}(\tau) - C_i}{m}$.
- $L(s)$ for real $s$ is defined as $\mathbf{L}(v(s))$.
- $M(s)$ for real $s$ is defined as $L(s) - ms$.

The purpose of these notations is that by Lemma 5, any vector is a minimal compliant vector if and only if it is $v(s)$ for some $s$ and $M(s) = 0$. This is true because $\mathbf{L}(\boldsymbol{v})$ does not depend on $i$ and thus can be viewed as a constant. Here we intend $s = \frac{\mathbf{L}(\boldsymbol{v})}{m}$. We will use the Intermediate Value Theorem from calculus to demonstrate that exactly one such $s$ exists.

**Lemma 2.1.** $L(s)$ *is continuous over* $\mathbb{R}$.

*Proof.* Let $\epsilon > 0$. Let $\delta_c = \frac{\epsilon}{m_0 - 1}$. Suppose $|x - x_0| < \delta_c$. Without loss of generality, assume $x < x_0$ (otherwise we can swap them.) Then $v(x)$ and $v(x_0)$ are of the form of $\boldsymbol{x}$ and $\boldsymbol{y}$, respectively, in Lemma 4, with $k = n$. Thus,

$$\mathbf{L}(v(x)) < \mathbf{L}(v(x_0))$$
$$\leq \mathbf{L}(v(x)) + \delta_c(m_0 - 1)$$
$$= \mathbf{L}(v(x)) + \epsilon$$

Thus, $|L(x) - L(x_0)| < \epsilon$. This proves that $L(s)$ is continuous.

It follows immediately from Lemma 2.1 that $M(s)$ is also continuous over $\mathbb{R}$, because $M(s)$ is the difference of two continuous functions.

**Lemma 2.2.** $M(s)$ *is strictly decreasing over* $\mathbb{R}$.

*Proof.* Suppose $x < y$. Then $v(x)$ and $v(y)$ are of the form of $\boldsymbol{x}$ and $\boldsymbol{y}$ in Lemma 4, with $k = n$ and $\delta = y - x$. Thus, $L(x) < L(y) \leq L(x) + (m_0 - 1)(y - x)$. Therefore,

$$M(y) - M(x)$$
$$= L(y) - L(x) - my - mx$$
$$\leq (m_0 - 1)(y - x) - m(y - x)$$
$$\leq -(y - x)$$
$$< 0$$

**Lemma 2.3.** $M(0) > 0$.

*Proof.* Observe $\forall i$

$$x_i = \frac{\mathbf{S}(\tau) - C_i}{m} \geq -\frac{C_i}{m}.$$

Therefore, because $C_i > 0$, $m > 1$, and $U_i \leq 1$,

$$x_i U_i + C_i \geq -\frac{U_i C_i}{m} + C_i > 0.$$

Therefore, $L(0) > 0$, because it is a sum of positive numbers. Thus, $M(0) = L(0) - 0 > 0$.

**Lemma 2.4.** $M(C_{sum} + \mathbf{S}(\tau)) < 0$, where $C_{sum}$ is the sum of the $m_0 - 1$ largest $C_i$ values.

*Proof.* We first compute an upper bound on $L(C_{sum} + \mathbf{S}(\tau))$. By definition, $L(C_{sum} + \mathbf{S}(\tau))$ is the sum of the largest $m_0 - 1$ components of the form

$$x_i U_i + C_i = \left( C_{sum} + \mathbf{S}(\tau) + \frac{\mathbf{S}(\tau) - C_i}{m} \right) U_i + C_i.$$

Because $U_i \leq 1$, this is upper bounded by

$$C_{sum} + \mathbf{S}(\tau) + \frac{\mathbf{S}(\tau) - C_i}{m} + C_i$$
$$= C_{sum} + \frac{m+1}{m} \mathbf{S}(\tau) + \frac{m-1}{m} C_i$$

The sum of $m_0 - 1$ such terms is upper bounded by

$$(m_0 - 1) C_{sum} + (m_0 - 1) \frac{(m+1)}{m} \mathbf{S}(\tau) + \frac{m-1}{m} C_{sum}$$
$$\leq \frac{m^2 - m}{m} C_{sum} + \frac{m-1}{m} C_{sum} + \frac{(m-1)(m+1)}{m} \mathbf{S}(\tau)$$
$$= \frac{m^2 - 1}{m} (C_{sum} + \mathbf{S}(\tau))$$

Thus,

$$M(C_{sum} + \mathbf{S}(\tau))$$
$$\leq \frac{m^2 - 1}{m} (C_{sum} + \mathbf{S}(\tau)) - m(C_{sum} + \mathbf{S}(\tau))$$
$$< 0.$$

We are now ready to prove that the minimal compliant vector is unique. By Lemma 2.1, Lemma 2.3, Lemma 2.4, and the Intermediate Value Theorem, $\exists s$ between 0 and $C_{sum} + \mathbf{S}(\tau)$ such that $M(s) = 0$. By Lemma 2.2, $t < s$ implies $M(t) > 0$ and $t > s$ implies $M(t) < 0$. Thus, the solution is unique, so $v(s)$ is the unique minimal (i.e., minimum) compliant vector.

## 3.2   Improved Computation Algorithm

The algorithm presented in [2] for computing a minimal compliant vector is iterative: it starts out with the (non-compliant) vector of all zeros and increases some $x_i$ during each iteration, terminating when the resulting vector becomes compliant. However, [2] was unable to provide a bound on the number of such iterations that may be needed; instead, heuristics were proposed for fixing the minimum amount by which some $x_i$ is increased during each iteration, in order to bound the total number of steps. As a consequence, the value of $x_i$ so computed may exceed the actual value by up to this step-size. An alternative iterative formulation that uses unbounded binary search is also possible, the inaccuracy in the values of the $x_i$'s decreasing exponentially with the number of iterations of the binary search algorithm.

In this section, we significantly improve on either of these 2 approaches from [2], by deriving an algorithm that computes the *exact* minimum compliant vector in time that is *polynomial* in the number of tasks in the task system.

By Lemma 5 above, we know the minimum compliant vector can be obtained by solving the following system of equations

$$
\begin{cases}
x_1 = \dfrac{\mathbf{L}(\boldsymbol{x}) + \mathbf{S}(\tau) - C_1}{m} \\
x_2 = \dfrac{\mathbf{L}(\boldsymbol{x}) + \mathbf{S}(\tau) - C_2}{m} \\
\cdots\cdots \\
x_n = \dfrac{\mathbf{L}(\boldsymbol{x}) + \mathbf{S}(\tau) - C_n}{m}
\end{cases}
$$

Defining $z$ as follows:

$$
z = \frac{\mathbf{L}(\boldsymbol{x}) + \mathbf{S}(\tau)}{m} \tag{14}
$$

our system of equations becomes

$$
\begin{cases}
x_1 = z - (C_1/m) \\
x_2 = z - (C_2/m) \\
\cdots\cdots \\
x_n = z - (C_n/m)
\end{cases} \tag{15}
$$

Applying this to the definition of $\mathbf{L}(\boldsymbol{x})$, we get

$$
\mathbf{L}(\boldsymbol{x}) \stackrel{\text{def}}{=} \sum_{(m_0-1)\ \text{largest}} \left( U_i \left( z - \frac{C_i}{m} \right) + C_i \right) \tag{16}
$$

By (14) and (16), we get the following equation in a single unknown $z$

$$
z = \left( \sum_{(m_0-1)\ \text{largest}} \left( U_i \left( z - \frac{C_i}{m} \right) + C_i \right) - \mathbf{S}(\tau) \right)/m \tag{17}
$$

Solving $z$ in the above equation and using (15) above, we can get the desired minimum compliant vector.

In the following we will show how to solve Equation (17). We observe that the RHS of (17) is a piece-wise linear function with respect to $z$, and its slope changes only at the points where the tasks involved in the summation (the $(m_0 - 1)$ largest sum term) change. To find out these points, we establish an equation for each pair of $(i, j), i \neq j$,

$$
U_i \left( z - \frac{C_i}{m} \right) + C_i = U_j \left( z - \frac{C_j}{m} \right) + C_j \tag{18}
$$

and the solution of the above equation, if it has one, is a potential point where the RHS of (17) may change its slope. For each solution, we use a tuple $\{z_{ij}, i, j\}$ to record the solution $z_{ij}$ and the index of the two tasks involved. Let *Points* denote the set of all these tuples, sorted in non-decreasing order of $z_{ij}$. It could be the case that more than two functions intersect with each other at the same point. For example, with task $\tau_i$, $\tau_j$ and $\tau_k$ we got three tuples $\{z_{ij}, i, j\}$, $\{z_{jk}, j, k\}$ and

$\{z_{ki}, k, i\}$, where $z_{ij} = z_{jk} = z_{ki}$. In this case, $z_{ij}$, $z_{jk}$ and $z_{ki}$ are kept as three distinguished points, and sorted in an arbitrary order in *Points*.

The algorithm to solve Equation (17) is shown in Algorithm 1. We use $z_0$ and $z_1$ to keep track of the two endpoints of current segment; at the beginning they are initialized as 0 and the first slope-changing point in *Points* respectively (line 1 and 2). Then we find out the $m_0 - 1$ tasks involved in the sum item in (17) in the first segment, recorded in $\Theta$ (line 3). Then we solve the equation with $\Theta$, i.e., assuming the solution falls in the segment $[z_0, z_1]$, and get the solution $z^*$ (line 5). We examine whether $z^* \in [z_0, z_1]$, if so, we have found the solution; if not, the assumption that the solution falls in the segment $[z_0, z_1]$ is incorrect, so we move to the next segment. We check whether we should replace $\tau_i$ by $\tau_j$ (or the other way around) in the next segment (line 9). A task $\tau_i$ is replaced by $\tau_j$ if it satisfies:

$$\tau_i \in \Theta \wedge \tau_j \notin \Theta \wedge U_j > U_i \tag{19}$$

Then we get the end point of the next segment (line 14), and repeat the above procedure with the next segment, until we find the solution. Note that by Theorem 2 the algorithm is guaranteed to terminate.

Now we discuss the complexity of Algorithm 1. The number of points in *Points* is at most $n^2$, and to sort them the complexity is $O(n^2 \log(n))$. The complexity to find out tasks in $\Theta$ is $O(n \log n)$. In each iteration of the while loop, the complexity of solving the equation with $\Theta$ is $O(m_0)$, which is bounded by $O(m)$. All other operations in the while loop are of constant complexity. So the overall complexity of the algorithm is $O(n^2 \log(n) + n^2 m)$.

---

1: $z_0 := 0$
2: $\{z_1, i, j\} :=$ the first element in *Points*
3: $\Theta :=$ the set of tasks involved in the $m_0 - 1$ largest sum item
    with $z = 0$
4: **while** (1) **do**
5:     $z^* :=$ the solution of

$$z = \Big( \sum_{\tau_i \in \Theta} \Big( U_i(z - \frac{C_i}{m}) + C_i \Big) - \mathbf{S}(\tau) \Big)/m$$

6:     **if** $z^* \in [z_0, z_1]$ **then**
7:         **return** $z^*$
8:     **else**
9:         **if** $(ShouldReplace(i, j) ==$ true$)$ **then**
10:             Update $\Theta$
11:         **end if**
12:     **end if**
13:     $z_0 := z_1$
14:     $\{z_1, i, j\} :=$ the next element in *Points*
        ($z_1 := +\infty$ if no more element in *Points*).
15: **end while**

**Algorithm 1.** An algorithm for solving Equation (17)

# 4 Conclusion

The bounded tardiness property of EDF [1] makes it an attractive choice of algorithm for systems which can tolerate bounded amounts of tardiness. However, previous work has either analyzed only implicit deadline systems (e.g. [1], [2]) or provided a highly general analysis ([4]). We have extended the result provided in [2] to the case in which deadlines do not need to equal periods. We have also proposed an improved algorithm for computing the tardiness bounds that can run in polynomial-time. These bounds provide the tightest known bounds for global EDF in the case in which deadlines need not equal periods. As in [2], we allow the tardiness for each task to be bounded individually.

# References

1. Devi, U.C., Anderson, J.H.: Tardiness bounds under global EDF scheduling on a multiprocessor. The Journal of Real-Time Systems 38(2), 133–189 (2008)
2. Erickson, J.P., Devi, U., Baruah, S.K.: Improved tardiness bounds for global EDF. In: Euromicro Conference on Real-Time Systems, pp. 14–23 (2010)
3. Leontyev, H.: Compositional Analysis Techniques For Multiprocessor Soft Real-Time Scheduling. PhD thesis, Department of Computer Science, The University of North Carolina at Chapel Hill (2010)
4. Leontyev, H., Anderson, J.H.: Generalized tardiness bounds for global multiprocessor scheduling. The Journal of Real-Time Systems 44(1), 26–71 (2010)
5. Erickson, J.P., Guan, N., Baruah, S.K.: Addendum to tardiness bounds for global EDF with deadlines different from periods. Technical Report TR10-014, The University of North Carolina at Chapel Hill, Computer Science department (September 2010), ftp://ftp.cs.unc.edu/pub/publications/techreports/10-014.pdf
6. Baruah, S.K., Mok, A.K., Rosier, L.E.: Preemptively scheduling hard-real-time sporadic tasks on one processor. In: IEEE Real-Time Systems Symposium, pp. 182–190 (1990)

# Cache-Aware Lock-Free Queues for Multiple Producers/Consumers and Weak Memory Consistency

Anders Gidenstam[1], Håkan Sundell[1], and Philippas Tsigas[2]

[1] School of Business and Informatics, University of Borås, Borås, Sweden
[2] Department of Computer Science and Engineering, Chalmers University of Technology, 412 96 Göteborg, Sweden

**Abstract.** A lock-free FIFO queue data structure is presented in this paper. The algorithm supports multiple producers and multiple consumers and weak memory models. It has been designed to be cache-aware and work directly on weak memory models. It utilizes the cache behavior in concert with lazy updates of shared data, and a dynamic lock-free memory management scheme to decrease unnecessary synchronization and increase performance. Experiments on an 8-way multi-core platform show significantly better performance for the new algorithm compared to previous fast lock-free algorithms.

## 1 Introduction

Lock-free implementation of data structures is a scalable approach for designing concurrent data structures. Lock-free data structures offer high concurrency but also immunity to deadlocks and convoying, in contrast to their blocking counterparts. Concurrent FIFO queue data structures are fundamental data structures that are key components in applications, algorithms, run-time and operating systems. This paper presents an efficient lock-free queue data structure for multiple producers and consumers. The algorithm is cache-aware in order to minimize its communication overhead. It works also on weak memory consistency models (due to out-of-order execution) without need for additional fence [4] instructions for reads and writes done in the algorithm towards the shared memory.

With the strongly emerging multi-core architectures for main-stream as well as high-performance computing, there is an increasing interest for efficient concurrent data structures that allow maximal exploitation of the available parallelism. With the evolving more complex multithreaded architectures of applications and systems, there is also likely to be an increasing need for stronger progress and safety guarantees of components in supporting frameworks, and consequently non-blocking synchronization would fit very well thanks to both its possible advantages in performance and its progress properties.

Two basic non-blocking methods have been proposed in the literature, *lock-free* and *wait-free* [3]. *Lock-free* implementations of shared data structures guarantee that at any point in time in any possible execution some operation will complete in a finite number of steps. In cases with overlapping accesses, some of them might have to repeat the operation in order to correctly complete it. However, real-time systems might have

C. Lu, T. Masuzawa, and M. Mosbah (Eds.): OPODIS 2010, LNCS 6490, pp. 302–317, 2010.

stronger requirements on progress, and thus in *wait-free* implementations each task is guaranteed to *correctly* complete any operation in a *bounded* number of its own steps, regardless of overlaps of the individual steps and the execution speed of other processes; i.e., while the lock-free approach might allow (under very bad timing) individual processes to starve, wait-freedom strengthens the lock-free condition to ensure individual progress for every task in the system.

Large efforts have been made on designing efficient concurrent queue data structures and blocking (or mixed with non-blocking techniques) implementations are available in most contemporary programming language frameworks supporting multithreading. In this paper, we focus only on strictly non-blocking queue algorithms as implementations being just "concurrent" (and possibly efficient as e.g. "lock-less") are still prune to problems as e.g. deadlocks. Absence of explicit locks does not imply any non-blocking properties, unless the latter are proven to be fulfilled. A large number of lock-free (and wait-free) queue implementations have appeared in the literature, e.g. [6][1][11][8][9][5] being the most influential or recent and most efficient results. These results all have a number of specialties or drawbacks as e.g. limitations in allowed concurrency, static in size, requiring atomic primitives not available on contemporary architectures, and scalable in performance but having a high overhead. This paper improves on previous results by combining the underlying approaches and designing the new algorithm cache-aware and tolerant to weak memory consistency models in order to maximize efficiency on contemporary multi-core platforms. The new lock-free algorithm has no limitations on concurrency, is fully dynamic in size, and only requires atomic primitives available on contemporary platforms. Experiments on an 8-way multi-core platform show significantly better performance for the new algorithm compared to previous lock-free implementations.

The rest of the paper is organized as follows. In Section 2, related work is discussed. Section 3 presents the new algorithm. The corresponding proofs and analysis are outlined in Section 4. In Section 5, some benchmark experiments are described. Finally, Section 6 concludes this paper.

## 2   Related Work

Lamport [6] presented a lock-free (actually wait-free) implementation of a queue based on a static array, with a limited concurrency supporting only one producer and one consumer. In this algorithm, synchronization is done via shared indices indicating the current head and tail array element. Giacomoni et al. [1] presented a cache-aware modification which instead synchronize directly on the array elements. Tsigas and Zhang [11] presented a lock-free extension of [6] where synchronization is done both directly on the array elements and the shared head and tail indices using $CAS$[1], thus supporting multiple producers and consumers. In order to avoid the ABA problem when updating the array elements, the algorithm exploits using two (or more) null values; the ABA

---

[1] The Compare-And-Swap (CAS) atomic primitive will update a given memory word, if and only if the word still matches a given value (e.g. the one previously read). CAS is generally available in contemporary systems with shared memory, supported mostly directly by hardware and in other cases in combination with system software.

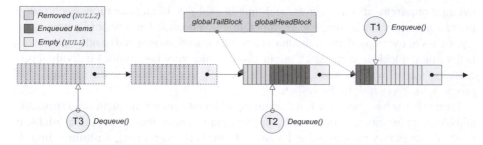

**Fig. 1.** A lock-free queue implemented using a linked list of arrays, where each thread is avoiding accesses to global pointers in order to reduce number of cache misses

problem is due to the inability of *CAS* to detect concurrent changes of a memory word from a value (A) to something else (B) and then again back to the first value (A). Moreover, for lowering the memory contention the algorithm alternates every other operation between scanning and updating the shared head and tail indices.

In resemblance to [6][1][11] the new algorithm uses arrays to store (pointers to) the items, and in resemblance to [11] it uses *CAS* and two null values. Moreover, shared indices [1] are avoided and scanning [11] is preferred as much as possible. In contrast to [6][1][11] the array is not static or cyclic, but instead more arrays are dynamically allocated as needed when new items are added, making our queue fully dynamic.

Michael and Scott [8] presented a lock-free queue based on a linked list, supporting multiple producers and consumers. Synchronization is done via shared pointers indicating the current head and tail node as well via the next pointer of the last node, all updated using *CAS*. The queue is fully dynamic as more nodes are allocated as needed when new items are added. The original presentation used unbounded version counters, and therefore required double-width *CAS* which is not supported on all contemporary platforms. The problem with the version counters can easily be avoided by using some memory management scheme as e.g. [7]. Moir et al. [9] presented an extension where elimination is used as a back-off strategy and increasing scalability when contention on the queue's head or tail is noticed via failed *CAS* attempts. However, elimination is only possible when the queue is close to be empty during the operation's invocation. Hoffman et al. [5] takes another approach to increase scalability by allowing concurrent *Enqueue* operations to insert the new node at adjacent positions in the linked list if contention is noticed during the attempted insert at the very end of the linked list. To enable these "baskets" of concurrently inserted nodes, removed nodes are logically deleted before the actual removal from the linked list, and as the algorithm traverses through the linked list it requires stronger memory management than [7] and a strategy to avoid long chains of logically deleted nodes.

In resemblance to [8][9][5] the new algorithm is dynamic, and in resemblance to [5] removed blocks are logically deleted, blocks are being traversed and creation of long chains are avoided. In contrast to [9][5] the new algorithm employs no special strategy for increasing scalability besides allowing disjoint *Enqueue* and *Dequeue* operations to execute in parallel.

---

**Program 1.** The functionality supported by the memory management scheme

```
1 node_t * NewNode(int size);
2 void DeleteNode(node_t *node);
3 node_t * DeRefLink(node_t **link);
4 void ReleaseRef(node_t *node);
5 bool CASRef(node_t **link, node_t *old, node_t *_new);
6 void StoreRef(node_t **link, node_t *node);
```

---

**Program 2.** Callback procedures for the memory management

```
1 void TerminateNode(block_t *node) {
2     StoreRef(&node->next,NULL);
3 }
4 void CleanUpNode(block_t *node) {
5     block_t *next = DeRefLink(&node->next);
6     block_t *next2 = DeRefLink(&globalTailBlock);
7     CASRef(&node->next, next, next2);
8 }
```

---

## 3   The New Algorithm

The underlying data structure that our algorithmic design uses is a linked list of arrays, and is depicted in Figure 1. In the data structure every array element contains a pointer to some arbitrary value. Both the *Enqueue* and *Dequeue* operations are using increasing array indices as each array element gets occupied versus removed. To ensure consistency, items are inserted or removed into each array element by using the *CAS* atomic synchronization primitive. To ensure that a *Enqueue* operation will not succeed with a *CAS* at a lower array index than where the concurrent *Dequeue* operations are operating, we need to enable the *CAS* primitive to distinguish (i.e., avoid the ABA problem) between "used" and "unused" array indices. For this purpose two null pointer values [11] are used; one (NULL) for the empty indices and another (NULL2) for the removed indices. As each array gets fully occupied (or removed), new array blocks are added to (or removed from) the linked list data structure. Two shared pointers, globalHeadBlock and globalTailBlock, are globally indicating the first and last active blocks respectively. These shared pointers are also concurrently updated using *CAS* operations as the linked list data structure changes. However, as these updates are done lazily (not atomically together with the addition of a new array block), the actually first or last active block might be found by following the next pointers of the linked list.

   As a successful update of a shared pointer will cause a cache miss to the other threads that concurrently access that pointer, the overall strategy for improving performance and scalability of the new algorithm is to avoid accessing pointers that can be concurrently updated [5]. Moreover, our algorithm achieves fewer updates by not having shared variables with explicit information regarding which array index currently being the next active for the *Enqueue* or *Dequeue*. Instead each thread is storing its own[2] pointers indicating the last known (by this thread) first and active block as well as active indices

---

[2] Each thread have their own set of variables stored in separate memory using thread-local storage (TLS).

**Program 3.** The block structure and auxiliary functions

```
 1 struct block_t : public node_t {
 2     void * nodes[BLOCK_SIZE];
 3     int head;
 4     int tail;
 5     bool deleted;
 6     block_t * next;
 7 };
 8 block_t * NewBlock() {
 9     block_t * block = NewNode(sizeof(block_t));
10     block->next = NULL;
11     block->head = 0;
12     block->tail = 0;
13     block->deleted = false;
14     for(int i=0;i<BLOCK_SIZE;i++) block->nodes[i]=NULL;
15     return block;
16 }
17 void InitQueue() {
18     block_t * block = NewBlock();
19     StoreRef(&globalHeadBlock,block);
20     StoreRef(&globalTailBlock,block);
21 }
22 void InitThread() {
23     threadHeadBlock = DeRefLink(&globalHeadBlock);
24     threadTailBlock = DeRefLink(&globalTailBlock);
25     threadHead = threadHeadBlock->head;
26     threadTail = threadTailBlock->tail;
27 }
28 // Shared variables
29 block_t * globalHeadBlock, globalTailBlock;
30 // Thread-local storage
31 block_t * threadHeadBlock, threadTailBlock;
32 int threadHead, threadTail;
```

for inserting and removing items. When a thread recognizes its own pointers to be inaccurate and stale, it performs a scan of the array elements and array blocks towards the right, and only resorts to reading the global pointers when it's beneficial compared to scanning. The *Dequeue* operation to be performed by thread T3 in Figure 1 illustrates a thread that has a stale view of the status of the data structure and thus needs to scan. As array elements are placed next to each other in memory, the scan can normally be done without any extra cache misses (besides the ones caused by concurrent successful *Enqueue* and *Dequeue* operations) and also without any constraint on in which order memory updates are propagated through the shared memory, thus allowing weak memory consistency models without the need for additional memory fence instructions.

For our implementation of the new lock-free queue algorithm, we have selected the lock-free memory management scheme proposed by Gidenstam et al. [2] which makes use of the *CAS* and *FAA* atomic synchronization primitives. The interface defined by the memory management scheme is listed in Program 1 and are fully described in [2]. Using this scheme we can assure that an array block can only be reclaimed when there is no next pointer in the linked list pointing to it and that there are no local references to it from pending concurrent operations or from pointers in thread-local storage. By supplying the scheme with appropriate callback functions, the scheme automatically reduces the length of possible chains of deleted nodes (held from reclamation by late threads holding a reference to an old array block), and thus enables an upper bound on

**Program 4.** The new Enqueue operation.

```
1  void Enqueue(void *item) {
2      int head = threadHead;
3      block_t *block = threadHeadBlock;
4      for(;;) {
5          if(head==BLOCK_SIZE) {
6              block_t *oldBlock = block;
7              block->head = head;
8              block = DeRefLink(&block->next);
9              if(block == NULL) {
10                 block = (queueblock_t *) NewBlock();
11                 while(globalHeadBlock != oldBlock && oldBlock->next==NULL) {
12                     queueblock_t *headBlock = DeRefLink(&globalHeadBlock);
13                     if(headBlock->next != oldBlock) break;
14                     if(CASRef(&globalHeadBlock,headBlock,oldBlock)) break;
15                 }
16                 if(CASRef(&oldBlock->next,NULL,block))
17                     CASRef(&globalHeadBlock,oldBlock,block);
18                 else {
19                     DeleteNode(block);
20                     block = DeRefLink(&oldBlock->next);
21                 }
22             }
23             else if(block->head==BLOCK_SIZE && block->next!=NULL)
24                 block = DeRefLink(&globalHeadBlock);
25             threadHeadBlock = block;
26             head = block->head;
27         }
28         else if(block->nodes[head]==NULL) {
29             if(CAS(&block->nodes[head],NULL,item)) {
30                 threadHead = head+1;
31                 return;
32             }
33         }
34         else head++;
35     }
36 }
```

the maximum memory usage for the data structure. The task of the callback function for breaking cycles, see the *CleanUpNode* procedure in Program 2, is to update the next pointer of a deleted array block such that it points to an active array block, in a way that is consistent with the semantics of the *Enqueue* and *Dequeue* operations. The *TerminateNode* procedure is called by the memory management scheme when the memory of an array block is possible to reclaim.

The specific fields of each array block are described in Program 3 as it is used in this implementation. Note that the linked list data structure always contains at least one array block. Note also that the additional fields head and tail in the array block are only used for indicating either fullness or emptiness of the whole array, and not any intermediate status. In order to simplify the description of our new algorithm, we have omitted some of the details of applying the operations of the memory management [2]. In actual implementations, *ReleaseRef* calls should be inserted at appropriate places whenever a variable holding a safe pointer goes out of scope or is reassigned.

The *Enqueue* operation is described in Program 4. After scanning for the first empty (i.e., an array element containing NULL) array index, it tries to insert the new item by updating the array element with *CAS*. If this fails (due to a concurrent successful

**Program 5.** The new Dequeue operation.

```
1  void * Dequeue() {
2      int tail = threadTail;
3      block_t *block = threadTailBlock;
4      for(;;) {
5          if(tail==BLOCK_SIZE) {
6              block_t *oldBlock = block;
7              block->tail = tail;
8              block=DeRefLink(&block->next);
9              if(block == NULL)
10                 return NULL;
11             else {
12                 if(!oldBlock->deleted) {
13                     while(globalTailBlock != oldBlock && !oldBlock->deleted) {
14                         block_t *tailBlock= DeRefLink(&globalTailBlock);
15                         if(tailBlock->next != oldBlock) continue;
16                         if(CASRef(&globalTailBlock,tailBlock,oldBlock))
17                             DeleteNode(tailBlock);
18                     }
19                     if(CAS(&oldBlock->deleted,false,true)) {
20                         if(CASRef(&globalTailBlock,oldBlock,block))
21                             DeleteNode(oldBlock);
22                     }
23                 }
24                 if(block->deleted)
25                     block=DeRefLink(&globalTailBlock);
26             }
27             threadTailBlock = block;
28             tail = block->tail;
29         }
30         else {
31             void *data = block->nodes[tail];
32             if(data==NULL2)
33                 tail++;
34             else if(data==NULL && CAS(&block->nodes[tail],NULL,NULL)) {
35                 threadTail = tail;
36                 return NULL;
37             }
38             else if(CAS(&block->nodes[tail],data,NULL2)) {
39                 threadTail = tail+1;
40                 return data;
41             }
42         }
43     }
44  }
```

Enqueue), it continues scanning until the end of the array. If the end of the array is reached, it first assures lock-freedom and accuracy of the global head pointer:

1. If the global head pointer is not pointing to the current block, the operation (after it verifies that the global head pointer is pointing to the previous block) updates the head pointer to do so by using a *CAS* operation.
2. If the global head pointer is pointing to the current array block, the algorithm tries to insert a new array block by updating the next pointer using a *CAS*. If this fails, this is due to some concurrent *Enqueue* operation having already added a new block, henceforth the operation continues scanning for an empty array index in that block.

The *Dequeue* operation is described in Program 5. After scanning for the first non-empty (i.e., an array element with neither NULL or NULL2) array index, it tries to remove the found item by updating the array element with a *CAS*. If this fails (due to a concurrent successful *Dequeue*), it continues scanning until the end of the array. If NULL is found during scanning, the queue is (after also ensuring the NULL value to be globally consistent using *CAS*[3] ) recognized to be empty and the operation returns an empty value. If the end of the array is reached, the algorithm first assures lock-freedom and accuracy of the global tail pointer:

1. If the global tail pointer is pointing to the current array block, it tries to logically mark the block as deleted using a *CAS*.
2. If the global tail pointer was not pointing to the current block, it is (after verified that it is pointing to the previous block) updated to do so using a *CAS*. Whenever the global tail pointer is successfully updated (either when helping or after a successful logical deletion), the previously global tail-block is sent for memory reclamation.

Whenever an array element is successfully updated with NULL2 using *CAS*, the found item is returned by the *Dequeue* operation.

## 4   Correctness and Analysis

In this section we show that the new queue algorithm is linearizable and lock-free. Line numbers given for actions in *Enqueue* operations refer to Program 4, while line numbers for actions in *Dequeue* operations refer to Program 5. Due to space limitations some of the detailed proofs have been omitted in this version of the paper.

**Assumption 1 (Memory order).** *All CAS operations are atomic.*

*A CAS operation behaves as a memory barrier for a thread's memory reads and writes. All reads and writes done before the CAS in program order are committed to memory before the CAS takes effect and none of the reads and writes following a CAS are visible in memory before the CAS takes effect.*

**Definition 1.** *The linearization point of an Enqueue operation is the successful CAS at line 29 in Enqueue.*

**Definition 2.** *The linearization point of a Dequeue operation is either:*
*i) the CAS at line 34 in Dequeue (Program 5) iff NULL is returned; or*
*ii) the successful CAS at line 38 in Dequeue otherwise.*

### 4.1   Properties of an Array Block

**Definition 3.** *A **full** array block is a block where all array elements have been changed from NULL to another value (i.e., there is no array element with value NULL). An array block is **marked full** when its head field is set to BLOCK_SIZE.*

**Definition 4.** *An **emptied** array block is a block where all array elements have been changed to NULL2. An array block is **marked emptied** when its block.tail field is set to BLOCK_SIZE.*

---

[3] On the platform we used for our experimental evaluation, CAS is faster than memory barriers for this purpose. On platforms with suitable memory barriers, these could be used instead.

**Lemma 1 (Block array element life cycle).** *An array element in a block can change value at most two times during the life time of the block in the following order: i) first from the initial value* NULL *to an item; and subsequently ii) from an item to* NULL2.

**Lemma 2 (Thread-local head lag).** *The thread-local static variable* threadHead *is never ahead of the true head index (i.e., the index of the first* NULL *value in the block) of the block at the starting point of an* Enqueue *operation.*

**Lemma 3 (Thread-local tail lag).** *The thread-local static variable* threadTail *is never ahead of the true tail position (i.e., the index after the last* NULL2 *in the block) of the block at the starting point of a* Dequeue *operation.*

## 4.2   Properties of the Chain of Array Blocks

**Definition 5.** *An **active** array block is a block that has been created, has been published in a shared variable (i.e., in* globalHeadBlock, globalTailBlock *or a next pointer) and not yet been marked as deleted by setting the block's deleted flag.*

**Definition 6.** *A **valid** array block is a block that has been created and has not (yet) become reclaimable.*

**Lemma 4 (Block next pointer).** *The next pointer in an active block initially contains* NULL *and can change at most once while the block is active, from* NULL *to a pointer to a new block.*

**Lemma 5 (Unique head block).** *At any time there is exactly one valid block that has a next pointer with the value* NULL.

**Lemma 6 (At least one active block).** *There is always at least one active block in the queue.*

**Lemma 7 (globalHeadBlock).** *The global variable* globalHeadBlock *always points to either: i) the block at the head of the chain of blocks; or ii) the block immediately before the head of the chain of blocks.*

**Lemma 8 (globalTailBlock).** *The global variable* globalTailBlock *always points to either: i) the first active block in the chain of blocks; or ii) the block immediately before the first active block in the chain of blocks.*

## 4.3   Linearizability

**Lemma 9 (Linearizability I).** *The operation* Enqueue *is linearizable with respect to other* Enqueue *and* Dequeue *operations with linearization points according to Definition 1 and Definition 2.*

*Proof.* First observe that from Lemma 5 and Lemma 6 there is always a well defined array block at the head of the chain of array blocks.

Consider two concurrent *Enqueue* operations $Enq_1(A)$ and $Enq_2(B)$, enqueuing the elements $A$ and $B$ respectively. According to Lemma 2 we can, without loss of generality, assume that both operations start with their threadHead variables set to 0. Both operations do a linear search for the first array element in the block at the head of the chain of blocks that contains NULL and will try to update that array element using CAS (line 29 in Program 4). Only one can succeed and that *Enqueue* will be linearized at that point. The other will retry from line 4.

Consider an *Enqueue* operation $Enq(A)$ and a concurrent *Dequeue* operation $Deq$. The critical case is when the queue is initially empty. According to Lemma 2 and Lemma 3 we can, without loss of generality, assume that the operations start with their threadHead and respectively threadTail variables set to 0. Assume towards a contradiction that $Deq$ returns $A$ despite being linearized before $Enq(A)$. The contradiction is obvious since there is no way that $Deq$ can return $A$ before $A$ is written into the array block, which occurs at the linearization point of $Enq(A)$ (line 29 in Program 4).

For the opposit case assume towards a contradiction that $Deq$ returns NULL despite being linearized after $Enq(A)$. To return NULL $Deq$ must traverse the array block until it finds NULL. In particular, it must have read the first index that contained NULL, which is where $Enq(A)$ will write $A$ using *CAS* (*Enqueue* line 29). Since *CAS* is atomic according to our assumption on memory order a read returning NULL must have occured before the *CAS*. Since this read is the linearization point of $Deq$ we have a contradiction with the assumption that $Deq$ was linearized after $Enq(A)$.    □

**Lemma 10 (Linearizability II).** *The operation Dequeue is linearizable with respect to other Dequeue and Enqueue operations with linearization points according to Definition 1 and Definition 2.*

*Proof.* Consider two *Dequeue* operations, $Deq_1$ and $Deq_2$ on a non-empty queue. The operations will first search the first active block, via their threadTailBlock variables and globalTailBlock, where the latter is guaranteed to point to the first active block or the block immediately before it by Lemma 8. Once a *Dequeue* has reached the first active block it will scan it, looking for an array element that is not NULL2. If such an array element is found the $Deq$ operation tries to change that element to NULL2 using *CAS* (line 38). Assume towards a contradiction that the $Deq_1$ returning $B$ is linearized before $Deq_2$ returning $A$ where $A$ was enqueued before $B$ (in the same array block). From Lemma 9 we know that $A$ is in an array element with lower index than $B$. Since *Dequeue* only scans past NULL2 values (line 32), $Deq_1$, which must have scanned past the index of $A$ to reach $B$, must have read NULL2 from $A$'s array element. According to our memory order assumption all local memory reads that precede a CAS must have occured before the *CAS*. Hence, $Deq_1$ read NULL2 from the array element of $A$ before its linearization point. From Lemma 1 we know that an array element can only change to NULL2 once which contradicts our assumption that $Deq_2$ which is linearized after $Deq_1$ returns $A$.

Consider two *Dequeue* operations, $Deq_1$ and $Deq_2$ on a queue containing exactly one item $A$. Assume towards a contradiction that $Deq_1$ returns NULL despite being linearized before $Deq_2$ returning $A$. As above by Lemma 1 NULL can only occur at a higher array element index than that of $A$ and consequently $Deq_1$ have to read NULL2 from that location before its *CAS* operation from NULL to NULL at line 34 succeeds giving a contradiction.

That *Dequeue* is linearizable with respect to concurrent *Enqueue* operations is shown in the proof of Lemma 9 above.

Note that the scan procedure in *Dequeue* is performing speculative reads that might have taken effect out of program order. If the scan was performing at least one search step, the preceding speculative reads in the steps before the last step must have read the NULL2 value (as line 34 must have been executed). These speculative NULL2 reads must have taken effect before the last atomic NULL read during the *CAS* at line 34, as the *CAS* implies a memory barrier and must have taken effect after the previous speculative reads.                                                                                          □

### 4.4  Lock-Freedom

**Lemma 11  (Lock-free I).** *The operation Enqueue is lock-free.*

*Proof.* The *Enqueue* operation contains two nested loops. There are three cases to consider:

First consider the case where threadHeadBlock points to a block that is not marked full. According to Lemma 2 the value of the threadHead variable will be smaller or equal to the index of the first NULL value in the block when the *Enqueue* operation starts. The operation will finish if it finds an array element in the block containing NULL and successfully puts its item there using a *CAS*. The index it looks at increases in each iteration, except when an unsuccessful *CAS* occurs, something that according to Lemma 1 can only happen once per array element. Thus the search index will reach the end of the block after at most 2*BLOCK_SIZE iterations and would find a free array element if there is any left. That is, progress is made unless concurrent operations fill the block first. If the block is found to be full the next iteration will mark the block full (line 7) and continue in one of the cases below.

Second, consider the case where threadHeadBlock points to a block that is marked full and has a next pointer that isn't NULL. Finding out that the block is full takes at most BLOCK_SIZE iterations. After that *Enqueue* will read the full block's next pointer into block (line 8). Since block isn't NULL the *Enqueue* operation tests if the new block is marked as full (line 23). If it is full and isn't the last block (i.e., block.next is not NULL) the *Enqueue* operation moves to the block that globalHeadBlock points to. According to Lemma 7 this is either the last or second last block of the chain. If block is full and is the last block the next iteration will enter case three below.

Third, consider the case where threadHeadBlock points to a block that is full (oldBlock) and has a next pointer that is NULL. This case proceeds as the second case until the *Enqueue* reads the oldBlock.next pointer to be NULL at line 8. When it does that, it enters the inner loop at line 11. To remain in the loop globalHeadBlock must not be equal to oldBlock and oldBlock must remain the last block in the chain. Further globalHeadBlock must point to the block before oldBlock at line 13 and not at line 14 since the *CAS* would succeed and exit the loop otherwise. With Lemma 7 in mind this can clearly only occur once since in the next iteration either globalHeadBlock is equal to oldBlock or, if globalHeadBlock has moved further, oldBlock.next is not NULL anymore. Past the inner loop the *Enqueue* tries to add a new block. Regardless of whether it succeeds or not the next iteration of the outer loop will be done on a new block.    □

**Lemma 12 (Lock-free II).** *The operation Dequeue is lock-free.*

*Proof.* The *Dequeue* operation contains two nested loops. There are three cases to consider:

First consider the case where threadTailBlock points to a block that has not been marked emptied. According to Lemma 3 the value of the threadTail will be smaller or equal to the index of the first value not equal to NULL2 in the block when the *Dequeue* operation starts. At worst the operation has to search from the beginning of the block (i.e., threadTail was 0). Each array element in the block is read (line 31 in *Dequeue*) and depending on the value found at the current array element the operation either moves to the next array element if the value was NULL2 (line 33), returns NULL (line 36) if the value was NULL and then verified to be NULL by the *CAS* at line 34, or attempts to change the value of the location to NULL2 using a *CAS* (line 38). If the *CAS* succeeds the removed item is returned, otherwise the *Dequeue* operation will do another iteration in which it will move to the next array element in the block (since according to Lemma 1 the only possible reason for the *CAS* to fail is that a concurrent *Dequeue* operation changed the value to NULL2). In all at most 2*BLOCK_SIZE iterations of the outer loop is required to either find and successfully dequeue an item or find the block emptied. If the block is found to be emptied the next iteration will mark the block emptied (line 7) and continue in one of the cases below.

Second, consider the case where threadTailBlock points to a block that has been marked emptied and has a next pointer that is NULL. In this case the queue is empty and NULL is returned (line 10).

Third, consider the case where threadTailBlock points to a block that has been marked emptied and has a next pointer that is not NULL. In this case the current block is referenced by oldBlock (line 6) and its next pointer is read into block (line 8). There are two cases depending on whether oldBlock is marked deleted or not (line 12). If oldBlock is marked deleted the *Dequeue* operation checks if the next block is also marked deleted (line 24) in which case it moves directly to the block that globalTailBlock points to, which according to Lemma 8 is the first active block of the queue or the block immediately before it. Otherwise the *Dequeue* moves the next block (which at least was active at line 24). If oldBlock is not marked deleted the *Dequeue* will enter the inner loop (line 13). To remain in this loop, the variable globalTailBlock has to be different from oldBlock and oldBlock must not be marked deleted. From Lemma 8 we know that globalTailBlock points to the first active block or the block immediately before that. At the time the inner loop is entered oldBlock is the first active block so globalTailBlock is the block immediately before oldBlock or else the loop would not be entered (since globalTailBlock would be equal to oldBlock). In this case the *CAS* at line 16 in this or a concurrent *Dequeue* can advance globalTailBlock to oldBlock and terminate the loop. Further, Lemma 8 shows that it is impossible to advance globalTailBlock past oldBlock without marking oldBlock deleted and thereby making sure the inner loop cannot continue.

Once clear of the inner loop the *Dequeue* tests if the next block is marked deleted (line 24) and acts as described above, continuing with either the next block or the block pointed to by globalTailBlock.                                                                    □

### 4.5  Concurrent FIFO Queue

**Theorem 1.** *The algorithm implements a lock-free and linearizable FIFO queue data structure.*

*Proof.* The minimal set of operations[4] necessary for implementing a FIFO queue is consisting of the *Enqueue* and *Dequeue* operations. Correspondingly, given by Lemmas 11 and 12 our implementation is lock-free, and given by Lemmas 9 and 10 our implementation is linearizable.

## 5  Experiments

We have evaluated the performance of our lock-free queue algorithm by the means of some custom micro-benchmarks. The purpose of these experiments is to help estimate how well the new algorithm compares with other known lock-free queues under high contention and increasing concurrency. The benchmarks are the following:

1. Random 50%/50%. Each thread is randomly (the sequence is decided in forehand) executing either an *Enqueue* or a *Dequeue* operation.
2. Random 50%/50% Bias 1000. Performed as the previous benchmark, besides that the queue is initialized with 1000 items.
3. 1 Producer / N-1 Consumers. Each thread (out of N) is either a producer or consumer, throughout the whole experiment. The producer is repeatedly executing *Enqueue* operations, whereas the consumers are executing *Dequeue*.
4. N-1 Producers / 1 Consumer. Same as the previous benchmark, with the producer and consumer distributions interchanged.

For comparison we have also implemented the dynamic lock-free queues by Michael and Scott [8], ditto with elimination [9], the baskets queue [5], and the static cyclic array lock-free queue presented in [11]. All dynamic queues (including the new algorithm) have been implemented to support queue sizes only limited by the system's memory, i.e., using lock-free management schemes [7] or [2] and lock-free free-lists where appropriate. For the new implementation, the size of the array block (BLOCK_SIZE) is chosen to fit within one cache line. All implementations are written in C and compiled with the highest optimization level. In our experiments, each concurrent thread is started at the very same time and each benchmark runs for one second for each implementation. Exactly the same sequence of operations was performed for all different implementations compared. A clean-cache operation was also performed just before each run.

---

[4] If required, operations as *Peek* and *IsEmpty* can be derived straight-forwardly out of the *Dequeue* algorithm by omitting the update part of the *CAS* operation in line 38 combined with other minor changes.

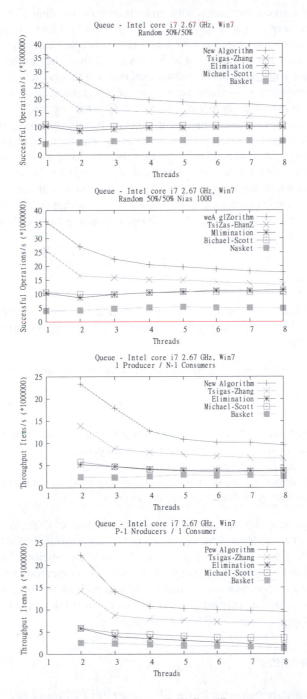

**Fig. 2.** Experiments on a 8-way Intel Core i7 processor system

The results from the experiments with up to 8 threads are shown in Figure 2. The benchmarks have been executed on an Intel Core i7 920 2.67 GHz with 6 GB DDR3 1333 MHz system running Windows 7 64-bit. This processor has 4 cores, capable of executing 2 threads each. The results of benchmarks 1-2 show the number of successful (failed *Dequeue*s are not counted) operations executed per second in the system in total. The results of benchmarks 3-4 show the number of items per second that have passed through the queue (i.e., the number of successful *Dequeue* operations). In all of the benchmarks, the two array-based implementations perform significantly better than the other implementations. The worse performance of the other implementations compared to the static array-based implementation can be explained to be mainly due to the costs of having dynamic allocation of nodes. Interestingly, the new dynamic implementation performs significantly better than the implementation with a static array. This can be explained by the benefits of the cache-awareness (also causing fewer shared updates) apparently being significantly higher than the corresponding costs of having dynamic allocation of arrays.

# 6   Conclusions

We have presented a new algorithm for implementing a lock-free queue data structure. To the best of our knowledge, this is the first lock-free queue algorithm with all of the following properties:

- Cache-aware algorithmic handling of shared pointers including lazy updates to decrease communication overhead.
- Linked-list of arrays as underlying structure for efficient dynamic algorithmic design.
- Exploitation of thread-local static storage for efficient communication.
- Fully dynamic in size via lock-free memory management.
- Lock-free design for supporting concurrency.
- Algorithmic support for weak memory consistency models, resulting in more efficient implementation on contemporary hardware.

The algorithm has been shown to be lock-free and linearizable. Experiments on a contemporary multi-core platform show significantly better performance for the new algorithm compared to previous state-of-the-art lock-free implementations. We believe that our implementation should be of highly practical interest to contemporary and emerging multi-core and multi-processor system thanks to both its high performance, its strong progress guarantees, and its support to weak memory consistency models. We are currently incorporating it into the NOBLE [10] library.

## Acknowledgments

This work was partially supported by the EU as part of FP7 Project PEPPHER (www.peppher.eu) under grant 248481 and the Swedish Research Council under grant number 37252706 and 13671-60582-29.

# References

1. Giacomoni, J., Moseley, T., Vachharajani, M.: Fastforward for efficient pipeline parallelism: a cache-optimized concurrent lock-free queue. In: Proceedings of the 13th ACM SIGPLAN Symposium on Principles and Practice of Parallel Programming (PPoPP 2008), pp. 43–52. ACM, New York (2008)
2. Gidenstam, A., Papatriantafilou, M., Sundell, H., Tsigas, P.: Efficient and reliable lock-free memory reclamation based on reference counting. IEEE Transactions on Parallel and Distributed Systems 20(8), 1173–1187 (2009)
3. Herlihy, M.: Wait-free synchronization. ACM Transactions on Programming Languages and Systems 11(1), 124–149 (1991)
4. Higham, L., Kawash, J.: Impact of instruction re-ordering on the correctness of shared-memory programs. In: Proceedings of the 8th International Symposium on Parallel Architectures, Algorithms and Networks, pp. 25–32. IEEE, Los Alamitos (December 2005)
5. Hoffman, M., Shalev, O., Shavit, N.: The baskets queue. In: Tovar, E., Tsigas, P., Fouchal, H. (eds.) OPODIS 2007. LNCS, vol. 4878, pp. 401–414. Springer, Heidelberg (2007)
6. Lamport, L.: Specifying concurrent program modules. ACM Trans. Program. Lang. Syst. 5(2), 190–222 (1983)
7. Michael, M.M.: Hazard pointers: Safe memory reclamation for lock-free objects. IEEE Transactions on Parallel and Distributed Systems 15(8) (August 2004)
8. Michael, M.M., Scott, M.L.: Simple, fast, and practical non-blocking and blocking concurrent queue algorithms. In: Proceedings of the Fifteenth Annual ACM Symposium on Principles of Distributed Computing, pp. 267–275. ACM Press, New York (1996)
9. Moir, M., Nussbaum, D., Shalev, O., Shavit, N.: Using elimination to implement scalable and lock-free fifo queues. In: Proceedings of the 17th Annual ACM Symposium on Parallelism in Algorithms and Architectures (SPAA 2005), pp. 253–262. ACM, New York (2005)
10. Sundell, H., Tsigas, P.: Noble: non-blocking programming support via lock-free shared abstract data types. SIGARCH Comput. Archit. News 36(5), 80–87 (2008)
11. Tsigas, P., Zhang, Y.: A simple, fast and scalable non-blocking concurrent FIFO queue for shared memory multiprocessor systems. In: Proceedings of the 13th Annual ACM Symposium on Parallel Algorithms and Architectures (SPAA 2001), pp. 134–143. ACM Press, New York (2001)

# An Adaptive Technique for Constructing Robust and High-Throughput Shared Objects

Danny Hendler[1], Shay Kutten[2], and Erez Michalak[2]

[1] Department of Computer-Science, Ben-Gurion University
[2] Department of Industrial Engineering and Management, Technion

**Abstract.** Shared counters are the key to solving a variety of coordination problems on multiprocessor machines, such as barrier synchronization and index distribution. It is desired that they, like shared objects in general, be robust, linearizable and scalable.

We present the first linearizable and wait-free shared counter algorithm that achieves high throughput without a-priori knowledge about the system's level of asynchrony. Our algorithm can be easily adapted to any other combinable objects as well, such as stacks and queues.

In particular, in an $N$-process execution $E$, our algorithm achieves high throughput of $\Omega(\frac{N}{\phi_E^2 \log^2 \phi_E \log N})$, where $\phi_E$ is $E$'s level of asynchrony. Moreover, our algorithm stands any constant number of faults. If $E$ contains a constant number of faults, then our algorithm still achieves high throughput of $\Omega(\frac{N}{\phi_E'^2 \log^2 \phi_E' \log N})$, where $\phi_E'$ bounds the relative speeds of any two processes, at a time that both of them participated in $E$ and none of them failed.

Our algorithm can be viewed as an adaptive version of the Bounded-Wait-Combining (BWC) prior art algorithm. BWC receives as an input an argument $\phi$ as a (supposed) upper bound of $\phi_E$, and achieves optimal throughput if $\phi = \phi_E$. However, if the given $\phi$ happens to be lower than the actual $\phi_E$, or much greater than $\phi_E$, then the throughput of BWC degraded significantly. Moreover, whereas BWC is only lock-free, our algorithm is more robust, since it is wait-free.

To achieve high throughput and wait-freedom, we present a method that guarantees (for some common kind of procedures) the procedure's successful termination in a bounded time, regardless of shared memory contention. This method may prove useful by itself, for other problems.

## 1 Introduction

A shared counter is a shared object that holds an integer and supports the *fetch&increment* (FAI) operation for atomically incrementing the counter and returning its previous value. It is desirable that algorithms for shared counters be *linearizable*, robust, and scalable. Linearizability [17] is the most widely-used correctness condition for shared objects. Intuitively, it requires that each operation appears to take effect instantaneously at some moment between its invocation and response. Regarding robustness - *Lock-freedom* is a global progress guarantee. It requires that some operation must be completed in a finite number

C. Lu, T. Masuzawa, and M. Mosbah (Eds.): OPODIS 2010, LNCS 6490, pp. 318–332, 2010.

of processes' steps. *Wait-freedom* [15] is a stronger guarantee - a process completes its own operation in a finite number of its own steps. Wait-freedom provides strong fault-tolerance [15]: no process can be prevented from completing an operation by undetected failures of other processes, or by arbitrary speed variations. We measure scalability in terms of shared objects throughput [12]. Intuitively, the throughput of a non-empty execution $E$ is the ratio between $E$'s duration and the number of completed FAI instances in $E$. For example, suppose the hardware supports a primitive of FAI. A counter implementation in which every process simply performs the FAI primitive on a shared base object is both linearizable and wait-free. However, this implementation is not scalable, since its throughput does not grow with $N$.

This paper presents the first linearizable and wait-free shared counter algorithm that achieves high throughput without any a-priori knowledge of the execution. Our algorithm can be easily adapted to work for any other combinable operation as well.

## 1.1   Related Work

To allow parallelism, researchers proposed highly-distributed coordination structures such as *counting networks* [1]. Though they are wait-free and scalable, the counting networks of [1] are not linearizable. Herlihy, Shavit, and Waarts demonstrated that counting networks can be adapted to implement linearizable counters [16]. However, the first counting network they present is not lock-free, while the others are not scalable, since each operation has to access $\Omega(N)$ base objects.

*Combining* is a well-established technique for highly parallel shared objects. Combining was introduced by Gottlieb et al. to be used in switches of a processor-to-memory network [8]. It reduces contention by merging several messages with the same destination. When a switch discovers several memory requests directed to the same memory location, a combined request is created to represent these requests. Separate responses to the original requests are created later from the reply to the combined request. Goodman et al. [7] introduced Combining Tree (CT), that was used to implement a linearizable and scalable counter, but their implementation is not lock-free. Shavit and Zemach [19] introduced *diffracting trees* to replace the static CT with a collection of randomly created dynamic trees. Diffracting trees were used to implement wait-free and scalable shared counters but they are not linearizable. Hoai Ha, Papatriantafilou, and Tsigas introduced another version of adaptive CT [11], but it is not linearizable either.

Chandra, Jayanti, and Tan [2] introduced a construction that implements a large class of objects they call *closed objects*. An object contains a set of operations, such that every operation in the set causes the object a state transition. The object is closed, if any two consecutive state transitions can be replaced with a single state transition that brings the object to the same state. A closed object may support the FAI operation. In their construction, processes combine operations over dynamically created trees. Their algorithm is wait-free and linearizable, but its throughput does not grow with $N$.

Ellen, Lev, Luchangco, and Moir ([6]) introduced *Scalable Non-Zero Indicator* (SNZI), a shared object that is related to a shared counter, but has weaker semantics. SNZI can replace a shared counter for some applications. They present a linearizable implementation that is scalable and lock-free, and can be fast in the absence of contention. However, their approach does not seem suitable for a full fledged shared counter.

Achieving fault-tolerance using bounded-time locking was proposed by Gray and Cheriton in [9], in the context of caching. They presented the *lease*, that grants its holder control over writes to a covered datum (during the term of the lease). Another fault-tolerance tool is Greenwald's two-handed-emulation [10], which uses the *double-compare-and-swap* (DCAS) primitive to construct lock-free implementations of shared objects. When some process $p$ calls the emulation to execute an operation $o$ of the object, $p$ tries to register a new instance of $o$ to the emulation. Only one operation instance may be registered to the emulation at any moment. Hence, first, $p$ has to assist executing the steps of the currently registered instance $o'$ (if such exists). When $o'$ is completed, the call of its creator process completes, and other process (possibly $p$) registers its operation instance. DCAS is used to ensure that exactly one process executes successfully the current step of the currently registered operation instance, while the other simultaneous writes have no effect.

Hendler and Kutten integrated combining, bounded waiting and two-handed-emulation, and proposed BWC, a linearizable and lock-free counter implementation [12]. Their algorithm receives as an input an argument $\phi$ and guarantees a high throughput of $\Omega(\frac{N}{\log N})$ in an $N$-process execution $E$ in which $\phi$ is an asymptotically tight upper bound on $E$'s level of asynchrony. Intuitively, BWC is a lock-free variation of CT, in which processes wait for each other, but only up to some bounded number of steps (determined using $\phi$). However, in an execution $E$ in which $\phi < \phi_E$, BWC achieves low throughput. When $\phi >> \phi_E$, BWC's throughput decreases by a factor of at least $\phi/\phi_E$ compared to the optimum.

The difficulty arising from an unknown level of asynchrony received a lot of attention. Dolev, Dwork and Stockmeyer presented consensus algorithms for a number of partial synchrony models with different timing requirements and failure assumptions [3]. In some of their models, a fixed upper bound on the relative processor speed (denoted by $\phi$) is not known in advance, and their protocols overcome this difficulty. However, their approach does not seem to provide a robust solution in our model, since a fixed $\phi$ may not exists. Even if it did, it is unclear how to broadcast it to all processes without high contention.

Dwork, Lynch and Stockmeyer suggested distributed consensus protocols that are tolerant to some number of failures in various models with different synchrony conditions [5]. They deal with point-to-point models and with models that allow multicasting to some of the processors in an atomic step. However, in their solutions, the level of asynchrony is known in advance.

Like in the case for counters, we are not aware of any linearizable and wait-free *deterministic* stack or queue algorithm that is also scalable. Hendler, Shavit, and Yerushalmi [14] presented an elimination-based *randomized* linearizable stack algorithm that is both lock-free and scalable in practice. Moir et al. [18] used ideas similar to [14] to obtain a queue algorithm that possesses the same properties.

The new algorithm presented in this paper can be viewed as an adaptive version of $BWC$, that gracefully adapts to the asynchrony bound. Unfortunately, we have not found a way to use $BWC$ in a modular manner, and our algorithm is therefore rather involved (see Sect. 3.3).

The rest of the paper is organized as follows. Section 2 provides the model of the shared system we use. Section 3 explains some key concepts in the new algorithm, by outlining some previous combining techniques. Section 4 describes the new algorithm. Section 5 presents a new technique we use to execute concurrent procedures, that may be useful by itself, for other problems. Section 6 outlines the analysis of the new algorithm. In Sect. 7, we conclude and present directions for future work.

## 2   Model

We consider a shared-memory system, in which a set of $N$ asynchronous processes communicate by applying to shared variables *read*, *write*, or *read-modify-write* primitives - *compare-and-swap* (CAS) and DCAS. To execute a primitive, a process performs an *invocation* of the operation, 0 or more *stall* steps and a *response* step. From the invocation step and until the response step, the operation is considered to be *pending*. If two or more processes have a pending write/read-modify-write operation on a shared variable $v$, then exactly one of them receives a response, and the rest are stalled. Similarly to [4], we assume that if a process $p$ has a pending operation on a shared variable $v$, then $p$ incurs a stall only if another process with a pending operation on $v$ receives a response. Pending operations receive their responses in the order of their invocations.

A *configuration* specifies the value of each shared variable and the state of each process. An *initial configuration* is a configuration in which all the shared variables have their initial values and all the processes are in their initial states. An *execution fragment* is a sequence of steps, in which processes take steps and change states according to their algorithm. A process' state may change based on the response it receives in a response step only. An *execution* is an execution fragment that starts from the initial configuration. We assign times to execution steps as follows. Assigned times constitute a non-decreasing sequence of integers starting from 0. Let $s$ be a step performed by process $q$, let $E = E_1 s E_2$, and let $s'$ denote $q$'s last step in $E_1$ (if any). The time assigned to $s$ in $E$ is denoted *time(E,s)*. If $s$ is a response step, then *time(E,s)* is set to *time(E,s')*. If $s$ is a stall step, then *time(E,s)* is set to *time(E,s')+1*. If $s$ is an invocation step, *time(E,s)* is set to the maximum between 0 and (if exists): (1) the time of the last step in $E_1$; (2) *time(E,s')* plus 1; (3) the time of the last step in $E_1$ that accesses the same variable as $s$ plus 1.

The *duration* of an execution $E$ is the time of the last step of $E$. The *throughput* of an execution $E$ is the ratio between the number of FAI operations that complete in $E$ and $E$'s duration.

Let $E$ be an execution fragment. $E$ is $\phi$-*synchronous* if, for any $E_0$, $E_1$, $E_2$ such that $E = E_0 E_1 E_2$, and for any two distinct processes $p$ and $q$, if $p$ has to invoke some primitive at the end of $E_0$ and $E_1$ contains $\phi + 1$ invocation steps by $q$, then $E_1$ contains at least one invocation step by $p$.

An integer $\phi$ is a *correct asynchrony bound* for an execution $E$ if it is an upper bound of the speed ratio between any two processes in $E$. $E$'s *level of asynchrony*, denoted by $\phi_E$, is the minimal corret asynchrony bound (or infinity if no finite bound exists). The *contention level* of an execution $E$ is the maximum number of consecutive stalls that are incurred by a process in $E$.

For simplicity of presentation, we assume in the following that $N$ is an integral power of 2. This assumption does not change our results: the solution for a set of $N$ processes is equivalent to the solution for a set of $2^{\lceil \log N \rceil}$ processes (in which only $N$ processes actually make requests).

## 3   Prior Art Combining Techniques

### 3.1   Combining Tree [7]

The CT algorithm [7] uses a full binary tree, in which each process owns one of the leaves and a unique *color*. A process $p$ stores a new request in $p$'s leaf, and climbs the tree towards the root. To climb to a new node $n$, $n$ must be uncolored and $p$ has to color $n$ with $p$'s own color. If $p$ succeeds in coloring $n$, then $p$ stays at $n$ for some $\epsilon$ steps.

Some other process $q$ may reach $n$ during that time and has to wait at $n$ (since $n$ is colored by $p$). We say that $p$ and $q$ are *buddies*, such that $p$ is the *climbing buddy*, $q$ is the *waiting buddy* and $n$ is their *meeting point*. The climbing buddy combines the requests of both buddies (stores at $n$ a union of the requests that are stored in both child nodes of $n$) and climbs to bring corresponding responses for both buddies. The waiting buddy waits at the meeting point for the responses.

Process $p$ may reach a node $n_1$ colored by some process $q_1$, but after $q_1$ had already finished forwarding requests into $n_1$. In that case, we say that $p$ *suffers*, since it has to wait at $n_1$ until $q_1$ returns and *uncolors* $n_1$ (as described later). Only then $p$ can color $n_1$ and proceed.

Eventually (assuming no faults occur), a union of requests that includes $p$'s request is stored at the root by some process $r$, who produces corresponding responses. For example, responses for $x$ FAI requests are an interval of $x$ consecutive numbers starting with the current value of a counter at the root (the counter is increased accordingly by $x$).

Then, $r$ descends towards its own leaf, uncoloring the nodes along the way. At every node $n_2$ which is a meeting point of $r$ with some waiting buddy $q_2$ who gave $y$ requests, $r$ gives $y$ responses to $q_2$. Both buddies then descend in the same manner, propagating responses to the rest of the waiting processes. When returning to the leaf with a response, a process returns that response.

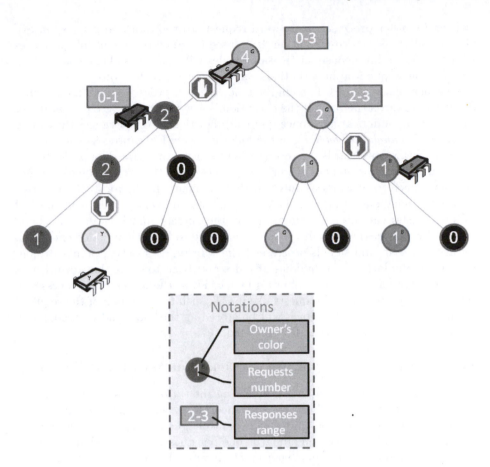

**Fig. 1.** Combining Tree

Figure 1 depicts an illustration of a CT example. In this example, three processes wait for responses at non-root nodes, and one process is at the root with a combination of four requests, propagating four responses (responses 0 and 1 to the left child; responses 2 and 3 to the right child).

## 3.2    Bounded Wait Combining [12]

CT is linearizable but not lock-free, and does not guarantee high throughput. BWC is a lock-free version of CT that guarantees high throughput in $\phi$-synchronous executions (assuming $\phi$ is known).

In CT, a deadlock can happen if a buddy $p$ waits for one of its buddies indefinitely. In BWC, waiting times are bounded as a function of $\phi$. Only if the execution is not $\phi$-synchronous, $p$'s bounded wait may expire, and then $p$ proceeds in an *asynchronous mode of operation* (continues independently, without any waiting for other processes, to forward requests to the root and get responses). Greenwald's two-handed-emulation [10] is used to synchronize multiple processes who operate simultaneously at the same nodes without locks. This makes BWC lock-free but sequential in executions that are not $\phi$-synchronous.

Another drawback of CT, is high latency when a process suffers many times in its way towards the root. If the execution is $\phi$-synchronous, BWC guarantees that a process suffers at most once (per call). For that, BWC arranges the calling processes in *bunches of buddies*, in the following way: First, processes just try to color their way up towards the root (without storing or forwarding requests yet, or waiting $\epsilon$ time for buddies). A process $r$ who colors the root becomes a *leader* of a new bunch (that contains only $r$ at that moment). In the root, $r$ waits a *root wait*, long enough as a function of $\phi$ to guarantee that every process that suffered (if such exists) catches a meeting point in the current bunch (waits on a node colored by a process who is in $r$'s bunch, or going to be). Then, $r$ descends back towards the leaf, and signals processes (who wait on $r$'s meeting points) to join $r$'s bunch. Similarly, these buddies of $r$ descend back towards their own leaves and signals other processes to join $r$'s bunch. From the leaves, the processes of the bunch start to forward requests like in CT, such that a union of the requests of all the processes in the bunch will be stored at the root, and corresponding responses will be produced and propagated.

### 3.3    Difficulties in Adapting to the Unknown Asynchrony Bound $\phi_E$

Recall that BWC assumes a known bound on the ratio of the speeds of different processes. A first, trivial (but probably incorrect) idea towards getting rid of this assumption seems to be (1) to make multiple guesses of that bound and (2) run multiple copies of BWC modularly, one per guess. We tried that direction, but, unfortunately, did not manage to make it work. Intuitively, this would have violated the linearizability or the counter semantics. In BWC, a process gives its request (for a counter value) to be combined with other requests. When the request is combined, it is not easy to locate it and cancel it. Hence, assume the process gives the request to one copy of BWC. Then other copies cannot handle the same request, otherwise the counter would have been increased several times. Moreover, if those hypothetical copies of BWC communicate with each other, the modularity is lost. We did use multiple guesses of the bound, but we had to do that in a less modular way. Moreover, we had to overcome difficulties in coordinating processes working on different guesses.

For example, we could not use two-handed-emulation as is done by BWC. As long as BWC worked in its synchronous modes, only one process can move requests into each node. In the new algorithm, various processes may access the same node, each working on a different guess. This could have caused starvation, had we stayed with the original two-handed-emulation. The *communal procedures* technique we developed to solve this problem (see Sect. 5) may be useful for future studies and is an independent contribution of this paper.

# 4    The New Algorithm

We describe our algorithm in two levels. First, we provide a high level description of the algorithm (Sect. 4.1). Then, we provide a more detailed description of the algorithm's modes and transitions (Sect. 4.2). For lack of space, detailed pseudo-code is provided in [13].

## 4.1    High Level Overview

In the new algorithm, a process $p$ holds a *guess* of $\phi_E$, according to which $p$ attempts to perform an algorithm that resembles BWC. If $p$ finds its guess insufficient, then $p$ *restarts* (doubles its guess and tries again). We call the progress of $p$ with a certain guess a *guess iteration*.

In the initial configuration, every node is uncolored and holds the dummy guess 1, while every process starts the algorithm with the minimum guess 2. When $p$ colors a node $n$, it updates $n$'s guess to $p$'s guess. To color a node $n$, a process $p$ must have a higher guess than $n$ (even if $n$ is already colored, $p$ paints over with its own color). As soon as a process $p_1$ detects that its color was removed from a node $n_1$, it restarts. As a result, requests or responses that belong to other buddies in $p_1$'s bunch are being left at $n_1$. We say that $p_1$'s bunch was *interrupted*.

First, a process injects a request to its leaf and begins the first guess iteration with a guess 2. Let $p$ be a process that begins a new guess iteration with a guess $\phi$. Denote $p$'s leaf by $L_p$ and $p$'s guess by $G_p$. First, $p$ checks whether a response for $p$'s request resides already at $L_p$ (as a result of a former guess iteration, if such exists). If so, then $p$ returns with that response. Otherwise, if $G_p$ is above our defined *guess threshold*, then $p$ shifts to an *asynchronous mode of operation*. In that mode, $p$ round trips from $L_p$ to the root and back (promoting requests and responses in its way towards their destination), until $p$ finds its response in $L_p$, and returns. Assume there is no response in $L_p$ and $\phi$ is lower than the guess threshold. Then, during the guess iteration, $p$ aims to join a bunch of buddies, who later combine together requests and propagate responses. If $\phi$ is an incorrect guess, then at some point, this bunch may be interrupted by some process with a higher guess. If $p$ detects such interruption, $p$ restarts. Additionally, $p$ bounds the number of steps it spends during the guess iteration with guess $\phi$. If $\phi$ is smaller than the actual $\phi_E$, then $p$ may reach the bound and restart. If $p$ restarts while some buddies wait for it, at some point they will restart too (either their bounded wait will expire or they will detect an interruption).

Let us describe the guess iteration progress, as long as $p$ does not restart from reasons mentioned above. To color nodes (see Sect. 3), $p$ climbs from $L_p$ towards the root. Here (unlike CT and BWC), $p$ succeeds to color a node $n$ only if $n$'s guess is lower than $G_p$. If so, $p$ stores its color and guess into $n$. Eventually, some process $r$ who guesses $\phi$ (either $p$ or some other process) colors the root with the guess $\phi$, becomes a bunch leader and begins a *root wait* (see Sect. 3.2). After the root wait, $r$ descends to *recruit* to $r$'s bunch processes who guess $\phi$ and wait on $r$'s colored nodes. The bunch is constructed recursively (every recruited process descends and recruits processes who guess $\phi$ and wait on its colored nodes). At every node $m$ in which a process $q_1$ recruited a process $q_2$, we call $m$ a *meeting point*, $q_1$ a *climbing buddy* and $q_2$ a *waiting buddy*.

Both in the case that $p$ is the bunch leader itself, and the case that $p$ was recruited by some other process, $p$ descends back to $L_p$. From its leaf, $p$ climbs up its colored nodes. At every non-leaf node $n$ along its way, $p$ forwards new requests from the children of $n$ in $p$'s bunch. If $n$ is a meeting point of $p$ with a waiting buddy $q$, and $p$ reaches $n$ before $q$, then $p$ waits until $q$ completes forwarding requests to their meeting point. Then, $q$ waits at $n$ for responses to all of its pending requests.

Eventually, the bunch leader $r$ reaches the root again and forwards $y$ new combined requests. The counter is increased from (say) $z$ to $z+y$, and $y$ responses (from $z$ to $z + y - 1$) are produced.

Then, $r$ descends towards $L_r$. At every non-leaf node $n$ along its way, $r$ propagates new responses to $n$'s children. In addition, $r$ uncolors $n$ and decrements $n$'s guess, to allow recoloring of $n$ later with the guess $\phi$. Recursively, every waiting buddy in $r$'s bunch (including $p$ if $p \neq r$) gets responses for all the pending requests in its meeting point, and proceeds similarly towards its leaf. Finally, every process in $r$'s bunch reaches its leaf and returns with the single response that resides there.

## 4.2   A More Detailed Description of the Algorithm

The algorithm is composed of *modes*. Below, we describe each mode of the algorithm and specify the transition between modes. Figure 2 depicts the mode transitions diagram of our algorithm.

- *START* - Process $p$ injects a new request to $L_p$, sets its own guess $G_p$ to 2 and colors $L_p$. Then, $p$ enters the first guess iteration, at the *INIT_RANK* mode.

- *INIT_RANK* - First, $p$ updates $L_p$'s guess to $G_p$. Second, $p$ remembers $L_p$ as its *top colored node* (the highest node $p$ has colored in $p$'s current guess iteration).
  If there is already a response for $p$'s request in $L_p$, $p$ switches to *PROP_RESPONSES* to return with the response. Else, if $p$'s guess surpasses the guess threshold, $p$ shifts to an *asynchronous mode of operation* (switches to *ASYNC_UP*).
  Otherwise, $p$ switches to *SLOCK_UP*, to join (or lead) a bunch with the guess $G_p$.

- *SLOCK_UP* - In switching to this mode, $p$ sets its *timer*. At the beginning of every *SLOCK_UP* iteration, $p$ decreases its timer and performs two tests: A *timer verification* (that $p$'s timer has not expired) and a *color verification* (that $p$'s current node is still colored by $p$). If a test fails, $p$ *restarts* (doubles its guess and switches back to *INIT_RANK*).

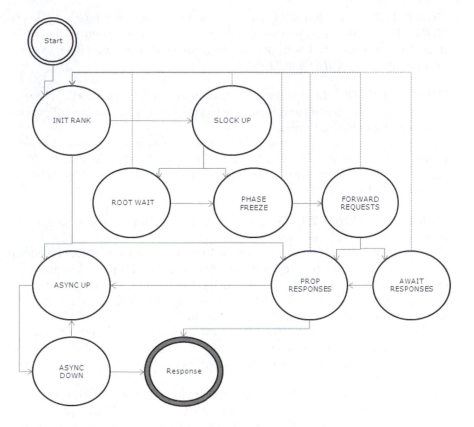

**Fig. 2.** Mode Transitions; Dashed arrows reflect a restart for a new guess iteration

Starting from $L_p$, at each non-root node $n$ that $p$ manages to color, $p$ stores two indications: First, that $n$ is ready to join a bunch with guess $G_p$. Second, that $p$ has not forwarded yet requests to $n$. Note, that even though $p$ has colored $n$ successfully, some other process (with a guess higher than $G_p$) may recolor $n$ at any time. Thus, $p$ updates $n$ only if $n$ is colored by $p$. Using DCAS, $p$ can detect an interruption (and restart if it finds any).

If $n$'s parent (say $np$) has a guess that is lower than $G_p$, then $p$ colors $np$ with $G_p$, updates $p$'s top colored node to be $np$ and ascends to $np$. Otherwise, if a signal (an indication that $p$ may join a bunch that is currently being formed) is written at $n$, then $p$ (a waiting buddy at the meeting point $n$) switches to *PHASE_FREEZE*.

If $p$ colors the root, $p$ becomes a new bunch leader. To give other processes (who guess $G_p$ too) time to climb up to a node that will be recruited to that bunch, $p$ switches to *ROOT_WAIT*.

If it did not switch to any other mode, $p$ stays for another *SLOCK_UP* iteration.

- *ROOT_WAIT* - In switching to this mode, $p$ sets its timer. At every *ROOT_WAIT* iteration, $p$ ensures that the root has not been recolored (if it is, then $p$ restarts). The root wait is completed when $p$'s timer expires, and $p$ switches to *PHASE_FREEZE*.

- *PHASE_FREEZE* - In switching to this mode, $p$ sets its timer. At every iteration, $p$ decreases its timer and performs color and timer verifications (like in the *SLOCK_UP* mode).
  Descending down from $p$'s top colored node towards $L_p$, at each non-leaf node $n$ along the way, $p$ signals each child $c$ of $n$ that is waiting for a signal to join a bunch with guess $G_p$. When $p$ reaches $L_p$, it switches to *FORWARD_REQUESTS*.

- *FORWARD_REQUESTS* - At every iteration, $p$ decrements its timer and performs color and timer verifications.
  At each non-leaf node $n$ along $p$'s way from $L_p$ to $p$'s top colored node, $p$ forwards requests up to $n$ from every child $c$ of $n$ who got a signal of $p$'s current bunch. If $c$ is colored by a waiting buddy $q$ of $p$, then $p$ waits at $n$ until $q$ indicates at $c$ the completion of requests forwarding to $c$. Similarly, $p$ indicates at $n$ when it completes forwarding requests to $n$.
  To handle simultaneous writes of different processes, this critical part is executed using a new technique (see Sect. 5). This technique, called *communal procedures*, guarantees successful completion after some known constant number of steps.
  At $p$'s *top colored* node, there may be two cases: (1) $p$ is at the root with responses for its bunch, and switches to *PROP_RESPONSES*; (2) $p$ is not at the root and switches to *AWAIT_RESPONSES* (to wait for responses from $p$'s climbing buddy).

- *AWAIT_RESPONSES* - In switching to this mode, $p$ sets its timer. At every iteration, $p$ decrements its timer and performs color and timer verifications. If $p$ gets all the required responses at its top colored node, $p$ switches to *PROP_RESPONSES*.

- *PROP_RESPONSES* - In switching to this mode, $p$ sets its timer. At every iteration, $p$ decrements its timer and performs color and timer verifications. Starting from its top colored node, at each non-leaf node $n$ in $p$'s way to $L_p$, $p$ propagates from $n$ to $n$'s children all the responses for the requests (again, using a communal procedure). In addition, $p$ uncolors $n$ and decrements $n$'s guess.
  Finally, $p$ reaches $L_p$ and returns with the single response that resides in it.

The *asynchronous mode of operation* consists of the following two modes:
- *ASYNC_UP* - $p$ forwards requests in the path from $L_p$ towards the root without waiting for any other process. At each non-leaf node $n$ along the way, $p$ forwards the requests of both $n$'s children (using a communal procedure). At the root, $p$ switches to *ASYNC_DOWN*.

– *ASYNC_DOWN* - $p$ propagates responses down the path from the root to $L_p$. Additionally, $p$ uncolors and resets the guess of the nodes along the way, thus enabling future operations to avoid using high guesses (to stabilize the system, if it becomes more synchronous later on).

During the asynchronous mode of operation new pending requests may arrive to a node at each moment. Therefore, unlike in *PROP_RESPONSES*, $p$ does not wait for responses to *all* of the pending requests at the node. Instead, $p$ descends from a node after responding to up to two pending requests elements. When $p$ reaches $L_p$, if a response resides at $L_p$, $p$ returns with that response. Otherwise, $p$ switches back to *ASYNC_UP*.

## 5   Communal Procedures

Different processes who call a procedure to forward requests to (or propagate responses from) the same node may try to write to the same variables simultaneously. Unlike in BWC, this may happen in the new algorithm not only when some process switches to an asynchronous mode of operation. Had we used two-handed-emulation as in BWC (see Sect. 3.2), starvation could happen and we could not guarantee wait-freedom and high throughput. The new algorithm requires that each call to such procedures is completed successfully after some finite number of steps (bounded by a known-in-advance constant). We present a technique that achieves all that, for some common kind of concurrent procedures we call *communal procedures*.

A procedure $f$ is *communal* if when called by multiple processes simultaneously, a single complete instance of $f$ yields the desired output for all the simultaneous calls. For example, a procedure to forward requests into a node from its children can be implemented as a communal procedure, since a single instance of this procedure can forward requests for all of the simultaneous calls.

A process executes a communal procedure $f$ according to $f$'s *state object*, that simply holds a *step number* variable (initialized to 0). The statements of $f$ are translated into numbered steps (starting from step 1). The state object of $f$ indicates the last step that was executed in the current instance of $f$. DCAS is used to execute the following step: one DCAS "hand" promotes the step number variable and the other "hand" performs the corresponding step. The DCAS ensures that only one of the simultaneous callers succeeds. When $f$'s instance reaches its last step, to start a new instance, a process allocates a fresh state object (with a step number 0), and tries to update $f$'s reference at the node to point to the new object (using DCAS that promotes the step number, such that only one update succeeds). Garbage collection (or equivalent techniques such as lock-free reference counting or dynamic lock-free objects, see [10]) guarantees that this object will not be recycled while any reference to it exists, and hence will be a unique token describing the particular procedure instance.

Many processes may wish to execute $f$ at the same time. Every process helps executing $f$'s instances only until it guarantees a full execution of at least one instance of $f$. For example, assume process $p$ calls $f$ at a node $n$, and finds

$f$'s instance at $n$ at step number $j$. If $j = 0$, then $p$ helps to execute $f$ only until $p$ detects a new instance of $f$ at $n$. Otherwise ($j > 0$), $p$ stays for another round to execute the steps of the new instance too, but only until $p$ detects a third instance of $f$ at $n$ (because this ensures that the second instance was fully executed during $p$'s call). Note, that the total number of primitives executed by $p$ is bounded by twice the number of steps in $f$.

We use two communal procedures at every node (detailed pseudo-code appears in [13]): (1) ForwardRequestsFromChildren() - specifically for the FAI operation, a process $p$ forwards requests to some node $n$ from $n$'s child $c$, by computing the number of additional requests to forward, and adding this number to $n$'s counter. Later, the propagation of the responses for the requests must be in the order of requests arrivals (to preserve linearizability). Therefore, $p$ enqueues an appropriate pending requests element to $n$'s pending requests queue to record this order. If $n$ is the root, then $p$ also enqueues a responses element to $n$'s responses queue. Specifically for the FAI operation, the responses element consists of the range of responses for the new requests (starting with the previous value of the root's requests counter). (2) PropagateResponsesToChildren() - a process $p$ propagates responses from some node $n$ to $n$'s child $c$, according to a pending requests element from $n$'s pending requests queue. Assume that such element requires $x$ responses to be propagated down to $c$. Process $p$ slices a range of $x$ responses from $n$'s responses queue and enqueues that range into $c$'s responses queue.

## 6  Algorithm Analysis Outline

For lack of space, the complete analysis is provided in [13]. First, let us define some additional terms. An *E-interval* is a sub-sequence of $E$. We say that $[t_0, t_1]$ is a *FAI-segment of $p$*, if it is an $E$-interval such that $p$ starts a FAI operation at $t_0$ and returns at $t_1$. Let $m$ be a mode of the algorithm. We say that $[t_0, t_1]$ is an *m-segment of $p$*, if $p$ switches into mode $m$ at $t_0$ and stays in this mode until it switches out of $m$ at $t_1$ (or returns with a response). An *ASYNC-segment of $p$* starts when $p$ switches from *INIT_RANK* to *ASYNC_UP* and ends when $p$ returns with a response (or infinite if $p$ does not return - though we prove this cannot happen). The outline of the analysis follows.

- Linearizability - First, we prove (Lemma 1) by induction that there are no overflows in the requests/responses queues in the nodes. This ensures that there are no interfering writes between a communal procedure who enqueues into such a queue and a communal procedure who dequeues from it. With that, we prove (Lemma 2) that no field changes its value between the moments a communal procedure reads from it and writes into it. Then, we prove (Lemma 3) that the critical procedures forward requests and propagate responses as expected. With that, we prove (Lemma 4) that if a process ends the algorithm, it returns with a response. Finally, we prove (Theorem 1) the linearizability of our algorithm, and that it preserves the counter semantics.

– Wait-freedom - First, we bound (Lemma 5) the number of steps performed by a process during a single iteration at *any* mode. Second, we prove (Lemmas 6-9) that an *ASYNC*-segment of execution of a process is finite and its length is a function of $N$. Finally, we prove (Theorem 2) the wait-freedom of our algorithm, by proving that a process $p$ either completes in a finite time or switches to the asynchronous mode of operation (for which, Lemma 6 proved the execution is finite).

– Scalability - Specifically, we prove that in an $N$-process execution $E$, our algorithm achieves the throughput of $\Omega(N/((\phi_E \log \phi_E)^2 \log N))$. If $\phi_E$ is greater or equal to the guess threshold, the proof is relatively straightforward. Otherwise (Assumption 1: $\phi_E$ is lower than the guess threshold), we prove that a process returns with a response after at most $\lceil \log_2 \phi_E \rceil$ guess iterations. For that, we prove (Lemma 23) that a process $p$ with a correct guess $G$ does not restart. Specifically, $p$ succeeds in every color and timer verification during that guess iteration, until it returns with a response. A color verification may fail only if some other process has a guess higher than $G$. Thus, we prove (Lemmas 10-21) $p$'s success at every timer verification, assuming no process guessed higher than $G$ (Invariant 1), and with that we prove (Lemma 22) that no process can be the first to guess higher than $G$ (i.e. Invariant 1 holds under Assumption 1). We bound the number of primitives $p$ applies during a FAI-segment (Lemma 24), and finally, prove (Theorem 3) the above throughput whether or not Assumption 1 holds.

– Additional properties - We prove (Theorem 4) that our algorithm achieves high throughput even after a constant number of process failures. When some process $p$ who guessed $G$ fails, it has no effect on processes who guessed higher than $G$ (because they don't wait on $p$'s colored nodes, and are not $p$'s bunch buddies). At the worst case, if $G$ is a correct guess, then processes who guessed $G$ would have to restart with guess $G + 1$, which is also a correct rank.

## 7 Conclusions

In this paper we presented the first linearizable and wait-free shared counter algorithm that achieves high throughput without any a-priori knowledge of the system's asynchrony level.

A communal procedure is a procedure that can be executed concurrently by multiple processes, such that a single complete execution of the procedure is sufficient for each participating process. Another contribution of our paper, which we believe is of independent interest, is the definition and efficient implementation of communal procedures.

The algorithm as described does not adapt when the level of asynchrony decreases during the execution. Such an adaptive version was developed, but is not presented here because of its considerable additional complication. One direction for future research, is to make the throughput of the algorithm adaptive to the number of participating processes. Another interesting research direction is to establish corresponding lower bounds on the throughput of algorithms with similar properties.

# References

1. Aspnes, J., Herlihy, M., Shavit, N.: Counting networks. STOC 23, 348–358 (1991)
2. Chandra, T.D., Jayanti, P., Tan, K.: A Polylog Time Wait-Free Construction for Closed Objects. PODC 17, 287–296 (1998)
3. Dolev, D., Dwork, C., Stockmeyer, L.: On the Minimal Synchronism Needed for Distributed Consensus. JACM 34, 77–97 (1987)
4. Dwork, C., Herlihy, M., Waarts, O.: Contention in shared memory algorithms. STOC 25, 174–183 (1993)
5. Dwork, C., Lynch, N., Stockmeyer, L.: Consensus in the Presence of Partial Synchrony. JACM 35, 288–323 (1988)
6. Ellen, F., Lev, Y., Luchangco, V., Moir, M.: SNZI: scalable NonZero indicators. PODC 26, 13–22 (2007)
7. Goodman, J.R., Vernon, M.K., Woest, P.J.: Efficient Synchronization Primitives for Large-Scale Cache-Coherent Multiprocessors. ASPLOS 3, 64–75 (1989)
8. Gottlieb, A., Grishman, R., Kruskal, C.P., McAuliffe, K.P., Rudolph, L., Snir, M.: The NYU ultracomputer - designing a MIMD, shared-memory parallel machine. ISCA 9, 239–254 (1998)
9. Gray, C., Cheriton, D.: Leases: an efficient fault-tolerant mechanism for distributed file cache consistency. SOSP 12, 202–210 (1989)
10. Greenwald, M.: Two-Handed Emulation: How to build Non-Blocking implementations of Complex Data-Structures using DCAS. PODC 21, 260–269 (2002)
11. Ha, P.H., Papatriantafilou, M., Tsigas, P.: Self-tuning Reactive Distributed Trees for Counting and Balancing. OPODIS 8, 213–228 (2004)
12. Hendler, D., Kutten, S.: Constructing shared objects that are both robust and high-throughput. DISC 20, 428–442 (2006)
13. Hendler, D., Kutten, S., Michalak, E.: An Adaptive Technique for Constructing Robust and High-Throughput Shared Objects - Technical Report, http://ie.technion.ac.il/~kutten/hkm2010.pdf
14. Hendler, D., Shavit, N., Yerushalmi, L.: A scalable lock-free stack algorithm. SPAA 16, 206–215 (2004)
15. Herlihy, M.: Wait-free synchronization. TOPLAS 13, 124–149 (1991)
16. Herlihy, M., Shavit, N., Waarts, O.: Linearizable Counting Networks. FOCS 32, 526–535 (1991)
17. Herlihy, M., Wing, J.M.: Linearizability: a correctness condition for concurrent objects. TOPLAS 12, 463–492 (1990)
18. Moir, M., Nussbaum, D., Shalev, O., Shavit, N.: Using elimination to implement scalable and lock-free FIFO queues. SPAA 17, 253–262 (2005)
19. Shavit, N., Zemach, A.: Diffracting trees. SPAA 6, 167–176 (1994)

# Efficient Lock Free Privatization

Yehuda Afek[1], Hillel Avni[1], Dave Dice[2], and Nir Shavit[1,2]

[1] Tel-Aviv University, Tel-Aviv 69978, Israel
[2] Sun Labs at Oracle, 1 Network Drive, Burlington MA 01803-0903
hillel.avni@gmail.com

**Abstract.** Working on shared mutable data requires synchronization through barriers, locks or transactional memory mechanisms. To avoid this overhead a thread may privatize part of the data and work on it locally. By privatizing a data item a thread is guaranteed that it is the only one accessing this data, i.e., that it accesses the data item in exclusion.

The most robust and yet lock-free privatization algorithms, are lock-free reference counting (LFRC). These algorithms attach a counter to each node, which counts the number of references to the node. However, these counters are shared by all threads in the system and thus are contention prone, and must be updated with expensive atomic operations such as CAS.

We present a new privatization algorithm, Public Guard (PG); an algorithm which eliminates most of the contention of LFRC algorithms, while maintaining their robustness and non blocking nature. Our evaluation shows that PG improves performance by up to 50% in many work loads.

Another problematic issue with LFRC, that we address in this paper, is that a counter of a private node, may be accessed by a slow thread. This may prevent LFRC from freeing memory to the system. In another contribution of this paper we suggest a method with minimal overhead to allow LFRC to reclaim memory.

## 1 Introduction and Related Work

A privatization algorithm is a technique that allows the thread to make part of the data it accesses private, guaranteeing that no other thread can access it. In the literature, privatization is usually part of the memory management algorithm: threads privatize buffers before freeing them back into the allocatable memory pool.

However, in many situations it is important to provide privatization that is unrelated to memory management. For example, to allow threads to operate on chucks of data without the overhead of synchronization, or to allow them to move records between data structures. Thus, in its general form, we will say that

**Definition 1.** *Privatization is the process of thread $\pi$ verifying that the data of an object $O$ is not accessed nor accessible by any thread in the system except $\xi$.*

C. Lu, T. Masuzawa, and M. Mosbah (Eds.): OPODIS 2010, LNCS 6490, pp. 333–347, 2010.

Meta data that is related to a private node may be accessed by the system. This meta data may reside inside or outside the private node.

We identify three privatization algorithms:

1. **Guards:** In this group each thread has a set of designated pointers (guards) that are pointing to all the nodes it is about to access. In privatization the algorithm scans all the guards to verify that no thread accesses the privatized node. Examples are Herlihy et al. PTB [1] and Michael hazard pointers [2].
2. **Epoch:** Here when a thread wants to privatize a node it deposits it in a to-be private group of nodes. Then the algorithm verifies that all the threads in the system passed through a code segment where they are guaranteed not to access any shared data. Then it frees all the nodes in the group. Hart EBR and QSBR [3] are members of this group. Dice et al. [4] created a transactional memory flavor of it.
3. **LFRC:** Count all references to each node. Both global, from the heap, and local from thread local stacks. When the count drops to zero, the node may be privatized. Representatives are Valois [5] and Detlef [6].

There are also hybrids of the above. Herlihy et al. [1] introduces SLFRC which combines guards with LFRC, and Gidenstam [7] mixes LFRC for global references with guards for local ones. Our new PG merges the robustness of LFRC with the thread designation of guards.

Privatization serves for two purposes; a thread which ends up privatizing a node n, can either free the node, or work on it exclusively, i.e., it serves both for privatization and for memory management.

Before we continue to the pros and cons of the above types, we make the following definition:

**Definition 2.** *Privatization is lock-free if a thread which does not accesses node n can not stop other threads from privatizing n.*

All epoch-based algorithms we encountered (e.g., [4,3]) are blocking. I.e a thread that never exits its critical section prevents privatization across the system. So, while having extremely low overhead in the optimistic case, epoch-based may hang the system in the general case.

Guards are not scalable to arbitrary data structures, due to two reasons. First, there might be a lot of guards in the system making the process of privatizing awkward. The second is that in real applications it is very tricky to tell what guards are necessary.

LFRC algorithms are both lock-free and scalable. If a node is not used by a thread, that thread can not prevent the node privatization, which makes LFRC lock-free. It is scalable because no matter how many references exist in the system, the process of privatizing an item has the same overhead.

However, LFRC comes with a price:

1. It has memory overhead as a reference count is appended to each data item.
2. Updating the reference count requires a CAS, which is an expensive operation, that may fail in the presence of contention.

3. Unlike epoch-based and guard types, the meta data in LFRC is embedded in the node. Thus there is always a point where a thread must touch the node, through touching its meta data, before it knows that the node is private.

Our PG is an LFRC algorithm, but it replaces the simple integer reference counter with a structure of combined counters. The structure has an integer counter for global references, an array of local integer counters for local references, and a seqlock to facilitate the privatization process.

PG reduces the overhead of updating the reference count, for local references, by each thread counting independently in its own local counter. This reduction improves the performance as it prevents retries and eliminates slow atomic operations like CAS and F&A.

As mentioned above, privatization is used frequently in the context of memory management. There LFRC has a big advantage: it is automatic. A thread does not need to free a node explicitly, and when the total reference count is zero the node is freed. In real software, it can be difficult to tell from the application when a node can be freed, thus requirement of explicit freeing is likely to create both memory leaks and other bugs.

When we talk about memory management we actually talk about two different things:

1. **Recycling:** Maintain a group of nodes, with the same size and same structure that can be used and reused but not freed to the OS.
2. **Reclamation:** Free memory to the OS so it can be used for any purpose.

When using LFRC privatization the application can recycle nodes but is not allowed to reclaim them. The reason is, that in a reclaimed node the reference count too may be reused for other purposes. When a thread reads it, not knowing that the node was privatized, it gets meaningless result or even a segmentation fault. One way to solve this problem is Herlihy [1] SLFRC, where they protect accesses to meta data with a guard. However, this method has a price per node that grows with the number of threads. In this paper we introduce an algorithm to reclaim nodes that are privatized with LFRC, with overhead that is amortized over multiple nodes. Although this algorithm uses epoch-based construction, we show the amount of unreclaimed memory is bounded, unlike all other pure epoch-based algorithms.

## 2   PG in a Nutshell

Here is the idea of the PG algorithm. In PG each node has three fields that are used for privatization:

1. g_cnt is used to count references from the heap to the node.
2. l_cnt is an array of integers with an entry for each thread in the system. Each thread uses its entry in the array to count references from its local stack to the node.
3. inc is a seqlock [8], which is locked when the node is private.

PG has a separate method and API for updating global and local references. When a reference from the heap is added, a thread must use CAS to update g_cnt. However, in the much more common case, when a thread updates a local reference count, it just uses regular reads and writes on its entry in l_cnt, without any possible contention.

The l_cnt access pattern is similar to TLRW-bytelock's [9], and reuses its slot allocation algorithm (though we did not find it beneficial to use byte size counters).

If, upon decrementing any counter, a thread finds that the global count equals zero, it automatically tries to privatize the node (either because it tries to privatize the node or for memory management).

Privatization has the following steps after finding the global count equals zero:

1. inc is sampled and, if locked, the node is already private, so done.
2. if g_cnt $\neq$ 0, done.
3. l_cnt is scanned and, if any entry is $\neq$ 0, done.
4. Try to lock inc by incrementing it with a CAS from sampled value, if failed done.
5. Privatization has succeeded.

Upon completing the above sequence, exactly one thread succeeds in privatizing the node, which makes the node actually private (assuming all thread well behave and follow the rules as below).

PG works under the assumption that threads are well behaved. This implies three rules that all threads must follow:

1. Before accessing any node, protect it by incrementing its associated local reference count, and when done decrement that counter.
2. When adding or removing a global reference, update the global counter with the corresponding API.
3. Global count in PG serves also as a flag. If a thread does not want to privatize a node, nor to release it then it should verify that global counter $\neq$ 0.

If a thread wants to delete a node n from a data structure and then work on it not in private, or insert it to another structure, it must verify g_cnt $\neq$ 0. This can be done by calling PgInc(n) before deletion and PgDec(n) after insertion.

Figure 1 demonstrates a situation where two processors want to access the same object at the same time. In LFRC both need to update the same address, i.e. the reference count, before and after the access, thus they need to use CAS, which is resource consuming and may fail. In PG, each processor updates its relative l_cnt entry with a simple write, an operation that is cheaper and uninterrupted.

In summary, accessing an object for read or write is generally far more common than handling it for memory management. LFRC involves a CAS operation for each such access, which PG manages to avoid. This is the root of PG better performance.

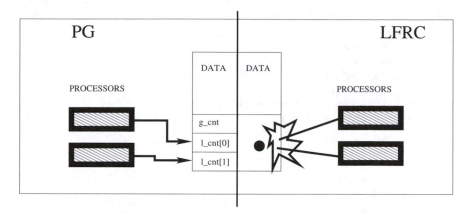

**Fig. 1.** PG vs. LFRC behavior for simple access

# 3    PG Privatization Algorithm

In this section we call Valois [5] algorithm simply LFRC, and let it represent all previous LFRC algorithms. To let the reader see where PG is better than LFRC, we explain LFRC and PG together.

Then we show how to let both LFRC and PG reclaim memory which they privatized, with minimal and amortizable overhead.

## 3.1    Overview of PG and LFRC

In Figure 2 we see the mandatory fields of a node which is used with LFRC or PG. LFRC has one field which holds both the total reference count and a claim bit, which is an indication to the status of the node, i.e., whether it is private. In PG this field is broken into three separate fields. Local counter (l_cnt) per thread for local references, a global counter (g_cnt) for global references and incarnation (inc). Incarnation is a seqlock, i.e., a combination of a counter and a lock. When incarnation is locked ($=$ odd value), the node is private. When a node is shared again, incarnation field is unlocked ($=$ even value) and incremented in one operation.

| Field | LFRC | PG | Description |
|-------|------|----|-------------|
| data | V | V | Application dependant. |
| links | V | V | References to other nodes. |
| p_cnt | V | X | Counts references and includes an indication the node is private. |
| l_cnt | X | V | An array of local reference counters per thread. |
| g_cnt | X | V | A counter of global references. |
| inc | X | V | Incarnation: a seqlock, that is used in the privatization process. |

**Fig. 2.** LFRC and PG node structures

```
int LfrcDecrementAndTAS(int *cnt)
LD01:     repeat
LD02:          old = *cnt
LD03:          new = old - 2
LD04:          if(new == 0)
LD05:              new = 1
LD06:     until (CAS(cnt, old,new) == TRUE)
LD07:     return ((old - new) & 1)
```

**Fig. 3.** LFRC decrement reference count of a node

```
void LfrcRelease(node *n)
LR01:     if(n==NULL)
LR02:          return
LR03:     if (LfrcDecrementAndTAS(&n->p_cnt)==0)
LR04:          return
LR05:     foreach (L in n.links)
LR06:          LfrcRelease(L)
LR07:     Privatize (n)
```

**Fig. 4.** LfrcRelease privatizes a node and decrements all its links reference counts

**Decrement Reference Count and Free.** When LFRC or PG remove a local or global reference, they decrement the reference count. LFRC has one reference count and PG maintains one for global references and an array of reference counters per thread in each node for local references. When reference count, or in PG, when global reference count, drops to zero the algorithm tries to privatize the node. That is why this section explains decrement and privatization together.

Figure 3 shows the actual decrement of the reference count of a node in LFRC. If the reference count drops to zero the LFRC tries to privatize it by setting the claim bit. Both decrementing and privatizing are done with a CAS in line LD06 of `LfrcDecrementAndTAS`. `LfrcDecrementAndTAS` is called from `LfrcRelease` in Figure 4.

PG uses two functions to perform reference count decrement. `PgDoneAccess` in Figure 5 decrements the local counter by a simple write. As most reference are local, and as LFRC uses CAS to decrement local count, this is a place where PG avoids many CAS. In Figure 6 `PgDec` uses CAS to decrement the global reference count of a node.

```
void PgDoneAccess(node *n)
PD01:     If(n==NULL)
PD02:          return
PD03:     decrement(n->l_cnt[SELF_ID])
PD04:     PgTryPrivatize(n)
```

**Fig. 5.** PG remove a local reference

```
void PgDec(node *n)
PC01:     if(n==NULL)
PC02:         return
PC03:     while (!CAS(n->g_cnt, n->g_cnt, (n->g_cnt-1)))
PC04:     PgTryPrivatize(n)
```

**Fig. 6.** PG PgDec removes a global reference and then tries to privatize the node

```
void PgTryPrivatize(node *n)
PT01:     retry:
PT02:         cur = n->inc
PT03:         if locked(cur)
PT04:             return
PT05:         if (n->g_cnt == 0)
PT06:             for(id = 1...MAXID)
PT07:                 if(n->l_cnt[id])
PT08:                     return
PT09:             if(CAS(&n->inc, cur, cur + 1))
PT10:                 tmp = n
PT11:                 Privatize n
PT12:                 foreach l in tmp->links
PT13:                     PgDec(l)
```

**Fig. 7.** PG Try to privatize the node

Both functions are calling `PgTryPrivatize` from Figure 7 which checks (PT05 and LD04) that global count and local count (PT06-PT08) are zero and only than tries to privatize by locking the incarnation field of the node with a CAS (PT09). Thus privatizing a node both in LFRC and PG involves a CAS, and incurs similar overheads.

**Safe Read.** An application that wants to access a node, must read its pointer with a safe read. This operation increments the reference count of the node and then verifies the node is still pointed from where it was pointed before.

In line LS05 of Figure 8, `LfrcSafeRead` uses `AtomicAdd` to increment a node reference count.

```
node *LfrcSafeRead(node **n)
LS01:     forever
LS02:         q = *n
LS03:         if (q == NULL)
LS04:             return NULL
LS05:         AtomicAdd(q->p_cnt, 2)
LS06:         if(q == *n)
LS07:             return q
LS08:         LfrcRelease(q)
```

**Fig. 8.** LFRC Safe read function

```
node *PgSafeRead(node **n)
PS01:    forever
PS02:        q = *n
PS03:        if (q == NULL)
PS04:            return NULL
PS05:        (q->l_cnt[ID])++
PS06:        if(q == *n)
PS07:            return q
PS08:        PgDoneAccess(q)
```

**Fig. 9.** PG Safe read function

PgSafeRead from Figure 9 uses a simple increment function (PS05) to add a local reference. As safe read is a very frequent operation this simpler command saves work for PgSafeRead.

**New Node Allocation.** The allocation operation returns a node from a free nodes pool, which might be global or local and hands it to the application. If the pool is global the algorithms must be aware of ABA risks, which makes the function less trivial. In this section we focus only on allocating from a global pool.

```
node *LfrcNew()
LN01:    forever
LN02:        p = SafeRead(&free_list)
LN03:        if (p == NULL)
LN04:            Assert(Out of memory)
LN05:        if(CAS(&free_list, p, p->next) == TRUE)
LN06:            ClearLowestBit(&p->p_cnt)
LN07:            return p
LN08:        LfrcRelease(p)
```

**Fig. 10.** LFRC allocation from a global pool

In line LN02 in Figure 10, LfrcNew reads the first object in the free list with a SafeRead. The reason is that after p was acquired with a SafeRead, p→next can not change as long as it is in the free pool, and p can not be freed again after it was allocated. If p was read without LfrcSafeRead / PgSafeRead, it could have been allocated and freed again and have a new p→next and now the old p→next might be pointing to a currently allocated node. This would cause a double allocation which is erroneous.

Here is the scenario, in a more formal way, how threads $T_1$ and $T_2$ allocate a node twice if not using SafeRead:

1. $T_1$ reads p from freelist and $p_{T_1}$=p→next.
2. $T_2$ allocates p .
3. $T_2$ allocates $p_{T_1}$.
4. $T_2$ frees p again with a new $p_{T_2}$=p→next.

5. $T_1$ arrives at line LN05 of LfrcNew and replaces freelist with $p_{T_1}$ which has been allocated already by $T_2$.
6. $T_2$ allocates $p_{T_1}$ again.

In line LR07 of LfrcRelease function and line PT11 of PgTryPrivatize a node may be recycled into a global shared pool or privatized, i.e., accessed without synchronization by the thread who holds a reference to it. As we recall in LFRC New had to call SafeRead so the node will not be freed again during a critical part of the function. In PG this is more complicated as a thread checks both g_cnt and l_cnt and incarnation before freeing a buffer so when extracting n from a free pool n→inc must be locked, at least one of n→l_cnt entries and n→g_cnt must be non zero.

Figure 11 is the PgNew which allocates from a global shared pool.

```
node *PgNew()
PN01:    forever
PN02:        q = PgSafeRead(&free_list)
PN03:        PgInc(q)
PN04:        if(p == NULL)
PN05:            Assert(Out of memory)
PN06:        if(CAS(&free_list, p, p->next) == TRUE)
PN07:            increment(p->inc)
PN08:            PgDoneAccess(p)
PN09:            return p
PN10:        PgDoneAccess(p)
PN11:        PgDec(p)
```

**Fig. 11.** PG allocation from a global pool

The function is using PgSafeRead in line PN02 to read from the free list. Then, in line PN03 it increments the global count as well. If it manages to extract the node in PN06 it returns it decrements the local count in PN08 and returns the new node in PN09. Otherwise it decrements the node counters in PN10-PN11 and retries.

## 3.2   PG and LFRC Usage for Reclamation

As explained in section 1 LFRC and PG can be used to recycle memory but not for reclamation (i.e., recycling but not reclamation). The following algorithm maintain the lock-free property of PG and LFRC, but gives them the ability to reclaim memory. To reclaim memory we need to know if there is a sleeping thread that is about to access a reference count of a free node. This can happen only in the SafeRead function, before checking if a node is still pointed from the same place. In order to reclaim, the system must know there is no thread in that section, so we make it as swift as possible and wrap it with a local seqlock (rec_epo]) which is incremented when we enter or exit that section. We also add the field rec_epo, as shown in Figure 12, in LFRC and PG nodes.

| Field | LFRC | PG | Description |
|-------|------|-----|-------------|
| rec_epo | V | V | Reclamation epoch: a seqlock, that is used for reclamation. |

**Fig. 12.** LFRC and PG with reclamation

When a thread has too many nodes in its local pool it scans all the seqlocks and waits for all threads to get out of that section. For this solution we modify `SafeRead` as depicted in Figure 13. In SR05, just before incrementing the reference count, the local rec_epo seqlock is incremented to locked state. After increment completion (SR10) or failure (SR08), rec_epo is incremented locally to unlocked state. Our tests show this method has literally the same performance as `SafeRead` that does not allow reclamation.

```
node *SafeRead(node **n)
SR01:     forever
SR02:         q = *n
SR03:         if (q == NULL)
SR04:             return NULL
SR05:         Lock rec_epo
SR06:         increment reference_count
SR07:         if(q == *n)
SR08:             Unlock rec_epo
SR09:             return q
SR10:         Unlock rec_epo
SR11:         decrement reference count
```

**Fig. 13.** Generic safe read function with reclamation for LFRC and PG

The spirit of this technique is like Herlihy's which uses a guard for that critical section, but the same absolute overhead which they have per node is here per all freed nodes.

In order to verify memory consumption is limited the free function counts the number, N, of nodes it freed locally. If N equals a threshold H, the thread takes a snapshot $S_1$ of all critical counts. If N>H the node is freed to a global pool and another snapshot $S_2$ is taken. If all threads which were in the critical section in $S_1$ made progress in $S_2$, all the local pool is reclaimed. $S_2$ is created only once, and if there is a thread which made no progress the scan stops and continues from where it stopped in the next free operation.

**Lemma 1.** *Assume J is the number of threads in the system, H is the threshold and M is maximal number of simultaneous used nodes, maximal unreclaimed nodes is less than $M+((H+1)*(J-1))$*

*Proof.* T allocates a node if its local pool and the global pool are empty. So the maximal number each other thread may hold is H+1, i.e., threshold and another one it privatized but did not free into global pool. So we have J-1 threads which hold H+1 nodes each and one which holds 0 nodes.

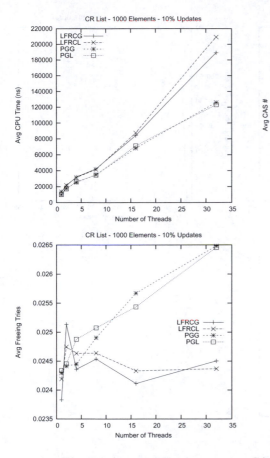

**Fig. 14.** Average time vs. average CAS on various threads number

## 4   Evaluation

We tested our algorithms on 8 processors Intel Core i7 Processor I7-920, running 64-bit RedHat Linux.

The graphs demonstrate the performance of two algorithms, PG and LFRC [5]. Both algorithms are run with local buffer pools (PGL and LFRCL) and with global buffer pool (PGG and LFRCG).

For each configuration/test (each in a separate figure) we generate three graphs. The first is the average time in nano-seconds taken per one operation, over 5 runs. The second is the average number of CAS operations during that time, which explains the overhead of LFRC. The third is the average number of times in the operation that it tries to privatize the node i.e., the number of times PG was in line PT09 of `PgTryPrivatize` or the number of times LFRC

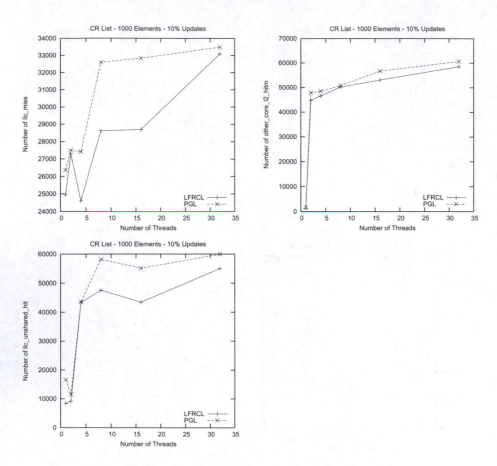

**Fig. 15.** Number of cache misses various threads number

was in line LD06 of `LfrcDecrementAndTAS`. This graph is the overhead created by PG. PG tries to free a node when the global references drop to zero, but it is not aware of local references. Thus it does more retries than LFRC which slows its operation.

The operations are on a list where insert and delete operations take a lock, while search operations are lock-free. In this test a lock-free search manipulates local references proportionally to the length of the list, but an insert or delete do exactly one local and one global reference. We show three tests:

1. Figure 14 demonstrates various thread numbers with 1K elements and 10% updates. Here, the amount of CAS is proportional to the number of operations and thus PG is better by a constant coefficient. The number of free retries is higher in PG but its absolute number is too small too make a difference.

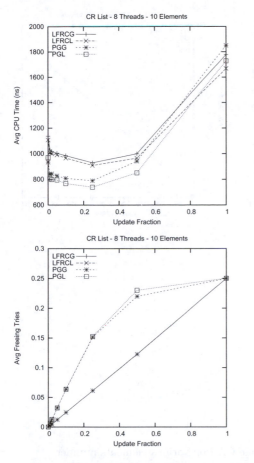

**Fig. 16.** Average time vs. average CAS on various update rates

For this test we added, in figure 15 cache misses as counted by VTune$^{TM}$ Performance Analyzer 9.1 for Linux. Although PG is better than LFRC for this work load on all thread counts, it has either equal or slightly more cache misses. This makes sense as everywhere LFRC gets a miss PG suffers one as well, while PG touches more memory locations.

2. In figure 16 there are 8 threads with 10 elements and various update rates. Up to about 80% updates we can see PG with local pools is best, and there LFRC with local pools passes it. The reason is that as the number of PG freeing tries grows and overshadows LFRC CAS number, which drops as the number of searches goes down.

3. Figure 17 shows 8 threads with 10% updates and various elements number. We see in 10% updates, the number of CAS, which in LFRC is proportional to the duration of the transaction, makes PG always better. The number of free tries is very small except in very small lists where PG is not better than LFRC.

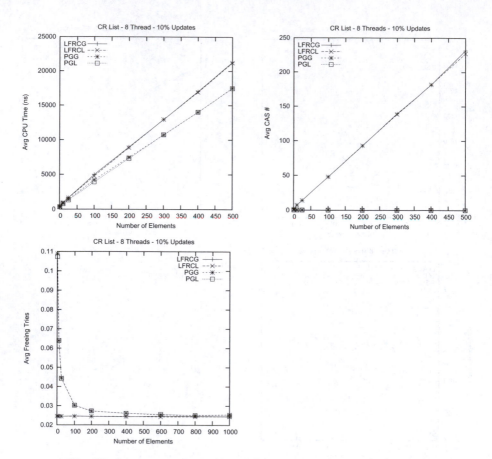

**Fig. 17.** Average time vs. average CAS on various elements number

## 5    Conclusion

In this paper we introduced PG, the first LFRC which does not use CAS [5] nor HP [7] for local references. We showed it has superior performance compared with previous LFRC although it has slightly more cache misses. This fact emphasizes that PG advantage is in the algorithmic reduction in the number of CAS operations. PG outperforms LFRC in all workloads that are not dominated by allocation and free operations.

Another contribution is a method to allow LFRC and PG free memory to the system, and bound the maximal amount of allocated memory. This is the first time LFRC can free memory without HP[7,1] that has significant overhead or DCAS [6] which is not common in today hardware.

## Acknowledgements

This paper was supported by European Union grant FP7-ICT-2007-1 (project VELOX).

# References

1. Herlihy, M., Luchangco, V., Martin, P., Moir, M.: Nonblocking memory management support for dynamic-sized data structures. ACM Trans. Comput. Syst. 23, 146–196 (2005)
2. Michael, M.M.: Hazard pointers: Safe memory reclamation for lock-free objects. IEEE Trans. Parallel Distrib. Syst. 15(6), 491–504 (2004)
3. Hart, T., McKenney, P., Brown, A.: Making lockless synchronization fast: performance implications of memory reclamation. In: International Parallel and Distributed Processing Symposium, p. 4 (2006)
4. Dice, D., Matveev, A., Shavit, N.: Implicit privatization using private transactions. In: TRANSACT (2010)
5. Valois, J.D.: Lock-free linked lists using compare-and-swap. In: Proceedings of the Fourteenth Annual ACM Symposium on Principles of Distributed Computing, PODC 1995, pp. 214–222. ACM, New York (1995)
6. Detlefs, D.L., Martin, P.A., Moir, M., Steele Jr., G.L.: Lock-free reference counting. In: Proceedings of the Twentieth Annual ACM Symposium on Principles of Distributed Computing, PODC 2001, pp. 190–199. ACM, New York (2001)
7. Gidenstam, A., Papatriantafilou, M., Sundell, H., Tsigas, P.: Efficient and reliable lock-free memory reclamation based on reference counting. IEEE Trans. Parallel Distrib. Syst. 20(8), 1173–1187 (2009)
8. Lameter, C.: Effective synchronization on linux/numa systems. In: Gelato Federation Meeting (2005)
9. Dice, D., Shavit, N.: Tlrw: return of the read-write lock. In: Proceedings of the 4th ACM SIGPLAN Workshop on Transactional Computing (2009)

# A Competitive Analysis for Balanced Transactional Memory Workloads

Gokarna Sharma and Costas Busch

Department of Computer Science, Louisiana State University
Baton Rouge, LA 70803, USA
{gokarna,busch}@csc.lsu.edu

**Abstract.** We consider transactional memory contention management in the context of *balanced workloads*, where if a transaction is writing, the number of write operations it performs is a constant fraction of its total reads and writes. We explore the theoretical performance boundaries of contention management in balanced workloads from the worst-case perspective by presenting and analyzing two new polynomial time contention management algorithms. The first algorithm Clairvoyant is $O(\sqrt{s})$-competitive, where $s$ is the number of shared resources. This algorithm depends on explicitly knowing the conflict graph. The second algorithm Non-Clairvoyant is $O(\sqrt{s} \cdot \log n)$-competitive, with high probability, which is only a $O(\log n)$ factor worse, but does not require knowledge of the conflict graph, where $n$ is the number of transactions. Both of these algorithms are greedy. We also prove that the performance of Clairvoyant is tight, since there is no polynomial time contention management algorithm that is better than $O((\sqrt{s})^{1-\epsilon})$-competitive for any constant $\epsilon > 0$, unless NP$\subseteq$ZPP. To our knowledge, these results are significant improvements over the best previously known $O(s)$ competitive ratio bound.

## 1 Introduction

The ability of multi-core architectures to increase application performance depends on maximizing the utilization of the computing resources provided by them and using multiple threads within applications. These architectures present both an opportunity and challenge for multi-threaded software. The opportunity is that threads will be available to an unprecedented degree, and the challenge is that more programmers will be exposed to concurrency related synchronization problems that until now were of concern only to a selected few. Writing concurrent programs is a non-trivial task because of the complexity of ensuring proper synchronization. Conventional lock based synchronization (i.e., mutual exclusion) suffers from well known limitations, so researchers considered non-blocking transactions as an alternative. Herlihy and Moss [16] proposed Transactional Memory (TM), as an alternative implementation of mutual exclusion, which avoids many of the drawbacks of locks, e.g., deadlock, reliance on the programmer to associate shared data with locks, priority inversion, and failures of threads while holding locks. Shavit and Touitou [24] extended this idea

C. Lu, T. Masuzawa, and M. Mosbah (Eds.): OPODIS 2010, LNCS 6490, pp. 348–363, 2010.

to Software-only Transactional Memory (STM) by proposing a novel software method for supporting flexible transactional programming of synchronization operations [15,12,13].

A transaction consists of a sequence of read and write operations to a set of shared system resources (e.g. shared memory locations). Transactions may conflict when they access the same shared resources. If a transaction $T$ discovers that it conflicts with another transaction $T'$ (because they share a common resource), it has two choices, it can give $T'$ a chance to commit by aborting itself, or it can proceed and commit by forcing $T'$ to abort; the aborted transaction then retries again until it eventually commits. To solve the transaction scheduling problem efficiently, each transaction consults with the *contention manager* module for which choice to make. Dynamic STM (DSTM) [15], proposed for dynamic-sized data structures, is the first STM implementation that uses a contention manager as an independent module to resolve conflicts between two transactions and ensure progress. Of particular interest are *greedy contention managers* where a transaction restarts immediately after every abort. As TM has been gaining attention, several (greedy) contention managers have been proposed in the literature [2,11,10,5,21,19]. which have been assessed formally and experimentally by specific benchmarks [20].

A major challenge in guaranteeing progress through transactional contention managers is to devise a policy which ensures that all transactions commit in the shortest possible time. The goal is to minimize the *makespan* which is defined as the duration from the start of the schedule, i.e., the time when the first transaction is issued, until all transactions commit. The makespan of the transactional scheduling algorithm can be compared to the makespan of an optimal off-line scheduling algorithm to provide a *competitive ratio*. The makespan and competitive ratio primarily depend on the *workload* − the set of transactions, along with their arrival times, duration, and resources they read and modify [3].

The performance of some of the contention managers has been analyzed formally in [3,2,11,10,21,23] (the detailed description is given in Section 1.2). The best known formal bound is provided in [2] where the authors give an $O(s)$ competitive ratio bound, where $s$ is the number of shared resources. When the number of resources $s$ increases, the performance degrades linearly. A difficulty in obtaining better competitive ratios is that the scheduling problem of $n$ concurrent transactions is directly related to the vertex coloring problem which is a hard problem to approximate [17]. A natural question which we address here is whether it is possible to obtain better competitive ratios. As we show below, it is indeed possible to obtain sub-linear competitive ratios for balanced transaction workloads.

### 1.1 Contributions

In this paper, we study contention management in the context of *balanced workloads* which have better performance potential for transactional memory. A balanced workload consists of a set of transactions in which each transaction has the following property: if the transaction performs write operations, then the number of writes it performs is a constant fraction of the total number of operations (read and writes) of the transaction. The *balancing ratio* $\beta$ expresses the ratio of write operations of a transaction to the overall operations of the transaction. The balancing ratio is bounded as $\frac{1}{s} \leq \beta \leq 1$, since a writing transaction writes to at least one resource. In balanced workloads $\beta = \Theta(1)$ for all the transactions which perform writes. Balanced workloads can also include read-only transactions, but we assume that there is at least one transaction that performs writes, since otherwise the scheduling problem is trivial (no conflicts).

Balanced transaction workloads represent interesting and practical transaction memory scheduling problems. For example balanced workloads represent the case where we have small sized transactions each accessing a small (constant) number of resources, where trivially $\beta = \Theta(1)$. Other interesting scenarios are transaction workloads which are write intensive, where transactions perform many writes, as for example in scientific computing applications where transactions have to update large arrays.

We present two new polynomial time contention management algorithms which are especially tailored for balanced workloads and analyze their theoretical performance boundaries from the worst-case perspective. The first algorithm, called Clairvoyant, is $O\left(\ell \cdot \sqrt{\frac{s}{\beta}}\right)$-competitive where $s$ is the number of shared resources, and $\ell$ expresses the logarithm ratio of the longest to shortest execution times of the transactions. (The transaction execution time is the time it needs to commit uninterrupted from the moment it starts.) For balanced transaction workloads where $\beta = \Theta(1)$, and when transaction execution times are close to each other, i.e. $\ell = O(1)$, Algorithm Clairvoyant is $O(\sqrt{s})$-competitive. This algorithm is greedy and has the pending commit property (where at least one transaction executes uninterrupted each time). However, it depends on assigning priorities to the transactions based on the explicit knowledge of the transaction conflict graph which evolves while the execution of the transactions progresses. It also assumes that each transaction knows how long is its execution time and how many resources it accesses.

The second algorithm, called Non-Clairvoyant, is $O\left(\ell \cdot \sqrt{\frac{s}{\beta}} \cdot \log n\right)$-competitive, with high probability (at least $1 - \frac{1}{n}$), where $n$ is the number of transactions concurrently executing in $n$ threads. For balanced transaction workloads, where $\beta = \Theta(1)$, and when transaction execution times are close to each other, i.e. $\ell = O(1)$, Algorithm Non-Clairvoyant is $O(\sqrt{s} \cdot \log n)$-competitive. This is only a $O(\log n)$ factor worse than Clairvoyant, but does not require explicit knowledge of the conflict graph. The algorithm is also greedy. This algorithm uses as a subroutine a variation of the RandomizedRounds scheduling algorithm by Schneider and Wattenhofer [21] which uses randomized priorities and doesn't require knowledge of the conflict graph.

The $O(\sqrt{s})$ bound of Algorithm Clairvoyant is actually tight. Through a reduction from the graph coloring problem, we show that it is impossible to approximate in polynomial time any transactional scheduling problem with $\beta = 1$ and $\ell = 1$ with a competitive ratio smaller than $O((\sqrt{s})^{1-\epsilon})$ for any constant $\epsilon > 0$, unless NP⊆ZPP. To our knowledge, these results are significant improvements over the best previously known bound of $O(s)$ for transactional memory contention managers. For general workloads (including non-balanced workloads), where transactions are equi-length ($\ell = O(1)$), our analysis gives $O(s)$ competitive worst case bound, since $\beta \geq 1/s$. This bound matches the best previously known bound of $O(s)$ for general workloads. The parametrization of $\beta$ that we provide gives more tradeoffs and flexibility for better scheduling performance, as depicted by the performance of our algorithms in balanced workloads.

## 1.2 Related Work

Almost 10 year after publishing the seminal paper [16] to introduce the new research area of transactional memory, Herlihy *et al.* [15] proposed Dynamic STM (DSTM) for dynamic-sized data structures. Later on, several other STM implementations have been proposed, such as TL2 [4], TinySTM [8], and RSTM [18] to name a few. Among them, DSTM is the first practical obstruction-free[1] implementation that seeks advice from the contention manager module to either wait or abort a transaction at the time of conflict.

Several contention managers have been proposed in STM and the performance of some of them has been analyzed formally in [3,2,11,10,21,23]. The first formal analysis of the performance of a contention manager is given by Guerraoui *et al.* [11] where they present the Greedy contention manager which decides in favor of older transactions using timestamps and achieves $O(s^2)$ competitive ratio. This bound holds for any algorithm which ensures the *pending commit* property (see Definition 1). Attiya *et al.* [2] improve the competitive ratio to $O(s)$, and prove a matching lower bound of $\Omega(s)$ for any deterministic *work-conserving* algorithm which schedules as many transactions as possible (by choosing a maximal independent set of transactions). The model in [2] is non-clairvoyant in the sense that it requires no prior knowledge about the transactions while they are executed.

Schneider and Wattenhofer [21] present a deterministic algorithm Commit-Bounds with competitive ratio $\Theta(s)$ and a randomized algorithm RandomizedRounds with makespan $O(C \log n)$ with high probability, for a set of $n$ transactions, where $C$ denotes the maximum number of conflicts among transactions (assuming unit execution time durations for transactions). Sharma *et al.* [23] study greedy contention managers for $M \times N$ *execution windows of transactions* with $M$ threads and $N$ transactions per thread and present and analyze two new randomized greedy contention management algorithms. Their first algorithm Offline-Greedy produces a schedule of length $O(\tau_{\max} \cdot (C + N \log(MN)))$ with high probability, where $\tau_{\max}$ is the execution time duration of the longest transaction in the system, and the second algorithm Online-Greedy produces a schedule of length $O(\tau_{\max} \cdot (C \log(MN) + N \log^2(MN)))$. The competitiveness of both of the algorithms is within a poly-log factor of $O(s)$. Another recent work

---

[1] A synchronization mechanism is obstruction-free if any thread that runs for a long time it eventually makes progress [14].

is Serializer [5] which resolves a conflict by removing a conflicting transaction $T$ from the processor core where it was running, and scheduling it on the processor core of the other transaction to which it conflicted with. It is $O(n)$-competitive and in fact, it ensures that two transactions never conflict more than once.

TM schedulers [3,6,25,1] offer an alternative approach to boost the TM performance. A TM scheduler is a software component which decides when a particular transaction executes. One proposal in this approach is Adaptive Transaction Scheduling (ATS) [25] which measures adaptively the contention intensity of a thread, and when the contention intensity increases beyond a threshold it serializes the transactions. The Restart and Shrink schedulers, proposed by Dragojević et al. [6], depend on the prediction of future conflicts and dynamically serialize transactions based on the prediction to avoid conflicts. The ATS, Restart, and Shrink schedulers are $O(n)$-competitive. Steal-On-Abort [1] is yet another proposal where the aborted transaction is given to the opponent transaction and queued behind it, preventing the two transactions from conflicting again.

Recently, Attiya et al. [3] proposed the BIMODAL scheduler which alternates between *writing epochs* where it gives priority to writing transactions and *reading epochs* where it gives priority to transactions that have issued only reads so far. It achieves $O(s)$ competitive ratio on bimodal workloads with equi-length transactions. A bimodal workload contains only early-write and read-only transactions.

*Outline of Paper.* The rest of the paper is organized as follows. We present our TM model and definitions in Section 2. We present and formally analyze two new randomized algorithms, Clairvoyant and Non-Clairvoyant, in Sections 3 and 4, respectively. The hardness result of balanced workload scheduling is presented in Section 5. Section 6 concludes the paper.

## 2   Model and Definitions

Consider a system of $n \geq 1$ threads $\mathcal{P} = \{P_1, \cdots, P_n\}$ with a finite set of $s$ shared resources $\mathcal{R} = \{R_1, \ldots, R_s\}$. We consider batch execution problems, where the system issues a set of $n$ transactions $\mathcal{T} = \{T_1, \cdots, T_n\}$ (*transaction workload*), one transaction $T_i$ per thread $P_i$. Each transaction is a sequence of actions (operations) each of which is either a read or write to some shared resource. The sequence of operations in a transaction must be *atomic*: all operations of a transaction are guaranteed to either completely occur, or have no effects at all. A transaction that only reads shared resources is called *read-only*; otherwise it is called a *writing* transaction. We consider transaction workloads where at least one transaction is writing.

After a transaction is issued and starts execution it either *commits* or *aborts*. A transaction that has been issued but not committed yet is said to be *pending*. A pending transaction can *restart* multiple times until it eventually commits. Concurrent write-write actions or read-write actions to shared objects by two or more transactions cause conflicts between transactions. If a transaction conflicts then it either aborts, or it may commit and force to abort all other conflicting transactions. In a *greedy schedule*, if a transaction aborts due to conflicts it then immediately restarts and attempts to commit again. We assume that the execution time advances synchronously for all threads and a preemption and abort require negligible time. We also assume that all transactions in the system are correct, i.e., there are no faulty transactions.[2]

**Definition 1 (Pending Commit Property [11]).** *A contention manager obeys the* pending commit *property if, whenever there are pending transactions, some running transaction T will execute uninterrupted until it commits.*

Let $\mathcal{R}(T_i)$ denote the set of resources used by a transaction $T_i$. We can write $\mathcal{R}(T_i) = \mathcal{R}_w(T_i) \cup \mathcal{R}_r(T_i)$, where $\mathcal{R}_w(T_i)$ are the resources which are to be written by $T_i$, and $\mathcal{R}_r(T_i)$ are the resources to be read by $T_i$.

**Definition 2 (Transaction Conflict).** *Two transactions $T_i$ and $T_j$ conflict if at least one of them writes on a common resource, that is, there is a resource $R$ such that $R \in (\mathcal{R}_w(T_i) \cap \mathcal{R}(T_j)) \cup (\mathcal{R}(T_i) \cap \mathcal{R}_w(T_j))$ (we also say that $R$ causes the conflict).*

From the definition of transaction conflicts we can define the *conflict graph* for a set of transactions. In the conflict graph, each node corresponds to a transaction and each edge represents a conflict between the adjacent transactions.

**Definition 3 (Conflict Graph).** *For a set of transactions $\mathcal{T}$, the* conflict *graph $G(\mathcal{T}) = (V, E)$ has as nodes the transactions, $V = \mathcal{T}$, and $(T_i, T_j) \in E$ for any two transactions $T_i, T_j$ that conflict.*

Let $\gamma(R_j)$ denote the number of transactions that write resource $R_j$. Let $\gamma_{\max} = \max_j \gamma(R_j)$. Denote $\lambda_w(T_i) = |\mathcal{R}_w(T_i)|$, $\lambda_r(T_i) = |\mathcal{R}_r(T_i)|$, and $\lambda(T_i) = |\mathcal{R}(T_i)|$, the number of resources which are being accessed by transaction $T_i$ for write, read, and both read and write. Let $\lambda_{\max} = \max_i \lambda(T_i)$. Note that in the conflict graph $G$ the maximum node degree is bounded by $\lambda_{\max} \cdot \gamma_{\max}$, and also there is a node whose degree is at least $\gamma_{\max}$.

For any transaction $T_i$ we define the *balancing ratio* $\beta(T_i) = \frac{|\mathcal{R}_w(T_i)|}{|\mathcal{R}(T_i)|}$ as the ratio of number of writes versus the total number of resources it accesses. For a read-only transaction $\beta(T_i) = 0$. For a writing transaction it holds $\frac{1}{s} \le \beta(T_i) \le 1$, since there will be at least one write performed by $T_i$ to one of the $s$ resources. We define the *global balancing ratio* as the minimum of the individual writing transaction balancing ratios: $\beta = \min_{(T_i \in \mathcal{T}) \wedge (\lambda_w(T_i) > 0)} \beta(T_i)$. We define *balanced transaction workloads* as follows (recall that we consider workloads with at least one writing transaction):

**Definition 4 (Balanced Workloads).** *We say that a workload (set of transactions) $\mathcal{T}$ is* balanced *if $\beta = \Theta(1)$.*

---

[2] A transaction is called faulty when it encounters an illegal instruction producing a segmentation fault or experiences a page fault resulting to wait for a long time for the page to be available [10].

In other words, in balanced transaction workloads the number of writes that each writing transaction performs is a constant fraction of the total number of resource accesses (for read or write) that the transaction performs.

Each transaction $T_i$ has execution time duration $\tau_i > 0$. The execution time is the total number of discrete time steps that the transaction requires to commit uninterrupted from the moment it starts. In our model we assume that the execution time of each transaction is fixed. Let $\tau_{max} = \max_i \tau_i$ be the execution time of the longest transaction, and $\tau_{min} = \min_i \tau_i$ be the execution time of the shortest transaction. We denote $\ell = \left\lceil \log \left( \frac{\tau_{max}}{\tau_{min}} \right) \right\rceil + 1$. We finish this section with the basic definitions of *makespan* and *competitive ratio*.

**Definition 5 (Makespan and Competitive Ratio).** *Given a contention manager $\mathcal{A}$ and a workload $\mathcal{T}$, makespan$_{\mathcal{A}}(\mathcal{T})$ is the total time $\mathcal{A}$ needs to commit all the transactions in $\mathcal{T}$. The competitive ratio is $CR_{\mathcal{A}}(\mathcal{T}) = \frac{makespan_{\mathcal{A}}(\mathcal{T})}{makespan_{opt}(\mathcal{T})}$, where* opt *is the optimal off-line scheduler.*

## 3 Clairvoyant Algorithm

We describe and analyze Algorithm Clairvoyant (see Algorithm 1). The writing transactions are divided into $\ell$ groups $A_0, A_1, \ldots, A_{\ell-1}$, where $\ell = \left\lceil \log \left( \frac{\tau_{max}}{\tau_{min}} \right) \right\rceil + 1$, in such a way that $A_i$ contains transactions with execution time duration in range $[2^i \cdot \tau_{min}, (2^{i+1} \cdot \tau_{min} - 1)]$, for $0 \leq i \leq \ell - 1$. Each group of transactions $A_i$ is then again divided into $\kappa$ subgroups $A_i^0, A_i^1, \ldots, A_i^{\kappa-1}$, where $\kappa = \lceil \log s \rceil + 1$, such that each transaction $T \in A_i^j$ accesses (for read and write) a number of resources in range $\lambda(T) \in [2^j, 2^{j+1} - 1]$, for $0 \leq j \leq \kappa - 1$. We assign an order to the subgroups in such a way that $A_i^j < A_k^l$ if $i < k$ or $i = k \wedge j < l$. Note that some of the subgroups may be empty. The read-only transactions are placed into a special group $B$ which has the highest order.

At any time $t$ the pending transactions are assigned a priority level which determines which transactions commit or abort. A transaction is assigned a priority which is one of: *high* or *low*. Let $\Pi_t^h$ and $\Pi_t^l$ denote the set of transactions which will be assigned high and low priority, respectively, at time $t$. In conflicts, high priority transactions abort low priority transactions. Conflicts between transactions of the same priority level are resolved arbitrarily. Suppose that $\widehat{A}_t$ is the lowest order subgroup that contains pending transactions at time $t$. Only transactions from $\widehat{A}_t$ can be given high priority, that is $\Pi_t^h \subseteq \widehat{A}_t$.

The priorities are determined according to the conflict graph for the transactions. Let $\mathcal{T}_t$ denote the set of all transactions which are pending at time $t$. (Initially, $\mathcal{T}_0 = \mathcal{T}$.) Let $\widehat{\mathcal{T}}_t$ denote the pending transactions of $\widehat{A}_t$ at time $t$. (Initially, $\widehat{\mathcal{T}}_0 = \widehat{A}_0$.) Let $\widehat{S}_t$ denote the set of transactions in $\widehat{\mathcal{T}}_t$ which are pending and have started executing before $t$ but have not yet committed or aborted. Let $\widehat{S}_t'$ denote the set of transactions in $\mathcal{T}_t$ which conflict with $\widehat{S}_t$. Let $\widehat{I}_t$ be a maximal independent set in the conflict graph $G(\widehat{\mathcal{T}}_t \setminus \widehat{S}_t')$. Then, the set of high priority transactions at time $t$ is set to be $\Pi_t^h = \widehat{I}_t \cup \widehat{S}_t$. The remaining transactions are

---

**Algorithm 1.** Clairvoyant

---

**Input**: A set $\mathcal{T}$ of $n$ transactions with global balancing ratio $\beta$;
**Output**: A greedy execution schedule;

- Divide writing transactions into $\ell = \lceil \log(\frac{\tau_{\max}}{\tau_{\min}}) \rceil + 1$ groups $A_0, A_1, \cdots, A_{\ell-1}$ in such a way that $A_i$ contains transactions with execution time duration in range $[2^i \cdot \tau_{\min}, (2^{i+1} \cdot \tau_{\min} - 1)]$; Read-only transactions are placed in special group $B$;
- Divide $A_i$ again into $\kappa = \lceil \log s \rceil + 1$ subgroups $A_i^0, A_i^1, \cdots, A_i^{\kappa-1}$ in a way that each subgroup $A_i^j$ contains transactions that access a number of resource in the range $[2^j, 2^{j+1} - 1]$;
- Order the groups and subgroups such that $A_i^j < A_k^l$ if $i < k$ or $i = k \wedge j < l$; special group $B$ has highest order;

**foreach** *time step* $t = 0, 1, 2, 3, \ldots$ **do**
 **Set Definitions:**
  $\mathcal{T}_t$: set of transactions that are pending; // $\mathcal{T}_0 \leftarrow \mathcal{T}$
  $\widehat{A}_t$: lowest order group that contains pending transactions;
  $\widehat{\mathcal{T}}_t$: set of transactions in $\widehat{A}_t$ which are pending; // $\widehat{\mathcal{T}}_0 \leftarrow \widehat{A}_0$
  $\widehat{S}_t$: set of transactions in $\widehat{\mathcal{T}}_t$ which were started before $t$;
  $\widehat{S}'_t$: set of conflicting transactions in $\mathcal{T}_t$ which conflict with $\widehat{S}_t$;
  $\widehat{I}_t$ : maximal independent set in the conflict graph $G(\widehat{\mathcal{T}}_t \setminus \widehat{S}'_t)$;
 **Priority Assignment:**
  High priority transactions: $\Pi_t^h \leftarrow \widehat{I}_t \cup \widehat{S}_t$;
  Low priority transactions: $\Pi_t^l \leftarrow \mathcal{T}_t \setminus \Pi_t^h$;
 **Conflict Resolution:**
  Execute all pending transactions;
  **On conflict** of transaction $T_u$ with transaction $T_v$:
   **if** $(T_u \in \Pi_t^h) \wedge (T_v \in \Pi_t^l)$ **then** $abort(T_u, T_v)$; **else** $abort(T_v, T_u)$;
   // $abort(T_u, T_v)$ aborts transaction $T_v$

---

given low priority, that is, $\Pi_t^l = \mathcal{T}_t \setminus \Pi_t^h$. Note that the transactions in $\Pi_t^h$ do not conflict with each other. The transactions $\Pi_t^h$ will remain in high priority in subsequent time steps $t' > t$ until they commit, since the transactions in $\widehat{S}_{t'}$ are included in $\Pi_{t'}^h$.

This algorithm is clairvoyant in the sense that it requires explicit knowledge of the various conflict relations at each time $t$. The algorithm is greedy, since at each time step each pending transaction is not idle. The algorithm also satisfies the pending commit property since at any time step $t$ at least one transaction from $\widehat{A}_t$ will execute uninterrupted until it commits. We have assumed above that each transaction knows its execution length and the number of resources it accesses. Clearly, the algorithm computes the schedule in polynomial time.

## 3.1   Analysis of Clairvoyant Algorithm

We now give a competitive analysis of Algorithm Clairvoyant. Define $\tau_{\min}^j = 2^i \cdot \tau_{\min}$ and $\tau_{\max}^j = (2^{i+1} \cdot \tau_{\min} - 1)$. Note that the duration of each transaction

$T \in A_i^j$ is in range $[\tau_{\min}^j, \tau_{\max}^j]$, and also $\tau_{\max}^j \leq 2\tau_{\min}^j$. Define $\lambda_{\min}^j = 2^j$ and $\lambda_{\max}^j = 2^{j+1} - 1$. Note that for each transaction $T \in A_i^j$, $\lambda(T) \in [\lambda_{\min}^j, \lambda_{\max}^j]$, and $\lambda_{\max}^j \leq 2\lambda_{\min}^j$.

In the next results we will first focus on a subgroup $A_i^j$ and we will assume that there are no other transactions in the system. We give bounds for the competitive ratio for $A_i^j$ which will be useful when we later analyze the performance for all the transactions in $\mathcal{T}$.

**Lemma 1.** *If we only consider transactions in subgroup $A_i^j$, then the competitive ratio is bounded by $CR_{Clairvoyant}(A_i^j) \leq 2 \cdot \lambda_{\max}^j + 2$.*

*Proof.* Let $\gamma_i^j(R_v)$ denote the number of transactions in a subgroup $A_i^j$ that write $R_v$, $1 \leq v \leq s$. Let $\gamma' = max_{v \in [1,s]} \gamma_i^j(R_v)$. Since there is only one subgroup, $\widehat{A}_t = A_i^j$. A transaction $T \in A_i^j$ conflicts with at most $\lambda_{\max}^j \cdot \gamma'$ other transactions in the same subgroup. If transaction $T$ is in low priority it is only because some other conflicting transaction in $A_i^j$ is in high priority. If no conflicting transaction is in high priority then $T$ becomes high priority immediately. Since a high priority transaction executes uninterrupted until it commits, it will take at most $\lambda_{\max}^j \cdot \gamma'$ time steps until all conflicting transactions with $T$ have committed. Thus, it is guaranteed that in at most $\lambda_{\max}^j \cdot \gamma' \cdot \tau_{\max}^j$ time steps $T$ becomes high priority. Therefore, $T$ commits by time $(\lambda_{\max}^j \cdot \gamma' + 1) \cdot \tau_{\max}^j$. Since $T$ is an arbitrary transaction in $A_i^j$, the makespan of the algorithm is bounded by:

$$makespan_{Clairvoyant}(A_i^j) \leq (\lambda_{\max}^j \cdot \gamma' + 1) \cdot \tau_{\max}^j.$$

There is a resource that is accessed by at least $\gamma'$ transactions of $A_i^j$ for write. All these transactions have to serialize because they all conflict with each other in the common resource. Therefore, the optimal makespan is bounded by:

$$makespan_{opt}(A_i^j) \geq \gamma' \cdot \tau_{\min}^j.$$

When we combine the upper and lower bounds we obtain a bound on the competitive ratio of the algorithm:

$$CR_{Clairv.}(A_i^j) = \frac{makespan_{Clairv.}(A_i^j)}{makespan_{opt}(A_i^j)} \leq \frac{(\lambda_{\max}^j \cdot \gamma' + 1) \cdot \tau_{\max}^j}{\gamma' \cdot \tau_{\min}^j} \leq 2 \cdot \lambda_{\max}^j + 2.$$

**Lemma 2.** *If we only consider transactions in subgroup $A_i^j$, then the competitive ratio is bounded by $CR_{Clairvoyant}(A_i^j) \leq 4 \cdot \frac{s/\beta}{\lambda_{\max}^j}$.*

*Proof.* Since the algorithm satisfies the pending-commit property, if a transaction $T \in A_i^j$ does not commit, then some conflicting transaction $T' \in A_i^j$ must commit. Therefore, the makespan of the algorithm is bounded by:

$$makespan_{Clairvoyant}(A_i^j) \leq |A_i^j| \cdot \tau_{\max}^j.$$

Each transaction in $T \in A_i^j$ accesses at least $\lambda_w(T)$ resources for write. Since we only consider transactions in $A_i^j$, $\lambda_w(T) \geq \beta \cdot \lambda_{\min}^j \geq \beta \cdot \lambda_{\max}^j / 2$. Consequently, by the pigeonhole principle, there will be a resource $R \in \mathcal{R}$ which is accessed by at least $\sum_{T \in A_i^j} \lambda_w(T)/s \geq |A_i^j| \cdot \beta \cdot \lambda_{max}^j/(2s)$ transactions for write. All these transactions accessing $R$ have to serialize because they conflict with each other. Therefore, the optimal makespan is bounded by:

$$makespan_{opt}(A_i^j) \geq \frac{|A_i^j| \cdot \beta \cdot \lambda_{max}^j}{2s} \cdot \tau_{min}^j.$$

When we combine the above bounds of the makespan we obtain the following bound on the competitive ratio of the algorithm:

$$CR_{Clairvoyant}(A_i^j) = \frac{makespan_{Clairvoyant}(A_i^j)}{makespan_{opt}(A_i^j)} \leq \frac{|A_i^j| \cdot \tau_{max}^j}{\frac{|A_i^j| \cdot \beta \cdot \lambda_{max}^j}{2s} \cdot \tau_{min}^j} \leq 4 \cdot \frac{s/\beta}{\lambda_{max}^j}.$$

From Lemmas 1 and 2, we obtain:

**Corollary 1.** *If we only consider transactions in subgroup $A_i^j$, then the competitive ratio of the algorithm is bounded by $CR_{Clairvoyant}(A_i^j) \leq 4 \cdot \min\left\{\lambda_{max}^j, \frac{s/\beta}{\lambda_{max}^j}\right\}$.*

We now continue to provide a bound for the performance of individual groups. This will help to provide bounds for all the transactions.

**Lemma 3.** *If we only consider transactions in group $A_i$, then the competitive ratio of the algorithm is bounded by $CR_{Clairvoyant}(A_i) \leq 32 \cdot \sqrt{\frac{s}{\beta}}$.*

*Proof.* Since $\lambda_{max}^j = (2^{j+1} - 1)$, Corollary 1 gives for each subgroup $A_i^j$ competitive ratio

$$CR_{Clairvoyant}(A_i^j) \leq 4 \cdot \min\left\{2^{j+1} - 1, \frac{s/\beta}{2^{j+1} - 1}\right\} \leq 8 \cdot \min\left\{2^j, \frac{s/\beta}{2^j}\right\}.$$

Let $\psi = \frac{\log(s/\beta)}{2}$. Note that $\min\left\{2^j, \frac{s/\beta}{2^j}\right\} \leq 2^j, \forall j \in [0, \lfloor\psi\rfloor]$; and $\min\left\{2^j, \frac{s/\beta}{2^j}\right\} \leq \frac{s/\beta}{2^j} = 2^{2\psi-j}, \forall j \in [\lfloor\psi\rfloor + 1, \kappa - 1]$. Group $A_i$ contains $\kappa$ subgroups of transactions. In the worst case, Algorithm Clairvoyant will commit the transactions in each subgroup according to their order starting from the lowest order subgroup and ending at the highest order subgroup, since that's the order that the transactions are assigned a high priority. Therefore,

$$CR_{Clairv.}(A_i) \leq \sum_{j=0}^{\kappa-1} CR_{Clairv.}(A_i^j)$$

$$= \sum_{j=0}^{\lfloor\psi\rfloor} CR_{Clairv.}(A_i^j) + \sum_{j=\lfloor\psi\rfloor+1}^{\kappa-1} CR_{Clairv.}(A_i^j)$$

$$\leq 8 \cdot \left(\sum_{j=0}^{\lfloor\psi\rfloor} 2^j + \sum_{j=\lfloor\psi\rfloor+1}^{k-1} 2^{2\psi-j}\right) \leq 8 \cdot \left(2 \cdot 2^\psi + 2 \cdot 2^\psi\right) = 32 \cdot \sqrt{\frac{s}{\beta}}.$$

**Theorem 1 (Competitive Ratio of** Clairvoyant**).** *For set of transactions* $\mathcal{T}$, *Algorithm* Clairvoyant *has competitive ratio* $CR_{Clairvoyant}(\mathcal{T}) = O\left(\ell \cdot \sqrt{\frac{s}{\beta}}\right)$.

*Proof.* As there are $\ell$ groups of transactions $A_i$, and one group $B$, in the worst case, Algorithm Clairvoyant will commit the transactions in each group according to their order starting from the lowest order group and ending at the highest order group. Clearly, the algorithm will execute the read-only transactions in group $B$ in optimal time. Therefore, using Lemma 3, we obtain:

$$CR_{Clairvoyant}(\mathcal{T}) \leq \sum_{i=0}^{\ell-1} CR_{Clairvoyant}(A_i) + CR_{Clairvoyant}(B)$$

$$\leq \sum_{i=0}^{\ell-1} 32 \cdot \sqrt{\frac{s}{\beta}} + 1 = 32 \cdot \ell \cdot \sqrt{\frac{s}{\beta}} + 1.$$

The corollary below follows immediately from Theorem 1.

**Corollary 2 (Balanced Workload).** *For balanced workload* $\mathcal{T}$ *($\beta = \Theta(1)$) and when* $\ell = O(1)$, *Algorithm* Clairvoyant *has competitive ratio* $CR_{Clairvoyant}(\mathcal{T}) = O(\sqrt{s})$.

## 4    Non-clairvoyant Algorithm

We present and analyze Algorithm Non-Clairvoyant (see Algorithm 2). This algorithm is similar to Clairvoyant given at Section 3 with the difference that the conflicts are resolved using priorities which are determined without the explicit knowledge of the conflict graph.

Similar to Algorithm Clairvoyant, the transactions are organized into groups and subgroups. Lower order subgroups have always higher priority than higher order subgroups. At each time step $t$, let $\widehat{A}_t$ denote the lowest order subgroup. Clearly, the transactions in $\widehat{A}_t$ have higher priority than the transactions in all other subgroups, and in case of conflicts only the transactions in $\widehat{A}_t$ win. When transactions in the same subgroup conflict, the conflicts are resolved according to random priority numbers. When a transaction starts execution it chooses uniformly at random a discrete number $r(T) \in [1, n]$. In case of a conflict of transaction $T_w$ with another transaction $T_x$ in the same subgroup with $r(T_x) < r(T_w)$, then $T_x$ aborts $T_w$, and otherwise $T_w$ aborts $T_x$. When transaction $T_w$ restarts, it cannot abort $T_x$ until $T_x$ has been committed or aborted. After every abort, the newly started transaction chooses again a new discrete number uniformly at random in the interval $[1, n]$. The idea of randomized priorities has been introduced originally by Schneider and Wattenhofer [21] in their Algorithm RandomizedRounds.

This algorithm is non-clairvoyant in the sense that it does not depend on knowing explicitly the conflict graph to resolve conflicts. The algorithm is greedy but does have the pending commit property. The groups and subgroups can be

---

**Algorithm 2.** Non-Clairvoyant

---

**Input**: A set $\mathcal{T}$ of $n$ transactions with global balancing ratio $\beta$;
**Output**: A greedy execution schedule;

- Divide transactions into $\ell = \lceil \log(\frac{\tau_{\max}}{\tau_{\min}}) \rceil + 1$ groups $A_0, A_1, \cdots, A_{\ell-1}$ in such a way that $A_i$ contains transactions with execution time duration in range $[2^i \cdot \tau_{\min}, (2^{i+1} \cdot \tau_{\min} - 1)]$; Read-only transactions are placed in special group $B$;
- Divide $A_i$ again into $\kappa = \lceil \log s \rceil + 1$ subgroups $A_i^0, A_i^1, \cdots, A_i^{\kappa-1}$ in a way that each subgroup $A_i^j$ contains transactions that access a number of resource in the range $[2^j, 2^{j+1} - 1]$;
- Order the groups and subgroups such that $A_i^j < A_k^l$ if $i < k$ or $i = k \wedge j < l$; special group $B$ has highest order;

**foreach** *time step* $t = 0, 1, 2, 3, \ldots$ **do**
  Execute all pending transactions; // at $t = 0$ issue all transactions
  **On (re)start** of transaction $T$:
    $r(T) \leftarrow$ random integer in $[1, n]$;
  **On conflict** of transaction $T_u \in A_i^j$ with transaction $T_v \in A_k^l$:
  **if** $A_i^j < A_k^l$ **then** $abort(T_u, T_v)$;
  **else if** $A_i^j > A_k^l$ **then** $abort(T_v, T_u)$;
    **else if** $r(T_u) < r(T_v)$ **then** $abort(T_u, T_v)$ ;        // The case $A_i^j = A_k^l$
    **else** $abort(T_v, T_u)$;
    // In case a transaction $T_u$ aborts $T_v$ because $r(T_u) < r(T_v)$,
    then when $T_v$ restarts it cannot abort $T_u$ until $T_u$
    commits or aborts

---

implemented in the algorithm since we assume that each transaction knows its execution time and the number of resources that it accesses. Clearly, the algorithm computes the schedule in polynomial time.

### 4.1   Analysis of Non-clairvoyant Algorithm

In the analysis given below, we study the properties of Algorithm Non-Clairvoyant and give its competitive ratios. We use the following adaptation of the response time analysis of Algorithm RandomizedRounds given in [21]. It uses the following Chernoff bound:

**Lemma 4 (Chernoff Bound).** *Let $X_1, X_2, \ldots, X_n$ be independent Poisson trials such that, for $1 \leq i \leq n$, $\mathbf{Pr}(X_i = 1) = pr_i$, where $0 < pr_i < 1$. Then, for $X = \sum_{i=1}^{n} X_i$, $\mu = \mathbf{E}[X] = \sum_{i=1}^{n} pr_i$, and any $0 < \delta \leq 1, \mathbf{Pr}(X < (1 - \delta)\mu) < e^{-\delta^2 \mu/2}$.*

**Lemma 5 (Adaptation from Schneider and Wattenhofer [21]).** *Given a transaction scheduling problem with $n$ concurrent transactions, where each transaction has execution time at most $\tau$, the time span a transaction $T$ needs from the moment it is issued until commit is $16 \cdot e \cdot (d_T + 1) \cdot \tau \cdot \ln n$ with probability at least $1 - \frac{1}{n^2}$, where $d_T$ is the number of transactions conflicting with $T$.*

*Proof.* Consider the respective conflict graph $G$ of the problem with the $n$ transaction. Let $N_T$ denote the set of conflicting transactions for $T$ (these are the neighbors of $T$ in $G$). Let $r(T)$ denote the random priority number choice of $T$ in range $[1, n]$. The probability that for transaction $T$ no transaction $T' \in N_T$ has the same random number is:

$$\mathbf{Pr}(\nexists T' \in N_T | r(T) = r(T')) = \left(1 - \frac{1}{n}\right)^{d_T} \geq \left(1 - \frac{1}{n}\right)^n \geq \frac{1}{e}.$$

The probability that $r(T)$ is at least as small as $r(T')$ for any transaction $T' \in N_T$ is $\frac{1}{d_T+1}$. Thus, the chance that $r(T)$ is smallest and different among all its neighbors in $N_T$ is at least $\frac{1}{e \cdot (d_T+1)}$. If we conduct $16 \cdot e \cdot (d_T + 1) \cdot \ln n$ trials, each having success probability $\frac{1}{e \cdot (d_T+1)}$, then the probability that the number of successes $Z$ is less than $8 \ln n$ becomes: $\mathbf{Pr}(Z < 8 \cdot \ln n) < e^{-2 \cdot \ln n} = 1/n^2$, using the Chernoff bound of Lemma 4. Since every transaction has execution time at most $\tau$, the total time spent until a transaction commits is at most $16 \cdot e \cdot (d_T + 1) \cdot \tau \cdot \ln n$, with probability at least $1 - 1/n^2$.

We now give competitive bounds for some subgroup $A_i^j$ and later extend the results to all the transactions in $\mathcal{T}$. The proofs are similar as in the analysis of Algorithm Clairvoyant and can be found in the full version of paper (See [22]).

**Lemma 6.** *If we only consider transactions in subgroup $A_i^j$, then the competitive ratio is bounded by $CR_{Non-Clairvoyant}(A_i^j) \leq 64 \cdot e \cdot \lambda_{\max}^j \cdot \ln n$ with probability at least $1 - \frac{|A_i^j|}{n^2}$.*

**Lemma 7.** *If we only consider transactions in subgroup $A_i^j$, then the competitive ratio is bounded by $CR_{Non-Clairvoyant}(A_i^j) \leq 64 \cdot e \cdot \frac{s/\beta}{\lambda_{\max}^j} \cdot \ln n$ with probability at least $1 - \frac{|A_i^j|}{n^2}$.*

From Lemmas 6 and 7, we obtain:

**Corollary 3.** *If we only consider transactions in subgroup $A_i^j$, then the competitive ratio of the algorithm is bounded by $CR_{Non-Clairvoyant}(A_i^j) \leq 64 \cdot e \cdot \min\left\{\lambda_{\max}^j, \frac{s/\beta}{\lambda_{\max}^j}\right\} \cdot \ln n$ with probability at least $1 - \frac{|A_i^j|}{n^2}$.*

We now provide a bound for the performance of individual groups which will help to provide bounds for all the transactions.

**Lemma 8.** *If we only consider transactions in group $A_i$, then the competitive ratio of the algorithm is bounded by $CR_{Non-Clairvoyant}(A_i) \leq 512 \cdot e \cdot \sqrt{\frac{s}{\beta}} \cdot \ln n$ with probability at least $1 - \frac{|A_i|}{n^2}$.*

**Theorem 2 (Competitive Ratio of Non-Clairvoyant).** *For a set of transactions $\mathcal{T}$, Algorithm Non-Clairvoyant has competitive ratio $CR_{Non-Clairvoyant}(\mathcal{T}) = O\left(\ell \cdot \sqrt{\frac{s}{\beta}} \cdot \log n\right)$ with probability at least $1 - \frac{1}{n}$.*

The corollary below follows immediately from Theorem 2.

**Corollary 4 (Balanced Workload).** *For balanced workload $\mathcal{T}$ ($\beta = \Theta(1)$) and when $\ell = O(1)$, Algorithm* Non-Clairvoyant *has competitive ratio $CR_{Non-Clairvoyant}(\mathcal{T}) = O(\sqrt{s} \cdot \log n)$ with probability at least $1 - \frac{1}{n}$.*

## 5    Hardness of Balanced Transaction Scheduling

In this section, we show that the performance of Clairvoyant is tight by reducing the graph coloring problem to the transaction scheduling problem.

A VERTEX COLORING problem instance asks whether a given graph $G$ is $k$-colorable [9]. A valid $k$-coloring is an assignment of integers $\{1, 2, \cdots, k\}$ (the colors) to the vertices of $G$ so that neighbors receive different integers. The chromatic number, $\chi(G)$ is the smallest $k$ such that $G$ has a valid $k$-coloring. We say that an algorithm approximates $\chi(G)$ with approximation ratio $q(G)$ if it outputs $u(G)$ such that $\chi(G) \leq u(G)$ and $u(G)/\chi(G) \leq q(G)$. Typically, $q(G)$ is expressed only as a function of $n$, the number of vertices in $G$. It is well known that known VERTEX COLORING is NP-complete. It is also shown in [7] that unless NP⊆ZPP, there does not exist a polynomial time algorithm to approximate $\chi(G)$ with approximation ratio $O(n^{1-\epsilon})$ for any constant $\epsilon > 0$, where $n$ denotes the number of vertices in graph $G$.

A TRANSACTION SCHEDULING problem instance asks whether a set of transactions $\mathcal{T}$ with a set of resources $\mathcal{R}$ has makespan $k$ time steps. We give a polynomial time reduction of the VERTEX COLORING problem to the TRANSACTION SCHEDULING problem. Consider an input graph $G = (V, E)$ of the VERTEX COLORING problem, where $|V| = n$ and $|E| = s$. We construct a set of transactions $\mathcal{T}$ such that for each $v \in V$ there is a respective transaction $T_v \in \mathcal{T}$; clearly, $|\mathcal{T}| = |V| = n$. We also use a set of resources $\mathcal{R}$ such that for each edge $e \in V$ there is a respective resource $R_e \in \mathcal{R}$; clearly, $|\mathcal{R}| = |E| = s$. If $e = (u, v) \in E$, then both the respective transactions $T_u$ and $T_v$ use the resource $R_e$ for write. Since all transaction operations are writes, we have that $\beta = 1$. We take all the transactions to have the same execution length equal to one time step, that is, $\tau_{\max} = \tau_{\min} = 1$, and $\ell = 1$.

Let $G'$ be the conflict graph for the transactions $\mathcal{T}$. Note that $G'$ is isomorphic to $G$. Node colors in $G$ correspond to time steps in which transactions in $G'$ are issued. Suppose that $G$ has a valid $k$-coloring. If a node $v \in G$ has a color $x$, then the respective transaction $T_v \in G'$ can be issued and commit at time step $x$, since no conflicting transaction (neighbor in $G'$) has the same time assignment (color) as $T_v$. Thus, a valid $k$-coloring in $G$ implies a schedule with makespan $k$ for the transactions in $\mathcal{T}$. Symmetrically, a schedule with makespan $k$ for $\mathcal{T}$ implies a valid $k$-coloring in $G$.

It is easy to see that the problem TRANSACTION SCHEDULING is in $NP$. From the reduction of the VERTEX COLORING problem, we also obtain that TRANSACTION SCHEDULING is $NP$-complete.

From the above reduction, we have that an approximation ratio $q(G)$ of the VERTEX COLORING problem implies the existence of a scheduling algorithm

$\mathcal{A}$ with competitive ratio $CR_{\mathcal{A}}(\mathcal{T}) = q(G)$ of the respective TRANSACTION SCHEDULING problem instance, and vice-versa. Since $s = |\mathcal{R}| = |E| \le n^2$, an $(\sqrt{s})^{1-\epsilon}$ competitive ratio of $\mathcal{A}$ implies at most an $n^{1-\epsilon}$ approximation ratio of VERTEX COLORING. Since, we know that unless NP$\subseteq$ZPP, there does not exist a polynomial time algorithm to approximate $\chi(G)$ with approximation ratio $O(n^{1-\epsilon})$ for any constant $\epsilon > 0$, we obtain a symmetric result for the TRANSACTION SCHEDULING problem:

**Theorem 3 (Approximation Hardness of Transaction Scheduling).** *Unless* NP$\subseteq$ZPP, *we cannot obtain a polynomial time transaction scheduling algorithm such that for every input instance with $\beta = 1$ and $\ell = 1$ of the* TRANSACTION SCHEDULING *problem the algorithm achieves competitive ratio smaller than* $O((\sqrt{s})^{1-\epsilon})$ *for any constant $\epsilon > 0$.*

Theorem 3 implies that the $O(\sqrt{s})$ bound of Algorithm Clairvoyant, given in Corollary 2 for $\beta = \Theta(1)$ and $\ell = O(1)$, is tight.

## 6    Conclusions

We have studied the competitive ratios achieved by transactional contention managers on balanced workloads. The randomized algorithms presented in this paper allow to achieve best competitive bound on balanced workloads. We also establish hardness results on the competitive ratios in our balanced workload model by reducing the well known NP-complete vertex coloring problem to the transactional scheduling problem.

There are several interesting directions for future work. As advocated in [15], our algorithms are conservative − abort at least one transaction involved in a conflict − as it reduces the cost to track conflicts and dependencies. It is interesting to look whether the other schedulers which are less conservative can give improved competitive ratios by reducing the overall makespan. First, our study can be complemented by studying other performance measures, such as the average response time of transactions under balanced workloads. Second, while we have theoretically analyzed the behavior of balanced workloads, it is interesting to see how our contention managers compare experimentally with prior transactional contention managers, e.g., [5,25,11,1].

## References

1. Ansari, M., Luján, M., Kotselidis, C., Jarvis, K., Kirkham, C., Watson, I.: Steal-on-abort: Improving transactional memory performance through dynamic transaction reordering. In: Seznec, A., Emer, J., O'Boyle, M., Martonosi, M., Ungerer, T. (eds.) HiPEAC 2009. LNCS, vol. 5409, pp. 4–18. Springer, Heidelberg (2009)
2. Attiya, H., Epstein, L., Shachnai, H., Tamir, T.: Transactional contention management as a non-clairvoyant scheduling problem. Algorithmica 57(1), 44–61 (2010)
3. Attiya, H., Milani, A.: Transactional scheduling for read-dominated workloads. In: OPODIS 2009, pp. 3–17. Springer, Heidelberg (2009)

4. Dice, D., Shalev, O., Shavit, N.: Transactional locking II. In: Dolev, S. (ed.) DISC 2006. LNCS, vol. 4167, pp. 194–208. Springer, Heidelberg (2006)
5. Dolev, S., Hendler, D., Suissa, A.: CAR-STM: scheduling-based collision avoidance and resolution for software transactional memory. In: PODC 2008, pp. 125–134 (2008)
6. Dragojević, A., Guerraoui, R., Singh, A.V., Singh, V.: Preventing versus curing: avoiding conflicts in transactional memories. In: PODC 2009, pp. 7–16 (2009)
7. Feige, U., Kilian, J.: Zero knowledge and the chromatic number. In: CCC 1996, pp. 278–287 (1996)
8. Felber, P., Fetzer, C., Riegel, T.: Dynamic performance tuning of word-based software transactional memory. In: PPoPP 2008, pp. 237–246 (2008)
9. Garey, M.R., Johnson, D.S.: Computers and Intractability; A Guide to the Theory of NP-Completeness. W. H. Freeman & Co., New York (1990)
10. Guerraoui, R., Herlihy, M., Kapalka, M., Pochon, B.: Robust Contention Management in Software Transactional Memory. In: SCOOL 2005 (2005)
11. Guerraoui, R., Herlihy, M., Pochon, B.: Toward a theory of transactional contention managers. In: PODC 2005, pp. 258–264 (2005)
12. Harris, T., Fraser, K.: Language support for lightweight transactions. In: OOPSLA 2003, pp. 388–402. ACM, New York (2003)
13. Harris, T., Marlow, S., Peyton-Jones, S., Herlihy, M.: Composable memory transactions. In: PPoPP 2005, pp. 48–60 (2005)
14. Herlihy, M., Luchangco, V., Moir, M.: Obstruction-free synchronization: Double-ended queues as an example. In: ICDCS 2003, pp. 522–529 (2003)
15. Herlihy, M., Luchangco, V., Moir, M., Scherer III, I.W.N.: Software transactional memory for dynamic-sized data structures. In: PODC 2003, pp. 92–101 (2003)
16. Herlihy, M., Moss, J.E.B.: Transactional memory: Architectural support for lock-free data structures. In: ISCA 1993, pp. 289–300 (1993)
17. Khot, S.: Improved inapproximability results for maxclique, chromatic number and approximate graph coloring. In: FOCS 2001, pp. 600–609 (2001)
18. Marathe, V.J., Spear, M.F., Heriot, C., Acharya, A., Eisenstat, D., Scherer III, W.N., Scott, M.L.: Lowering the overhead of software transactional memory. Tech. Rep. TR 893, Computer Science Department, University of Rochester (2006)
19. Ramadan, H.E., Rossbach, C.J., Porter, D.E., Hofmann, O.S., Bhandari, A., Witchel, E.: Metatm/txlinux: Transactional memory for an operating system. IEEE Micro. 28(1), 42–51 (2008)
20. Scherer III, W.N., Scott, M.L.: Advanced contention management for dynamic software transactional memory. In: PODC 2005, pp. 240–248 (2005)
21. Schneider, J., Wattenhofer, R.: Bounds on contention management algorithms. In: Dong, Y., Du, D.-Z., Ibarra, O. (eds.) ISAAC 2009. LNCS, vol. 5878, pp. 441–451. Springer, Heidelberg (2009)
22. Sharma, G., Busch, C.: A competitive analysis for balanced transactional memory workloads. CoRR abs/1009.0056 (2010)
23. Sharma, G., Estrade, B., Busch, C.: Window-based greedy contention management for transactional memory. In: Lynch, N.A., Shvartsman, A.A. (eds.) DISC 2010. LNCS, vol. 6343, pp. 64–78. Springer, Heidelberg (2010)
24. Shavit, N., Touitou, D.: Software transactional memory. In: PODC 1995, pp. 204–213 (1995)
25. Yoo, R.M., Lee, H.H.S.: Adaptive transaction scheduling for transactional memory systems. In: SPAA 2008, pp. 169–178 (2008)

# Fast Local-Spin Abortable Mutual Exclusion with Bounded Space

Hyonho Lee

Department of Computer Science
University of Toronto, Toronto, ON, Canada, M5S 3G4
hlee@cs.toronto.edu

**Abstract.** Abortable mutual exclusion is a variant of mutual exclusion, where processes are allowed to abort their invocations while waiting to enter the critical section. In this paper, we present an FCFS abortable mutual exclusion algorithm with bounded time and space, in which each invocation performs $O(k^2)$ RMAs if at most $k$ processes abort. We define an object type, *S-HAD*, from which it is easy to construct local-spin abortable mutual exclusion algorithms. Our main contribution is a wait-free implementation of an S-HAD object. We also develop a new, wait-free memory reclamation method, which generalizes reference counting, to achieve bounded space. The resulting algorithm uses $O(N^2)$ shared variables, each with $O(\log N)$ bits, where $N$ is the number of processes.

## 1 Introduction

Abortable mutual exclusion [13] is a variant of classical mutual exclusion [5], in which a process performing a *trying protocol* to enter the *critical section* is allowed to stop waiting for the critical section to become available by performing an *abort protocol*, which returns the process to the *remainder section* within a bounded number of steps. Abortable mutual exclusion can be useful in real-time applications or in parallel database systems because, in these systems, users may want to abort any operation that takes too long [13].

In shared memory models, processes communicate with each other only via shared variables, so waiting processes must keep accessing shared variables until they stop waiting. Such *busy-waiting* may cause processes to perform an unbounded number of steps during the trying protocol. In the *distributed shared memory (DSM)* and *cache-coherent (CC)* models, the cost for a process to access its own local shared memory or cache is considered to be much less than the cost to access memory located remotely. Hence, in these models, counting only *remote memory accesses (RMAs, also known as remote memory references)* is a good measure of the time complexity of an algorithm. To achieve a bounded number of RMAs, many papers about mutual exclusion have considered *local-spin* algorithms. In such algorithms, each process accesses only a bounded number of RMAs while busy-waiting. In this paper, we restrict attention to local-spin algorithms.

C. Lu, T. Masuzawa, and M. Mosbah (Eds.): OPODIS 2010, LNCS 6490, pp. 364–379, 2010.

In some classical mutual exclusion algorithms, such as the Bakery algorithm [10], which are not local-spin, all waiting processes wait for the same shared variable to change. Then a process can abort by simply announcing that it is no longer trying. Scott and Scherer [13] proposed two first-come-first-served (FCFS) local-spin mutual exclusion algorithms that allow waiting processes to abort. In their first algorithm, each process waits for a change in a certain shared variable associated with its predecessor (the last process that was enqueued before it). This algorithm is local-spin in the CC model. In their second algorithm, each waiting process first announces itself to its predecessor and then waits for a certain locally stored variable to change value. This algorithm is local-spin in the DSM model. In these algorithms, processes in the trying protocol form a queue and each waits for a signal from its predecessor. In Scott and Scherer's algorithms, each process enters the critical section within $O(1)$ RMAs when no process aborts. However, their abort protocol contains a waiting period in which an aborting process performs handshakes with its predecessor and successor in the queue, so it may not terminate the abort protocol within a bounded number of steps.

Later, Scott [12] eliminated this waiting period in the abort protocol: He presented two FCFS local-spin abortable mutual exclusion algorithms in which a process aborts within a bounded number of its own steps. When no processes abort, each invocation performs only a constant number of RMAs in the trying protocol. However, when two or more processes repeatedly abort without removing themselves from the queue of waiting processes and then re-enter the trying protocol, the length of the queue may become unbounded. Hence, these algorithms use unbounded space. The number of RMAs a process performs in the trying protocol can be as large as the number of consecutive times processes began the trying protocol immediately beforehand and subsequently aborted [12,9]. This can be arbitrarily large, since a process can repeatedly enter the trying protocol and abort. However, the bad situation is only achieved when each invocation that aborts decides to do so before its predecessor begins the abort protocol.

In Section 2.4 of [12], Scott described a simple abortable mutual exclusion algorithm with $\Theta(N)$ space. This algorithm also uses a queue. When a process starts its trying protocol, it enqueues an element, and waits for the value of its predecessor in the queue to change. When a process aborts, it changes the value of the element it last enqueued. If this process re-enters the trying protocol, then it checks whether the element it last enqueued has been accessed and, if not, it reclaims this element, instead of enqueuing a new element. However, in this algorithm, a process can perform an unbounded number of RMAs in the trying protocol. For example, suppose process $p$ is the predecessor of another process $q$ in the queue, and $q$ is waiting for the value of $p$'s element to be changed. When $p$ aborts, it changes the value of its element. If it re-enters the trying protocol and reclaims the same element, it changes the value of the element back to its previous value. Even if $q$ did not notice $p$'s abort, $q$'s next read of $p$'s element generates a cache miss. Thus, if $p$ aborts and re-enters the trying protocol, reclaiming the same element an unbounded number of times, $q$ may perform an unbounded number of RMAs while waiting. Thus this algorithm is not local-spin or FCFS.

There are two previously known FCFS local-spin abortable mutual exclusion algorithms with bounded space in which each process performs a bounded number of RMAs for each entry to the critical section: Jayanti [9] uses registers and LoadLinked/StoreConditional (LL/SC), and Danek and Lee [3] use only registers. Jayanti's algorithm performs $\Theta(\min(k, \log N))$ and Danek and Lee's algorithm performs $\Theta(N)$ RMAs for each entry to the critical section, where $N$ is the number of processes and $k$ is the contention, i.e., the number of processes that are trying to enter the critical section at the same time. Danek and Lee also presented a local-spin abortable mutual exclusion algorithm with $\Theta(\log N)$ RMAs that does not satisfy FCFS. Since any mutual exclusion algorithm using only registers and comparison primitives, such as COMPARE_AND_SWAP or LL/SC, requires $\Omega(\log N)$ RMAs in the worst case for each entry to the critical section [2] and, since mutual exclusion is a special case of abortable mutual exclusion, both Jayanti's algorithm and Danek and Lee's $\Theta(\log N)$ algorithm are optimal.

In the worst case, each process performs fewer RMAs in Jayanti's algorithm than in Scott's local-spin algorithms. However, if the number of consecutive aborts is $o(\log N)$, then Scott's algorithms are better in terms of the number of RMAs. A natural question is whether there exists a local-spin abortable mutual exclusion algorithm that preserves all of the merits of Scott's algorithms, but uses only bounded space and performs a bounded number of RMAs in the worst case.

In this paper, we present a new FCFS local-spin abortable mutual exclusion algorithm for the CC model. It uses $O(N^2)$ space and a process performs $O(k^2)$ RMAs to enter the critical section, where $k$ is the number of processes that began the trying protocol immediately beforehand and subsequently aborted. The worst case is only achieved when each invocation returns to the remainder section and re-enters the trying protocol before its predecessor begins the abort protocol.

For modularity, we first define an object type, *S-HAD*, from which it is easy to construct a local-spin abortable mutual exclusion algorithm. S-HAD is a sequence that supports Head, Append, and Delete, but with two restrictions: Each process can own at most one element in the sequence at a time and only the owner of an element can perform these three operations on it.

We give two wait-free implementations of an S-HAD object. Our first implementation has $O(N^2)$ RMA complexity but uses unbounded space. Then, we extend it, using a generalization of reference counts, to achieve $O(N^2)$ space complexity as well. Our new memory reclamation method is wait-free and very efficient in terms of RMAs. It uses only standard operations (TEST_AND_SET, FETCH_AND_ADD, FETCH_AND_STORE, READ and WRITE) on $O(\log N)$ bit words, and each process performs $O(1)$ RMAs for recycling a record. In contrast, Detlefs et al.'s reference counting method [4] uses DOUBLE_COMPARE_AND_SWAP, which is not available in most systems, and Valois's reference counting method [14] allows processes to access a freed record or a recycled record, which would cause a significant increase in the RMA complexity of our algorithm. With *hazard pointers* [11], to reuse a record, a process must read the hazard pointers of all other processes, which takes $\Theta(N)$ RMAs. Herlihy et al. [7] proposed a reference counting method similar to hazard pointers. Their method also takes $\Theta(N)$ RMAs. Since we want each process to perform a small number of RMAs if aborts are rare, we needed to develop a new memory reclamation method.

Section 2 formally defines abortable mutual exclusion and describes the system model. Section 3 defines S-HAD, gives an abortable mutual exclusion algorithm based on S-HAD, and proves the correctness of the algorithm. Section 4 presents our unbounded space implementation of S-HAD, and Section 5 presents our bounded space implementation of S-HAD. Complete proofs of correctness of the algorithms in Sections 4 and 5 appear in the full paper.

## 2  Preliminaries

In an abortable mutual exclusion algorithm, processes that want to access the critical section first execute the trying protocol. After completing the trying protocol, a process enters the critical section. When it finishes the critical section, it then performs the *exit protocol*, and finally returns to the remainder section. If a process must wait in the trying protocol and wants to abort, it performs the abort protocol, and then returns to the remainder section. We assume no process failures.

An algorithm solves the abortable mutual exclusion problem, if it satisfies the following properties:

**Mutual Exclusion:** At most one process is in the critical section at any time.
**Lockout Freedom:** If a process $p$ starts executing the trying protocol and keeps taking steps in the trying protocol without aborting, then it will eventually enter the critical section.
**Bounded Exit:** If a process starts executing the exit protocol, then it returns to the remainder section within a bounded number of its own steps.
**Bounded Abort:** If a process starts executing the abort protocol, then it returns to the remainder section within a bounded number of its own steps.

The *First-Come-First-Served (FCFS)* property [10] is a strong fairness condition in which processes enter the critical section in roughly the same order they enter the trying protocol. Although it is not a requirement of abortable mutual exclusion, most mutual exclusion algorithms in which each process performs $O(1)$ RMAs to enter the critical section satisfy this property.
**FCFS:** The doorway is a bounded section of code that begins the trying protocol. If a process $p$ finishes executing the doorway before a process $q$ begins executing the doorway, and $p$ does not abort, then $p$ enters the critical section before $q$ does.

In this paper, we consider the asynchronous cache-coherent (CC) model with $N$ processes [1]. The CC model is a shared memory model in which each process has its own local cache. In this model, processes perform atomic operations on shared variables. We divide all atomic operations into two classes: *trivial* operations, which cannot change the value of a shared variable, and *non-trivial* operations, which may change the value of a shared variable. READ is an example of a trivial operation. WRITE, FETCH_AND_STORE and COMPARE_AND_SWAP are examples of non-trivial operations.

When a process $p$ performs a trivial operation on a shared variable, it first checks its own cache. If $p$ has a valid cached copy of the shared variable (i.e. no other process has performed a non-trivial operation on the shared variable since $p$ last accessed the shared variable and copied it to its cache), the trivial operation does not generate an RMA. If $p$ does not have a valid cached copy of the shared variable (either because $p$ has not accessed the shared variable before or because another process has performed a non-trivial operation on the shared variable after $p$'s last access of the shared variable), then $p$ accesses the shared variable from remotely located shared memory and copies the variable to its own cache. This generates an RMA. When $p$ performs a non-trivial operation on a shared variable, even if the value of the variable does not change, the system invalidates all other cached copies of the variable, which generates an RMA.

A *passage* is the sequence of steps performed by a process from when it begins the trying protocol until it next returns to the remainder section by finishing the exit or abort protocol. Our complexity measure is the worst case number of RMAs performed in the trying, exit, and abort protocols in any passage.

## 3   S-HAD and Abortable Mutual Exclusion

An S-HAD is a sequence of elements, each owned by a different process. A process can perform the following operations on an element that it owns:

Head($R$): returns TRUE if element $R$ is at the beginning of the sequence.
Append($R$): appends element $R$ to the end of the sequence.
Delete($R$): deletes element $R$ from the sequence.

Append($R$) may be called only when $R$ is not in the sequence, and Delete($R$) may be called only when $R$ is in the sequence. Thus, element $R$ occurs in the sequence if and only if Delete($R$) has not been performed since Append($R$) was last performed. Head($R$) is TRUE if and only if $R$ occurs in the sequence and each element $X$ that was appended before $R$ has been deleted from the sequence.

We can easily build an abortable mutual exclusion algorithm using a linearizable implementation of an S-HAD object. When a process tries to enter the critical section, it appends a new element to the S-HAD object. Then the process keeps performing Head until its element is at the head of the S-HAD object. When Head returns TRUE, the process enters the critical section. When the process finishes the critical section or wants to abort, it deletes the appended element from the S-HAD object. The detailed algorithms TryingProtocol, Exit-Protocol and AbortProtocol appear in Figure 1. GetNewElement is a function that returns a new element. This may be a system call that allocates a memory location for an element or a function that returns an element from a free list.

TryingProtocol()

  T1:  $R :=$ GetNewElement()

  T2:  Append($R$)

  T3:  **while** $\neg$ Head($R$) **do**

  T4:    if the process wants to abort, perform AbortProtocol()

    **end while**

ExitProtocol() / AbortProtocol()

  E1:  Delete(R)

**Fig. 1.** Abortable Mutual Exclusion Algorithm

To prove the correctness of this abortable mutual exclusion algorithm, we show that only the process whose element is at the head of the sequence enters the critical section. We also show that any appended element eventually becomes the head of the sequence if it is not deleted. If process $p$ gets an element $R$ on line T1, we say $owner(R) = p$.

**Observation 1.** *If a process $p$ is in the critical section, then the element $R$ at the head of the sequence is owned by $p$.*

An operation is *wait-free* if a process performs the operation within a bounded number of its own steps. Since any abortable mutual exclusion algorithm must satisfy the bounded exit and bounded abort properties, Delete must be wait-free. If GetNewElement, Append and Head are also wait-free, then the while loop starting on line T3 is the only waiting period. In this case, the algorithm in Figure 1 is an FCFS abortable mutual exclusion algorithm.

**Theorem 1.** *Given wait-free implementations of an S-HAD object and Get-NewElement, the algorithm in Figure 1 is an FCFS abortable mutual exclusion algorithm.*

*Proof.* The mutual exclusion property follows from Observation 1. Since Delete is wait-free, the algorithm satisfies the bounded abort and bounded exit properties. To prove lockout freedom, suppose that there exists an infinite execution $E$ in which some set of processes, $\mathcal{P}$, keep performing TryingProtocol without entering the critical section or performing AbortProtocol.

Since GetNewElement and Append are wait-free, each process $p$ in $\mathcal{P}$ eventually gets a new element, $R_p$, on line T1 of its last invocation of TryingProtocol, and finishes performing Append($R_p$) on line T2. Since $p$ does not perform Exit-Protocol or AbortProtocol after its last invocation of TryingProtocol, $p$ does not subsequently perform Delete($R_p$). Let $\mathcal{R} = \{R_p | p \in \mathcal{P}\}$. Let $X$ be the element in $\mathcal{R}$ appended earliest, and let $p \in \mathcal{P}$ be the process that performed Append($X$).

By definition, any invocation that last appended an element $S$ before $X$ either eventually enters the critical section and performs ExitProtocol, or eventually performs AbortProtocol. Hence, the invocation eventually performs Delete($S$). Thus, eventually, $X$ becomes the head of the sequence and Head($X$) returns

TRUE. Since $p$ keeps performing TryingProtocol without performing Abort-Protocol, it performs Head($X$) infinitely many times. Thus, $p$ will eventually enter the critical section. This contradicts the assumption that $p \in \mathcal{P}$, so the algorithm satisfies lockout freedom.

Since GetNewElement and Append are wait-free, each process performs lines T1 and T2 within a bounded number of its own steps. Let the doorway be lines T1 and T2. If process $p$ finishes Append($R$) before process $q$ starts an invocation of GetNewElement that returns $R'$, then $R$ is appended before $R'$. Thus, if $p$ does not abort, $R$ reaches the head of the sequence before $R'$. Then, by Observation 1, $p$ enters the critical section before $q$. Hence, the resulting abortable mutual exclusion algorithm satisfies the FCFS property.                                   □

In some systems, allocating a memory location may not be wait-free. However, the algorithm in Figure 1 still solves abortable mutual exclusion if GetNewElement satisfies the following properties: a process that invokes GetNewElement eventually completes GetNewElement and a process that invokes GetNewElement but wants to return to the remainder section before it completes can do so within a bounded number of its own steps. These properties are required for lockout freedom and bounded abort, respectively.

For this algorithm to be local-spin, Head must be implemented carefully. The RMA complexity of one passage is the sum of the RMAs performed during one execution of each of GetNewElement, Append and Delete, and an unbounded number of executions of Head. Thus, in the DSM model, if Head contains even a single RMA, then the resulting algorithm is not local-spin. However, in the CC model, when a process reads a shared variable, it copies its value to its local cache. Hence, even if Head contains remote memory reads, subsequent calls of Head by process $p$ do not generate RMAs unless another process performs a non-trivial operation on a shared variable $p$ reads in Head.

In the next two sections, we present wait-free, linearizable implementations of an S-HAD object shared by $N$ processes such that any number of calls of Head($R$) between a call of Append($R$) and the subsequent call of Delete($R$) generate only a bounded number of RMAs in the CC model. Moreover, if each element is deleted only when it is at the head of the sequence, this number of RMAs is bounded above by a small constant. Our first implementation is simpler but uses unbounded space, and our second implementation uses bounded space.

## 4   A Simple Implementation of S-HAD

In this section, we present a simple implementation of an S-HAD object. Detailed pseudo-code is given in Figure 2. Note that the lines are not consecutively numbered. This is so each line has the same number as in the bounded space implementation in Section 5.

We begin by explaining the overall structure of the implementation. An S-HAD object is represented by an intree of records, one per element, each with a pointer, *pred*, which is either NIL or points to another record, and a flag, *del*. The root of the tree is a *dummy* record, which is never deleted, whose *del* field is always 'head' and whose *pred* field is always NIL. For every other record $R$, the field *R.del* indicates whether the element it represents is in the S-HAD sequence or has been *logically deleted*. The initial value of *R.del* is FALSE, and it becomes TRUE when the owner of $R$ performs line D1 of Delete($R$). The field *R.pred* points to another record that was appended before $R$. Thus, the records form an acyclic graph rooted at the dummy record. There is a FETCH_AND_STORE (or SWAP) object, *Tail*, that initially points to the root. To perform Append($R$), a process atomically reads *Tail* and updates *Tail* to point to $R$ on line A2 of Append($R$). Hence, *Tail* always points to the record that was appended most recently.

When a process wants to know whether the element represented by its record $R$ is at the head of the sequence, it repeatedly updates *R.pred* until it points to a record that has not been deleted. This is done by Update($R$). Then, the element represented by $R$ is at the head of the sequence if and only if *R.pred* points to the dummy record.

A process logically deletes its record $R$ by setting *R.del* to TRUE. Then it calls Update($R$) one more time to ensure that $R$ does not point to another logically deleted record. This is necessary because, otherwise, a sequence with two records that are preceded by arbitrarily many logically deleted records between them and the dummy record can be created by repeatedly deleting the second last record and then appending a new record.

At any point during an execution, the state of the S-HAD object is the sequence of records $R$ for which line A2 of Append($R$) has been performed and *R.del* = FALSE. This sequence is ordered by the time at which line A2 of Append($R$) was performed. All records that represent elements in the S-HAD sequence are on the same path to the root and the one that is closest to the root is at the head of the sequence.

We define the linearization point of Append($R$) to be when line A2 is performed. Immediately afterwards, *R.del* = FALSE. Hence, by performing line A2 of Append($R$), the element represented by $R$ is appended to the end of the sequence. The element represented by $R$ is removed from the sequence when *R.del* is set to TRUE on line D1. We define this to be the linearization point of Delete. We define the linearization point of Head($R$) to be when Update($R$) returns on line H1, which is when line U2 of Update($R$) is performed with $(*mypred).del \neq$ TRUE. The correctness of the implementation in Figure 2 follows from the next two results.

**Observation 2.** *At the linearization point of Head($R$), let $S$ be the record pointed to by R.pred and let $d = S.del$. Then Head($R$) returns TRUE if and only if $S$ is the dummy record. If Head($R$) returns FALSE, then $d$ = FALSE.*

**Lemma 3.** *Head($R$) returns TRUE if and only if the element represented by $R$ is at the head of the sequence at the linearization point of Head($R$).*

Append($R$) is wait-free, since it consists of only two atomic operations. Similarly, Head($R$) and Delete($R$) are wait-free if Update($R$) is wait-free. The following lemma shows that Update is wait-free.

*shared variables:*

> type Record ( *pred*: pointer to a record ∪ { NIL }, initially NIL
>                     *del*: { TRUE, FALSE, 'head' }, initially FALSE)
> Record *Dummy* = (NIL, 'head')
> *Tail*: pointer to a record, initially points to *Dummy*

*private variables:*

> *mypred, ppred*: pointer to a record

Head(*R* :Record) % *Precondition:* R.del = FALSE, R.pred ≠ NIL
% *Postcondition:* returns TRUE, if R is the head of the list; otherwise, returns FALSE

  H1: Update(*R*)
  H2: *mypred* := *R.pred*
  H3: return ((\**mypred*).*del* = 'head')

Append(*R* :Record) % *Precondition:* R.del = FALSE, R.pred = NIL

  A2: *mypred* := FETCH_AND_STORE(*Tail*, &*R*)
  A3: *R.pred* := *mypred*

Delete(*R* :Record) % *Precondition:* R.del = FALSE, R.pred ≠ NIL

  D1: *R.del* := TRUE
  D2: Update(*R*)

Update(*R* :Record) % *Precondition:* R.pred ≠ NIL

  U1: *mypred* := *R.pred*
  U2: **while** (\**mypred*).*del* = TRUE **do**
  U3:     *ppred* := (\**mypred*).*pred*
  U5:     *R.pred* := *ppred*
  U9:     *mypred* := *ppred*
       **end while**

**Fig. 2.** An Implementation of S-HAD

**Lemma 4.** *If no record is appended more than once, then the while loop of Update(R) is not performed forever.*

*Proof sketch.* In each execution of the while loop in Update(*R*), *R.pred* is updated. Each time *R.pred* is updated, *R.pred* points to a record that was appended earlier than the record it previously pointed to. Since the number of records that were appended earlier than *R* is bounded, *R.pred* is updated a bounded number of times.                                                    □

Hence, the implementation in Figure 2 is wait-free. If GetNewElement is a wait-free system call that always returns a new record, then, by Theorem 1, the algorithm in Figure 1 using the implementation in Figure 2 is a correct FCFS abortable mutual exclusion algorithm.

While *R.pred* does not change, any sequence of calls to Head(*R*) generates at most three RMAs in the CC model: the first time *owner(R)* reads *R.pred* and (\**R.pred*).*del*, and when (\**R.pred*).*del* changes from FALSE to TRUE. If all records are deleted in the same order as they are appended, which is the case for our abortable mutual exclusion when no aborts occur, then $R.pred$ changes only once. Hence, in the abortable mutual exclusion algorithm using this implementation of S-HAD, each process performs $O(1)$ RMAs if no aborts occur.

We say that a record $R$ was *deleted prematurely* if $R.pred$ did not point to the dummy record when line D1 of Delete($R$) was performed. In the abortable mutual exclusion algorithm in Figure 1, each invocation that is aborted corresponds to a prematurely deleted record. If $k$ is the number of processes that delete records prematurely, then we prove that the *pred* pointer of every record changes $O(k^2)$ times.

**Lemma 5.** *Let $R'$ be the last record that was appended prior to element $R$, but was not prematurely deleted. If $k'$ is the number of different processes that appended records between $R'$ and $R$ inclusive, then the while loop of Update($R$) was performed at most $k'(k' + 3)/2$ times between beginning Append($R$) and completing Delete($R$).*

*Proof sketch.* A record $X$ is *active* if and only if the first line of Append($X$) has been performed, but the last line of Delete($X$) has not yet been performed. Note that, if the element represented by $X$ is in the sequence, then $X$ is also active, but the converse may not hold after line D1 of Delete($X$) has been performed. After record $X$ becomes inactive, $X.pred$ does not change.

Suppose there is a sequence of records, $W_1, W_2, \ldots, W_{j-1}, W_j$ such that $W_{i+1}.pred$ points to $W_i$ for $1 \le i < j$. In this case, we say that there is a *path* from $W_j$ to $W_1$. If all of $W_1, \ldots, W_j$ are inactive, then $W_i$ was active when $W_{i+1}$ became inactive. Hence, Delete($W_{i+1}$) was completed before Delete($W_i$). In particular, Delete($W_j$) was completed before Delete($W_1$). If $owner(W_1) = owner(W_j)$, then Delete($W_1$) was completed before $W_j$ was appended and hence, before Delete($W_j$) was completed. Thus, in this case, at least one of $W_1, \ldots, W_{j-1}$ is active.

Since there are $k'$ different processes that appended an element between $R'$ and $R$ inclusive during Update($R$), the path from $R$ to $R'$ contains $\ell \le k'$ active records, the first of which is $R$. If $Y_i$ is the $i$th active record between $R'$ and $R$, for $i = 1, \ldots, \ell$, then the records between $Y_i$ and $Y_{i+1}$ are all inactive. It follows from the previous paragraph that each inactive record on the path from $Y_{i+1}$ to $Y_i$ has a different owner. Hence, the path from $Y_{i+1}$ to $Y_i$ contains at most $k' - i$ records. Also, the subpath from $Y_1$ to $R'$ contains at most $k'$ records. Thus, the number of records on any path from $R$ to $R'$ is at most $(k' - \ell + 1) + (k' - \ell + 2) + \ldots + (k' - 1) + k' + \ell = \ell(2k' - \ell + 3)/2 \le k'(k' + 3)/2$. Since $R'$ was not prematurely deleted, $R'.pred$ pointed to the dummy record when $R'.del$ was set to TRUE. Hence, immediately after $R'$ was logically deleted, the path from $R$ to the dummy record contained at most $k'(k'+3)/2+1$ records. Therefore, $R.pred$ was updated at most $k'(k' + 3)/2$ times.                    □

Excluding Update, each process performs $O(1)$ RMAs during Append, Head, and Delete. Lemma 5 implies that a process performs $O(k^2)$ RMAs between beginning Append($R$) and completing Delete($R$). Thus, if GetNewElement takes $O(1)$ RMAs, the algorithm in Figure 1 using the implementation in Figure 2 has $O(k^2)$ RMA complexity.

**Theorem 2.** *Suppose GetNewElement takes $O(1)$ RMAs. In the algorithm in Figure 1 using the implementation in Figure 2, each process performs $O(k^2)$ RMAs per passage, where $k$ is the number of processes that began the trying protocol immediately beforehand and subsequently aborted.*

Since $k$ is bounded by $N$, the worst case RMA complexity is $O(N^2)$. This worst case can occur, but only if $\Theta(N)$ processes perform particular sequences of Appends and Deletes. A specific execution that generates the worst case is described in the full paper.

# 5   An Implementation of S-HAD with Bounded Space

In the previous algorithm, even though records have been logically deleted from the S-HAD object, processes can still access them to find out that they have been deleted. Also, each time a process performs Append, it uses a new record. Because logically deleted records are not deallocated, that algorithm uses unbounded space. However, eventually, a logically deleted record is no longer accessed, and we can safely reclaim the memory used by that record. To determine when a logically deleted record is no longer accessed, we use a generalization of *reference counts*. If the generalized reference count for a record becomes zero, then the record can be physically deleted, since no process will subsequently access the information in the record.

In simple reference counting, each record contains a counter and a record can be physically deleted when its counter is zero. If record $R$ points to record $X$, record $S$ points to $Y$, and a process wants to change $R$ to point to the same place as $S$, then it reads $\&Y$ from $S$, increments $Y$'s counter, sets $R$ to $\&Y$, and finally decrements $X$'s counter. However, if $Y$'s counter becomes zero and $Y$ is physically deleted between the first two steps, then the owner of $R$ will not notice this and may access the location in memory from which $Y$ was deleted. One way to prevent this is to perform the first two steps atomically using DOUBLE_COMPARE_AND_SWAP (Detlefs et al. [4]). Unfortunately, DOUBLE_COMPARE_AND_SWAP is not available in most systems.

Another approach is for the process to read $S$ after it increments $Y$'s counter and, if $S$'s pointer has been changed, it decrements $Y$'s counter instead of changing $R$ [14]. In this case, no other information in $Y$ is accessed. However, this method still allows access to the counter of a physically deleted record, so the memory it occupies cannot be reclaimed by the system. Physically deleted records can be put into a free list and reused in the future. However, when processes access the counter of a free or recycled record, extra RMAs are generated. Using this method in our algorithm increases the worst case RMA complexity from $\Theta(N^2)$ to $\Theta(N^4)$.

*shared variables:*

  type Record ( *rc*: a pair of integers $(orc, drc)$, where $0 \leq orc < N$ and
        $-N < drc < N$, initially $(0, 0)$

      *pred*: a pair $(predptr, prc)$, where *predptr* is NIL or a pointer to
          a record and $0 \leq prc < N$ is an integer, initially (NIL, 0)

      *del*: { TRUE, FALSE, 'head' }, initially FALSE

      *done*: { TRUE, FALSE}, initially FALSE)

  Record *Dummy* = ((0,0), (NIL, 0), 'head', 0)

  *Tail*: pointer to a record, initially points to *Dummy*

*private variables:*

  *mypred*, *ppred*: pointer to a record

  *myprc*, *x*, *y*: integer

Head($R$ :Record) % *Precondition:* $R.del$ = FALSE, $R.pred \neq$ (NIL, $-$)
% *Postcondition:* returns TRUE, if $R$ is the head of the list; otherwise, returns FALSE

  H1: Update($R$)

  H2: $(mypred, -) := R.pred$

  H3: return $((*mypred).del = $ 'head')

Append($R$ :Record) % *Precondition:* $R.del$ = FALSE, $R.pred$ = (NIL, 0)

  A1: $R.rc := (1, 1)$

  A2: $mypred := $ FETCH_AND_STORE($Tail$, &$R$)

  A3: $R.pred := (mypred, 0)$

Delete($R$ :Record) % *Precondition:* $R.del$ = FALSE, $R.pred \neq$ (NIL, $-$)

  D1: $R.del := $ TRUE

  D2: Update($R$)

  D3: Remove($R$)

Update($R$ :Record) % *Precondition:* $R.pred \neq$ (NIL, $-$)

  U1: $(mypred, -) := R.pred$

  U2: **while** $(*mypred).del$ = TRUE **do**

  U3:   $(ppred, -) := $ FETCH_AND_ADD($(*mypred).pred, (0, 1)$)

  U4:   FETCH_AND_ADD($(*ppred).rc, (1, 0)$)

  U5:   $(-, myprc) := $ FETCH_AND_STORE($R.pred, (ppred, 0)$)

  U6:   $(x, y) := $ FETCH_AND_ADD($(*mypred).rc, (-1, myprc - 1)$)

  U7:   **if** $(x, y) = (1, 1 - myprc)$ **then**
        % Note that $(*mypred).rc = (0, 0)$

  U8:     Remove($*mypred$)
      **end if**

  U9:   $mypred := ppred$
    **end while**

Remove($R$ :Record)

  R1: **if** TEST_AND_SET($R.done$) = TRUE **then**

  R2:   $(mypred, myprc) := $ FETCH_AND_STORE($R.pred, $ (NIL, 0))

  R3:   $(x, y) := $ FETCH_AND_ADD($(*mypred).rc, (-1, myprc - 1)$)

  R4:   recycle($R$)

  R5:   **if** $(x, y) = (1, 1 - myprc)$ **then**
        % Note that $(*mypred).rc = (0, 0)$

  R6:     Remove($*mypred$)
      **end if**
    **end if**

**Fig. 3.** An Implementation of S-HAD with bounded space

Due to their weaknesses, instead of adopting previous methods, we devise a new reference counting method for our algorithm. In our new memory reclamation method, each record has a pointer *predptr*, and an *original reference counter* (*orc*), which stores an upper bound on the number of pointers in shared memory that point to it. In addition to *orc*, each record also has two more counters, a *proactive reference counter* (*prc*) and a *distributed reference counter* (*drc*). Both *prc* and *drc* are used to keep track of pointers that have been read and may be written to shared memory in the future.

$R.prc$ stores the number of times $R.predptr$ has been read since $R.predptr$ was last updated. This value is transferred to $S.drc$ when $R.predptr$ is changed from pointing to $S$ to pointing to another record. In general, for any record $S$, the sum of the *prc*'s of all records that point to $S$ plus $S.drc$ is bounded above by the number of times a pointer to $S$ has been read minus the number of times a pointer to $S$ has been overwritten.

$R.drc$ is stored together with $R.orc$ in a single variable $R.rc$, so that they can be accessed together. The range of *orc* is from 0 to $N-1$ and the range of *drc* is from $1-N$ to $N-1$. Hence, $rc = (drc, orc)$ can be represented using $O(\log N)$ bits in a single word of memory. FETCH_AND_ADD$(rc, (m, n))$ can be simulated by FETCH_AND_ADD$(rc, m \cdot 2^{\lceil \log_2 N \rceil} + n)$.

$R.prc$ is stored together with $R.predptr$ in a single variable $R.pred$. Associating a pointer with a counter was also done in [6] and [8]. In [8], a wait-free implementation of a pointer requires a complicated atomic operation. However, in our algorithm, processes perform only READ, WRITE, FETCH_AND_ADD and FETCH_AND_STORE operations on pointers. Pointers in [6] are similar to ours, but are stored together with two integers.

Since the range of *prc* is from 0 to $N-1$, *pred* can be represented using $\lceil \log_2 N \rceil$ bits in addition to the bits used for the pointer, all stored in one word. Since we use only $O(N^2)$ records, a pointer can be represented using $O(\log N)$ bits. FETCH_AND_ADD$(pred, k)$ adds $k$ to *prc*. In our algorithm, whenever *predptr* is set to point to a record $X$, *prc* becomes zero, which can be accomplished by FETCH_AND_STORE$(pred, (\&X, 0))$.

Pseudo-code for the algorithm is presented in Figure 3. Head$(R)$ is essentially the same as in the previous algorithm. In Append$(R)$, the owner of $R$ sets $R.rc$ to $(1,1)$ before appending $R$ to the end of the sequence. Most of the differences are inside the while loop of Update$(R)$. Unlike the previous algorithm, $R.pred$ can now be changed by processes other than the owner of $R$, on lines U3 and R2, but only after $R$ has been logically deleted. This does not affect the RMA complexity of Head$(R)$, which is only performed while $R$ is in the sequence.

To see how Update$(R)$ was modified, consider the situation when process $p$, which owns record $R$, wants to update $R$'s predecessor pointer to point to the predecessor of its predecessor, i.e. $R.predptr := (*R.predptr).predptr$. Suppose $X$ is $R$'s predecessor, $Y$ is $X$'s predecessor, and $p$'s local variable *mypred* points to $X$. To change $R$ to point from $X$ to $Y$, process $p$ performs line U3, in which $p$ atomically reads $X.predptr$ and increments $X.prc$ using FETCH_AND_ADD. This indicates that $R$ will reference $Y$ and it learned about $Y$ from $X$. Next, $p$ increments

$Y.orc$ on line U4. On line U5, $p$ atomically changes $R.predptr$ to $Y$, reads $R.prc$ into its local variable $myprc$, and resets $R.prc$ to 0, using FETCH_AND_STORE. Hence, $myprc$ stores the number of processes that have accessed $R.predptr$ between the last two updates of $R.predptr$. Finally, on line U6, $p$ atomically decrements $X.orc$ and adds $myprc - 1$ to $X.drc$, using FETCH_AND_ADD. The distributed reference count is decremented, since $R$ is no longer pointing to $X$. The value that had been stored in $R.prc$ before it was reset is transferred to $X.drc$. Both of these are accomplished by adding $myprc - 1$ to $X.drc$.

When a process tries to physically delete a record $R$, it calls function Remove($R$). $R$ can be physically deleted only when no record points to $R$, no records will point to $R$, and lines D1 and D2 of Delete($R$) have been completed. $R.orc = 0$ indicates that no record currently points to $R$, and $R.drc = 0$ indicates that no record will point to $R$. Hence, when $R.rc = (0,0)$ and Delete($R$) is completed, $R$ can be physically deleted. To ensure that both conditions are met, Remove($R$) is called twice: one by the owner of $R$ at the end of Delete($R$) (line D3) and the other by a process who finds that $R.rc = (0,0)$ during Update (line U8) or Remove (line R6). Only the later of these two calls physically deletes $R$ by calling recycle($R$) on line R4. recycle($R$) can be either a system call that deallocates $R$ from memory or some function that moves $R$ into a free list.

Remove($R$) is called from exactly one of line U8 or line R6, so Remove($R$) is called exactly twice. To ensure that only the later call physically deletes $R$, we use a TEST_AND_SET object, $R.done$, and only perform the rest of Remove if it returns TRUE. Note that a record $R$ can be physically deleted by any process, although Delete($R$) can be called only by the owner of $R$.

Remove is called recursively if physically deleting a record causes another record's reference counts to become $(0,0)$. When a process physically deletes a record $R$, it also removes its pointer, $R.predptr$. If $R.predptr$ pointed to another record $S$, then $S$'s reference counts must be updated. This may cause $S.rc$ to become $(0,0)$ and, if it is, Remove($S$) is called recursively on line R6. These recursive calls add only $O(k^2)$ RMAs in total, if $k$ is the number of processes that appended a record before $R$ and deleted it prematurely. Hence, it does not affect the overall asymptotic RMA complexity of the algorithm.

Unlike the reference counting in [14], our algorithm allows each record to be reclaimed by the system, provided the system calls for memory allocation and deallocation each take $O(1)$ RMAs. In this case, GetNewRecord in Figure 1 is a system call for memory allocation and recycle($R$) on line R4 of Figure 3 is a system call for memory deallocation.

Alternatively, we can use a free list of length at most $3N$ for each process. The reason $3N$ records per process suffice is discussed in the full paper. Each process, $p$, maintains a Boolean array of size $3N$, which indicates which records are available. To get a new record, process $p$ keeps checking each element of the array until it finds a true bit. If the $i$th bit in the array is TRUE, $p$ sets it to FALSE and uses its $i$th record. When some process recycles the $i$th record of $p$, it sets the $i$th element of $p$'s array to TRUE. Since $p$ is the only process that sets

**Table 1.** Local-spin abortable mutual exclusion algorithms

| | Scott [12] | Jayanti [9] | Danek and Lee [3] | New Algorithm |
|---|---|---|---|---|
| Atomic operations used besides READ and WRITE | FETCH_AND_STORE, COMPARE_AND_SWAP | LL/SC | - | TEST_AND_SET, FETCH_AND_ADD, FETCH_AND_STORE |
| Local-spin on CC | Yes | Yes | Yes | Yes |
| Local-spin on DSM | Yes | Yes | Yes | No |
| RMAs / passage if no aborts | $O(1)$ | $\Theta(\log N)$ | $\Theta(\log N)$ ; $\Theta(N)$ | $O(1)$ |
| RMAs / passage | unbounded | $\Theta(\log N)$ | $\Theta(\log N)$ ; $\Theta(N)$ | $O(N^2)$ |
| space | unbounded | $\Theta(N)$ | $\Theta(N)$ | $\Theta(N^2)$ |
| FCFS | Yes | Yes | No ; Yes | Yes |

elements of its array to FALSE, no RMA is generated when $p$ reads FALSE. Therefore, both GetNewRecord in Figure 1 and recycle($R$) on line R4 of Figure 3 generate only $O(1)$ RMAs.

The resulting algorithm uses only $O(N^2)$ space. It also has the same RMA complexity, $O(N^2)$, as the previous algorithm. Therefore, the abortable mutual exclusion algorithm in Figure 1 using this implementation of S-HAD is local-spin, uses $O(N^2)$ space, has $O(N^2)$ RMA complexity, and each process performs $O(k^2)$ RMAs per passage, where $k$ is the number of processes that began the trying protocol immediately beforehand and subsequently aborted.

## 6    Conclusions

We presented a local-spin abortable mutual exclusion algorithm with $O(N^2)$ space, in which each process performs $O(1)$ RMAs for each entry to the critical section when no processes abort, and each process performs $O(k^2)$ RMAs when aborts occur in the CC model, where $k$ is the number of processes that abort. Table 1 compares our algorithm with previous abortable mutual exclusion algorithms.

Our algorithm performs more RMAs per passage than Jayanti's and Danek and Lee's in the worst case, but fewer when no aborts occur. If $k = o(\sqrt{\log N})$ processes began the trying protocol immediately before process $p$ and subsequently aborted, then $p$ performs $o(\log N)$ RMAs per passage in our algorithm, which is better than Jayanti's or Danek and Lee's algorithms. It would be interesting to compare the experimental performance of our algorithm with the other algorithms.

It is open whether $\Omega(N^2)$ space and RMAs are necessary in the CC model, if each process performs a constant number of RMAs when no processes abort. It is also open whether there exists a local-spin abortable mutual exclusion algorithm in the DSM model with bounded space and RMAs, in which each process performs a constant number of RMAs when no processes abort.

# Acknowledgements

I would like to thank Professor Faith Ellen and the anonymous reviewers for numerous helpful suggestions and careful corrections.

# References

1. Anderson, J.H., Kim, Y.-J., Herman, T.: Shared-Memory Mutual Exclusion: Major Research Trends Since 1986. Distributed Computing (2002)
2. Attiya, H., Hendler, D., Woelfel, P.: Tight RMR Lower Bounds for Mutual Exclusion and Other Problems. In: Proceedings of the 40th Annual ACM Symposium on Theory of Computing, pp. 217–226 (2008)
3. Danek, R., Lee, H.: Brief Announcement: Local-Spin Algorithms for Abortable Mutual Exclusion and Related Problems. In: Taubenfeld, G. (ed.) DISC 2008. LNCS, vol. 5218, pp. 512–513. Springer, Heidelberg (2008)
4. Detlefs, D.L., Martin, P.A., Moir, M., Steele Jr., G.L.: Lock-Free Reference Counting. In: The 20th Annual ACM Symposium on Principles of Distributed Computing, pp. 190–199 (2001)
5. Dijkstra, E.W.: Solution of a Problem in Concurrent Programming Control. Communications of the ACM 8(9), 569 (1965)
6. Goldberg, B.: Generational Reference Counting: A Reduced-communication Distributed Storage Reclamation Scheme. In: Proceedings of the ACM SIGPLAN 1989 Conference on Programming Language Design and Implementation (1989)
7. Herlihy, M., Luchangco, V., Martin, P., Moir, M.: Brief Announcement: Dynamic-sized lock-free data structures. In: Proceedings of the 21st Annual Symposium on Principles of Distributed Computing (2002)
8. Herlihy, M., Luchangco, V., Moir, M.: Space and Time Adaptive Non-blocking Algorithms. Electronic Notes in Theoretical Computer Science 78, 260–280 (2003)
9. Jayanti, P.: Adaptive and Efficient Abortable Mutual Exclusion. In: Proceedings of the 22th Annual ACM Symposium on Principles of Distributed Computing (July 2003)
10. Lamport, L.: A New Solution of Dijkstra's Concurrent Programming Problem. Communications of the ACM 17(8), 453–455 (1974)
11. Michael, M.M.: Hazard Pointers: Safe Memory Reclamation for Lock-Free Objects. IEEE Transactions on Parallel and Distributed Systems 15(6) (2004)
12. Scott, M.L.: Non-blocking Timeout in Scalable Queue-based Spin Locks. In: The 21st Annual Symposium on Principles of Distributed Computing (July 2002)
13. Scott, M.L., Scherer III, W.N.: Scalable Queue-based Spin Locks with Timeout. In: The 8th ACM Symposium on Principles and Practice of Parallel Programming (June 2001)
14. Valois, J.D.: Lock-Free Linked Lists Using Compare-and-Swap. In: Proceedings of the 14th Annual ACM Symposium on Principles of Distributed Computing, pp. 214–222 (1995)

# Turning Adversaries into Friends: Simplified, Made Constructive, and Extended

Eli Gafni[1] and Petr Kuznetsov[2]

[1] Computer Science Department, UCLA
[2] Deutsche Telekom Laboratories/TU Berlin

**Abstract.** A *liveness contract* is an agreement between the specifier of a system and a task to solve, and the programmer who makes her living by delivering protocols. In a shared-memory system, a liveness contract specifies infinite suffixes of executions in which the programmer is required to solve a distributed task. If the behavior of the system does not comply with the specification, no output is required. A convenient way to describe a large class of liveness contracts was recently proposed by Delporte et al. For a system $\Pi$ of $n$ processes, an *adversary* is a set $\mathcal{A}$ of subsets of $\Pi$. The system is required to make progress only in executions in which the set of correct processes is in $\mathcal{A}$.

Given an adversary $\mathcal{A}$ and a task $T$, should the programmer sign the contract? Can she deliver?

In this paper, we give a very simple resolution of this question for colorless tasks that contrasts with more involved arguments of the original paper of Delpote et al. More importantly, our resolution is constructive — it tells the programmer how to use $\mathcal{A}$ to solve $T$, when it is solvable.

Our framework naturally generalizes to systems enriched with more powerful objects than read-write registers. We determine necessary and sufficient conditions for an adversary $\mathcal{A}$ to solve consensus using $j$-process consensus objects and read-write registers, which resolves an open question raised recently by Taubenfeld.

## 1 Introduction

Distributed computing is about overcoming asynchrony and failures. Wait-free system, a system where we make no assumptions about some synchrony or correctness of some processes, can solve only few interesting tasks. To solve more interesting tasks, we should make more assumptions about the system behavior.

Recently [7], Delporte et al. proposed a class of assumptions that they called *adversaries*. In their view, an *adversary* controls *sets* of processes that may fail in a given execution, regardless of the time when they fail. Put differently, an adversary is defined as a collection $\mathcal{A}$ of sets of processes, and they only consider executions where some element in $\mathcal{A}$ is exactly the set of correct processes. Following [7], in this paper, we explore the ability of such adversaries to enhance solvability of *distributed tasks*, defined in terms of inputs the processes receive, outputs the processes produce, and a binary relation that maps inputs to the sets of possible outputs.

C. Lu, T. Masuzawa, and M. Mosbah (Eds.): OPODIS 2010, LNCS 6490, pp. 380–394, 2010.

Why are adversaries interesting to look at? In a shared-memory system, it is straightforward to ensure that the outputs a protocol solving a task provides are always correct [10,22]. However, ensuring that the outputs can indeed be eventually produced is sometimes tricky. Therefore, an adversary can be viewed as a *liveness* property, that specifies under which condition the correct processes are expected to produce outputs.

Given a task $T$ and an adversary $\mathcal{A}$, can the task be solved? It is known that this question is in general undecidable [12,16], and Delporte et al. [7] reduced it to the question of $k$-*resilient* solvability, i.e., assuming an adversary that consists of all sets of $n - k$ or more processes, restricted to *colorless* tasks (also called convergence tasks [4]). The resolution proposed in [7] is not easy to follow, and moreover, it is not constructive — it does not tell the programmer of the protocol how to use the adversary $\mathcal{A}$ to solve a colorless task $T$, when $T$ is solvable.

In this paper, we give a simpler constructive resolution of the question.

How to use such a condition $\mathcal{A}$ is shown below on a "back of an envelope" example. The paper is just a detailed elaboration of the envelope.

Consider a system of four processes, $p$, $q$, $r$, and $s$, and consider the *obstruction-free* adversary $\mathcal{A}_{OF}$ defined as the set of all singletons $\{\{p\}, \{q\}, \{r\}, \{s\}\}$.[1] Thus, $\mathcal{A}_{OF}$ stipulates that an algorithm solving a task is only required to make progress if some process is eventually forever running solo. It is immediate that $\mathcal{A}_{OF}$ allows for solving consensus [8]: a sequence of commit-adopt [10] instances, where the first instance is called with the input value, every next instance is called with the value returned by the previous instance, and the first committed value is returned. Thus, the *set consensus power* [13] of $\mathcal{A}_{OF}$, i.e., the smallest $k$ such that $k$-set agreement can be solved in the presence of $\mathcal{A}_{OF}$, is 1.

In general, under which condition an adversary $\mathcal{A}$ allows for solving consensus? In this paper, we show that $\mathcal{A}$ provides consensus if for all $S \in \mathcal{A}$ all subsets of $S$ that belong to $\mathcal{A}$ have a non-empty intersection. Intuitively, a correct process in the intersection acts as a leader in a classical eventual leader-based consensus protocol [6].

What if we weaken $\mathcal{A}_{OF}$ by adding one more allowed set of correct processes: $\mathcal{A}'_{OF} = \{p, q, r, s, pqrs\}$: either some process eventually runs solo, or no process fails? What is the set consensus power of $\mathcal{A}'_{OF}$? It is easy to observe that $\mathcal{A}'_{OF}$ allows for solving 2-set agreement: As $\{p, q, r, s\}$ can do consensus and $\{pqrs\}$ can do consensus, run both in parallel.

But can we solve consensus with $\mathcal{A}'_{OF}$? The answer is "no". Indeed, by assuming the converse, that there exists a read-write protocol $P$ that, under $\mathcal{A}'_{OF}$, solves consensus, we can derive a read-write consensus protocol for 2 processes violating [9,20], as follows.

We take 2 simulators $s_0$ and $s_1$ that mimic a run of $P$ in $\mathcal{A}'_{OF}$ using *BG-agreement* [3,4] to make sure that every step in $P$ is simulated consistently across the simulators. Initially, $s_0$ tries to start $P$ with all 0s and $s_1$ with all 1s as input values of $p$, $q$, $r$, and $s$. Recall that BG-agreement is allowed to block forever if one of the simulators fails in the middle of it. Steps of $P$ are simulated

---

[1] For brevity, we simply write $\{p, q, r, s\}$ in the following.

in a round-robin fashion on the codes of $p, q, r, s$ until a decision value is output in the simulated run or one code blocks because of an unresolved BG-agreement. This unresolved agreement may block the code of either $p$ or $q$ but not both (we do not need other singleton sets in $\mathcal{A}'_{OF}$ for the simulation). Say the code of $p$ is blocked. Thus, a live simulator, say $s_0$, picks $q$ and simulates just it, as long as the unresolved BG-agreement on the code of $p$ stays unresolved. In case it does resolve, $s_0$ resumes again to continue round-robin on the codes of $p$, $q$, $r$ and $s$. Thus, if no BG-agreement remains unresolved forever, the codes of all processes $p, q, r, s$ accept infinitely many simulated steps. Otherwise, an eventually solo execution of $p$ or $q$ is simulated. Thus, the correct processes in the simulated execution of $P$ are $\{pqrs\}$ and $P$ should output, else, $p$ or $q$ continue forever solo, and $P$ again should output. Thus, set consensus number of $\mathcal{A}'_{OF}$ is 2.

Further, imagine that we want to boost the power of $\mathcal{A}'_{OF} = \{pqrs, p, q, r, s\}$ using objects that solve consensus among two or more processes [18]. A simple extension of the argumentation above shows that $j$-process consensus objects are necessary and sufficient for solving consensus with $\mathcal{A}'_{OF}$, where $j$ is the maximum of the hitting set size of $S$ in $\mathcal{A}'_{OF}$, over all $S \in \mathcal{A}'_{OF}$.[2] In our case, the hitting set size of $\{pqrs\}$ in $\mathcal{A}'_{OF}$ is 4 and, thus, we need 4-process consensus. But if we restrict ourselves to the adversary $\mathcal{A}''_{OF} = \{pqrs, p, q\}$, then we would need only 2-process consensus.

In this paper, we generalize the observations made above for the special case of $\mathcal{A}'_{OF}$, to any adversary of [7]. We introduce an alternative definition of the set consensus power of an adversary $\mathcal{A}$, a positive one as we view adversaries as helpful entities: The smallest $k$ such that $k$-set agreement can be solved in the presence of $\mathcal{A}$.[3] Then we provide a simple characterization of the set consensus power of an adversary. Our characterization is self-consistent and, unlike the definition given in [7] does not involve reductions to $k$-resilience.

Our simulations allow us to derive a more general result: every two adversaries that have the same set consensus power $k$ agree on the set of colorless tasks they are able to solve. Informally, colorless tasks allow every process to adopt an input or output value from any other participating process. Thus, every colorless task is equivalent to some level of set agreement. Our technique is based on simple direct simulations and it does not employ failure detector-based reductions of [7].

Recently, following [7], Herlihy and Rajsbaum [17], and a concurrent paper [14], considered a restricted set of adversaries that are closed under superset: for every $S \in \mathcal{A}$, every its superset $S' \subseteq \Pi$ is also in $\mathcal{A}$. Informally, such adversaries say what sets of processes are expected to be live, but do not say which sets of processes are supposed to fail. By employing elements of modern combinatorial topology, [17] derives the characterization of colorless tasks with respect to superset-closed adversaries. In [14], we derive this result employing a very simple

---

[2] The hitting set size of $S$ in $\mathcal{A}$ is the size of the minimum-cardinality subset of $S$ that meets every element of $\mathcal{A}$ subset of $S$.

[3] More precisely, [7] talks about the *disagreement power* of $\mathcal{A}$ which is the largest $d$ such that $d$-set agreement cannot be solved in the presence of $\mathcal{A}$. The disagreement power of $\mathcal{A}$ is the set consensus power of $\mathcal{A}$ minus one.

simulation algorithm, a precursor to the one in this paper. Indeed, our paper [14] generalizes naturally to unrestricted adversaries. We suspect that doing the same with [17] is a major undertaking.

Imbs et al. [18] and Taubenfeld [21] considered special classes of *progress conditions* in the context of shared-memory systems enriched with consensus objects shared by subsets of $j < n$ processes. We observe that, with respect to colorless tasks, progress conditions of [18,21] are in fact special cases of adversaries [7]. Then we reconstruct the characterization of the power of *leveled* adversaries [21] to solve consensus using $j$-process consensus objects and extend the result to general adversaries, closing a question left open in [21].

This paper provides therefore a purely algorithmic characterization of adversaries that neither involves "esoteric" (for the distributed community) topological arguments, as [17], nor does it rely upon weakest failure detector results, as [7]. Neither it is stuck in the 80's resorting at these days and age to bivalency arguments [18,21]. Overall, this supports the contention that beyond dealing with sub-consensus tasks [15], topology is the analogue of plowing your field with an F16 fighter rather than a simple tractor — the F16 may do the job faster, but it takes years to master and you are liable to crash because of the low altitude flying and sharp turns plowing requires. Bivalency is the Ox. You can go with it so far but no more. The golden path, between bivalency and topology, is the BG simulation [3,4]. A tractor - simple, yet powerful and exactly suitable for the job.

The rest of the paper is organized as follows. Section 2 briefly describes our system model. Section 3 defines the notion of the power of a general adversary. Section 4 presents our characterization of adversaries with respect to colorless tasks. Section 5 extends our characterization to other computing models. Section 6 overviews the related work and concludes the paper.

## 2    Model

We adopt the conventional read-write shared memory model and only describe necessary details.

*Processes and objects.* We consider a distributed system composed of a set $\Pi$ of $n$ processes $\{p_1, \ldots, p_n\}$ $(n \geq 2)$. Processes communicate by applying atomic operations on a collection of *shared objects*. In the most of this paper, we assume that the shared objects are registers that export only atomic read-write operations. The shared memory can be accessed using atomic snapshot operations [1]. An *execution* is a pair $(I, \sigma)$ where $I$ is an initial state and $\sigma$ is a sequence of process ids. A process that takes at least one step in an execution is called *participating*. A process that takes infinitely many steps in an execution is said to be *correct*, otherwise, the process is *faulty*.

*Distributed tasks.* A *task* is defined through a set $\mathcal{I}$ of input $n$-vectors (one input value for each process, where the value is $\bot$ for a non-participating process), a set $\mathcal{O}$ of output $n$-vectors (one output value for each process, $\bot$ for non-terminated processes) and a total relation $\Delta$ that associates each input vector with a set of possible output vectors. A protocol *wait-free* solves a task $T$ if in every execution, every correct process eventually outputs, and all outputs respect the specification of $T$.

*Correct sets and adversaries.* The *correct set* of an execution $e$, denoted $correct(e)$ is the set of processes that appear infinitely often in $e$. An *adversary* [7] is a collection of subsets of $\Pi$. We say that an execution $e$ is $\mathcal{A}$-*compliant* if $correct(e) \in \mathcal{A}$.

*Hitting sets.* Given a set system $(\Pi, \mathcal{A})$ where $\mathcal{A}$ is a set of subsets of $\Pi$, a set $H \subseteq \Pi$ is a *minimum cardinality hitting set of* $(\Pi, \mathcal{A})$ if it is a minimum cardinality subset of $\Pi$ that meets every set in $\mathcal{A}$. The *hitting set size* of $(\Pi, \mathcal{A})$, i.e., the size of a minimum cardinality hitting set of $(\Pi, \mathcal{A})$, is denoted by $h(\mathcal{A})$. Obviously, if $h(\mathcal{A}) = 1$, then $\forall \mathcal{A}' \subseteq \mathcal{A}$, $\mathcal{A} \neq \emptyset$, $h(\mathcal{A}') = 1$. Finding the hitting set size is NP-complete [19].

*Colorless tasks.* In *colorless* task (also called *convergence* tasks [4]) processes are free to use each others' input and output values, so the task can be defined in terms of input and output *sets* instead of vectors.

Formally, let $val(U)$ denote the set of non-$\perp$ values in a vector $U$. In a colorless task, for all input vectors $I$ and $I'$ and all output vectors $O$ and $O'$, such that $(I, O) \in \Delta$, $val(I') \subseteq val(I)$, $val(O') \subseteq val(O)$, we have $(I', O) \in \Delta$ and $(I, O') \in \Delta$.

*The Commit-Adopt protocol.* The *commit-adopt* abstraction (CA) [10] exports one operation $propose(v)$ that returns $(commit, v')$ or $(adopt, v')$, for $v', v \in V$, and guarantees that (a) every returend value is a proposed value, (b) if only one value is proposed then this value must be committed, (c) if a process commits on a value $v$, then every process that returns adopts $v$ or commits $v$, and (d) every correct process returns. The commit-adopt abstraction can be implemented wait-free.

*The BG-simulation technique.* BG-simulation is a technique by which $k + 1$ processes $s_1, \ldots, s_{k+1}$, called *simulators*, can wait-free simulate a $k$-resilient execution of any asynchronous $n$-process protocol $A$ [3,4]. The simulation guarantees that each simulated step of every process $p_j$ is either agreed on by all simulators, or one less simulator participates further in the simulation for each step which is not agreed on.

The central building block of the simulation is the *BG-agreement* protocol. The protocol is safe—every decided value was previously proposed, and no two different values are decided— but not necessarily live. If a simulator slows down in the middle of BG-agreement, the protocol's execution at other correct simulators may "block" until the slow simulator finishes the protocol. If the simulator is faulty, no simulator is guaranteed to decide.

Suppose the simulation tries to promote $m > k$ codes in a fair (e.g., round-robin) way. As long there is a live simulator, at least $m - k$ simulated processes accept infinitely many steps of $A$ in the simulated execution.

## 3    Set Consensus Power of $\mathcal{A}$

Let $\mathcal{A}$ be an adversary and take any set $S \subseteq P$. $\mathcal{A}_S$ denotes the adversary that consists of $S$ and all elements of $\mathcal{A}$ that are subsets of $S$. E.g., for $\mathcal{A} = \{pq, qr, q, r\}$ and $S = qr$, $\mathcal{A}_S = \{qr, q, r\}$.

Let $S \in \mathcal{A}$ and take $a \in S$. Then $\mathcal{A}_{S,a}$ denotes the adversary that consists of all elements of $\mathcal{A}_S$ that *do not* include $a$. E.g., for $\mathcal{A} = \{pq, qr, q, r\}$, $S = qr$, and $a = q$, $\mathcal{A}_{S,a} = \{r\}$. Note that if the hitting set size of $(\Pi, \mathcal{A}_S)$ is 1, then for every $a \in S$ that meets every set in $\mathcal{A}_S$, we have $\mathcal{A}_{S,a} = \emptyset$. Thus:

**Lemma 1.** $h(\mathcal{A}_S) > 1$ *if and only if* $\forall a \in S : \mathcal{A}_{S,a} \neq \emptyset$.

**Definition 1.** *The quantity denoted* $setcon(\mathcal{A})$, *which will later be shown to be the set consensus power of* $\mathcal{A}$, *is defined as follows:*

- *If* $\mathcal{A} = \emptyset$, *then* $setcon(\mathcal{A}) = 0$
- *Otherwise,* $setcon(\mathcal{A}) = \max_{S \in \mathcal{A}} \min_{a \in S} setcon(\mathcal{A}_{S,a}) + 1$

Thus, $setcon(\mathcal{A})$, for a non-empty adversary $\mathcal{A}$, is determined as $setcon(\mathcal{A}_{\bar{S}, \bar{a}}) + 1$ where $\bar{S}$ is an element of $\mathcal{A}$ and $\bar{a}$ is a process in $\bar{S}$ that "max-minimize" $setcon(\mathcal{A}_{S,a})$. Note that for $\mathcal{A} \neq \emptyset$, $setcon(\mathcal{A}) \geq 1$.

We say that $S \in \mathcal{A}$ is *proper* if it is not a subset of any other element in $\mathcal{A}$. Let $proper(\mathcal{A})$ denote the set of proper elements in $\mathcal{A}$. Note that since for all $S' \subset S$, $\min_{a \in S'} setcon(\mathcal{A}_{S',a}) \leq \min_{a \in S} setcon(\mathcal{A}_{S,a})$, we can replace $S \in \mathcal{A}$ with $S \in proper(\mathcal{A})$ in Definition 1.

For example, for $\mathcal{A} = \{pqr, pq, pr, p, q, r\}$, we have $setcon(\mathcal{A}) = 2$: for $S = pqr$ and $a = p$, we have $\mathcal{A}_{S,a} = \{q, r\}$ and $setcon(\mathcal{A}_{S,a}) = 1$. Intuitively, in an execution where the correct set belongs to $\mathcal{A} - \mathcal{A}_{S,a} = \{pqr, pq, pr, p\}$, process $p$ can act as a leader for solving consensus. If the execution's correct set belongs to $\mathcal{A}_{S,a} = \{q, r\}$ (either $q$ or $r$ eventually runs solo) then $q$ and $r$ can solve consensus using an obstruction-free algorithm. Running the two algorithms in parallel, we obtain a solution to 2-set agreement. The reader can easily verify that any other choice of $a \in pqr$ results in larger values of $setcon(\mathcal{A}_{S,a})$.

As another example, consider the $t$-resilient adversary $\mathcal{A}_{t\text{-}res} = \{S \subseteq \Pi, |S| \geq n - t\}$. It is easy to verify recursively that $setcon(\mathcal{A}_{t\text{-}res}) = t + 1$: at each level $1 \leq j \leq t + 1$ of recursion we consider a set $S$ of $n - j + 1$ elements, pick up a process $p \in S$ and delegate the set of $n - j$ processes that do not include $p$ to level $j+1$. At level $t+1$ we get a set of size $n - t$ and stop. Thus, $setcon(\mathcal{A}_{t\text{-}res}) = t+1$.

More generally, consider *superset-closed* adversaries $\mathcal{A}$ [14]: for every $S \in \mathcal{A}$, every its set $S'$ such that $S \subseteq S' \subseteq \Pi$ is also in $\mathcal{A}$.

**Theorem 1.** *For all superset-closed adversaries* $\mathcal{A}$, $setcon(\mathcal{A}) = h(\mathcal{A})$.

*Proof.* By definition, for $\mathcal{A} = \emptyset$, $setcon(\mathcal{A}) = h(\mathcal{A}) = 0$. By induction, suppose that for all $0 \leq j < k$ and all superset-closed adversaries $\mathcal{A}'$ with $h(\mathcal{A}') = j$, we have $setcon(\mathcal{A}') = j$.

Consider a superset-closed adversary $\mathcal{A}$ such that $h(\mathcal{A}) = k$. The only proper element of $\mathcal{A}$ is the whole set of processes $\Pi$. Thus, $setcon(\mathcal{A}) = \min_{a \in \Pi}$

Initially:
$$\forall j, \; \mathcal{A}^j = \emptyset$$

$PartitionAdv(\mathcal{A})$

1    $partition(\mathcal{A}, 1)$

$partition(\mathcal{B}, j)$

2    **while** $\mathcal{B} \neq \emptyset$ **do**

3        $(B, b) := args \; \max_{S \in proper(\mathcal{B})} \min_{a \in S} setcon(\mathcal{B}_{S,a})$

4        $\mathcal{A}^j := \mathcal{A}^j \cup (\mathcal{B}_B - \mathcal{B}_{B,b})$

5        $partition(\mathcal{B}_{B,b}, j + 1)$

6        $\mathcal{B} := \mathcal{B} - \mathcal{B}_B$

**Fig. 1.** Partitioning an adversary with $setcon = k$

$setcon(\mathcal{A}_{\Pi,a}) + 1$. Since $h(\mathcal{A}) = k$, by removing all elements that include $a$ we obtain an adversary $\mathcal{A}_{\Pi,a}$ such that $h(\mathcal{A}_{\Pi,a}) \geq k - 1$. (Otherwise, there is a hitting set of $\mathcal{A}$ of size less than $k$.) By picking up $a$ in a hitting set of $\mathcal{A}$ of size $k$ we obtain, by the induction hypothesis, $h(\mathcal{A}_{\Pi,a}) = setcon(\mathcal{A}_{\Pi,a}) = k - 1$ and, thus, $setcon(\mathcal{A}) = k$.

For general adversaries, for convenience, we first consider the special case of set consensus power 1. Definition 1 and Lemma 1 imply:

**Lemma 2.** $setcon(\mathcal{A}) = 1$ *if and only if* $\forall S \in \mathcal{A}, \; h(\mathcal{A}_S) = 1$

We show below that the elements of every adversary $\mathcal{A}$ with $setcon(\mathcal{A}) = k$ can be split into $k$ sub-adversaries such that $setcon$ of every sub-adversary is 1.

**Theorem 2.** *Let $\mathcal{A}$ be an adversary, and let $setcon(\mathcal{A}) = k$. Then there exists $\mathcal{A}^1, \ldots, \mathcal{A}^k$, a partitioning of $\mathcal{A}$, such that, for all $1 \leq j \leq k$, $setcon(\mathcal{A}^j) = 1$.*

*Proof.* Let $\mathcal{A}$ be an adversary such that $setcon(k)$. Our goal is to partition $\mathcal{A}$ into $k$ sub-adversaries $\mathcal{A}^1, \ldots, \mathcal{A}^k$ such that $\forall j = 1, \ldots, k, \; \forall S \in \mathcal{A}^j, \; h(\mathcal{A}^j_S) = 1$. We construct the desired partitioning of $\mathcal{A}$ using procedure $PartitionAdv(\mathcal{A})$ described in Figure 1.

Suppose that at a level $j \in \{1 \ldots, k\}$, we have $\mathcal{B} \subseteq \mathcal{A}$, a set of elements of $\mathcal{A}$ which were not yet assigned a level. We recursively assign elements of $\mathcal{B}$ to levels $j$ or more using procedure $partition(\mathcal{B}, j)$.

Let $B$ and $b \in B$ max-minimize $setcon(\mathcal{B}_{S,a})$ over all $S \in \mathcal{B}$ and $a \in S$ (ties broken deterministically). Then we assign $\mathcal{B}_B - \mathcal{B}_{B,b}$ to level $j$ and recursively partition $\mathcal{B}_{B,b}$ on level $j + 1$ by calling $partition(\mathcal{B}, j + 1)$. When we are done, i.e., all elements of $\mathcal{B}_B$ are assigned to levels $j$ or more, we proceed to assigning the remaining elements of $\mathcal{B} - \mathcal{B}_B$ to level $j$ or more, and we repeat this until we exhaust $\mathcal{B}$.

We observe first that this procedure recursively explores all elements in $\mathcal{A}$, i.e., every element $S \in \mathcal{A}$ is assigned to some level $j \geq 1$. By construction, each

$A^j$ only contains sets $S$ with the hitting set size 1, namely, all $S' \in A_B$ that contain $b$ (chosen in line 3). All other elements of $A_S$ are delegated to levels $j+1$ or more.

By Definition 1 and Lemma 1, if we start from the whole set $A$ at level 1 (line 1), and $setcon(A) = k$, exactly levels $1, \ldots, k$ are populated.

Finally, by construction, for all $j$ and all $S \in proper(A^j)$, $h(A_S^j) = 1$. By Lemma 2, for all $j$, $setcon(A^j) = 1$.

Before we characterize the ability of adversaries to solve generic colorless tasks, we consider the special case of adversaries of $setcon = 1$.

**Theorem 3.** *If $setcon(A) = 1$, then $A$ solves consensus.*

*Proof.* Recall that if $setcon(A) = 1$, then, by Lemma 2, $\forall S \in A$, $h(A_S) = 1$. The consensus algorithm is presented in Figure 2. This is a rotating coordinator-based algorithm inspired by the consensus algorithm by Chandra and Toueg [6].

The algorithm proceeds in rounds. In each round $r$, every process $p_i$ first tries to commit its proposal in a new instance of commit-adopt. If $p_i$ succeeds, then the committed value is written in the "decision" register $D$ and returned. Otherwise, $p_i$ adopts the returned value as its current estimate and writes it in $R_i$ equipped with the current round number $r$. Then $p_i$ takes snapshots of $\{R_1, \ldots, R_n\}$ until either a set $S \in A$ reaches round $r$ or a decision value is written in $D$ (in which case the process returns the value from $D$). If no decision is taken yet, then $p_i$ checks if the coordinator of this round, $p_{r \bmod n}$, is in $S$. If so, $p_i$ adopts the value written in $R_{r \bmod n}$ and proceeds to the next round.

Safety of the algorithm follows from the properties of commit-adopt. Indeed, the first round in which some process commits on some value $v$ in line 14 locks the value for all subsequent rounds.

For liveness, suppose, by contradiction, that the algorithm never terminates in some $A$-compliant execution $e$. Recall that we only consider executions in which some set in $A$ is exactly the set of correct processes. Therefore, every correct process goes through infinitely many rounds.

Let $\bar{S} \in A$ be the set of correct processes in $e$. After a round $r'$ when all processes outside $\bar{S}$ have failed, every element of $A$ evaluated by a correct process in line 16 is a subset of $\bar{S}$. Finally, since the hitting set size of $A_{\bar{S}}$ is 1, all these elements of $A$ overlap on some correct process $p_j$.

Consider round $r = mn + j \geq r'$. In this round, $p_j$ not only belongs to all sets evaluated by the correct processes, but it is also the coordinator ($j = r \bmod n$). Thus, the only value that a process can propose to commit-adopt in round $r+1$ is the value previously written by $p_j$ in $R_j$. Thus, every process that returns from commit-adopt in round $r+1$ commits—a contradiction. Hence, no read-write protocol can solve $T'$ in the presence of $A$.

Theorems 2 and 3 imply the following:

**Corollary 1.** *Let $A$ be an adversary such that $setcon(A) = k$. Then the adversary can solve $k$-set agreement.*

---

Shared variables:
 $D$, initially $\bot$
 $R_1, \ldots, R_n$, initially $\bot$

$propose(v)$
7  $est := v$
8  $r := 0$
9  $S := P$
10  **repeat**
11   $r := r + 1$
12   $(flag, est) := CA_r.propose(v)$
13   **if** $flag = commit$ **then**
14    $D := est; return(est)$     {Return the committed value}
15   $R_i := (est, r)$
16   **wait until** $\exists S \in \mathcal{A}, \forall p_j \in S: R_j = (v_j, r_j)$ where $r_j \geq r$ **or** $D \neq \bot$
             {Wait until a set in $\mathcal{A}$ moves}
17   **if** $p_{r \bmod n+1} \in S$ **then**
18    $est := v_{r \bmod n+1}$     {Adopt the estimate of the current leader}
19  **until** $D \neq \bot$
20  $return(D)$

---

**Fig. 2.** Consensus with a "one-level" adversary $\mathcal{A}$, $setcon(\mathcal{A}) = 1$

*Proof.* First we apply Theorem 2 to partition $\mathcal{A}$ into $k$ classes $\mathcal{A}^1, \ldots, \mathcal{A}^k$ such that, for all $j = 1, \ldots, k$, $setcon(\mathcal{A}^j) = 1$. Then every process runs $k$ parallel consensus algorithms established by Theorem 3, one for each $\mathcal{A}^j$, proposing its input value in each of these consensus instances (the idea originally appeared in [2]). Since the set of correct processes in every $\mathcal{A}$-compliant execution belongs to some $\mathcal{A}^j$, at least one consensus instance returns. The process decides on the first such returned value. Moreover, at most $k$ different values are decided and each returned value was previously proposed.

The next section shows that no read-write protocol can solve $(k-1)$-set agreement under an adversary $\mathcal{A}$ such that $setcon(\mathcal{A}) = k$.

## 4 Characterizing Colorless Tasks

In this section, we show a more general result: the set of colorless tasks that can be solved with an adversary $\mathcal{A}$ with $setcon(\mathcal{A}) = k$ is exactly the set of colorless tasks that can be solved $(k-1)$-resiliently, but not $k$-resiliently. The proof is based on two simple applications of BG simulation [3,4].

First, we show that $\mathcal{A}$ solves every $(k-1)$-resiliently solvable colorless task $T$ by presenting an algorithm that, in every $\mathcal{A}$-compliant execution, simulates a $(k-1)$-resilient execution of a protocol solving $T$.

Second, we show that $\mathcal{A}$ cannot solve a colorless task $T'$ that is not $(k-1)$-resiliently solvable by presenting an algorithm that $(k-1)$-resiliently simulates any protocol that solves $T'$ in every $\mathcal{A}$-compliant execution.

Local variables:

$B_1, \ldots, B_k$, initially $\perp$          {Set of currently simulated elements of $\mathcal{A}$}

$b_1, \ldots, b_k$, initially $\perp$          {Set of currently blocked processes}

$L$, initially 1          {The current level of simulation}

Code for every simulator $s_i$, $i = 1, \ldots, n$

21    $B_1 :=$ the first element $S \in proper(\mathcal{A})$ such that $setcon(\mathcal{A}_S) = k$
         {In some deterministic order}

22    $L := 1$

23    **repeat forever**

24      $\ell := 1$

25      **while** $\ell < L$ **and** the current step of $b_\ell$ is still blocked **do** $\ell = \ell + 1$

26      **if** $\ell < L$ **then** $L := \ell$          {Return to level $\ell$ if the step of $b_\ell$ is resolved}

27      let $p_j \in B_L$ be the process with the least number of simulated steps

28      run BG-agreement for the next step of $p_j$

29      **if** the step of $p_j$ is blocked and $L < k$ **then**

30          $b_L := p_j$

31          $B_{L+1} :=$ the first set in $proper(\mathcal{A}_{B_L, b_L})$ with power $\geq k - L$
               {Such a set exists, since $setcon(\mathcal{A}_{B_L}) \geq k - L + 1$}

32          $L := L + 1$

**Fig. 3.** Simulating an $\mathcal{A}$-compliant execution

**Theorem 4.** *Let $\mathcal{A}$ be an adversary such that $setcon(\mathcal{A}) = k$ and $T$ be a color-less task. $\mathcal{A}$ solves $T$ if and only if $T$ is $(k-1)$-resiliently solvable.*

*Proof.* Let $\mathcal{A}$ be an adversary such that $setcon(\mathcal{A}) = k$.

Let $T$ be a colorless $(k-1)$-resiliently solvable task. By Corollary 1, $\mathcal{A}$ can implement $k$-set agreement. Then we apply the generic algorithm of [11] that solves every $(k-1)$-resilient colorless task using a solution to $k$-set agreement as a black box. Thus, $\mathcal{A}$ solves $T$.

For the other direction, suppose that $\mathcal{A}$ solves a colorless task $T'$ that is *not* solvable $(k-1)$-resiliently, and let $Alg$ be the corresponding algorithm.

We describe below a simulation protocol (summarized in Figure 3) that allows $n$ simulators, $s_1, \ldots, s_n$, to $(k-1)$-resiliently simulate an $\mathcal{A}$-compliant execution of $Alg$.

Essentially, the protocol builds upon BG-simulation, except that the *order* in which steps of $Alg$ is not fixed in advance. Instead, the order is determined online, based on the currently observed set of participating processes.

Let $B_1$ be an element of $proper(\mathcal{A})$ such that $setcon(\mathcal{A}_{B_1}) = k$ (by the definition of $setcon$ such a set exists). Initially, every simulator proceeds by simulating steps of processes in $B_1$ in a round-robin fashion. If simulating a step blocks—some other simulator stopped in the middle of the BG agreement protocol of some process $b_1 \in B_1$—the simulator proceeds to simulating steps of the processes in $B_2$, the "next" not yet blocked element of $\mathcal{A}_{B_1}$ such that $setcon(\mathcal{A}_{B_2}) \geq k - 1$. Indeed, by Definition 1, for all $b \in B$, $setcon(\mathcal{A}_{B_1, b}) \geq$

$setcon(\mathcal{A}_{B_1}) - 1 = k - 1$. Thus, such a set $B_2 \in \mathcal{A}_{B_1, b_1}$ exists. The procedure is then repeated for $B_2$: steps of processes in $B_2$ are simulated as long as no process in $B_2$ is blocked. As soon as a blocked process $b_2$ is observed, the simulator proceeds to simulating $B_3$, an element of $\mathcal{A}_{B_2, b_2}$ that has consensus power at least $k - 2$, etc. Inductively, since $setcon(\mathcal{A}) = k$, if the simulation reaches level $k$, then $B_k \neq \emptyset$.

Every simulator periodically checks if some of the previously blocked agreements are resolved (line 25). If so, the simulator jumps back to the smallest level with a resolved agreement (line 26).

Note that, since every step of $Alg$ is agreed upon using the BG-agreement protocol, the simulation constructs a correct execution of $Alg$ [3,4]. Now we show that the produced execution is indeed $\mathcal{A}$-compliant, and thus $Alg$ must terminate.

First, we observe that no line in the pseudo-code presented in Figure 3 is blocking. Thus, every correct simulator proceeds through infinitely many rounds in lines 24-29. Consider level $\ell$ and suppose that some correct process never observed the currently simulated step of $b_\ell$ being resolved (it is blocked forever by a faulty simulator). Since simulators explore the simulated sets in a deterministic order starting from level 1, every correct process eventually blocks on the same step of $b_\ell$.

Now let $\ell$ be the lowest level in which no step in $B_\ell$ is observed blocked forever. Since there are at most $k - 1$ faulty simulators, and a faulty simulator cannot block more than one simulated process, $\ell \leq k$. Thus, every correct process simulates infinitely many steps of $B_\ell$, and, eventually, every simulated step belongs to a process in $B_\ell$. By construction, $B_\ell \in \mathcal{A}$ and, thus, the simulated run of $Alg$ is $\mathcal{A}$-compliant. Therefore, $Alg$ must terminate in the simulated execution and we obtain a $(k - 1)$-resilient solution to $T'$ — a contradiction.

The *set consensus power* of an adversary $\mathcal{A}$ is the smallest $k$ such that $\mathcal{A}$ can solve $k$-set agreement. Theorem 4 implies:

**Corollary 2.** *The set consensus power of $\mathcal{A}$ is $setcon(\mathcal{A})$.*

By Theorem 1, determining $setcon(\mathcal{A})$ may boil down to determining the hitting set size of $(\Pi, \mathcal{A})$, and thus, by [19]:

**Corollary 3.** *Determining the set consensus power of an adversary is NP-complete.*

The *disagreement power* of an adversary $\mathcal{A}$ [7], denoted $d(\mathcal{A})$, is the largest $d$ such that $d$-set agreement *cannot* be solved in the presence of $\mathcal{A}$. By Corollary 2, $d(\mathcal{A}) = setcon(\mathcal{A}) - 1$.

## 5    Extension to Other Models

In a recent paper [18], Imbs et al. considered *asymmetric* progress conditions that allow for modeling different progress guarantees for different processes. An asymmetric progress condition associates each process $p_i$ with a set $\mathcal{P}_i$ of process subsets that contain $p_i$. Process $p_i$ is expected to make progress (e.g., output a value in a task solution) only if the current set of correct processes is in $\mathcal{P}_i$.

It is easy to see that with respect to the solvability of colorless tasks, the asymmetric progress conditions of [18] can be modeled as adversaries of [7]. Indeed, for each progress condition $\mathcal{P} = \{\mathcal{P}_1, \ldots, \mathcal{P}_n\}$, we can construct an adversary $\mathcal{A}^{\mathcal{P}} = \cup_i \mathcal{P}_i$. Since to solve a colorless task, it is sufficient to make sure that at least one process decides, every $\mathcal{P}$-resilient solution to a colorless task implies an $\mathcal{A}^{\mathcal{P}}$-resilient solution, and vice versa.

**Observation 5.** *A colorless task $T$ is solvable with a progress condition $\mathcal{P}$ if and only if it is solvable with the adversary $\mathcal{A}^{\mathcal{P}}$.*

In an even more recent paper [21], Taubenfeld focused on a special case of *leveled* adversaries that only specify the *sizes* of correct sets. Such an adversary $\mathcal{L}$ can be specified as a sequence of number in $\{1, \ldots, n\}$: for each $j \in \mathcal{L}$, the adversary contains all process sets of size $j$. The paper shows, among other things, that consensus can be solved with $\mathcal{L}$ using $j$-process consensus objects (i.e., objects that can solve consensus among up to $j$ processes) if and only if $j \geq width(\mathcal{L})$, where $width(\mathcal{L}) = \max(\mathcal{L}) - \min(\mathcal{L}) + 1$.

Note that $width(\mathcal{L})$ is exactly $h(\mathcal{L}_S)$ for any $S \in proper(\mathcal{L})$. Indeed, we need exactly $width(\mathcal{L})$ processes to meet every set of $\min(\mathcal{L})$ processes that is subset of an element of $proper(\mathcal{L})$ (a set of $\max(\mathcal{L})$ processes).

**Theorem 6.** *A leveled adversary $\mathcal{L}$ such that $width(\mathcal{L}) = k$ can wait-free solve consensus using $j$-process consensus objects if and only if $j \geq k$.*

*Proof.* (Sketch) $\mathcal{L}$ can solve consensus using $k$-process consensus and read-write registers as follows. As in the consensus algorithm in Figure 2, every process alternates between instances of commit-adopt and a leader-based reconciliation protocol. The first committed value is written in a decision register and returned. Instead of a single coordinator in a hitting set of size 1, we now select a "coordinator group" of size $k$. Thus, there are $n$ choose $k$ coordinator groups, and we place them in a deterministic order: $C_0, \ldots, C_{\binom{n}{k}-1}$. Now process $p_i$ considers itself a coordinator of a round $r$ if $p_i \in C_{r \bmod \binom{n}{k}}$. Furthermore, every round $r$ is associated with a $k$-process consensus object $cons_r$ that can only be accessed by processes in $C_{r \bmod \binom{n}{k}}$.

In Figure 4, we give an update of lines 15-18 of the consensus algorithm in Figure 2, the rest of the algorithm remains unchanged. As in the proof of Theorem 3, eventually, there will be a round $r'$ when only a subset of processes of some $S \in \mathcal{L}$ of size $\max(\mathcal{L})$ take steps and $C_{r' \bmod \binom{n}{k}}$ is a hitting set of $\mathcal{L}_S$ (the adversary that consists of $S$ and all its subsets in $\mathcal{L}$). Thus, in round $r'$, every correct process $p_i$ will adopt the estimate value agreed upon by the processes in $C_{r' \bmod \binom{n}{k}}$: every element of $\mathcal{L}$ evaluated by $p_i$ in line 35 should include at least one process in $C_{r' \bmod \binom{n}{k}}$. Thus, every correct process accesses the instance of commit-adopt in round $r' + 1$ with the same value and decides.

33    **if** $p_i \in C_{r \mod \binom{n}{k}}$ **then** $est := cons_r.propose(est)$

34    $R_i := (est, r)$

35    **wait until** $\exists S \in \mathcal{L}, \forall p_j \in S: R_j = (v_j, r_j)$ where $r_j \geq r$ **or** $D \neq \perp$
        {Wait until a set in $\mathcal{L}$ moves}

36    **if** $\exists p_j \in C_{r \mod \binom{n}{k}} \cap S$ **then**

37        $est := v_j$                {Adopt the estimate of the current coordinator group}

**Fig. 4.** Solving consensus with $\mathcal{L}$ and $k$-process consensus objects: replacing lines 15–18 in Figure 2

Now, by contradiction, suppose that we can solve consensus with $\mathcal{L}$ using $(k-1)$-process consensus objects, and let $Alg$ be the corresponding algorithm. We establish a contradiction by presenting a wait-free 2-process consensus algorithm.

It is straightforward to extend our simulation in Figure 3 to simulate a protocol that, in addition to read-write registers, uses $(k - 1)$-process consensus objects. Indeed, let two simulators simulate steps of $Alg$ of a set $B_1 \in \mathcal{L}$ of $\max(\mathcal{L})$ processes in a round-robin fashion. A simulator that fails while simulating a step that accesses a $(k - 1)$-process consensus object can block a set $S$ of up to $k - 1$ simulated codes that are in process of accessing this object. But since $h(\mathcal{L}_{B_1}) = k$, we still have at least one set $B_2$ of $\min(\mathcal{L})$ processes that are not blocked. By applying the logic used in the proof of Theorem 4, we obtain an $\mathcal{L}$-compliant execution of $Alg$. The simulated execution of $Alg$ must terminate — a contradiction.

Theorem 6 is mildly surprising in the sense that the ability of $j$-consensus objects to boost the power of $\mathcal{L}$ to solve consensus has nothing to do with the exact structure of $\mathcal{L}$, but depends only on the size of the hitting set of $\mathcal{L}_S$ for some $S \in \mathcal{L}$ of the maximal size. Indeed, notice that our argumentation has nothing to do with "sequences" or "width," it only uses the hitting set size of $\mathcal{L}_S$ for $S \in proper(\mathcal{L})$. A straightforward extension of Theorem 6 resolves an open question raised in [21].

**Theorem 7.** *An adversary $\mathcal{A}$ can wait-free solve consensus using $j$-process consensus objects if and only if $j \geq \max_{S \in \mathcal{A}}(h(\mathcal{A}_S))$.*

## 6    Concluding Remarks

An adversary, as defined by Delporte et al. [7], is in fact a special case of an environment of [5] that determines which sets of processes are allowed to fail without specifying the timing of failures. Thus, we can rephrase the statement "task $T$ can be solved with adversary $\mathcal{A}$", as "task $T$ can be solved in environment $\mathcal{A}$ using the dummy failure detector". (The output of the dummy failure detector does not depend on the failure pattern.) It is shown in [13] that, with respect to colorless tasks, failure detectors can be split into $n$ equivalence classes, and each class $j$ agrees on the set of tasks it can solve: namely, tasks that can be solved $(j-1)$-resiliently and not $j$-resiliently. Therefore, by applying [13], we conclude that each adversary belongs to one of such equivalence class. This characterization is however a brute-force solution and it does not give us an explicit algorithm to compute the class to which a given adversary belongs.

The approach taken in [7] is based on a three-stage simulation. First, it is shown how an adversary can simulate any *dominating* adversary, where the domination is defined through involved recursive inclusion properties. Second, it is shown that every adversary that does not dominate the $k$-resilient adversary is strong enough to implement the anti-$\Omega_k$ failure detector that, in turn, can be used to solve $k$-set agreement [23]. Finally, it is shown that anti-$\Omega_k$ can be used to solve any colorless task that can be solved $k$-resiliently.

Instead, this paper proposes a self-consistent, constructive and simple characterization of general adversaries of [7], and sketches an extension of the characterization to models that use $j$-process consensus objects [18,21].

**Acknowledgment.** The first author is grateful to Gadi Taubenfeld for a few nights of pillow talk about [18,21].

# References

1. Afek, Y., Attiya, H., Dolev, D., Gafni, E., Merritt, M., Shavit, N.: Atomic snapshots of shared memory. Journal of the ACM 40(4), 873–890 (1993)
2. Afek, Y., Gafni, E., Rajsbaum, S., Raynal, M., Travers, C.: Simultaneous consensus tasks: A tighter characterization of set-consensus. In: Chaudhuri, S., Das, S.R., Paul, H.S., Tirthapura, S. (eds.) ICDCN 2006. LNCS, vol. 4308, pp. 331–341. Springer, Heidelberg (2006)
3. Borowsky, E., Gafni, E.: Generalized FLP impossibility result for t-resilient asynchronous computations. In: STOC, pp. 91–100. ACM Press, New York (May 1993)
4. Borowsky, E., Gafni, E., Lynch, N.A., Rajsbaum, S.: The BG distributed simulation algorithm. Distributed Computing 14(3), 127–146 (2001)
5. Chandra, T.D., Hadzilacos, V., Toueg, S.: The weakest failure detector for solving consensus. Journal of the ACM 43(4), 685–722 (1996)
6. Chandra, T.D., Toueg, S.: Unreliable failure detectors for reliable distributed systems. Journal of the ACM 43(2), 225–267 (1996)
7. Delporte-Gallet, C., Fauconnier, H., Guerraoui, R., Tielmann, A.: The disagreement power of an adversary. In: Keidar, I. (ed.) DISC 2009. LNCS, vol. 5805, pp. 8–21. Springer, Heidelberg (2009)
8. Fich, F.E., Luchangco, V., Moir, M., Shavit, N.: Obstruction-free algorithms can be practically wait-free. In: Fraigniaud, P. (ed.) DISC 2005. LNCS, vol. 3724, pp. 493–494. Springer, Heidelberg (2005)
9. Fischer, M.J., Lynch, N.A., Paterson, M.S.: Impossibility of distributed consensus with one faulty process. Journal of the ACM 32(2), 374–382 (1985)
10. Gafni, E.: Round-by-round fault detectors (extended abstract): Unifying synchrony and asynchrony. In: Proceedings of the 17th Symposium on Principles of Distributed Computing (1998)
11. Gafni, E., Guerraoui, R.: Generalizing state machine replication. Technical report, EPFL (2010), http://infoscience.epfl.ch/record/150307

12. Gafni, E., Koutsoupias, E.: Three-processor tasks are undecidable. SIAM J. Comput. 28(3), 970–983 (1999)
13. Gafni, E., Kuznetsov, P.: On set consensus numbers. In: Keidar, I. (ed.) DISC 2009. LNCS, vol. 5805, pp. 35–47. Springer, Heidelberg (2009)
14. Gafni, E., Kuznetsov, P.: L-resilient adversaries and hitting sets. CoRR, abs/1004.4701 (2010) (to appear in ICDCN 2011), http://arxiv.org/abs/1004.4701
15. Gafni, E., Rajsbaum, S., Herlihy, M.: Subconsensus tasks: Renaming is weaker than set agreement. In: Dolev, S. (ed.) DISC 2006. LNCS, vol. 4167, pp. 329–338. Springer, Heidelberg (2006)
16. Herlihy, M., Rajsbaum, S.: The decidability of distributed decision tasks (extended abstract). In: STOC, pp. 589–598 (1997)
17. Herlihy, M., Rajsbaum, S.: The topology of shared-memory adversaries. In: PODC (2010)
18. Imbs, D., Raynal, M., Taubenfeld, G.: On asymmetric progress conditions. In: PODC (2010)
19. Karp, R.M.: Reducibility among combinatorial problems. Complexity of Computer Computations, 85–103 (1972)
20. Loui, M.C., Abu-Amara, H.H.: Memory requirements for agreement among unreliable asynchronous processes. Advances in Computing Research 4, 163–183 (1987)
21. Taubenfeld, G.: The computational structure of progress conditions. In: Lynch, N.A., Shvartsman, A.A. (eds.) DISC 2010. LNCS, vol. 6343, pp. 221–235. Springer, Heidelberg (2010)
22. Yang, J., Neiger, G., Gafni, E.: Structured derivations of consensus algorithms for failure detectors. In: Proceedings of the 17th ACM Symposium on Principles of Distributed Computing, pp. 297–306 (1998)
23. Zieliński, P.: Anti-omega: the weakest failure detector for set agreement. In: PODC (August 2008)

# Quasi-Linearizability:
# Relaxed Consistency for Improved Concurrency

Yehuda Afek, Guy Korland, and Eitan Yanovsky

Computer Science Department
Tel-Aviv University, Israel
{afek,guykorla,eitanyan}@post.tau.ac.il

**Abstract.** Linearizability, the key correctness condition that most optimized concurrent object implementations comply with, imposes tight synchronization between the object concurrent operations. This tight synchronization usually comes with a performance and scalability price. Yet, these implementations are often employed in an environment where a more relaxed linearizability condition suffices, where strict linearizability is not a must.

Here we provide a quantitative definition of limited non-determinism, a notion we call Quasi Linearizability. Roughly speaking an implementation of an object is quasi linearizable if each run of the implementation is at a bounded "distance" away from some linear run of the object. However, as we show the limited distance has to be relative to some operations but not all.

Following the definition we provide examples of quasi concurrent implementations that out perform state of the art standard implementations due to the relaxed requirement. Finally we show that the Bitonic Counting Network non-deterministic behavior can be quantified using our Quasi Linearizable notion.

## 1 Introduction

*Linearizability*[5] is a useful and intuitive consistency correctness condition that is widely used to reason and prove common data structures implementations. Intuitively it requires each run to be equivalent in some sense to a serial run of the algorithm. This equivalence to some serial run imposes strong synchronization requirements that in many cases results in limited scalability and synchronization bottlenecks. In order to overcome this limitation, more relaxed consistency conditions have been introduced. Such alternative consistency conditions for concurrency programming include Sequential consistency[7], Quiescent consistency[6], Causal consistency[2], Release consistency[3], Eventual consistency[10] and Timed consistency[9]. But, the semantics of these relaxed conditions is less intuitive and the results are usually unexpected from a layman point of view. In this paper we offer a relaxed version of linearizability that preserves some of the intuition, provides a flexible way to control the level of relaxation and supports the implementation of more concurrent and scalable data structures.

C. Lu, T. Masuzawa, and M. Mosbah (Eds.): OPODIS 2010, LNCS 6490, pp. 395–410, 2010.
© Springer-Verlag Berlin Heidelberg 2010

For example, SEDA[11], the motivating and initiating reason for the current research is a common design pattern for highly concurrent servers, which heavily relies on thread pools. Such thread pools are composed from two elements (i) a set of threads ready to serve tasks and (ii) a task queue from which the threads consume their tasks. For the task queue, state of the art concurrent queue of Michael and Scott[8] is usually used. It is based on the fact that enqueue and dequeue may happen concurrently while threads trying to enqueue should race. Meaning such queue, which is not part of the server logic in a highly concurrent system, can become by itself a bottleneck limiting the overall SEDA system utilization. One can claim however, that more than often a thread pool does not need a strict FIFO queue, what is required is a queue with relaxed linearizability, i.e., that does not allow one task to starve, meaning bypassed by more than a certain number of tasks.

Another common pattern is the shared counter, which in many applications may become a bottleneck by itself. In order to trim down this contention point Aspnes et al.[6] offered a *counting network* which reduces the contention while maintaining a relaxed consistency condition called quiescent consistency. Such a relaxed counter can be used for example as an id generator, the output of this algorithm is a unique id for each requesting thread while a strict order is not required. This counter may also match other design patterns for example a "Statistical Counter". Modern servers expose many statistical counters, mainly for administration and monitoring. These counters count "online" every operation done on the server. Due to their run time nature these counters by themselves may easily become a contention point. However, sometimes there is no real need for accurate numbers but to capture the general trend. On the other hand the main drawback of the counting network algorithm is also its relaxed consistency, such relaxation does not provide any upper bound for the "inconsistency". We show in Section 6 that the upper bound is $N * W$ where $N$ is the number of working threads, and $W$ is the width of the counting network.

Two more common examples for widely used data structures are the Hash Table and the Stack. While there is a broad range of highly concurrent implementations for a Hash Table as with the former examples the need for a linearizable implementation is often too strict. A very common use case for a Hash Table is a Web Cache. In this case a cache miss while the data is in the Cache might not be a desirable behavior but can be sacrificed for a better scalability. More than that, even getting a stale data for a short while might not be a problem. A similar thing commonly happens with a Stack, a linearizable LIFO implementation can ordinarily be replaced with an almost LIFO implementation for a better scalability.

The above examples have motivated us to provide a quantitative definition of the limited non-determinism that the application requirements might allow. We define a consistency condition which is a relaxed linearizability condition with an upper bound on the non-determinism. Each operation must be linearizable at most at some bounded distance from its strict linearization point. For example, tasks may be dequeued from a queue not in strict FIFO order. That is, a task

$t$ may be dequeued if no task that has been enqueued $k$ tasks or more before $t$ in a linearization order, has not yet been dequeued. Our definition is strong and flexible enough to define at the same time (continuing the above example) that a dequeue that returns empty may not be reordered, i.e., it has to be in its strict linearizable order. In this paper we introduce a formal definition of *quasi-linearizability* condition which captures this condition. This condition introduces some degree of non-determinism, but is useful to prove the quasi-linearizability of different implementations as exemplified in later sections.

## 1.1   Other Relaxed Consistency Conditions

Many models were offered as weaker alternatives to Linearizability two of them are *Quiescent consistency*[6] and *Eventual consistency*[10].

**Quiescent consistency** provides high-performance at the expense of weaker constraints satisfied by the system. This property has two conditions:

1. Operations should appear in some sequential order (legal for each object).
2. Operations whose occurrence is separated by a quiescent state should appear in the order of their occurrence. An object is in a quiescent state if currently there is no pending or executing operation on that object.

**Eventual consistency** this is a specific form of weak consistency; e.g. a storage system guarantees that if no new updates are made to the object, eventually all accesses will return the last updated value. The most popular system that implements eventual consistency is the Internet DNS (Domain Name System). Updates to a name are distributed according to a configured pattern and in combination with time-controlled caches; eventually, all clients will see the last update.

Both models, in most cases, allows better concurrency but on the other hand do not provide any strict upper bound or an adaptive way to determine the "inconsistency" gap when compared to Linearizability.

The contributions of this paper are, first a formal definition of *Quasi Linearizability*, second, two implementations of a FIFO queue that utilize this definition. Third, we empirically evaluate these implementations showing much better scalability relative to the state of the art implementations. Finally we show that a Bitonic[$W$] Counting Network is in fact quasi-linearizable such that its count operation quasi factor is bounded by $N * W$, where $N$ is the number of working threads, and $W$ is the width of the counting network.

## 2   Quasi Linearizable, Definition

### 2.1   Lineariazablity Review

**Definition 1. History:** *Following [5] a history is a list of events which are ordered according to the time line in which they occurred, each event represents* *either a method invocation or a method response, a method invocation event is represented by the tuple $< O.method(args), T >$, where $O$ is the object the invocation operates on, method is the invoked method, args are the invocation arguments and $T$ is the thread that started this invocation. Method invocation response is represented by the tuple $< O : t(results), T >$, where $t$ is either OK or an exception name and results are the invocation result set. A response matches a prior invocation if it has the same object and thread, and no other events of $T$ on object $O$ appear between them.*

History $H$ is called sequential if the first event of $H$ is an invocation, and each invocation, except possibly the last, is immediately followed by a matching response.

An invocation is pending in $H$ if no matching response follows the invocation. An extension of history $H$ is a history constructed by appending zero or more responses matching the pending invocation of $H$. *Complete($H$)* is the sub-sequence of $H$ consisting of all matching invocation and responses, thus, removing all pending invocations from $H$. $H|T$ is a history consisting of all and only the events of thread $T$ in history $H$, two histories $H$ and $H'$ are *equivalent* if for each thread $T$, $H|T = H'|T$.

**Definition 2. Linearizability** *A history $H$ is linearizable if it has an extension $H'$ and there is a legal sequential history $S$ such that:*

1. *Complete($H'$) is equivalent to $S$.*
2. *If method invocation $m_0$ precedes method invocation $m_1$ in $H$, then the same is true in $S$.*

We first define *quasi-sequential specification* and then define what a *quasi-linearizable history* is and finally define a *quasi-linearizable* data structure. The definition follows the notations and standard model as in "The Art of multiprocessor programming" [4].

For example, consider the following sequential history of a queue: $H = enq(1)$, $enq(2)$, $deq()=2$, $enq(3)$, $deq()=1$, $deq()=3$. This sequential history is not legal for a queue, however, it is not "far" from being legal, by exchanging $enq()=2$ with $enq()=1$, one can get a legal sequential history. To formally define this reordering of $H$ and to express how "far" is $H$ from a legal sequential history, we introduce the quasi-linearizable concept.

A *sequential* history is an alternating sequence of invocations and responses, starting with an invocation, and each response matches the preceding invocation. We substitute these two matching events by a single event: $< O.method(args), t(results) >$ (We ignore the thread executing this method call since it is redundant in a sequential history).

Unless specified otherwise all the *sequential* histories in the sequel are condensed in that way. Each event in such a history represents the tuple corresponding to both the invocation and the matching response.

A *sequential specification* is the set of all possible sequential runs of an object, each of these runs can be represented as a sequential history. The term *legal sequential history* specifies that a sequential history is part of the sequential specification of the object that generated that history.

**Definition 3.**   – *For each event $e$ in a sequential history $H$, we define $H[e]$ to be its index in the history, clearly for two events, $e$ and $e'$, $H[e'] < H[e]$ iff $e'$ is before $e$ in $H$.*
  – *$H_i$ is the ith element in $H$. I.e., $H_{H[e]} = e$.*
  – *Events$(H)$ is the set of all the events in $H$.*
  – *Distance$(H', H)$ the distance between two histories $H$ and $H'$, such that $H'$ is a permutation of $H$, is $\max_{e \in Events(H)}\{|H'[e] - H[e]|\}$. .*

Notice that by definition, the distance is defined only for histories which are a permutation of each other.

**Definition 4. Object domain:** *The set of all possible operations that are applicable to an object. We distinguish between operations that have different arguments or different returned values. For example, for $O$ =stack, Domain$(O)$ = $\{< O.push(x), void >, < O.pop()\ x > |x \in X\} \cup \{< O.pop(), \phi >\}$, where $X$ is the set of all the possible elements in the stack.*

A sequential history $H|D$, is the projection of history $H$ on a subset $D$ of the events, i.e., $H$ after removing from it all the events which are not in $D$. $H|O = H|Domain(O)$.

We extend the sequential specification of an object $O$ to a larger set that contains sequential histories which are not legal but are at a bounded "distance" from a legal sequential history. In other words, a sequential history $H$ is in this set if there is some legal sequential history that its "distance" from $H$ is bounded by some specified bound. We define that bound using a function that we name the *Quasi-linearization factor*. It is a function that operates on subsets of the object domain, mapping each subset to its "quasi factor", which is the upper bound on the relative movement among the operations in the subset that turn it into a legal sequential history. Formally,

**Definition 5. Quasi-linearization factor:** *A function $Q_O$ of an object $O$ defined as $Q_O : D \to \mathbb{N}^1$. $D$ is the set containing subsets of the object's domain, formally $D = \{d_1, d_2, \ldots\} \subset Powerset(Domain(O))^2$*

**Definition 6.** $Q_O$**-Quasi-Sequential specification:** *is a set of all sequential histories that satisfy the "distance" bound implied by the quasi-linearization factor $Q_O$ of an object $O$. Formally, for each sequential history $H$ in the set, there is a legal sequential history $S$ of $O$ such that $H$ is a prefix of some history $H'$ which is a permutation of $S$ and $\forall$ subset $d_i \in D$: Distance$(H'|d_i, S|d_i) \leq Q_O(d_i)$*

**Definition 7.** *Let Objects$(H)$ be the set of all the objects that $H$ involves with.*

---

[1] The quasi-linearization factor range $\mathbb{N}$ is extended to a more robust set in the sequel.
[2] $d_1, d_2, \ldots$ are not necessarily disjoint sets, The quasi factor for operations that do appear in $D$ is unbounded.

**Definition 8. Q-Quasi-Linearizable history:** *A history $H$ is Q-Quasi-Linearizable if it has an extension $H'$ and there is a sequential history $S'$ such that:*

1. *$Q = \bigcup_{O \in Objects(H)} Q_O.$[3]*
2. *Complete($H'$) is equivalent to $S'$.*
3. *If method invocation $m_0$ precedes method invocation $m_1$ in $H$, then the same is true in $S'$.*
4. *$\forall O \in Objects(H) : S'|O$ is member of the $Q_O$-Quasi-Sequential specification.*

*We notice that a linearizable history $H$ has Q-quasi-linearizable factor 0 for all of the domains of the objects that appear in it, i.e., for each object $O$ in $H$, $Q(Domain(O)) = 0$.*

**Definition 9. Q-quasi-linearizable object:** *An object implementation $A$ is Quasi-Linearizable with $Q$ if for every history $H$ of $A$ (not necessarily sequential), $H$ is Q-Quasi-Linearizable history of that object.*

For example, consider the following quasi-linearization factor for a blocking queue implementation which is $Q_{queue}$-Quasi-Linearizable:
$D_{enq} = \{< O.enq(x), void > | x \in X\}$, $D_{deq} = \{< O.deq(), x > | x \in X\}$
Domain(Queue)=$D_{enq} \cup D_{deq}$

- $Q_{queue}(D_{enq}) = k$
- $Q_{queue}(D_{deq}) = 0$

Practically it means that an enqueue operation can bypass at most k preceding enqueue operations (and an arbitrary number of dequeue operations occurring in between). This quasi-linearizable queue specifications may be used as the task queue in the SEDA[11] system described in the Introduction.

**H is a prefix of some history?** Consider the following history for a concurrent counter:
$H =< getAndInc(), 3 >, < getAndInc(), 1 >$ This history can never be reordered to a legal sequential history since the event $< getAndInc(), 2 >$ is missing. However, it is reasonable for an execution of a quasi-linearizable implementation of a counter to create such a history because the execution can be stopped at any time. By appending the missing response and invocation $< getAndInc(), 2 >$ at the end of H we can reorder this history to a legal sequential history. This addition of unseen future events is described in the definition by adding a sequential history $H$ to the quasi-sequential specification of the counter object if it is a prefix of *some* history which that history is equivalent to a legal sequential history, the *some* history is $H \cup < getAndInc(), 2 >$. If we do not allow completion of unseen events, hence do not place $H$ in the quasi-sequential specification of the counter, we reduce the definition strength

---

[3] $Q$ is a union of all the different object quasi-linearizable factors, each object has its own separate domain even for objects of the same type.

since any implementation would have been forced to return the entire range of getAndInc() results for not yet terminated operations (without skipping any numbers as the return value) in order to be quasi-linearizable, which in fact makes it similar to quiescent consistent, for instance, a single thread operating on the object has to get a fully linearizable contract from the implementation. It is important to notice that by adding non existing future events, it is not possible to make any history quasi-linearizable. For instance, if the quasi factor for the getAndInc() operation is 5, the following history $H =< getAndInc(), 8 >$ can never be transformed to a legal sequential history only by adding any future events, that is because no matter what unseen future events are added, the first event will need to be moved at least by distance 7 in a legal sequential history (because there are 7 events that must occur before it in any legal sequential history).

**Distance measured on each subset of the domain separately.** The distance is measured only on the projection of the entire history on a subset of the domain, this is done intentionally since some operations may have no effect on others and we do not want to take them into account when we calculate the distance, For instance:
$H = enq(1),\ size()=1,\ size()=1,\ldots,\ size()=1,\ enq(2),\ deq()=2,\ deq()=1.$
If we measure the distance on the enqueue operation and consider the $size()=1$ operations between $enq(1)$ and $enq(2)$, then the distance is unbounded, since an unbounded number of $size$ operations may be executed (in this case one should consider a subset containing all possible $enq$ operations separately). Another notion is that the subsets of the domain that has a quasi factor are not necessarily disjoint, which can be used to define a more generic quasi state. For instance it may be interesting to disallow reordering between $size$ and $enqueue$ operations, but to allow a reorder between $enqueue$ and $dequeue$ operations.

**Extend Quasi Linearizable factor bound.** In the definition we have specified that the bound for each domain subset is a constant number, however, in some cases (as shown later on the Bitonic Counting Network), the bound can vary depending on different parameters such as configurable implementation parameters or different use cases (i.e., the number of threads accessing the object concurrently or different system properties). This is addressed by providing the ability to specify a custom function as the bound instead of a constant bound, and that function arguments take the parameters mentioned above. Formally, instead of having $Q_O : D \to \mathbb{N}$ we change the function as follows: $Q_O : D \to F^N$ where $F^N$ is the set of all functions into $\mathbb{N}^4$. This way, a function $f$ that represents a bound of a domain subset can receive the above variables as its parameters.

**Predicting the future?** Consider the following history for a concurrent queue:
$H = enq(1),\ enq(2),\ deq()=3,\ enq(3),\ deq()=1, deq()=2.$
By the definition of quasi-linearizability, this history is quasi-linearizable for $Q(D_{enq}) \leq 2$, however, it may seem weird that we consider this history legal

---

[4] $F^N = \{f - f$ is a function and $Range(f) = \mathbb{N}\}$.

because the first dequeue operation returns an element which has not yet been enqueued. However, practically speaking, if there was an implementation of a concurrent queue that this history represents an execution of it, it would mean that the implementation is predicting the future, which obviously is not feasible. The only type of such implementation would be one that returns a random value on its dequeue operation. However, for a data structure implementation to satisfy quasi-linearizability, *all* of its possible execution histories must be quasi-linearizable and given an implementation that returns a random result, we can easily schedule one execution example which may never be transformed to a legal sequential history while keeping the quasi distance boundaries.

**Locality (Composition).** Following [5], a property P of a concurrent system is said to be local if the system as a whole satisfies P whenever each individual object satisfies P. As shown in [5], linearizability is a local property, that is a history $H$ is linearizable if and only if, $\forall O \in Objects(H) : H|O$ is linearizable.

**Theorem 1.** *H is Q-Quasi-Linearizable if and only if, $\forall O \in Objects(H) : H|O$ is $Q_O$-quasi linearizable.*

*Sketch of Proof:* It is easy to see that if the entire history $H$ is quasi linearizable, then the projection of it on each object is quasi-linearizable by definition. In order to prove the other direction we need to show that given a history $H$, such that $\forall O \in Objects(H) : H|O$ is $Q_O$-quasi-linearizable, $H$ is Q-quasi-linearizable. For each object $O$, we denote $H'_O$ as the extension of $H|O$ implied by the quasi-linearizable definition of $H|O$ and $S'_O$ as the sequential history that it is part of the $Q_O$-Quasi-sequential specification of $O$ such that $Complete(H'_O)$ is equivalent to $S'_O$. By definition, $S'_O$ is a prefix of some history which is a permutation of a legal sequential history, we denote that legal sequential history by $S_O$. We define $H' = H \bigcup_{O \in Objects(H)} H'_O$, clearly $H'$ is an extension of $H$. We construct $S$ by replacing all of the objects sub-histories $Complete(H'|O)$ with $S'_O$, clearly $S$ is equivalent to $Complete(H')$ and the order of method invocations is kept between the two histories. We need to show that $\forall O \in Objects(H) : S|O$ is part of the $Q_O$-Quasi-Sequential specification of $O$, we get the above since $S|O = S'_O$ by construction.    □

Composition is important in order to be able to use Quasi-Linearizable objects in a bigger system while keeping the Quasi-Linearizable property of the entire system. For instance, consider a system that keeps track of some internal components and operations and at some point needs to calculate the total number of operations executed on the system. Normally, such a system uses a linearizable shared counter that counts each of the operations occurrences, and a combined display counter that represents the total number of operations that is calculated by summing up all operation counters. Assume we have 2 counters for 2 different operations, we get the total number of operations by adding this two counters, assume this counters have a $k_1$, and $k_2$ respectively constant quasi-linearizable bounds for their *add* method. From the composition derives that the quasi bound for the combined counter is $k_1 + k_2$ since the bound is kept for each counter upon composition. (If $k_1 = k_2 = 0$ we get a fully linearizable combined counter).

**Operations that are invisible.** The definition treats all operation as equal, however, operations that are invisible (i.e, do not change the state of the object) can pose difficulties on actual implementations if they affect the distance equally as visible operation since the implementation will probably need to update some internal state for each of these operation in order to comply with the distance bounds. For instance, consider a queue that supports 3 operations: enqueue, dequeue and size, size in this case is considered as an invisible operation. There are a few natural ways to define such a quasi-linearizable queue, one would be to put the enqueue and size operations in the same domain subset in the quasi-linearization factor specification, as well as the dequeue and size operations, thus disabling the queue to return the wrong size value at any stage. However, this boundaries take size operations into consideration when calculating the distance of reordered enqueue operations. An alternative would be to put size in a separate domain subset, however, this will result in legal quasi-linearizable implementations that return size that was legal at some state. Intuitively, the distance between two visible operations should not be affected by invisible operation executed between the two. On the hand, there is still a need for a bound on the reordering distance of invisible operation, otherwise one cannot pose any limitations for this type of operations in a quasi-linearizable object. In order to address this, we can extend the *Distance* of two histories $H$ and $H'$ in the following way:

- Let $VEvent(H)$ be all the events that appear in $H$ that are visible.
- Let $IEvent(H)$ be all the events that appear in $H$ that are invisible.
- $Event(H) = VEvent(H) \cup IEvent(H)$
- $VDistance(H, H') = \max_{e \in VEvents(H)} \{|H'|VEvents(H)[e] - H|VEvents(H)[e]|\}$.
- $NDistance(H, H') = \max_{e \in IEvents(H)} \{|H'[e] - H[e]|\}$.
- $Distance(H, H') = \max\{NDistance(H, H'), VDistance(H, H')\}$.

Using this upgraded distance definition, the enqueue and size operations can be placed together in the same subset and also the dequeue and size operations, while we consider size to be an invisible operation.

**Timed consistency comparison.** Timed consistency[9] adds the notion of time to the occurrences of events and not just order, roughly speaking, timed consistency models require that if a write operation is executed at time $t$, it must be visible to all processes by time $t + \Delta$. In that sense it has some similarity to the quasi-linearizable model, however, the concept of time is not equivalent to the concept of distance in the quasi-linearizable model. Specifically the timed consistency model does allow reordering of events. For example, consider the quasi-linearizable queue mentioned before, in a scenario where each enqueue and dequeue operations are invoked with a time interval of $\Delta$ in between (no concurrent invocations), in order for an implementation to be timed consistent, it will have to return the serial order of events like a regular linearizable implementation. On the other hand, a quasi linearizable implementation does not, since the distance is not affected by time. In the sequel we show implementations which are quasi-linearizable but not timed consistent. Another difference is that timed consistency split operations into two groups, write or read. While quasi-linearizable separates operations according to their logical meaning.

## 3     Random Dequeued Queue

We offer a simple quasi linearizable non blocking queue implementation, illustrated in Figure 1, that behaves as follows: dequeue operation may return results not in the precise order they were enqueued (up to a constant bound) but when an empty (null) dequeue result is returned, there are no enqueued items in the queue. Formally, we describe this behavior with the following quasi-linearizable factor:

- $Q_{nb-queue}(D_{deq} \cup \{< deq(), null >\}) = 0$ (no reordering of dequeue operations is allowed).
- $Q_{nb-queue}(D_{enq}) = k$ (we reorder enqueue operations up to distance $k$, which dictates the actual dequeue order).
- $\forall x \in X : Q(\{< enq(x), void >, < deq(), null >\}) = 0$ (enqueue operation can not be reordered over an empty dequeue operation)

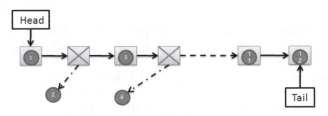

**Fig. 1.** Random Dequeue Queue

The idea is to spread the contention of the dequeue method by allowing to dequeue an element which is not at the head of the queue, but not more than $k$ places away from the head. We base our quasi queue implementation on [8] which is based on a linked list, in fact our enqueue operation is exactly the same. We change the dequeue operation to pick a random index between 0 and $k$ (the quasi factor), if the picked index is larger than 0 it iterates over the list from the head to the item at the specified index, it attempts to dequeue it by doing a single CAS(compare and set) which attempts to mark it as deleted. If failed it retries a few times and eventually falls back to the scenario as if index 0 is picked. If it succeeds, this is the dequeued item. If the selected number is 0, the operation iterates over the list from the head until it finds a node which has not yet been dequeued. While iterating it attempts to remove all encountered dequeued nodes by attempting to advance the head of the list using a CAS each step. The implementation's code and proof of its quasi-lineariazbility property is omitted due to space limitations, for the code and a sketch of proof see [1] for online version of this paper.

## 4   Segmented Queue

The previous implementation of a quasi linearizable queue only reduces contention on the dequeue operation while the *enqueuer* threads still compete over the tail reference trying to enqueue new elements. Additionally, a dequeue operation iterates over its randomly selected number of nodes, while it may traverse over a node that it can dequeue along the way. In the following section we present an algorithm, illustrated in Figure 2, that scatters the contention both for dequeue and enqueue operations and in the normal case, iterates over less nodes while still keeping a constant quasi factor.

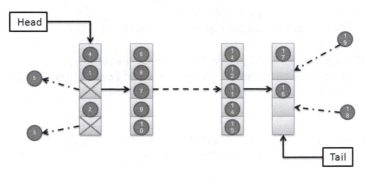

**Fig. 2.** Segmented Queue

The general idea is that the queue maintains a linked list of segments, each segment is an array of nodes in the size of the quasi factor (specified by *quasiFactor+1* in the implementation details), and each node has a deleted Boolean marker, which indicates if it has been dequeued. Each enqueuer iterates over the last segment in the linked list in some random permutation order; When it finds an empty cell it performs a CAS operation attempting to enqueue its new element. In case the entire segment has been scanned and no available cell is found (implying that the segment is full), then it attempts to add a new segment to the list.

The dequeue operation is similar, the dequeuer iterates over the first segment in the linked list in some random permutation order. When it finds an item which has not yet been dequeued, it performs a CAS on its deleted marker in order to "delete" it, if succeeded this item is considered dequeued. In case the entire segment was scanned and all the nodes have already been dequeued (implying that the segment is empty), then it attempts to remove this segment from the linked list and repeats the process on the next segment. If there's no next segment, the queue is considered empty.

Based on the fact that most of the time threads do not add or remove segments, most of the work is done in parallel on different cells in the segments. This ensures a controlled contention depending on the segment size, which is the quasi factor.

```
public void enq(Object value){
  AtomicReference<Node>[] lastSegment = getLast();
  Node newNode = new Node(value);
  if (lastSegment == null){
    //Queue has no segments, create a new segment
    lastSegment = createLast(null);
  }
  while(true){
    int[] permutation = getRandomPermutation();
    for(int i = 0; i <= quasiFactor; ++i){
      final int index = permutation[i];
      //Cell is not empty, continue
      if (lastSegment[index].get() != null)
        continue;
      //Found empty cell, try to enqueue here
      if (lastSegment[index].compareAndSet(null, newNode))
        return;
    }

    //If reached here, no available position, create a new segment
    lastSegment = createLast(lastSegment);
  }
}
```

The method *createLast(lastSegment)* creates and adds a new last segment only if the current last segment is the provided method argument (lastSegment). The result of the method will be the current last segment which was either created by this invocation or another invocation if the current last segment is different than the method argument.

```
public Object deq(){
  AtomicReference<Node>[] firstSegment = getFirst();
  while(true){
    boolean hadNullValue = false;
    if (firstSegment == null)
      //Queue is empty
      return null;
    int[] permutation = getRandomPermutation();
    for(int i = 0; i <= quasiFactor; ++i){
      Node node = firstSegment[permutation[i]].get();
      //Check if this cell is empty, which means
      //an element can be enqueued to this cell in the future
      if (node == null) {
        hadNullValue = true;
        continue;
      }
```

```
    //Check if can dequeue node at index
    if (node.deleted.compareAndSet(false, true))
      return node.value;
  }
  //scanned the entire segment without finding a candidate
  //to dequeue

  //If there was an empty cell, the queue is considered empty
  if (hadNullValue)
    return null;
  //All nodes have been dequeued, we can safely remove the
  //first segment
  firstSegment = removeFirst(firstSegment);
  }
}
```

The method *removeFirst(firstSegment)* removes the first segment only if the current first segment is the provided method argument (firstSegment). The result of the method will be the current first segment.

The proof of the quasi-lineariazbility property of the implementation is omitted due to space limitations, for a sketch of proof see [1] for online version of this paper.

### 4.1   Segmented Stack

The above algorithm can be adapted in order to implement a stack. The difference is that when a push operation needs to add a new segment, it is added to the head of the list instead of the tail and a pop operation is similar to a dequeue operation.

## 5   Performance Evaluation

We evaluated the performance of our new algorithms on a Sun UltraSPARC T2 Plus multicore machine. This machine has 2 chips, each with 8 cores running at 1.2 GHz, each core with 8 hardware threads, so 64 way parallelism on a processor and 128 way parallelism across the machine. There is obviously a higher latency when going to memory across the machine (a two fold slowdown).

We can see from Figure 3 that the Segmented queue implementation out perform both Michael and Scott[8] and Random Dequeue when the number of threads increases, which is reasonable since it spreads both enqueue and dequeue contention. However as we increase the quasi factor, the overhead of scanning an entire segment just to realize the enqueuer needs to create a new segment or the dequeuer needs to remove the first segment, is increasing. On the other hand the Random Dequeue behaves very similar to the Michael and Scott algorithm when the quasi factor is low, but on high number of threads it improves if we increase the quasi factor, which is because the contention is reduced on the dequeue operation.

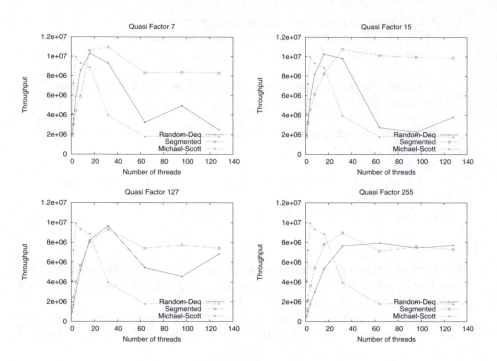

**Fig. 3.** Concurrent queue benchmark results

# 6    Bitonic[W] Counting Network

Next we show that the Bitonic[W] Counting Network (W the network width) is Q-quasi-linearizable, $Q(D_{inc} = \{< O.getAndInc(), n > |n \in \mathbb{N}\}) \leq N * W$ (where $N$ is the number of working threads). For example for the Bitonic[4] Counting Network showed in Figure 4 with $N = 4$ we show that $Q(D_{inc}) \leq 16$.

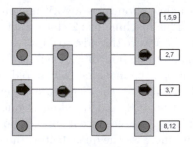

**Fig. 4.** Bitonic[4] Counting Network

For the Bitonic Counting Network we choose the quasi linearization points when a thread is increasing the counter at the exit from the network, we denote this history as $S'$.

**Lemma 1.** *For any $n$, in $S'$ all the operations $\{< O.getAndInc(), m > | m \leq n - (N * W)\}$ precede $< O.getAndInc(), n >$.*

**Proof:** *Assume in contradiction that the lemma does not hold, denote by $< O.getAndInc(), m >$, $m < n - (N * W)$ as the missing operation. From the quiescent consistency property of Bitonic network, we know that if we will now schedule an execution that lets all the threads that are currently traversing the network (executing getAndInc operation) to finish their current operation, and prevent new threads or the same threads to reenter the network, the network must be at a legal state, that is, all the values up to $n$ have been returned. From the algorithm of the network, we know that if $< O.getAndInc(), m >$ has not been executed yet, so does $\forall i > 0 :< O.getAndInc(), m + W * i >$ (because this operations are diverted to the same counter). From that we get that $< O.getAndInc(), m >$, $< O.getAndInc(), m + W >, \ldots, < O.getAndInc(), m + W * (N - 1) >$ have not happened by the time that $< O.getAndInc(), n >$ occurred. Since there are at most $N - 1$ threads that are pending execution completion, they can never fill in the missing $N$ operations to close the gap, in contradiction to the quiescent consistency property of the Bitonic network.*

From the lemma we know that each getAndInc() operation had bypassed at most $N * W$ other getAndInc() operation, therefore we can find a legal sequential history $S$ which satisfies $Distance(S'|D_{inc}, S|D_{inc}) \leq N * W$. In this part we have shown an upper bound of $N * W$ for getAndInc operation, but this is not necessarily a tight bound.

# 7   Conclusions

In this paper we have shown a more relaxed concurrent model for linearizability and a few actual implementations which take advantage of the new model and are more concurrent than the equivalent linearizable implementation. We have demonstrated this with a queue that supports enqueue and dequeue operations, we can see how this definition can be adapted to a queue that also supports a peek operation. One way to define its quasi-linearizablity, is by specifying the quasi factor parameters as follows: $Q(D_{enq} \cup D_{peek}) = k$ and $\forall x : Q(\{< deq(), x >, < peek(), x >\})=0$, meaning that a dequeue and peek operations may return an item at distance k from the head of the queue, but a peek can not return an item which has been already dequeued. Additionally we have shown that the already known Bitonic counting network implementation is quasi linearizable. This model can be applied to specify other quasi linearizable objects, such as, stack, heap etc., and thus allows a more concurrent implementation of these objects.

*Acknowledgements.* We would like to thank Adam Morrison, Nir Shavit and Maria Natanzon for very productive and helpful discussions. This paper was supported in part by grants from Sun Microsystems, Intel Corporation, as well as a grant 06/1344 from the Israeli Science Foundation and European Union grant FP7-ICT-2007-1 (project VELOX).

# References

1. Afek, Y., Korland, G., Yanovsky, E.: Quasi-linearizability: Relaxed consistency for improved concurrency, http://sites.google.com/site/gkorland/research
2. Ahamad, M., Hutto, P.W., Neiger, G., Burns, J.E., Kohli, P.: Causal memory: Definitions, implementation and programming. Technical Report GIT-CC-93/55, Georgia Institute of Technology (1994)
3. Gharachorloo, K., Lenoski, D., Laudon, J., Gibbons, P., Gupta, A., Hennessy, J.: Memory consistency and event ordering in scalable shared-memory multiprocessors. In: Proceedings of the 17th Annual International Symposium on Computer Architecture, ISCA 1990, pp. 15–26. ACM, New York (1990)
4. Herlihy, M., Shavit, N.: The art of multiprocessor programming. Morgan Kaufmann, San Francisco (2008)
5. Herlihy, M.P., Wing, J.M.: Linearizability: a correctness condition for concurrent objects. ACM Trans. Program. Lang. Syst. 12(3), 463–492 (1990)
6. Aspnes, M.H.J., Shavit, N.: Counting networks. Journal of the ACM 41(5), 1020–1048 (1994)
7. Lamport, L.: How to make a multiprocessor computer that correctly executes multiprocess program. IEEE Trans. Comput. 28(9), 690–691 (1979)
8. Michael, M.M.; Scott, M.L.: Simple, fast, and practical non-blocking and blocking concurrent queue algorithms. In: Proceedings of the Fifteenth Annual ACM Symposium on Principles of Distributed Computing, PODC 1996, pp. 267–275. ACM, New York (1996)
9. Torres-Rojas, F.J., Ahamad, M., Raynal, M.: Timed consistency for shared distributed objects. In: Proceedings of the Eighteenth Annual ACM Symposium on Principles of Distributed Computing, PODC 1999, pp. 163–172. ACM, New York (1999)
10. Vogels, W.: Eventually consistent. Communications of the ACM 52(1), 40–44 (2009)
11. Welsh, M., Culler, D., Brewer, E.: Seda: an architecture for well-conditioned, scalable internet services. SIGOPS Oper. Syst. Rev. 35(5), 230–243 (2001)

# A Token-Based Distributed Algorithm for the Generalized Resource Allocation Problem

Hirotsugu Kakugawa[1,*] and Sayaka Kamei[2,**]

[1] Osaka University, Suita, Osaka, Japan
[2] Hiroshima University, Higashihiroshima, Hiroshima, Japan

**Abstract.** The resource allocation problem is one of the fundamental problems for conflict resolution in distributed systems. In this paper, we consider the $(n, m, k, d)$-resource allocation problem, proposed by Joung [Joung, Distributed Computing (2010)], which is a generalization of the $k$-mutual exclusion problem and the group mutual exclusion problem. We propose a fully distributed solution based on tokens for the $(n, m, k, d)$-resource allocation problem for asynchronous message passing distributed systems. Previous works allow a process to request only one resource at a time. In our algorithm, the amount of resources that a process may request is arbitrary. The algorithm is designed based on leader-follower scheme, and permission to access resources is granted by tokens. The first process that requests resources becomes the leader, and then, other processes become followers and they are granted their requests by the leader. The message complexity of the proposed algorithm is bounded by $O(|Q|)$ in the worst case and $O(1)$ in the best case, where $|Q|$ is the size of quorums of a coterie that the algorithm uses.

## 1 Introduction

The resource allocation problem is one of fundamental problems for conflict resolution in distributed systems. The mutual exclusion problem, which has been studied extensively, is an example of such a problem. A process wishing to exclusively access a shared resource must obtain permission before its access. Mutual exclusion is necessary for consistent update of a shared resource for example. The resource allocation problem arises when processes in a distributed system require shared resources so they need to cooperate and compete with each other for their activities.

Because recent distributed systems grow huge and become complex, resource allocations of various types are necessary for various kind of network applications. Below, we use a term "critical section" in which each process accesses a shared resource in its program. The restriction of entry to critical section is

---

* This work is supported in part by Grant-in-Aid for Scientific Research ((B)20300012) of JSPS.
** This work is supported in part by Grant-in-Aid for Young Scientists ((B)22700074) of JSPS.

C. Lu, T. Masuzawa, and M. Mosbah (Eds.): OPODIS 2010, LNCS 6490, pp. 411–426, 2010.

considered as a problem of resource allocation because an entry to a critical section at a process can be considered as an allocation of resources to the process. Hence, we use terms "entry to critical section" and "allocation of resource" interchangeably in this paper.

Various generalizations of the mutual exclusion problem have been proposed so far. The $k$-mutual exclusion problem is a problem such that at most $k$ processes can enter critical section at the same time [1]. The distributed $k$-mutual exclusion problem is a generalization on the axis of the amount of resources that is accessed by processes at the same time. The group mutual exclusion problem is a problem such that any number of processes in the same group can enter critical section, but no two processes in different groups can enter critical section at the same time [2]. Hence, the distributed group mutual exclusion problem is a generalization on the axis of the type of resource (specified by group) that is accessed by processes at the same time. Therefore these two generalizations are orthogonal.

The $(n, m, k, d)$-resource allocation problem is a generalization of the $k$-mutual exclusion problem and the group mutual exclusion problem proposed in [3]. The access to shared resources is restricted as follows. The number of processes is $n$ and the number of groups (resource types) is $m$. The number of groups that can be accessed simultaneously is at most $k$, and, for each group, at most $d$ processes can enter critical section. The problem setting of [3] allows each process to request only one unit of resource. Hence, the $(n, 1, 1, 1)$-resource allocation problem is coincident with the mutual exclusion, the $(n, 1, 1, k)$- and the $(n, n, k, 1)$-resource allocation problems are coincident with $k$-mutual exclusion, and the $(n, m, 1, n)$-resource allocation problem is coincident with group mutual exclusion.

An application of the $(n, m, k, d)$-resource allocation problem is the following. Consider a peer-to-peer video streaming service over the Internet. There are $n$ nodes and $m$ possible channels. However, by limitation of bandwidth, at most $k$ channel can be served simultaneously and at most $d$ users can view each channel simultaneously. The $(n, m, k, d)$-resource allocation problem models such an access control in a distributed system.

The distributed algorithm mentioned in [3] adopts a communication structure called quorums, specifically, it adopts $(m, k, d)$-coterie which is a quorum structure specially designed to solve the problem, and it is a permission-based algorithm. Generally speaking, a quorum is a subset of processes in a distributed system. The quorums have some intersection relation with each other that is specified by the problem to be solved. A permission-based distributed resource allocation (mutual exclusion) algorithm that uses quorums follows the following scheme. A process that requests a resource selects a quorum and sends a request message to each process in a quorum for permission. When a process receives a request message, it grants the request under the constraint that it does not grant more than one requesting process at a time. When the requesting process obtains a grant (permission) from each process in a quorum, it may access a resource.

In this paper, we propose a new distributed algorithm for the $(n, m, k, d)$-resource allocation problem in asynchronous message passing environment. Contribution of our work is threefold: (1) full concurrency in resource access, (2) low message complexity, and (3) each process may specify an amount of resources it requests. Let us explain these points below.

Firstly, our algorithm achieves full concurrency in resource access. Because the $(n, m, k, d)$-resource allocation problem allows several processes to access shared resources simultaneously, increasing concurrency of resource access is an important issue. The existing algorithm proposed in [3], unfortunately, may not achieve full concurrency allowed by the $(n, m, k, d)$-resource allocation problem depending on timing of resource requests and on quorum selections. Our algorithm proposed in this paper solves this problem and it allows full concurrency, that is, as long as a request does not deviate the constraint of the $(n, m, k, d)$-resource allocation problem, the request is granted.

Secondly, our algorithm achieves a low message complexity. Although the existing algorithm proposed in [3] uses quorums for communication structure, a quorum selected by a requesting process may be used by other processes already. When this is the case, the requesting process must retry to find a free quorum to obtain permission; otherwise, it must wait other processes release a quorum. Retrying to find a free quorum may increase concurrency of resource access, however, it increases message complexity unfortunately. Our algorithm proposed in this paper solves this problem, and it does not require retry to find a quorum. Hence it achieves a low message complexity with full concurrency in resource access.

Lastly, in our algorithm, a requesting process can specify an amount of resources (of the same type) it requests. In our problem setting, the $(n, m, k, d)$-resource allocation problem is generalized in such a way that the total number of allocated resources is at most $d$ for each group. In the existing algorithm proposed in [3], an amount of resources that a requesting process can request is limited only one. Hence, our algorithm further generalizes allocation of resources.

To achieve the three points listed above, our algorithm is token-based for obtaining the privilege to access a shared resource. That is, a process that holds a token can access a shared resource. Our observation is that the drawback of the algorithm in [3] is the usage of a $(m, k, d)$-coterie which is a quorum structure specially designed to solve the $(n, m, k, d)$-resource allocation problem. The combinatorial structure of a $(m, k, d)$-coterie is theoretically interesting, however, algorithms based on it do not perform well. By using tokens, the drawbacks are resolved. On the other hand, our algorithm uses quorums for propagating a resource request between processes to reduce message complexity. Quorums used by our algorithm is called an ordinary coterie which is a simple quorum structure well studied [4–9].

Organization of this paper is as follows. In section 2, we review related works. In section 3, we present the computational model we assume and the problem statement. In section 4, we propose our resource allocation algorithm. In section 5, we show proof of correctness and performance analysis of the proposed algorithm. In section 6, we give concluding remarks.

## 2    Related Works

There are many papers on distributed mutual exclusion and its extension for asynchronous message passing distributed systems. There are two paradigms for distributed mutual exclusion and resource allocation algorithms: *permission-based* algorithm and *token-based* algorithm [10]. In a permission-based algorithm, a process obtains permission from processes for access to resources, whereas in a token-based algorithm, a process accesses resources only when it obtains a privilege token.

The distributed mutual exclusion problem is a problem to control accesses to shared resources in such a way that at most one process can access a shared resource at a time. In [11, 12], permission-based mutual exclusion algorithms are proposed with $O(n)$ message complexity. In [4], permission-based mutual exclusion algorithm is proposed. To reduce the message complexity, a coterie is adopted for communication. The message complexity of the algorithm is $O(|Q|)$, where $|Q|$ is the (maximum) size of quorum of a coterie used by the algorithm. $|Q|$ is typically $O(\sqrt{n})$ where $n$ is the total number of processes in a distributed system, by adopting a coterie based on the finite-projective plane. When a process makes a request, it contacts with only $|Q|$ processes, and hence its message complexity is reduced. In [13], a token-based algorithm is proposed with message complexity $O(n)$.

A coterie is a communication structure for distributed systems that reduces message complexity and improves availability. Intuitively, a coterie $C$ is a set of quorums $Q_1, Q_2, ...$, where $Q_i \subseteq V$ and $V$ is the set of processes in a distributed system, such that any two quorums have at least one common process. The idea of a permission-based mutual exclusion algorithm with a coterie is that a process must obtain permission from each process in a quorum to enter critical section and each process in a quorum givens permission at most the one process at a time. Because any two quorums have at least one common process, no two processes can enter critical section at a time. Since a coterie is an important communication structure for reducing message complexity and improvement of availability, there are many results, e.g., [4–9, 14].

The distributed $k$-mutual exclusion problem requires to control access to shared resources in such a way that (1) each process can request a unit of shared resource, and (2) at most $k$ units of shared resource is allocated to processes. i.e., at most $k$ processes can access shared resources at a time. This problem limits total number of processes that access shared resource. In [1], a permission-based distributed $k$-mutual exclusion problem is proposed that uses a $k$-coterie, which is an extension of a coterie, for communication structure.

The $h$-out of-$k$ mutual exclusion problem requires to control access to shared resources in such a way that (1) each process requests $h$ ($1 \leq h \leq k$) units of shared resource, and (2) at most $k$ units of shared resource is allocated to processes at a time. This problem is a generalization of $k$-mutual exclusion. In [15], $k$-arbiter is proposed as a generalization of a coterie, and a permission-based distributed algorithm for the $h$-out of-$k$ resource allocation problem is proposed that adopts $k$-arbiter as a communication structure.

The group mutual exclusion problem requires to control access to shared resources in such a way that (1) each process selects its group (resource type), and (2) no two processes in different groups access shared resource. Hence, any number of processes can access a shared resource as long as they are in the same group. This problem limits type of shared resource accessed by processes. In [16], a permission-based distributed algorithm is proposed with message complexity $O(|Q|)$, where $|Q|$ is the size of quorum of a coterie used by the algorithm. In [17], a token-based distributed algorithm with message complexity $O(n)$ is proposed. In [18], a distributed algorithm is proposed with message complexity $O(|Q|)$, where $|Q|$ is the size of quorum used by the algorithm. Their algorithm is based on the *leader-follower* scheme which is a hybrid of permission-based and token-based schemes. The first process that allocates a resource becomes the leader, and other processes, called followers, are allocated resources by the leader process. In [19], a token-based distributed algorithm with message complexity $O(|Q|)$ is proposed. This algorithm also takes the leader-follower scheme, and a coterie is used to propagate resource requests for reducing message complexity. It is shown in [19] that it performs better in message complexity and concurrency than the algorithm proposed in [18] by simulation.

In [3], further generalization to the distributed mutual exclusion, called the $(n, m, k, d)$-resource allocation problem, is proposed. In [3], a $(m, 1, d)$-coterie and a $(m, k, d)$-coterie are proposed as extensions of a coterie. The distributed algorithm proposed in [4] solves the $(n, m, 1, d)$-resource allocation problem if the algorithm uses a $(m, 1, d)$-coterie instead of a coterie for communication structure. Further generalization a $(m, k, d)$-coterie is for the $(n, m, k, d)$-resource allocation problem, and construction of a $(m, k, d)$-coterie from a $(m, 1, d)$-coterie is proposed. Unfortunately, simple combination of the distributed algorithm proposed in [4] with a $(m, k, d)$-coterie does not solve the $(n, m, k, d)$-resource allocation problem fully. Specifically, as we mentioned in the previous section, there is a case in which concurrent access to shared resources is impossible even if allocation is allowed by definition of the $(n, m, k, d)$-resource allocation problem.

## 3   Preliminary

### 3.1   The Computational Model

In this paper, we assume that a distributed system consists of a set of $n$ processes $V = \{P_0, P_1, ..., P_{n-1}\}$. We assume that the network is fully connected. Each process has a unique identifier selected from a set of integers $\{0, 1, ..., n-1\}$. A distributed system is *asynchronous*, i.e., there is no common global clock. Information exchange between processes is done by asynchronous message passing. Each communication channel is FIFO, and each message sent is delivered within finite time. We assume that there is no upper bound on message delivery time. We assume that the system is error-free.

## 3.2    The Generalized Resource Allocation Problem

There are $m$ groups (types) of resources, and let $G = \{0, 1, ...., m - 1\}$ be a set of groups (resource types). Each process selects a group $g \in G$, and makes a request for critical section entries. We model such a behavior of each process $P_i$ as follows.

> **while** (**true**) **do** {
>     Select a group $g \in G$ and amount of resources $r \in \{1, ..., d\}$;
>     $Request(g, r)$;    – *Entry protocol.*
>     Critical Section
>     $Release$;    – *Exit protocol.*
> }

Formally, the (generalized) problem of $(n, m, k, d)$-*resource allocation* is defined as follows, which is a generalization of [3].

**Definition 1.** *The* $(n, m, k, d)$-resource allocation problem *is a problem to control execution of processes to satisfy the following three conditions.*

- Safety (total group number) : *At most $k$ different group of processes are in critical section simultaneously.*
- Safety (group size) : *For each group, the total amount of allocated resources is at most $d$ at any time.*
- Liveness: *Any requesting process eventually enters its critical section.*    □

## 3.3    Coterie

In this paper, we use a coterie for communication between processes. It is defined formally as follows.

**Definition 2.** *(Coterie* [14]*) Let* $U = \{P_0, P_1, ..., P_{n-1}\}$ *be a set. A set $C$ of subsets of $U$ is a* coterie *under $U$ if and only if the following three conditions are satisfied.*

1. Non-emptiness: *For each $Q \in C$, $Q$ is not empty and $Q \subseteq U$,*
2. Intersection property: *For any $Q, Q' \in C$, $Q \cap Q'$ is not empty, and*
3. Minimality: *For any $Q, Q' \in C$, $Q$ is not a proper subset of $Q'$.*

*An element of $C$ is called a* quorum.    □

## 3.4    Performance Measures

We define performance measures of distributed resource allocation algorithms.

**Definition 3.** Message complexity *is the number of messages exchanged per request for critical section.* Maximum concurrency *is the maximum number of processes that can enter critical section simultaneously.*    □

# 4   The Proposed Algorithm

In this section, we describe the proposed algorithm for the $(n, m, k, d)$-resource allocation problem. The proposed algorithm here is based on the algorithm proposed in [19] for the group mutual exclusion algorithm. To solve the $(n, m, k, d)$-resource allocation problem, we introduced a new resource management scheme.

The main idea of the proposed algorithm TQGRA is as follows. Our algorithm TQGRA uses two types of tokens, the *main-token* and *sub-tokens*. Initially, process $P_0$ holds the main-token. The number of the main-token is exactly one in the network at any time, while the number of sub-tokens varies. A sub-token is generated by the holder of the main-token on request of other process. When no process is in critical section, i.e., no process is accessing any resource, a process must obtain the main-token to enter critical section. In this case, the holder of the main-token changes. When there is a process in a critical section, a process obtains a sub-token to enter critical section. The constraint of the resource allocation is managed by the holder of the main-token. The holder of the main-token issues a sub-token only when the constraint is not violated.

Outline of behavior of each process is described as follows.

- When a process $P_i$ makes a request some resources:
  - In case $P_i$ knows the holder of the main-token, it sends a request message directly sent to the holder.
  - Otherwise, it selects a quorum $q \in C$, and it sends a request message to each process $P_j \in q$. At least one process $P_j \in q$ knows the holder of the main-token, whose mechanism is described later, and such $P_j$ transfers the request message to the holder of the main-token.
- When the holder of the main-token receives a request message, it takes the following actions.
  1. If no process uses any resource (i.e., there is no process is in critical section), the holder transfers the main-token to $P_i$ by a token message.
  2. If some processes are accessing resources, the holder issues a sub-token and sends it to $P_i$ by a subtoken message.
- When $P_i$ receives a token or subtoken message, it can access the resource.
  - If it receives a token message, it selects a quorum $q \in C$ and sends an acquired message to each process in $q$. By this message, at least one process in any quorum $q' \in C$ knows the holder of the main-token.
- When $P_i$ releases resource:
  - If $P_i$ is granted by token message, it decrements the amount of resources currently in use that is maintained by a token object.
  - If $P_i$ is granted by subtoken message, it sends a release to the holder of the main-token.

The major change of the proposed algorithm from the one in [19] is resource management scheme. The algorithm in [19] issues a sub-token if the type of requested resource is the same as the resource type currently accessed. The proposed algorithm modifies this resource management scheme as follows. The main-token maintains the amount of resources currently allocated for each type of resource. The holder of the main-token issues a sub-token only when a request does not violate the constraint of the $(n, m, k, d)$-resource allocation problem, that is, the number of active resource type does not exceed $k$, and the total amount of allocated resources does not exceed $d$ for each resource type.

## 4.1   Local Variables at Each Process

Important variables maintained at each process $P_i$ are as follows.

- $mode_i \in \{\text{IDLE}, \text{TRYING}, \text{INCS}\}$ — Current status of process $P_i$.
- $ts_i$ — Timestamp for each request, which is incremented by one when $P_i$ makes a new request. We assume that this variable is implicitly maintained by the well-known protocol by Lamport [11]; hence we do now explicitly show its update protocol in our algorithm description.
- $grp_i$ — Group name of resource for the current request.
- $nres_i$ — The amount of resources for the current request.
- $type_i \in \{\bot, \text{MAIN}, \text{SUB}\}$ — The type of token that $P_i$ is holding.
- $holder_i \in \{P_0, ..., P_n, \bot\}$ — Process name that holds the main-token to the best knowledge of $P_i$. Its value is $\bot$ if $P_i$ does not know.
- $home_i \in \{P_0, ..., P_n, \bot\}$ — The process name to which $P_i$ should return a sub-token when its has a sub-token.
- $tmpQ_i$ — A temporary queue for requests that $P_i$ receives to forward to the holder of the main-token. A request is queued only when $P_i$ does not know the holder of the main-token. When $P_i$ knows the holder of the main-token by a token message, it forwards all the requests in the queue to the holder.
- $token_i$ — The token object that corresponds to the main-token. $P_i$ holds a token object only when it is the holder of the main-token.

## 4.2   Structure of the Token Object

The holder of the main-token manages allocation of the resources, and the token object maintains enough information to manage resource allocation. The token object $tok$ contains the following items.

- $tok.actGrp[g]$ — The amount of resources allocated to group $g$. The protocol maintains the following two conditions.
  - $0 \le tok.actGrp[g] \le d$ for each $g$, and
  - $|\{g : tok.actGrp[g] > 0\}| \le k$.

  The first condition is the constraint that there are $d$ instance for each resource type. The second condition is the constraint that the number of allocated resource types is at most $k$.
- $tok.reqQ$ — A queue of requests currently pending.
- $tok.tsReq[j]$ — Timestamp value for the latest request from $P_j$. Because requests are forwarded by processes in a quorum, the holder of the main-token may receive the same request more than once. This value is used to ignore duplicate requests.

**on initialization;**

1.1   $mode_i :=$ IDLE;

1.2   $ts_i := 0;\ \ grp_i := \bot;\ \ nres_i := 0;$

1.3   $type_i := \bot;\ \ \ holder_i := \bot;$

1.4   $home_i := \bot;\ \ \ leaving_i :=$ **false**;

1.5   $acqs_i := \emptyset;\ \ acks_i := \emptyset;$

1.6   $tmpQ_i :=$ **new**$(Queue);$

1.7   **if** $(P_i = P_0)$ {     – $P_0$ is the initial holder of the main-token.

1.8     $tok_i :=$ **new**$(Token);$

1.9     $tok_i.actGrp[g] := 0$ **for each** group $g$;

1.10    $tok_i.reqQ :=$ **new**$(Queue);$

1.11    $tok_i.tsReq[j] := 0$ **for each** $j = 0..n - 1$;

1.12    $q_i :=$ select a quorum in coterie $C$;

1.13    $acqs_i := q_i - \{P_i\};$

1.14    **send** $\langle$acquired$\rangle$ **to each** $P_j \in acqs_i$;

1.15  } **else** {

1.16    $tok_i := \bot;$

1.17  }

**on event** $requestEvent(g_i, r_i);$      // Request for CS

2.1   $mode_i :=$ TRYING;

2.2   $ts_i := ts_i + 1;\ \ grp_i := g_i;\ \ nres_i := r_i$

2.3   **if** $(tok_i \neq \bot)$ **then** {     – case A1 ($P_i$ is the holder of the main-token.)

2.4     $tok_i.tsReq[i] := ts_i;$

2.5     **enqueue**$(tok_i.reqQ, \langle P_i, ts_i, grp_i, nres_i\rangle);$

2.6     **call** $handlePendingRequests;$

2.7   } **else if** $(holder_i \neq \bot)$ {     – case A2 ($P_i$ knows the holder of the main-token.)

2.8     **send** $\langle$request, $P_i, ts_i, grp_i, nres_i\rangle$ **to** $holder_i$;

2.9   } **else** {     – case A3 (Otherwise, send requests to quorum members.)

2.10    $q_i :=$ select a quorum in coterie $C$;

2.11    **send** $\langle$request, $P_i, ts_i, grp_i, nres_i\rangle$
          **to each** $P \in (q_i - \{P_i\});$

2.12    **if** $(P_i \in q_i)$ {

2.13      **delete** $\langle P_i, *, *, *\rangle$ **from** $tmpQ_i$;

2.14      **enqueue**$(tmpQ_i, \langle P_i, ts_i, grp_i, nres_i\rangle);$

2.15    }

2.16  }

**on event** $releaseEvent;$      // Exit from CS

3.1   **if** $(type_i =$ MAIN$)$ **then** {     – case B1 ($P_i$ holds the main-token.)

3.2     $tok_i.actGrp[grp_i] := tok_i.actGrp[grp_i] - nres_i;$

3.3     $type_i := \bot;\ \ mode_i :=$ IDLE;

3.4     $grp_i := \bot;\ \ nres_i := 0;$

3.5     **call** $handlePendingRequests;$

3.6   } **else** {     – case B2 ($P_i$ holds a sub-token)

3.7     **send** $\langle$release, $grp_i, nres_i\rangle$ **to** $home_i$;

3.8     $type_i := \bot;\ \ mode_i :=$ IDLE;

3.9     $grp_i := \bot;\ \ nres_i := 0;$

3.10    $home_i = \bot;$

3.11  }

3.12  **trigger event** $releaseDone_i$      // Exit done

**Fig. 1.** Description of TQGRA for $P_i$ (1/3)

**procedure** *handlePendingRequests*

4.1     **if** $\neg leaving_i \wedge (tok_i \neq \bot) \wedge \neg$**empty**$(tok_i.reqQ)$
            $\wedge (\forall g' : tok_i.actGrp[g'] = 0)$ {     $-$ *No resource is in use.*

4.2         $\langle P_j, t, g, r \rangle :=$ **peek**$(tok_i.reqQ)$;     // *peek the top item*

4.3         **if** $(P_j = P_i)$ {     $-$ $P_i$ *has a priority to use the main-token.*

4.4             **dequeue**$(tok_i.reqQ)$;     // *discard the top item*

4.5             $tok_i.actGrp[grp_i] := nres_i$;

4.6             $type_i :=$ MAIN;     $mode_i :=$ INCS;

4.7             **trigger event** $requestDone_i$;         // Enter CS

4.8         } **else** {     $-$ $P_j \neq (P_i)$ *has a priority to use the main-token.*

4.9             **if** $(acqs_i \neq \emptyset)$ {     $-$ *If this is the case, holder$_j$ must be reset.*

4.10                **call** *beginTokenTransfer*;     $-$ *Start preparation to transfer the main-token.*

4.11            } **else** {     $-$ *Otherwise, the main-token is transferred immediately.*

4.12                **dequeue**$(tok_i.reqQ)$;     // *discard the top item*

4.13                **send** $\langle$token, $tok_i \rangle$ **to** $P_j$;   $tok_i := \bot$;

4.14            }

4.15        }

4.16    }

4.17    **while** $\neg leaving_i \wedge (tok_i \neq \bot) \wedge \neg$**empty**$(tok_i.reqQ)$ {
            $-$ *Grant a request in the queue if the resource constraint is not violated.*

4.18        $\langle P_j, t, g, r \rangle :=$ **peek**$(tok_i.reqQ)$;

4.19        **if** $(tok_i.actGrp[g] = 0) \wedge (|\{h : tok_i.actGrp[h] > 0\}| \geq k)$

4.20            **break**;     $-$ *Cannot allocate new group.*

4.21        **if** $(tok_i.actGrp[g] + nres_i > d)$

4.22            **break**;     $-$ *The requested group is full.*

4.23        **dequeue**$(tok_i.reqQ)$;

4.24        $tok_i.actGrp[g] := tok_i.actGrp[g] + nres_i$;

4.25        **send** $\langle$subtoken$\rangle$ **to** $P_j$;

4.26    }

**on receipt of** $\langle$token, $tok\rangle$;

5.1     $tok_i := tok$;

5.2     $q_i :=$ select a quorum in coterie $C$;

5.3     $acqs_i := q_i - \{P_i\}$;

5.4     **send** $\langle$acquired$\rangle$ **to each** $P_j \in acqs_i$;

5.5     $type_i :=$ MAIN;     $mode_i :=$ INCS;

5.6     $tok_i.actGrp[grp_i] := nres_i$;

5.7     **trigger event** $requestDone_i$;         // Enter CS

5.8     **while** $\neg$**empty**$(tmpQ_i)$ {
            $-$ *Enqueue pending requets into the queue of the main-token.*

5.9         $\langle P_j, t, g, r \rangle :=$ **dequeue**$(tmpQ_i)$;

5.10        **if** $(tok_i.tsReq[j] < t)$ {

5.11            $tok_i.tsReq[j] := t$;

5.12            **enqueue**$(tok_i.reqQ, \langle P_j, t, g, r \rangle)$;

5.13        }

5.14    }

5.15    **call** *handlePendingRequests*;

**on receipt of** $\langle$subtoken$\rangle$ **from** $P_\ell$

6.1     $type_i :=$ SUB;     $mode_i :=$ INCS;   $home_i := P_\ell$;

6.2     **trigger event** $requestDone_i$;         // Enter CS

**Fig. 2.** Description of TQGRA for $P_i$ (2/3)

**on receipt of** ⟨acquired⟩ **from** $P_\ell$;

7.1    $holder_i := P_\ell$;    – $P_i$ is notified the holder of the main-token.

7.2    **while** ¬ **empty**$(tmpQ_i)$ {    – Forward temporarily enqueued requests.

7.3        $\langle P_j, t, g, r \rangle :=$ **dequeue**$(tmpQ_i)$;

7.4        **send** ⟨request, $P_j, t, g, r$⟩ **to** $holder_i$;

7.5    }

**on receipt of** ⟨request, $P_\ell, t, g, r$⟩;

8.1    **if** $(tok_i \neq \bot)$ {    – case C1 ($P_i$ is the holder of the main-token.)

8.2        **if** $(tok_i.tsReq[\ell] < t)$ {

8.3            $tok_i.tsReq[\ell] := t$;

8.4            **enqueue**$(tok_i.reqQ, \langle P_\ell, t, g, r \rangle)$;

8.5            **call** $handlePendingRequests$;

8.6        }

8.7    } **else if** $(holder_i \neq \bot)$ {    – case C2 ($P_i$ knows the holder of the main-token.)

8.8        **send** ⟨request, $P_\ell, t, g, r$⟩ **to** $holder_i$;
            – Forward the request to the holder of the main-token.

8.9    } **else** {    – case C3 (Otherwise, enqueue the request temporarily.)

8.10        **delete** $\langle P_\ell, *, * \rangle$ **from** $tmpQ_i$;

8.11        **enqueue**$(tmpQ_i, \langle P_\ell, t, g, r \rangle)$;

8.12    }

**on receipt of** ⟨release, $g, r$⟩;

9.1    $tok_i.actGrp[g] := tok_i.actGrp[g] - r$;

9.2    **call** $handlePendingRequests$;

**procedure** $beginTokenTransfer$;

10.1    $leaving_i :=$ **true**;    – The main-token is going to leave $P_i$.

10.2    **send** ⟨leave⟩ **to each** $P_j \in acqs_i$ $(= q_i - \{P_i\})$;

10.3    $acqs_i := \emptyset$; $acks_i := q_i - \{P_i\}$;    – Wait for ack for each leave.

**on receipt of** ⟨leave⟩ **from** $P_\ell$;

11.1    $holder_i := \bot$;    – Holder of the main-token is unknown.

11.2    **send** ⟨ack⟩ **to** $P_\ell$;

**on receipt of** ⟨ack⟩ **from** $P_\ell$;

12.1    $acks_i := acks_i - \{P_\ell\}$;

12.2    **if** $(acks_i = \emptyset)$ {    – ack is received for each leave.
            – Preparation to transfer the main-token is done.

12.3        $leaving_i :=$ **false**;

12.4        $\langle P_j, t, g \rangle :=$ **dequeue**$(tok_i.reqQ)$;

12.5        **if** $(P_j = P_i)$ {    – Priority of $P_i$ is the highest.
                – Transfer of the main-token is not necessary. $P_i$ keeps it.

12.6            $q_i :=$ select a quorum in coterie $C$;

12.7            $acqs_i := q_i - \{P_i\}$;

12.8            **send** ⟨acquired⟩ **to each** $P_j \in acqs_i$;

12.9            $type_i :=$ MAIN; $mode_i :=$ INCS;

12.10            $tok_i.actGrp[grp_i] := 1$;

12.11            **call** $handlePendingRequests$;

12.12            **trigger event** $requestDone_i$;    // Enter CS

12.13        } **else** {    – Priority of $P_j (\neq P_i)$ is higher.

12.14            **send** ⟨token, $tok_i$⟩ **to** $P_j$; $tok_i := \bot$;

12.15        }

12.16    }

**Fig. 3.** Description of TQGRA for $P_i$ (3/3)

## 4.3    Description of the Proposed Algorithm

Let us explain the proposed algorithm TQGRA. A formal description of the action of each process $P_i$ is shown in Figures 1, 2 and 3.

- Initialization (Figure 1). First, $P_i$ initializes each local variable accordingly. Initially, $P_0$ is the holder of the main-token, and it creates and initializes the main-token.
- When $P_i$ makes a request for $r_i$ resources of group $g_i$ (Figure 1).
    - Case A1. If $P_i$ does not have the main-token, it first enqueues the request in *tok.reqQ* and then it handles the request by procedure *handlePendingRequests*.
    - Case A2. If $P_i$ knows the current holder of the main-token, it sends a request message directly to the holder.
    - Case A3. Otherwise, $P_i$ selects a quorum, and it sends a request message to each process in the quorum. The request is eventually enqueued in *tmpQ* of each process in the quorum.
- When $P_i$ releases resource (Figure 1).
    - Case B1. If $P_i$ holds the main-token, it decrements the amount of resources in use.
    - Case B2. Otherwise, it sends a release message to the holder of the main-token.
- Procedure *handlePendingRequests* (Figure 2). This procedure is called when $P_i$ receives a new request. First, we consider a case when $P_i$ is the holder of the main-token and no process is accessing any resource (line 4.1). If $P_i$ has the priority among pending requests, it allocates the requested resource and immediately accesses it (lines 4.5–4.7). Otherwise, the main-token is transferred to the process with the highest priority (lines 4.9–4.13). Procedure *beginTokenTransfer*, which will be explained shortly, is responsible for the transfer of the main-token without losing any request. Next, if $P_i$ holds the main-token, each request is granted, in the order of the priority defined by timestamp, as long as the constraint of the resource allocation is not violated (lines 4.18–4.25). $P_i$ sends a subtoken to grant a request. In other words, resource is allocated only when the amount of resource for each group does not exceed the limit $d$, and the number of active groups does not exceed the limit $k$.
- When $P_i$ receives a token message (Figure 2). By this message, the main-token is transferred to $P_i$. $P_i$ notifies, by an acquired message, that it is the holder of the main-token to each process in a quorum. Then, it allocates requesting resource and immediately accesses it. Pending requests are enqueued in the queue of the main-token, and procedure *handlePendingRequests* is called to handle them.
- When $P_i$ receives a subtoken message (Figure 2). $P_i$ is granted to access requesting resource. It sets local variable $home_i$ to which a release message should be sent when $P_i$ releases resource.

- When $P_i$ receives an **acquired** message (Figure 3). $P_i$ knows the holder of the main-token by this message. If there are pending requests, they are forwarded to the holder of the main-token.
- When $P_i$ receives a **request** message (Figure 3).
  - Case C1. If $P_i$ is the holder of the main-token, the request is handled by procedure *handlePendingRequests*.
  - Case C2. If $P_i$ does not hold the main-token but it knows the holder, it forwards the **request** message to the holder.
  - Case C3. Otherwise, an older request from the same process is discarded, and $P_i$ temporarily enqueues the request in its local queue.
- When $P_i$ receives a **release** message (Figure 3). $P_i$ receives this message when it is the holder of the main-token. It decrements the amount of resources in use, and then, calls procedure *handlePendingRequests* to handle pending requests.
- Procedure *beginTokenTransfer* (Figure 3). This procedure is called when $P_i$ is the holder of the main-token and the main-token is going to transfer to other process. Before the main-token is transferred, *holder* variable at each process $P_j$ in a quorum is cleared to defer forwarding request messages at $P_j$. This is achieved by a protocol with **leave** and **ack** messages.
- When $P_j$ receives a **leave** message (Figure 3). $P_j$ receives this message from the holder $P_i$ of the main-token. Local variable $holder_j$ becomes $\perp$ and while this is the case, request messages are kept in local queue $tmpQ_j$ and forwarding them is deferred until new holder of the main-token is informed.
- When $P_i$ receives an **ack** message (Figure 3). The holder $P_i$ of the main-token receives this message as a response to a **leave** message. If $P_i$ receives an **ack** message from each process in a quorum, it can transfer the main-token to other process without losing any request messages. When $P_i$ has the priority for the next holder (lines 12.4 and 12.5), it holds the main-token again and allocates resource for itself. Otherwise, the main-token is transferred by **token** message.

## 5    Proof of Correctness and Performance Analysis

**Lemma 1.** *The number of the main-token is one at any time.*

*Proof.* When the system is initialized, only $P_0$ creates the main-token. The main-token is sent by lines 4.13 and 12.14, however, the sender erases the copy at the same time ($tok_i := \perp$). Hence the number of the main-token is one at any time. $\square$

**Lemma 2.** *For any execution, the safety of resource allocation is maintained.*

*Proof.* The holder, say $P_i$, of the main-token maintains the amount of resources allocated to processes. $P_i$ grants a resource allocation request only when the safety is not violated at line 4.18–4.25, that is, resource is allocated only when the following two conditions are not violated.

- $0 \leq tok.actGrp[g] \leq d$ for each $g$, and
- $|\{g : tok.actGrp[g] > 0\}| \leq k$.

That is, resource is allocated only when the amount of resources for each group does not exceed the limit $d$, and the number of active groups does not exceed the limit $k$. By lemma 1, only $P_i$ issues subtoken message. Hence the safety of resource allocation is globally maintained.    □

**Lemma 3.** *Each request is eventually granted.*

*Proof.* Suppose that there exists an execution in which there exists a process, say $P_i$, for which request is never granted. If the request of $P_i$ is enqueued in $tok.reqQ$, it is eventually granted. Hence we assume that the request is never enqueued in $tok.reqQ$. In addition, we can assume that $P_i$ never holds the main-token forever.

- If $holder_i \neq \perp$ holds, the request is forwarded to the holder of the main-token. If the holder of the main-token does not change, the request of $P_i$ is eventually enqueued in $tok.reqQ$. If the holder of the main-token changes, by messages leave and ack, the request of $P_i$ is eventually enqueued in $tok.reqQ$ before the main-token is transferred. Hence, the request is forwarded to the holder of the main-token.
- Otherwise, i.e., if $holder_i = \perp$ holds, $P_i$ sends a request message to each process in a quorum. By the intersection property of quorums (a coterie), there is a process, say $P_j$, in a quorum that (eventually) knows the holder of the main-token. If $holder_j \neq \perp$ holds when $P_j$ receives the request, the discussion of the above case applies for $P_j$, and the request is forwarded to the holder of the main-token. Otherwise, i.e., if $holder_j = \perp$ holds, the request is enqueued in $tmpQ_j$ and, when $P_j$ is notified a new holder of the main-token, the request is forwarded to the holder. Also in this case, the discussion of the above case applies for $P_j$, and the request is forwarded to the holder of the main-token.    □

**Theorem 1.** *The proposed distributed algorithm TQGRA solves the $(n, m, k, d)$-resource allocation problem.*    □

**Theorem 2.** *The message complexity of TQGRA is $O(5|Q| + 1)$, and the maximum concurrency is $n$.*

*Proof.* The worst case scenario for the message complexity is as follows. A requesting process, say $P_i$, sends a request message to each process in a quorum, and each process forwards the request message to the holder, say $P_j$, of the main-token. $P_j$ sends leave message to each process in a quorum, and an ack message is replied by each process in a quorum. Then, a token message is sent from $P_j$ to $P_i$, and $P_i$ sends an acquired message to each process in a quorum.

Since the holder of the main-token keeps track of the amount of resources currently in use, any request is granted as long as the constraint of the $(n, m, k, d)$-resource allocation problem is not maintained. Hence, at most $n$ process can access resource at the same time. In addition, all the resource can be accessed at the same time.    □

Note that, in the best case, no message is exchanged. This case occurs when a requesting process is the holder of the main-token.

## 6   Conclusion

In this paper, we proposed a distributed algorithm for the $(n, m, k, d)$-resource allocation problem. Our solution is fully distributed, and it is the first distributed algorithm that achieves full concurrency to resource access. It is based on the idea of privilege tokens and the leader-follower scheme. To reduce the message complexity, a coterie is used as a communication structure. Although our algorithm is designed for the $(n, m, k, d)$-resource allocation problem, we believe that it can be extended for further generalized distributed resource allocation problems. Study of further generalization is an interesting problem. Open problem is to device a distributed algorithm based only on quorums, not on tokens, that achieves full concurrency to resource access.

## References

1. Kakugawa, H., Fujita, S., Yamashita, M., Ae, T.: A distributed k-mutual exclusion algorithm using k-coterie. Information Processing Letters 49(2), 213–218 (1994)
2. Joung, Y.J.: Ashynchronous group mutual exclusion. Distributed Computing 13, 189–206 (2000)
3. Joung, Y.J.: On quorum systems for group resources allocation. Distributed Computing 22, 197–214 (2010)
4. Maekawa, M.: A $\sqrt{N}$ algorithm for mutual exclusion in decentralized systems. ACM Transactions on Computer Systems 3(2), 145–159 (1985)
5. Barbara, D., Garcia-Molina, H.: The reliability of voting mechanisms. IEEE Transactions on Computers C-36(10), 1197–1208 (1987)
6. Agrawal, D., Abbadi, A.E.: An efficient and fault-tolerant solution for distributed mutual exclusion. ACM Transactions on Compututer Systems 9(1), 1–20 (1991)
7. Kumar, A.: Hierarchical quorum consensus: a new algorithm for managing replicated data. IEEE Transactions on Computers 40(9), 996–1004 (1991)
8. Neilsen, M.L., Mizuno, M.: Coterie join algorithm. IEEE Transactions on Parallel and Distributed Systems 3(5), 582–590 (1992)
9. Peleg, D., Wool, A.: Crumbling walls: a class of practical and efficient quorum systems. Distributed Computing 10, 87–97 (1997)
10. Raynal, M.: A simple taxonomy for distributed mutual exclusion algorithms. ACM Operating Systems Review 25(2), 47–51 (1991)
11. Lamport, L.: Time, clocks, and the ordering of events in a distributed system. Communications of the ACM 21(7), 558–565 (1978)
12. Ricart, G., Agrawala, A.K.: An optimal algorithm for mutual exclusion in computer network. Communications of the ACM 24(1), 9–17 (1981)
13. Suzuki, I., Kasami, T.: A distributed mutual exclusion algorithm. ACM Transactions on Computer Systems 3(4), 344–349 (1985)
14. Garcia-Molina, H., Barbara, D.: How to assign votes in a distributed system. Journal of the ACM 32(4), 841–860 (1985)

15. Manabe, Y., Baldoni, R., Raynal, M., Aoyagi, S.: $k$-Arbiter: A safe and general scheme for $h$-out of-$k$ mutual exclusion. Theoretical Computer Science 193(1-2), 97–112 (1998)
16. Joung, Y.J.: Quorum-based algorithms for group mutual exclusion. IEEE Transactions on Parallel and Distributed Systems 14(5), 463–476 (2003)
17. Mittal, N., Mohan, P.K.: A priority-based distributed group mutual exclusion algorithm when group access is non-uniform. Journal of Parallel and Distributed Computing 67, 797–815 (2007)
18. Atreya, R., Mittal, N., Peri, S.: A quorum-based group mutual exclusion algorithm for a distributed system with dynamic group set. IEEE Transactions on Parallel and Distributed Systems 18(10), 1345–1360 (2007)
19. Kakugawa, H., Kamei, S., Masuzawa, T.: A token-based distributed group mutual exclusion algorithm with quorums. IEEE Transactions on Parallel and Distributed Systems 19, 1153–1166 (2008)

# On the Message Complexity of Global Computations

Doron Nussbaum and Nicola Santoro

School of Computer Science, Carleton University,
1125 Colonel By Drive, Ottawa, Ontario, Canada, K1S 5B6
{nussbaum,santoro}@scs.carleton.ca

**Abstract.** It is well known that, for most non-trivial problems (such as *Election, Spanning-Tree Construction, Traversal, Broadcast*, etc.), any generic solution requires at least $\Omega(m)$ messages in the worst case, where $m$ is the number of links among the $n$ entities. However, all the existing proofs of this fact assume that the network size (i.e., the parameters $n$ and $m$) are *not* known to the protocol.

A natural question arises whether this rather strong assumption, which is crucial for the proofs, is truly necessary for establishing a lower bound to these problems.

In this paper we answer this question and prove that the $\Omega(m)$ bound is inherent for all these problems, as well as many more. In fact, we consider the class of *global* problems, that is those whose solution requires the involvement of every entity in the communication (sending or receiving messages). The relationship between $n$ and $m$ plays an important role in establishing the lower bound. We show that for most networks (where $m \leq \frac{1}{2}(n-2)(n-3)+1$) a generic solution for any problem in this class requires at least $m$ messages even if $n$, $m$, and the degree of each node are known. This result holds for almost all values of $m$ (e.g., when $\frac{1}{2}(n-2)(n-3)+1 < m \leq \frac{1}{2}(n-1)(n-2)+1$ the number of required messages is $m-1$), even if there is a single initiator and the entities have distinct identifiers, and both these facts are known. Moreover, the results hold even if the protocol can maintain a global view of the network.

As the networks become more dense, namely the network approaches a complete graph, the number of required messages is gradually reduced. For extreme values of $m$ ( i.e., $m = \frac{1}{2}n(n-1)-c > \frac{1}{2}(n-1)(n-2)+1$), where $c \geq 0$ is constant, the lower bound gradually approaches $\Omega(n)$; this is understandable since we establish it for the single initiator scenario. However, we prove that in networks of such a size, single initiators problems such as *Broadcast* and *Traversal* can be solved with precisely that order of magnitude. This means that for those problems the knowledge of $n$ and $m$ generates a significant and sudden complexity *drop* from $\Theta(n^2)$ to $\Theta(n)$.

**Keywords:** distributed algorithms, global computations, generic protocols, asynchronous systems, message complexity.

C. Lu, T. Masuzawa, and M. Mosbah (Eds.): OPODIS 2010, LNCS 6490, pp. 427–442, 2010.
© Springer-Verlag Berlin Heidelberg 2010

# 1   Introduction

In this paper we are interested in the communication complexity of problems in message-passing systems. In particular, we are concerned with the amount of communication (i.e., message transmissions) required in the worst case for a problem $\mathcal{P}$ to be solved by a collection of computational entities connected through a network $\mathcal{G}$ and communicating by exchanging messages with their neighbours in $\mathcal{G}$. As usual, we model $\mathcal{G}$ as an undirected edge-labelled graph $(G, \lambda)$ where $G = (V, E)$ describes the communication topology of $\mathcal{G}$ and $\lambda$ describes the port labelling. We make no temporal assumptions on computational or communication delays, other than they are finite; in other words, the systems we consider are asynchronous.

The message complexity of a problem $\mathcal{P}$ obviously depends on many different factors: not only on the nature of $\mathcal{P}$ but also on the properties of the system $\mathcal{G}$ in which it is being solved, as well as (and perhaps more importantly) on the amount and type of knowledge about $\mathcal{G}$ held by the entities.

Rather than in topology-specific solutions, we are interested, in this paper, in solution protocols that are *generic*, that is that can be employed in any system $\mathcal{G}$, provided the network $(G, \lambda)$ satisfies some predefined solvability assumptions (e.g., $G$ is bi-connected). Notice that a generic solution, while not requiring (even an unlabelled) map of $G$, might still need and use metric information about $G$ (e.g., the number $n$ of nodes, the number $m$ of links, etc.). Furthermore, we assume that the task must be explicitly performed within finite time; that is, within finite time, the system must reach a configuration where every entity is in a terminal state (in which no further action will be performed) and the problem is solved. Notice that this is the case of most tasks; e.g., in Election, one node must become *leader* and all others must become *defeated*; in Wake-Up, all nodes must become *awake*. It corresponds to a *local termination* requirement: each node knows when its own part in the execution has terminated, but not necessarily the status of the computation at the other nodes. Thus, in the following, the message complexity of a problem $\mathcal{P}$ will be measured only over generic solutions that explicitly terminate.

In general, different problems have different complexities, and the methods and tools to determine their complexity are usually specific for that problem. However, sometimes, even very different problems share some commonality that can be used to determine a common bound for all of them (e.g. [6, 8–10, 12]), *Broadcast* (e.g. [2, 16], *Traversal* (e.g., [1, 3–5]), and *Wake-Up* (e.g., [11]). These problems are very different from each other. For example, *Broadcast* and *Traversal* are single initiator problems: by definition their resolution is started by a single entity; on the other hand, the resolution of *Election* and *Wake-Up* is started by an arbitrary (a priori unknown) number of entities, making them multiple initiators problems. *Election* requires for its solvability that to each entities is associated a distinct value from a totally ordered set; the other three problems can instead be solved in totally anonymous systems. They have however something in common: any generic solution for any of them requires the active participation of all computational entities of the system; more precisely, they require each entity to be involved in communication during the execution of the solution. Let us properly characterize this commonality

We call *global* a distributed computation $\mathcal{C}$ if every entity sends or receives at least one message during the computation. A protocol $\mathcal{P}$ is global if every execution is a global computation, and a problem $\mathcal{P}$ is global if its solution is necessarily global. Clearly, the four problems above are global. Interestingly, the generic solution of almost all non-trivial problems and tasks requires a global computation. In other words, all these problems are necessarily global. This implies the existence of a common lower bound on the message complexity of all these problems in spite of their apparent differences.

## 1.1   Lower Bounds

By definition, any global computation $\mathcal{C}$ requires $\Omega(n)$ messages, where $n$ is the number of entities in the system (i.e., the number of nodes in $G$); this trivial lower bound is however too low. It is commonly known that for the class of problems considered here, (i.e. those whose generic solution is global and must terminate within finite time) indeed $\Omega(m)$ are needed, where $m = |E|$ is the number of communication links in the systems (i.e., the number of edges in $G$). For example, the $\Omega(m + n \log n)$ lower bound for generic election and (min-cost) spanning tree construction is just the addition of the $\Omega(n \log n)$ lower bound for rings to the above mentioned $\Omega(m)$ lower bound (e.g. see [13, 14]). Some explicit proofs for specific problems (e.g. Election or Broadcast) can also be found in textbooks on distributed algorithms (e.g. see [15, 17]).

However, the existing (problem-specific) $\Omega(m)$ lower-bounds are derived under the strong assumption that neither $n$ nor $m$ is known to the entities (and thus to the solution protocol). In fact, the adversary-based proofs use precisely this lack of knowledge to generate a contradiction: the adversary substitutes the original graph $G$ with another graph $G'$ that has more nodes and edges and in which the supposedly correct protocol is shown to fail. For example, the proof of [15] that $m$ messages are needed for broadcasting uses a new graph $G'$ with $n + 1$ nodes and $m + 1$ edges; the proof of [17] that $m$ messages are needed for election uses a new graph with $2n$ nodes and $2m$ edges.

This immediately raises several important questions: is this lack of knowledge determinant for the complexity of these problems? Namely, if $n$ and $m$ are known, can these problems be solved with (order of magnitude) fewer messages? Or is it irrelevant, in the sense that the complexity of these problems is the same regardless of additional metric information? In other words, what is the real impact of this (lack of) metric information on the message complexity of a global problem?

To answer these questions, it is necessary to study what happens to global problems if both $n$ and $m$ are indeed known. This is precisely what we do in this paper.

## 1.2   Our Results

We prove that, for all but very large values of $m$, every global problem requires in the worst case the transmission of at least $m$ messages for its solution. This result holds even if both $n$ and $m$ are known, the network is not anonymous (that is even if the entities have unique identifiers), there is a single-initiator and this fact is known. In other words we prove that the lower bound on the message complexity of global problems is unaffected by the additional knowledge of $n$ and $m$.

We summarize our results for general graphs in Table 1. For most graphs the lower bound of $\Omega(m)$ holds. More over in many configurations of $m$ and $n$ at least $m$ messages are required.

Regular graphs are special instance and we show that for most regular graphs the $\Omega(m)$ also holds. We summarize our results for regular graphs in Table 2.

In the case of very large values of $m$ (i.e., when $G$ is almost a complete graph where $m > \frac{1}{2}(n-1)(n-2)+1$, the lower bound is slowly approaching $\Omega(n)$. This is understandable since we establish it for the single initiator scenario. Interestingly, in networks of such a size, single initiators problems such as *Broadcast* and *Traversal* can be solved with precisely that order of magnitude. This means that for those problems the knowledge of $n$ and $m$ generates a significant and sudden complexity *gap*, actually a *drop* from $\Theta(n^2)$ to $\Theta(n)$.

**Table 1.** Summary of results when $n$ and $m$ are known

| Network Size | Lower Bound |
|---|---|
| $m \le \frac{1}{2}(n-2)(n-3)+1$ | $m$ |
| $\frac{1}{2}(n-2)(n-3)+1 < m \le \frac{1}{2}(n-1)(n-2)+1$ | $m-1$ |
| $\frac{1}{2}(n-1)(n-2)+1 < m = \frac{1}{2}(n-1)(n-2)+k$ where $2 \le k < \frac{1}{2}(n-1)$ | $\frac{1}{2}(n-2k)(n+2k-3)$ |
| $\frac{1}{2}(n-1)(n-2)+k = m < \frac{1}{2}(n)(n-1)$, where $\frac{1}{2}(n-1) \le k < n-1$ | $2n-k-2$ |
| $m = \frac{1}{2}n(n-1)$, | $n-1$ |

**Table 2.** Summary of results for regular graphs when $n$ and $d = deg(u)$ are known

| Network Degree | Lower Bound |
|---|---|
| $2 \le \deg(u) < \frac{1}{2}n$ | $m - \frac{d^2-d}{2}$ |
| $\frac{1}{2}n \le \deg(u) \le n-1$ | $\frac{1}{2}(n-d)(3d-n+1)$ |

## 1.3   Paper Organization

The paper is organized as follows: In Section 2 we present the distributed model used in the paper. In Section 3 we present lower bounds for general networks followed by lower bounds for regular graphs in Section 4. We conclude and discuss future work in Section 5.

## 2   Model and Strategy

In this section we present the overall strategy in proving the lower bounds for the communication complexity. Our strategy is to employ an adversary in order to prove the lower bounds on the number of messages that a global protocol operating on a generic network requires to send to achieve global computation. In Section 2.1 we present the network model and in Section 2.2 we present the adversary schema.

As stated earlier the protocol $\mathcal{P}$ is global and generic. Namely, when $\mathcal{P}$ completes its work each node in the network must send or receive at least one message. Moreover, $\mathcal{P}$ is a generic protocol and as such it must operate on any network where its decisions are only based on knowledge acquired during its execution.

### 2.1   Network Model

The networks we consider are edge-labelled connected graphs $(G, \lambda)$ where $G = (V, E)$ is a simple graph and $\lambda$ is the local orientation, i.e. the labeling that locally assigns a distinct label to all edges incident on the same node. Without loss of generality, we assume that the edges incident on $x \in V$ are locally labeled with the integers $1, 2, ...d(x)$, where $\deg(x)$ denotes the degree of $x$. We used the term port to define the local labelling of edges at a give node (a node has $x$ ports if $deg(v) = x$). We denote by $\lambda(v, i), 1 \le i \le \deg(v)$ the neighbour of $v$ accessible through port $i$. For example, given two nodes $u, v \in V$ where $(u, v) \in E$ then $\lambda(u, i) = v$ and $\lambda(v, j) = u$ means that edge $(u, v)$ is connected in the network via ports $i$ and $j$ at $u$ and $v$ respectively.

In our network model we relax the often used restriction of anonymity. Namely, the network nodes are not anonymous and each node is assigned an id from $v_0$ to $v_{n-1}$.

### 2.2   Adversary Model

Next we show the relationship between the state of the network during the execution of a global protocol $\mathcal{P}$.

The lower bound proofs on the minimum number of messages, which $\mathcal{P}$ must send to achieve invocation of all the nodes, are based on the ability to construct a network (graph) for which any given global protocol that attempts to use less messages will fail. The global protocol $\mathcal{P}$ has almost complete knowledge about the network with the exception of the network connectivity topology. Namely, $\mathcal{P}$ knows, prior to starting its execution, the total number of nodes, the total number of edges, the starting node/vertex, the target vertex or vertices, and the degree of each node in the network. The network connectivity is determined by an adversary program $\mathcal{A}$ that executes in parallel to the protocol. The adversary attempts to prevent the execution of protocol $\mathcal{P}$ from reaching one or more nodes until a certain threshold number of messages, denoted by $M$, were sent.

Initially, $\forall v \in V, \lambda(v, i) = \infty$, where $1 \le i \le \deg(v)$. During the execution of the protocol $\mathcal{P}$ the adversary determines the connectivity and assigns the values of $\lambda(v, i), 1 \le i \le \deg(v)$. For example, if $\lambda(v, i) = u$ then node $v$ is connected to node $u$ via port $i$.

In general forcing the global protocol $\mathcal{P}$ to send at least $M$ messages during its execution is done as follows: the adversary $\mathcal{A}$ chooses a vertex $v \in E$ as the initiator and one or more vertices ,e.g., $u \in V$, that will receive a message for the first time only if at least $M$ message were sent. Since the system is asynchronous, $\mathcal{A}$ adjusts the transmission delays provided that every transmitted message is delivered within finite time; it does so to ensure that, during the execution, at most one message being transmitted at any time instant $t$. Let $t_0$ be the time when the execution starts, and let $t_1, t_2, \ldots$ be the time instants when a message transmission operation is performed; let us denote by $T$ the (possibly infinite) ordered set of all these time units.

Once the number of nodes $n$ and the number of edges $m$ is given, the adversary constructs a graph $G = (V, E)$ on which $\mathcal{P}$ will execute. The adversary determines a starting vertex, e.g., $v$, and the target vertex or vertices, e.g., $u$. Once the graph construction is completed the adversary provides $\mathcal{P}$ with information about the size of the graph, $n$ and $m$, the starting vertex $v$, the target vertex $u$, and the degree of each vertex in the graph.

Since $\mathcal{P}$ is global, during its execution the initiator must send at least a message, and every other node must receive at least a message. The only interactions between $\mathcal{P}$ and $\mathcal{A}$ during the execution of $\mathcal{P}$ occur when a message transmission is performed: at that time, $\mathcal{P}$ specifies the sender node and the port number, and $\mathcal{A}$ reveals which vertex receives the message and in particular whether or not this message will reach $u$.

## 3    Complexity Lower Bound in General Networks

In this section we prove our main result, that for almost all $m$, there are graphs where any global protocol must send at least $m$ messages, even if both $n$ and $m$ (and much more) are known to the protocol. This result holds even if there is a sole initiator, the nodes have unique identifiers, and both these facts are common knowledge.

The adversary strategy depends on the size of the network (on $n$ and $m$). Namely, depending on the relationships between $n$ and $m$, the adversary determines the number of target nodes and constructs a network on which $\mathcal{P}$ executes. In order to establish the lower bounds we use the following relationships between $n$ and $m$.

1. $m = n - 1$ - *tree*
2. $(n - 1) < m \le \frac{1}{2}(n - 2)(n - 3) + 1$ - *not-dense network*
3. $\frac{1}{2}(n - 2)(n - 3) + 1 < m \le \frac{1}{2}(n - 1)(n - 2) + 1$ - *dense network*
4. $\frac{1}{2}(n - 1)(n - 2) + 1 < m \le \frac{1}{2}(n - 1)(n - 2) + k$ where $2 \le k < \frac{1}{2}(n - 1)$ - *extremely dense network*
5. $m = \frac{1}{2}n(n - 1)$ - *complete graph*

Cases 1 and 5 above are special cases of the network and represent opposite end of the classification spectrum. In each of these case any invocation algorithm would require $n - 1$ messages and therefore these cases are omitted due to their simplicity.

## 3.1   Not-Dense Network

In this section we look at communication lower bound when the network is a typical network. Given a connected network with $n$ nodes and $m$ edges we term the network a *typical network* if $(n-1) < m \leq \frac{1}{2}(n-2)(n-3)+1$. Here we show that $\mathcal{P}$ must send at least $m$ messages before reaching all the $n$ nodes in the network. Figure 1 illustrates the proof of Theorem 1.

---

**Algorithm 1.** Adversary strategy for typical networks at time $t_u$ for a given node $x$ and port $k$

```
 1: if λ(x, k) ≠ ∞ then
 2:     the message will reach y = λ(x, k)
 3: else {λ(x, k) = ∞}
 4:     if |H| > 1 then
 5:         select an H_{i,j} from H where v_i = x
 6:         λ(x, k) = v_j
 7:         find l such that λ(v_j, l) = ∞
 8:         λ(v_j, l) = x
 9:         H = H \ {H_{i,j}} {delete H_{i,j} from H }
10:         the message will reach v_j
11:     else {H only contains H_{i,j} (|H| = 1)}
12:         if (x, v_0) ∈ E_{i,j} then
13:             λ(x, k) = v_0
14:             λ(v_0, 1) = x
15:             the message will reach v_0
16:         else {(x, v_1) ∈ E_{i,j}}
17:             λ(x, k) = v_1
18:             λ(v_1, 1) = x
19:             the message will reach v_1
20:         end if
21:     end if
22: end if
```

---

**Theorem 1.** *Given $n$ and $(n-1) < m \leq \frac{1}{2}(n-2)(n-3)+1$ there exists a typical network with $n$ nodes and $m$ edges upon which every global protocol $\mathcal{P}$ requires to send at least $m$ messages in order to invoke all the nodes in the graph.*

*Proof.* Given $n$ and $m$ the adversary $\mathcal{A}$ constructs a set of typical graphs $\mathcal{H}$ as follows: First $\mathcal{A}$ constructs a graph $G' = (V', E')$ where $V' = V - \{v_0, v_1\}$ and $|E'| = m - 1$. The edges are constructed as follows: $\mathcal{A}$ adds $n-2$ edges to form a ring $\{(v_i, v_{i+1}), 2 \leq i < n-1\} \cup \{(v_2, v_{n-1})\}$. The remaining $m - n + 1$ edges are added as follows: $\mathcal{A}$ arbitrarily chooses $m - n + 1$ distinct pairs of vertices from $V'$ and insert an edge between each pair of nodes. Next, using $G'$, $\mathcal{A}$ creates a set $\mathcal{H}$ of *typical graphs*, where $|\mathcal{H}| = m - 1$ as follows: for each edge $(v_i, v_j) \in E'$ (note $2 \leq i, j \leq n-1$) create a graph $H_{i,j} = (V_{i,j}, E_{i,j}) \in \mathcal{H}$ where $V_{i,j} = V$ and $E_{i,j} = [E' \backslash (v_i, v_j)] \cup [\{(v_0, v_i), (v_1, v_j)\}]$. In this construction $(\deg(v_0) = \deg(v_1) = 1$ and $\deg(v_i) \geq 2, 2 \leq i \leq n-1)$. It then provides $\mathcal{P}$ with the following data: a. starting vertex is $v_2$, b. target vertices are $v_0$ and $v_1$, c.

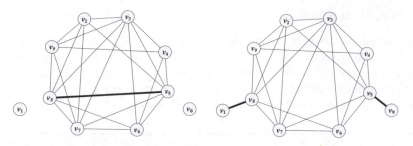

**Fig. 1.** Illustration of the proof of Theorem 1 using a graph with 10 vertices. The graph on the left shows the last graph in $\mathcal{H}$ (when $|\mathcal{H}| = 1$). The edge $(v_5, v_8)$, which is in bold, is the edge that was not used yet. The graph on the right shows how $v_0$ and $v_1$ are connected to create the required graph $G$.

$n$, d. $m$, e. degree of each node and f. that the graph $G$ in which the execution takes place is an element of $\mathcal{H}$.

$\mathcal{P}$ is global, and therefore, during its execution $v_2$ must send at least one message and every node must receive at least one message. In particular $v_0$ and $v_1$ must receive a message. When $\mathcal{P}$ wants to send a message at time $t_u$ from node $x \in V$ along port $k$ it asks $\mathcal{A}$ for the destination node. $\mathcal{A}$ uses Algorithm 1 to determine the destination node. The strategy of $\mathcal{A}$ is simple. If the port $k$ was used before then the $\mathcal{A}$ provides the previous destination port (Line 2). If the port was not used then $\mathcal{A}$ uses the set $\mathcal{H}$ , which consists of choices for a graph $G$ left to the adversary at time $t_u$. A message will be received by either $v_0$ or $v_1$ only if $|\mathcal{H}| = 1$ (Lines 11-21). Initially, $|\mathcal{H}| = m - 1$ and each time that $\mathcal{P}$ wants to send a message from a node $x$ along a port $k$ that was never used before and $|\mathcal{H}| > 1$, $\mathcal{A}$ provides a destination node and reduces the size of $\mathcal{H}$ by one (Lines 3-11). Therefore, before a message can either reach $v_0$ or $v_1$ at least $|\mathcal{H}| - 1 = m - 2$ messages must be sent. Adding to them two more messages (one to $v_0$ and one to $v_1$) we obtain that $m$ messages must be sent.    □

### 3.2    Dense Network

In this section we look at the communication lower bound when the network is a dense network. Given a connected network with $n$ nodes and $m$ edges we term the network a *dense network* if $\frac{1}{2}(n-2)(n-3) + 1 < m \leq \frac{1}{2}(n-1)(n-2) + 1$. Here we show that $\mathcal{P}$ must send at least $m - 1$ messages before reaching all the $n$ nodes in the network.

**Theorem 2.** *Given $n$ and $\frac{1}{2}(n-2)(n-3) + 1 < m \leq \frac{1}{2}(n-1)(n-2) + 1$ there exists a dense network with $n$ nodes and $m$ edges upon which every global protocol $\mathcal{P}$ requires to send at least $m$ messages in order to invoke all the nodes in the graph.*

### 3.3    Extremely Dense Network

In this section we look at the communication lower bound when the network is an extremely dense network. Given a connected network with $n$ nodes and $m$

---

**Algorithm 2.** Adversary strategy for dense networks at time $t_u$ for a given node $x$ and port $k$

---

```
 1: if λ(x, k) ≠ ∞ then
 2:     the message will reach y = λ(x, k)
 3: else {λ(x, k) = ∞}
 4:     if |H| > 1 then
 5:         select an H_{i,j} from H where v_i = x
 6:         λ(x, k) = v_j
 7:         find l such that λ(v_j, l) = ∞
 8:         λ(v_j, l) = x
 9:         H = H \ {H_{i,j}} {delete H_{i,j} from H }
10:         the message will reach v_j
11:     else {H only contains H_{i,j} (|H| = 1)}
12:         λ(x, k) = v_0
13:         if λ(v_0, 1) ≠ ∞ then
14:             λ(v_0, 1) = x
15:         else
16:             λ(v_0, 2) = x
17:         end if
18:         the message will reach v_0
19:     end if
20: end if
```

---

edges we term the network an *extremely dense network* if $\frac{1}{2}(n-1)(n-2)+1 < m = \frac{1}{2}(n-1)(n-2) + k$ where $2 \le k < \frac{1}{2}(n-1)$. Note, that in this case we assume that $deg(v_i) < (n-1), 0 \le i \le n-1$.

In this case we show how $\mathcal{A}$ determines a graph $G$ that require $\mathcal{P}$ to send at least $M = \frac{1}{2}(n-2k)(n+2k-3)$ messages during its execution on $G$. In this case the strategy is somewhat similar to that given in Section 3.2. First $\mathcal{A}$ constructs a graph $G' = (V', E')$ where $V' = V - \{v_0\}$ and $|E'| = \frac{1}{2}(n-1)(n-2)$ edges. The graph $G'$ is a complete network.

In this case $(deg(v_0) = 2k$ and $deg(v_i) = n-2, 1 \le i \le n-1$. The adversary $\mathcal{A}$ provides $\mathcal{P}$ with the following data: a. starting vertex $v_1$, b. target vertex $v_0$, c. $n$, d. $m$, e. degree of each node.

We say that a node $x \in V$ is *exhausted* during the execution of a global protocol $\mathcal{P}$ if each of its incident edges was used for communication (either to send or to receive a message). We denote by $E^*$ the set of edges that were used by $\mathcal{P}$ during its execution and by $V^E$ the set of vertices of $G'$ that were exhausted by $\mathcal{P}$. We also denote by $deg^*(v)$ the number of edges incident to $v$ that were used to send or receive a message from or to $v$ respectively. Note, throughout the algorithm $deg^*(v) \le deg(v)$.

**Theorem 3.** *Given $n$ and $\frac{1}{2}(n-1)(n-2)+1 < m = \frac{1}{2}(n-1)(n-2)+k$ where $2 \le k < \frac{1}{2}(n-1)$ there exists an extremely dense network with $n$ nodes and $m$ edges upon which every global protocol $\mathcal{P}$ requires to send at least $\frac{1}{2}(n-2k)(n+2k-3)$ messages in order to invoke all the nodes in the graph.*

**Algorithm 3.** Adversary strategy for extremely dense networks at time $t_u$ for a given node $x$ and port $k$

---

1: **if** $\lambda(x,k) \neq \infty$ **then**
2:     the message will reach $y = \lambda(x,k)$
3: **else** $\{\lambda(x,k) = \infty\}$
4:     **if** $|E^*| < \frac{1}{2}(n-2k)(n+2k-3) - 1$ **then**
5:         find a vertex $v_j \notin V^E$ such that $\lambda(v_j,l)_{1\leq l\leq n-2} \neq x$ and
            $\deg^*(v_j) = \min_{1\leq i\leq n-2}\{\deg^*(v_i)\}$
6:         $\lambda(x,k) = v_j$
7:         find $l$ such that $\lambda(v_j,l) = \infty$
8:         $\lambda(v_j,l) = x$
9:     **else** $\{|E^*| = \frac{1}{2}(n-2k)(n+2k-3)\}$
10:         $\lambda(x,k) = v_0$
11:         the message will reach $v_0$
12:     **end if**
13: **end if**

---

*Proof.* Given $n$ and $m$ the adversary $\mathcal{A}$ constructs the extremely dense graph during run time. The strategy of the adversary is somewhat similar to that given in Section 3.2. First $\mathcal{A}$ constructs a complete graph $G' = (V', E')$ where $V' = V - \{v_0\}$ and $|E'| = \frac{1}{2}(n-1)(n-2)$ edges. The graph $G'$ is a complete network where $\deg(v_i) = n-2, 1 \leq i \leq n-1$. In this case the adversary will construct during run time a graph G where $(\deg(v_0) = 2k$ and $\deg(v_i) = n-2, 1 \leq i \leq n-1$. The adversary $\mathcal{A}$ provides $\mathcal{P}$ with the following data: a. starting vertex $v_1$, b. target vertex $v_0$, c. $n$, d. $m$, e. degree of each node. Since $\mathcal{P}$ is global then during its execution $v_1$ must send at least one message and every node must receive at least one message. In particular $v_0$ must receive a message. When $\mathcal{P}$ wants to send a message at time $t_u$ from node $x \in V$ along port $k$ it asks $\mathcal{A}$ for the destination node. $\mathcal{A}$ uses Algorithm 3 to determine the destination node. The strategy of $\mathcal{A}$ is simple. If the port $k$ was used before then the $\mathcal{A}$ provides the previous destination port (Line 2). Otherwise $\mathcal{A}$ counts how many edges were used to send or received messages. If this number has reached $\frac{1}{2}(n-2k)(n+2k-3)$ then it sends a message to $v_0$ (Line 4). If less than $\frac{1}{2}(n-2k)(n+2k-3)$ were used to send or receive a message then $\mathcal{A}$ sends the message to a node with a minimum assigned connectivity so far.

What is left to show is that $v_0$ can be connected to the remaining vertices to form graph $G$. The maximum number of messages that can be sent before a message reaches $v_0$ is $\frac{1}{2}(n-2k)(n+2k-3)$. Thus at any time, prior to sending $\frac{1}{2}(n-2k)(n+2k-3)$ messages there are at most $n-2k$ exhausted nodes. When $\mathcal{A}$ routes the message to $v_0$ (Line 9 there are at most $n - 2k + 1$ exhausted nodes. It is sufficient to show that an edge matching of size $k - 1$ can be found among the remaining non exhausted nodes (the number of un exhausted nodes is greater than $2k - 1$). The number of edges that were not used is

$$|E| - |E^*| = [\frac{1}{2}(n-1)(n-2)] - [\frac{1}{2}(n-2k)(n+2k-3)]$$

$$|E| - |E^*| = 2k^2 - 3k + 1.$$

Assuming that a matching of size $k-1$ cannot be found then the maximum edge matching in the remaining non exhausted edges and unused edges is of size $k-2$. According to Erdös and Gallai [7], the maximum number of edges $\nu$ that a graph with $n$ vertices whose maximum edges matching is $\rho$ is bounded by

$$\nu = max\{\binom{2\rho+1}{2}, \rho(n-\rho) + \binom{\rho}{2}\}.$$

Using Erdös' formula we obtain that the maxim number of edges in a graph with $k-2$ maximal edge matching is

$$\nu = max\{\binom{2(k-2)+1}{2}, (k-2)(n-(k-2)) + \binom{k-2}{2}\}.$$

Note that $\binom{2(k-2)+1}{2} < 2k^2 - 3k + 1$ for $k > 1$, therefore we are only concerned with the second term $(k-2)(n-(k-2)) + \binom{k-2}{2}$, which depends on the number of nodes in the graph. In order to avoid a matching of size $k-1$ the number of un-exhausted nodes must be $|V| - |V^x| \geq \frac{5k^2-9k+4}{2k-4}$, where $V^x \subseteq V$ is the set of exhausted nodes. A construction that yields an edge matching of size $k-2$ consists of two subsets of vertices: (i) a set $S^1$ with at most $k-2$ nodes, which form a complete graph; and (ii) a set $S^2$ with the remaining $\frac{5k^2-9k+4}{2k-4} - (k-2) = \frac{3k^2-k-4}{2k-4}$ nodes. The nodes in $S^2$ are only connected to nodes in $S^1$. This implies that the $deg^*$ of $k-2$ nodes is much smaller than the remaining nodes in the graph (about a factor $k$). This cannot occur because in Line 5, $\mathcal{A}$ chooses a node with a minimum $deg^*$. □

When the number of edges exceeds $\frac{1}{2}(n-1)(n-1)$ then at least one node in the network is connected to all other nodes. Therefore the adversary will try to prevent the $\mathcal{P}$ from reaching such a node.

Before we present the lower bound in this case we first prove several properties related to the state of the network when all nodes in the network received at least one message. We start by addressing the state of a single vertex that was used to send a message to a designated vertex. We say that a node $x \in V$ is *exhausted* during the execution of a global protocol $\mathcal{P}$ if all its incident edges were used for communication (to send or receive a message). Let $\mathcal{H} = \{\mathcal{H}_1, ...\}$ be a set of all possible labeling of $G$.

**Lemma 1.** *Consider a graph $G = (V, E)$ and let $(u, v) \in E$. Let $\mathcal{P}$ be a generic protocol that requires in its execution that $u$ sends a message directly to $v$ through $(u, v)$, Then there exists an execution of $\mathcal{P}$ in $H \in \mathcal{H}$ where $u$ becomes exhausted. This result holds even if $\mathcal{P}$ knows $G$.*

*Proof.* Consider an execution of $\mathcal{P}$ in $H = (G, \lambda) \in \mathcal{H}$. Let $nu(H) \leq deg(u)$ be the number of unused edges incident on $u$ when $\mathcal{P}$ decides to send a message from $u$ to $v$. The lemma trivially holds if $nu(H) = 1$. If $nu(H) = k > 1$, it is suffices to show that, for any $1 < k \leq deg(u)$ there exist a $H' \in \mathcal{H}$ such that $nu(H') = k - 1$. Let $nu(H) = k > 1$; then there is at least one edge

$(u, w)$, $w \neq v$, on which no message was sent. Let $\lambda(u, i) = v, \lambda(u, j) = w$ and let $H' = (G, \lambda') \in \mathcal{H}$, be the network obtained from $(G, \lambda)$ by simply setting $\lambda'(u, i) = w, \lambda'(u, j) = v$. The same execution of $\mathcal{P}$ is possible in $H'$; however in this case, the message will arrive at $w$ and not at $v$ resulting in $nu(H') = k - 1$ in this execution.                                                                 □

**Corollary 1.** *Consider a graph $G = (V, E)$ and let $v_1, \ldots, v_k \in V, k \geq 2$ be neighbouring vertices of $u$ $((u, v_i) \in E, 1 \leq i \leq k$ where $l < k$ vertices have a special property $z$. Let $\mathcal{P}$ be a generic protocol that requires in its execution that $u$ sends a message directly to $v \in v_1, \ldots, v_k \in V, k \geq 2$ through $(u, v)$, where $v$ has the property $z$. Then there exists an execution of $\mathcal{P}$ in $H \in \mathcal{H}$ where $u$ must send $k - l + 1$ messages in order to guarantee that a vertex with property $z$ will receive the message. This result holds even if $\mathcal{P}$ knows $G$ which vertices have the property $z$ but not the topology of $G$.*

Using this properties we can now prove our result for cases where the number of edges exceeds $\frac{1}{2}(n-1)^2$.

**Theorem 4.** *Given $n$ and $\frac{1}{2}(n-1)(n-2) + k = m < \frac{1}{2}(n)(n-1)$ where $\frac{1}{2}(n-1) \leq k < (n-1)$ there exists an extremely dense network with $n$ nodes and $m$ edges upon which every global protocol $\mathcal{P}$ requires to send at least $2n - k - 2$ messages in order to invoke all the nodes in the graph.*

### 3.4    Tightness

We have seen that, for high values of $m$ (i.e., when $m = \frac{1}{2}n(n-1) - k$, where $1 < k < n$), the lower bound we established on the message complexity of global computations is only $\Omega(n)$. This is not surprising because we are only focusing on computations started by a single initiator. However, for several single-initiator problems, including *Broadcast* and *Traversal*, this lower bound in actually tight in order of magnitude, as we will now show.

Let $\tau(n) = \frac{1}{2}n(n-2)$.

**Theorem 5.** *For any $n > 1$ and $m > \tau(n)$, if $n$ and $m$ are known then* Broadcast *can be performed using at most $3n - 6$ messages, and* Traversal *can be performed using at most $4n - 7$ messages*

*Proof.* If $m > \tau(n)$ then, by the pigeon principle, there is at least one node of degree $n - 1$. Let $u$ be the initiator.

In case of *Broadcast*, if $deg(u) = n - 1$, $u$ just sends a message to each of its neighbours. Otherwise, it sequentially searches for a neighbour of degree $n - 1$; once located, that neighbour $v$ will forward the broadcast message to all its neighbours. This approach uses at most $2 \, deg(u) - 1 + n - 1 \leq 3n - 6$ messages.

Similarly, if $deg(u) = n - 1$ *Traversal* can be performed in just $2(n - 1)$ messages; otherwise, a neighbour $v$ of degree $n - 1$ must first be found; the cost would then be at most $2 \, deg(u) - 1 + 2(n - 1) \leq 4n - 7$ messages.                    □

---

**Algorithm 4.** Adversary strategy for typical regular networks at time $t_u$ for a given node $x$ and port $k$

---

1: **if** $\lambda(x,k) \neq \infty$ **then**
2:     the message will reach $y = \lambda(x,k)$
3: **else** $\{\lambda(x,k) = \infty\}$
4:     **if** $|\mathcal{H}| > 1$ **then**
5:         select an $H_{i,j}$ from $\mathcal{H}$ where $v_i = x$
6:         $\lambda(x,k) = v_j$
7:         find $l$ such that $\lambda(v_j, l) = \infty$
8:         $\lambda(v_j, l) = x$
9:         $\mathcal{H} = \mathcal{H} \setminus \{H_{i,j}\}$ {delete $H_{i,j}$ from $\mathcal{H}$ }
10:         the message will reach $v_j$
11:     **else** $\{\mathcal{H}$ only contains $H_{i,j}$ $(|\mathcal{H}| = 1)\}$
12:         Allow $\mathcal{P}$ to reach all vertices of $V''$ using $|V''|$ messages
13:     **end if**
14: **end if**

---

The tight $\Theta(n)$ complexity bound for very large $m$ applies not only to *Broadcast* and *Traversal*. In fact, let $\mathcal{LT}$ denote the set of global single-initiator problems that can be solved with a linear number of messages in tree networks. Then, Theorem 5 can be generalized as follows:

**Theorem 6.** *Let $G$ be an arbitrary connected network with $n > 1$ nodes and $m > \tau(n)$ edges. Every problem $\mathcal{P} \in \mathcal{LT}$ can be solved in $G$ with $\Theta(n)$ messages provided $n$ and $m$ are known..*

## 4   Complexity Lower Bound in Regular Networks

In the previous sections, the graphs, which were considered, were arbitrary in topology. Clearly, for restricted classes of graphs, better bounds can be hoped if the protocol is aware of the restrictions and properties of the class. In this section we consider the class of *regular* graphs, that is where every node has the same degree deg. We show that perhaps surprisingly, global computations require a large number of messages for a large number of graphs in this class, even if the protocol is aware of the regularity and of the size of the graph, and there is a single initiator.

Here we provide the algorithm $Invoke_{NW}$ with the degree of each vertex. Namely, here we are looking at uniform graph where the $\forall v \in V, \deg(v) = d$. Since the graphs are regular graphs the relationship between $n$ and $m$ cannot be set arbitrarily. Rather, one must ensure that a graph can be legally constructed from the given $n$ and $d$ by obeying $\frac{nd}{2} = \lfloor \frac{nd}{2} \rfloor$.

There are two cases to consider:

1. $2 < d < \frac{n}{2}$ - in this case the network connectivity level if typical.
2. $d \geq \frac{n}{2}$ - in the case the network has a very high level of connectivity (the network is a dense).

---

**Algorithm 5.** Adversary strategy for regular dense networks at time $t_u$ for a given node $x$ and port $k$

---

1: **if** $\lambda(x, k) \neq \infty$ **then**
2:     the message will reach $y = \lambda(x, k)$
3: **else** $\{\lambda(x, k) = \infty\}$
4:     **if** $|V^e| = n - d - 1$ and $x$ is connected to $d - 1$ vertices in $G$ **then** $\{x$ is the $n - d$ vertex to become exhausted$\}$
5:         $\lambda(x, k) = v_0$
6:         $\lambda(v_0, 1) = x$
7:         the message will reach $v_0$
8:     **else** $\{$after this processing $|V^e| < n - d - 1\}$
9:         **if** $x \in V^R$ **then**
10:             **if** $\exists v_j \in V^R$ and $(x, v_j) \notin E$ **then**
11:                 let $v_j = v \in V^R$ where $(x, v_j) \notin E$
12:             **else**
13:                 let $v_j = v \in V^B \cup V^G$ where $(x, v) \notin E$ and $v^c = \min_{u \in V^B \cup V^G}(u^c)$
14:             **end if**
15:         **else**
16:             **if** $x \in V^B$ **then**
17:                 **if** $\exists v_j \in V^B$ and $(x, v_j) \notin E$ **then**
18:                     let $v_j = v \in V^B$ where $(x, v_j) \notin E$
19:                 **else**
20:                     let $v_j = v \in V^R \cup V^G$ where $(x, v) \notin E$ and $v^c = \min_{u \in V^R \cup V^G}(u^c)$
21:                 **end if**
22:             **else** $\{x \in V^G\}$
23:                 let $v_j = v \in V^B \cup V^R$ where $(x, v) \notin E$ and $v^c = \min_{u \in V^B \cup V^R}(u^c)$
24:             **end if**
25:         **end if**
26:         $\lambda(x, k) = v_j$
27:         find $l$ such that $\lambda(v_j, l) = \infty$
28:         $\lambda(v_j, l) = x$
29:         $E = E \cup \{(x, v_j)\}$ $\{$add the new edge to $E\}$
30:         the message will reach $v_j$
31:         **if** $x$ is exhausted **then**
32:             $V^e = V^e \cup \{x\}$
33:         **end if**
34:     **end if**
35: **end if**

---

## 4.1 Regular Typical Networks

In this section we present a communication lower bound when the graph is typical regular graph. A graph $G = (V, E)$ is a typical regular graph when $\forall v \in V, 2 < deg(v) < \frac{n}{2}$.

**Theorem 7.** *Given $n$ and $m = \frac{nd}{2}, 2 < d < \frac{n}{2}$ there exists a typical regular network with $n$ nodes and $m$ edges upon which every global protocol $\mathcal{P}$ requires to send at least $m - \frac{d^2 - d}{2}$ messages in order to invoke all the nodes in the graph.*

## 4.2   Regular Dense Network

In this section we look at communication lower bound when the network is a dense regular network. Given a connected network with $n$ nodes and $m = \frac{nd}{2}$ edges we term the network a *dense regular network* if $\frac{n}{2} < d \leq n - 1$.

**Theorem 8.** *Given $n$ and $m = \frac{nd}{2}, \frac{n}{2} < d < n - 1$ there exists a dense regular network with $n$ nodes and $m$ edges upon which every global protocol $\mathcal{P}$ requires to send at least $(n-d)d - \frac{(n-d)(n-d-1)}{2}$ messages in order to invoke all the nodes in the graph.*

## 5   Conclusions

Global distributed computation in a network with $n$ nodes and $m$ communication links between the nodes (edges) requires that all nodes participate in the computation (namely, each node in the network must send or receive at least one message). An $\Omega(m)$ lower bound on the number of messages that must be send and receive during any global distributed computation was weak because it assumes no knowledge about the network. In this paper we provide a strong lower answer and prove that $\Omega(m)$ bound is inherent for all these problems, as well as many more. We show that there exist networks where any generic global distributed solution is requiring to send $\Omega(m)$ messages even if the size of the network ($n$, and $m$) and the degree of each node are known in advance. The relationship between $n$ and $m$ plays an important role in establishing the lower bound. We showed that for most networks (where $m \leq \frac{1}{2}(n-2)(n-3)+1$) a generic solution for any problem in this class requires at least $m$ messages. This result holds for almost all values of $m$ (e.g., when $\frac{1}{2}(n-2)(n-3)+1 < m \leq \frac{1}{2}(n-1)(n-2)+1$ the number of required messages is $m-1$), even if there is a single initiator and the entities have distinct identifiers, and both these facts are known. Moreover, the results hold even if the protocol can maintain a global view of the network.

As the networks become more dense, namely the network approaches a complete graph, the number of required messages is gradually reduced. For extreme values of $m$ ( i.e., $m = \frac{1}{2}n(n-1) - c > \frac{1}{2}(n-1)(n-2) + 1$), where $c \geq 0$ is constant, the lower bound gradually approaches $\Omega(n)$; this is understandable since we establish it for the single initiator scenario. However, we prove that in networks of such a size, single initiators problems such as *Broadcast* and *Traversal* can be solved with precisely that order of magnitude. This means that for those problems the knowledge of $n$ and $m$ generates a significant and sudden complexity *drop* from $\Theta(n^2)$ to $\Theta(n)$.

## References

1. Awerbuch, B.: A new distributed depth-first search algorithm. Information Processing Letters 20, 147–150 (1985)
2. Awerbuch, B., Goldreich, O., Peleg, D., Vainish, R.: A Trade-off between Information and Communication in Broadcast Protocols. Journal of the ACM 37(2), 238–256 (1990)

3. Chang, E.J.H.: Echo algorithms: Depth parallel operations on general graphs. IEEE Transactions on Software Engineering SE-8(4), 391–401 (1982)
4. Cheung, T.Y.: Graph traversal techniques and the maximum flow problem in distributed computation. IEEE Transactions on Software Engineering 9, 504–512 (1983)
5. Cidon, I.: Yet another distributed depth-first search algorithm. Information Processing Letters 26, 301–305 (1987)
6. Dobrev, S.: Leader election using any sense of direction. In: 6th International Colloquium on Structural Information and Communication Complexity, Lacanau, pp. 93–104 (July 1999)
7. Erdös, P., Gallai, T.: On maximal paths and circuits of graphs. Acta Mathemtica Hunarica 10(3-4), 337–356 (1959)
8. Gafni, E.: Improvements in the time complexity of two message-optimal election algorithms. In: 4th ACM Symposium on Principles of Distributed Computing, Minaki, pp. 175–185 (August 1985)
9. Gallager, R.G., Humblet, P.A., Spira, P.M.: A distributed algorithm for minimum spanning tree. ACM Transactions on Programming Languages and Systems 5(1), 66–77 (1983)
10. Garay, J.A., Kutten, S., Peleg, D.: A sublinear time distributed algorithm for minimum-weight spanning trees. SIAM Journal on Computing 27(1), 302–316 (1998)
11. Israeli, A., Kranakis, E., Krizanc, D., Santoro, N.: Time-Messages tradeoffs for the weak unison problem. Nordic Journal of Computing 4, 317–329 (1997)
12. Korach, E., Kutten, S., Moran, S.: A modular technique for the design of efficient distributed leader finding algorithms. ACM Transactions on Programming Languages and Systems 12(1), 84–101 (1990)
13. Lynch, N.: A hundred impossibility proofs for distributed computing. In: Proceedings of the Eighth Annual ACM Symposium on Principles of Distributed Computing, pp. 1–28 (1989)
14. Santoro, N.: On the message complexity of distributed problems. Int. Journal of Computer and Information Sciences 13(3), 131–147 (1984)
15. Santoro, N.: Design and Analysis of Distributed Algorithms. John Wiley, Chichester (2007)
16. Segall, A.: Distributed network protocols. IEEE Transactions on Information Theory IT-29(1), 23–35 (1983)
17. Tel, G.: Introduction to Distributed Algorithms. Cambridge University Press, Cambridge (1994)

# Optimizing Regenerator Cost
## in Traffic Grooming[*]
### (Extended Abstract)

Michele Flammini[1], Gianpiero Monaco[2], Luca Moscardelli[3],
Mordechai Shalom[4], and Shmuel Zaks[5]

[1] Dipartmento di Informatica, Università degli Studi dell'Aquila, L'Aquila, Italy
flammini@di.univaq.it
[2] Mascotte Project, INRIAI3S(CNRS/UNSA), Sophia Antipolis, France
gianpiero.monaco@sophia.inria.fr
[3] Dipartmento di Scienze, Università degli Studi di Chieti-Pescara, Pescara, Italy
moscardelli@sci.unich.it
[4] Tel Hai Academic College, Upper Galilee, 12210, Israel
cmshalom@telhai.ac.il
[5] Department of Computer Science, Technion, Haifa, Israel
zaks@cs.technion.ac.il

**Abstract.** In optical networks regenerators have to be placed on lightpaths in order to regenerate the signal. In addition, grooming enables the use of the same regenerator by several lightpaths. In this work we consider the problem of minimizing the number of regenerators used in traffic grooming in optical networks. We deal with the case in which a regenerator has to be placed at every internal node of each lightpath. Up to $g$ (the grooming factor) lightpaths can use the same regenerator. Starting from the 4-approximation algorithm of [7] that solves this problem for a path topology, we provide an approximation algorithm with the same approximation ratio for the ring and tree topologies. We present also a technique based on matching that leads to the same approximation ratio in tree topology and can be used to obtain approximation algorithms in other topologies. We provide an approximation algorithm for general topology that uses this technique.

**Keywords:** Optical Networks, Wavelength Division Multiplexing(WDM), Regenerators, Traffic Grooming, Tree Networks.

## 1 Introduction

In modern optical networks, high-speed signals are sent through optical fibers using WDM (Wavelength Division Multiplexing) technology. Currently deployed networks carry around 80 wavelengths per fiber, whereas networks with a few

---

[*] This work was partially supported by the Israel Science Foundation grant No. 1249/08, by British Council Grant UKTELHAI09, and by the PRIN 2008 research project COGENT (COmputational and GamE-theoretic aspects of uncoordinated NeTworks), funded by the Italian Ministry of University and Research.

C. Lu, T. Masuzawa, and M. Mosbah (Eds.): OPODIS 2010, LNCS 6490, pp. 443–458, 2010.

hundred wavelengths per fiber are being used in testbeds. The decrease in the energy of the signal with the traveled distance raises the requirement of optical amplifiers at every (almost) fixed distance. However, optical amplifiers introduce noise into the signal, thus after a certain number of amplifications, the optical signal needs to be regenerated. In the current technology, the signal is regenerated by first using a ROADM (Reconfigurable Optical Add-Drop Multiplexer) to extract a set of wavelengths from the optical fiber. Then, for each extracted wavelength, an optical regenerator is needed to regenerate the signal carried by that wavelength. That is, at a given optical node, one needs as many regenerators as wavelengths one wants to regenerate.

Nowadays the cost of a regenerator is considerably higher than the cost of an ROADM. Moreover, as described above, the regenerator cost is per wavelength, as opposed to ROADM cost that is paid once per several wavelengths. Therefore the *total* number of regenerators is an important cost parameter to be minimized. Another possible criterion is to minimize the number of *locations* (that is, the number of nodes) in which optical regenerators are placed. This measure is the one assumed in [6], which makes sense when the dominant part of the cost is the set-up of new optical nodes, or when the equipment to be placed at each node is the same for all nodes. In this work we consider the total number of regenerators as the cost function.

A logical path formed by a signal travelling from its source to its destination using a unique wavelength is termed a *lightpath*. Let $d$ be the maximum number of hops a lightpath can make without meeting a regenerator. Then, for each lightpath $\ell$, we need to place one regenerator every $d$ consecutive vertices in $\ell$, to get an optimal solution. However the problem becomes harder when the *traffic grooming* comes into the picture.

*Traffic grooming*: The network usually supports traffic that is at rates which are lower than the full wavelength capacity, and therefore the network operator has to be able to put together (= groom) low-capacity connections into the high capacity lightpaths. In graph-theoretic terms, we associate a path in the graph with each connection, and the problem can viewed as assigning wavelengths to these paths so that at most $g$ of them using the same wavelength ($g$ being the *grooming factor*) can share one edge. Thus, all paths (i.e. connections) that get the same color and form a connected subgraph correspond to grooming of these connections into one lightpath.

In this work we concentrate on the special case $d = 1$ and general $g$. It is expected that the techniques presented in our work will be carried out to similar studies for higher values of $d$.

## 1.1   Related Work

Various variants of regenerator placement problems were studied in [1, 4, 5, 10, 12, 13, 15, 16]. Most of these results concentrate in heuristics and simulations and do not consider traffic grooming.

In [6] theoretical results (upper bounds and lower bounds) are presented for some variants of this problem. This work considers the number of regenerator locations (as opposed to the total number of regenerators) as the cost measure, and does not consider traffic grooming. On the other hand [11] uses the same cost measure but still does not consider traffic grooming.

The problem is shown to be NP-hard in other contexts such as fiber minimization in [14] and is also implied by the proof of a similar result in [8].

When the underlying graph is a path the problem is equivalent to a machine scheduling problem studied in [7]. Several approximation algorithms are presented in this work for this scheduling problem and its special cases.

## 1.2    Our Contribution

In this work we consider the traffic grooming problem to minimize the number of regenerators used. We consider only the case $d = 1$, i.e. the case that a regenerator has to be placed at every internal node of every lightpath. Our starting point is a 4-approximation algorithm of [7] that solves a closely related problem for a path topology. We prove that the same algorithm can be used for our problem and show that it has the same approximation ratio not only for path topology, but also for ring topology. We present a greedy 4-approximation algorithm for tree networks. We also show a general technique using matchings that can lead to approximation algorithms in other topologies. We use this technique and show an $\lfloor \frac{L+7}{2} \rfloor$-approximation algorithm for general topology, where $L$ is the maximum load (i.e. number of paths that share a common edge) in the input.

In Section 2 we present preliminary results and definitions, including the above mentioned algorithm for path networks and extension of its analysis to the case of ring topology. In Section 3 we present an algorithm with the same performance for tree topology. In Section 4 we present the matching technique and its use for general toplogies. We summarize the results and suggest open research directions in Section 5. Due to the lack of space, some proofs and figures have been removed. For a full version of the paper see [9].

## 2    Preliminaries

### 2.1    Definitions and Problem Statement

An instance of the *Regenerators Grooming Problem* is a triple $(G, \mathcal{P}, g)$ where $G = (V, E)$ is a graph modeling the optical network, $\mathcal{P}$ is a set of simple paths in $G$ and $g$ is a positive integer, namely the grooming factor.

A coloring (or wavelength assignment) of $(G, \mathcal{P})$ is a function $w : \mathcal{P} \mapsto \mathbb{N}$. For a coloring $w$ and color $\lambda$, $\mathcal{P}_\lambda^w$ is the subset of paths from $\mathcal{P}$ colored $\lambda$ by $w$, i.e. $\mathcal{P}_\lambda^w \stackrel{def}{=} \{P \in \mathcal{P} | w(P) = \lambda\}$. When there is no ambiguity on the coloring $w$ under consideration, we omit the superscript $w$ and use $\mathcal{P}_\lambda$.

For an edge $e$, $\mathcal{P}_e$ denotes the subset of paths of $\mathcal{P}$ using the edge $e$. For every $e \in E$ we define $load(\mathcal{P}, e) \stackrel{def}{=} |\mathcal{P}_e|$ and $load(\mathcal{P}) \stackrel{def}{=} \max_{e \in E} load(\mathcal{P}, e)$. A *valid* coloring (or wavelength assignment) $w$ of $(G, \mathcal{P}, g)$ is a coloring of $\mathcal{P}$ in which for any edge $e$ at most $g$ paths using $e$ are colored with the same color, i.e. for every color $\lambda$ we have $load(\mathcal{P}_\lambda^w) \leq g$.

We denote by $INT(P)$ the set of intermediate nodes, i.e. of all the nodes not being endpoints, of a path $P$ in $G$, and $int(P) \stackrel{def}{=} |INT(P)|$. For a set $\mathcal{P}$ of paths we define

$$SPAN(\mathcal{P}) \stackrel{def}{=} \bigcup_{P \in \mathcal{P}} INT(P),$$

$$span(\mathcal{P}) \stackrel{def}{=} |SPAN(\mathcal{P})|,$$

$$len(\mathcal{P}) \stackrel{def}{=} \sum_{P \in \mathcal{P}} int(P).$$

A set of paths is called a *no-split instance* or shortly an NSI if the union of its paths (as sets of edges) induces a graph of maximum degree 2. Due to technological constraints, paths using a same wavelength and going through a same edge of the network can be routed only to another unique edge, and therefore every set of paths with the same color has to be an NSI.

The number of regenerators operating at wavelength $\lambda$ is $span(\mathcal{P}_\lambda^w)$; in fact, at each node being an intermediate node of some path in $\mathcal{P}_\lambda^w$ a regenerator operating at this wavelength is needed.

We are now ready to give a formal definition of our problem.

**Input:** An instance $(G, \mathcal{P}, g)$, where $G = (V, E)$ is a network, $\mathcal{P} = \{P_1, P_2, ..., P_n\}$ is a set of simple paths in $G$, and $g$ is the grooming factor.

**Output:** A valid coloring $w : \mathcal{P} \mapsto \mathbb{N}$ of the paths such that, for every $\lambda$, $\mathcal{P}_\lambda$ is an NSI (*no splitting condition*).

**Measure:** The cost of a solution is given by the total number of regenerators $REG^w \stackrel{def}{=} \sum_\lambda span(\mathcal{P}_\lambda^w)$.

**Objective:** The goal is to minimize the total number of regenerators $REG^w$.

$OPT(G, \mathcal{P}, g)$ denotes the cost of any optimal coloring and $ALG(G, \mathcal{P}, g)$ denotes the cost of the coloring returned by some algorithm $ALG$ on instance $(G, \mathcal{P}, g)$. As the cost function depends only on the partition of the paths induced by the coloring, with some abuse of notation, a coloring $w$ denotes also the equivalence class of colorings that induce the same partition as $w$.

## 2.2   Lower Bounds

We have the following trivial lower bounds for the cost of any coloring $w$, in particular for an optimal coloring.

- The grooming bound:

$$REG^w \geq \frac{len(\mathcal{P})}{g}.$$

- The span bound:

$$REG^w \geq span(\mathcal{P}).$$

The grooming bound holds because a regenerator can be used by a maximum of $g$ intermediate nodes of paths. The span bound holds because at least one regenerator is needed on any node that is an intermediate node of some path.

## 2.3   Path and Ring Networks

Now we focus on ring and path networks. We adapt Theorem 2.1 in [7] to our problem and generalize it to the case of ring networks. Specifically, we show that the *FirstFit* algorithm presented in [7] is a 4-approximation algorithm for our problem. The proof goes along the same lines, and we bring it here for sake of completeness; the main difference is in Lemma 2, whose proof required modifications of the proof of the corresponding claim in [7] in order to assure correctness for the case of ring topology.

Notice that when $G$ is a ring or a path, all subsets of $\mathcal{P}$ constitute an NSI.

Algorithm *FirstFit* colors the paths greedily by considering them one after the other, from longest to shortest. Each path is assigned the lowest possible color for it.

---

**Algorithm 1.** $FirstFit(G, \mathcal{P}, g)$ with $G$ being a path or a ring

---

1: Sort the paths in non-increasing order of length, i.e., $int(P_1) \geq int(P_2) \geq \ldots \geq int(P_n)$.
2: Consider the paths by the above order: assign to the next path, $P_j$, the first possible color $\lambda$ that will not violate the load condition. Namely, find the minimum value $\lambda \geq 1$ such that, for every edge $e$ of $P_j$, $load(\mathcal{P}_\lambda, e) \leq g - 1$ and $w(P_j) \leftarrow \lambda$.

---

The upper bound proof is based on the observation stated in the following lemma, and depicted in Figure 1.

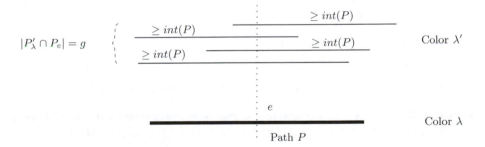

**Fig. 1.** Basic observation

**Lemma 1.** *Let $w$ be the coloring returned by $FirstFit$. Let $P$ be a path colored $\lambda$, i.e. $P \in \mathcal{P}_\lambda^w$, for some $\lambda \geq 2$. Then for any $\lambda' < \lambda$, (a) there is an edge $e \in P$ such that $load(\mathcal{P}_{\lambda'}^w, e) = g$, (b) each path $P' \in \mathcal{P}_{\lambda'} \cap \mathcal{P}_e$ is no shorter than $P$.*

We use the above properties stated in Lemma 1 in order to show the following claim, which will be crucial in order to prove the desired result.

**Lemma 2.** *For any $\lambda > 1$, $len(\mathcal{P}_{\lambda-1}) \geq \frac{g}{3} span(\mathcal{P}_\lambda)$.*

*Proof.* For every path $P \in \mathcal{P}_\lambda$, we choose arbitrarily an edge $e$ of $P$ among the edges whose existence is guaranteed by Lemma 1. Let $b(P) \stackrel{def}{=} \mathcal{P}_{\lambda-1} \cap \mathcal{P}_e$ be the *blocking* paths of $P$. By Lemma 1, $|b(P)| = g$ and $len(b(P)) \geq g \cdot int(P)$. Let $\overline{\mathcal{P}}$ be the set of all *blocking* paths defined as above, i.e. $\overline{\mathcal{P}} \stackrel{def}{=} \cup_{P \in \mathcal{P}_\lambda} b(P)$. Clearly $\overline{\mathcal{P}} \subseteq \mathcal{P}_{\lambda-1}$.

Now, we consider a blocking path $P' \in \overline{\mathcal{P}}$ (see Figure 3 in [9]). Consider the set of all paths in $\mathcal{P}_\lambda$ *blocked* by $P'$. With a little abuse of notation we denote them by $b^{-1}(P)$. Consider a node $v \in SPAN(b^{-1}(P'))$. It is in some path $P'' \in \mathcal{P}_\lambda$ which is no longer than $P'$ and intersects with $P'$, therefore there exists an intermediate node of $P'$ which is at distance to $v$ at most $int(P'') \leq int(P')$. As $G$ is a path or a cycle the number of such nodes $v$ is at most $3 \cdot int(P')$. We conclude that $span(b^{-1}(P')) \leq 3 \cdot int(P')$. Summing up for all the paths in $\overline{\mathcal{P}}$ we get

$$\sum_{P' \in \overline{\mathcal{P}}} span(b^{-1}(P')) \leq 3 \sum_{P' \in \overline{\mathcal{P}}} int(P') = 3 \cdot len(\overline{\mathcal{P}}).$$

Consider a node $v \in SPAN(\mathcal{P}_\lambda)$. It is an intermediate node of at least one path $P \in \mathcal{P}_\lambda$, which in turn is blocked by at least $g$ paths of $\overline{\mathcal{P}}$. Therefore $v \in SPAN(b^{-1}(P'))$ for at least $g$ paths $P'$ of $\overline{\mathcal{P}}$, in other words $v$ contributes at least $g$ to the sum in the left hand side above. Thus we have $\sum_{P' \in \overline{\mathcal{P}}} span(b^{-1}(P')) \geq g \cdot span(\mathcal{P}_\lambda)$. Therefore,

$$3 \cdot len(\mathcal{P}_{\lambda-1}) \geq 3 \cdot len(\overline{\mathcal{P}}) \geq \sum_{P' \in \overline{\mathcal{P}}} span(b^{-1}(P')) \geq g \cdot span(\mathcal{P}_\lambda). \qquad \square$$

We are now ready to prove the following theorem providing an upper bound to the approximation ratio of the $FirstFit$ algorithm. The proof exploits arguments similar of the one of the corresponding theorem in [7].

**Theorem 1.** *If $G$ is a path or a ring, then for any instance $(G, \mathcal{P}, g)$, $FirstFit(G, \mathcal{P}, g) \leq 4 \cdot OPT(G, \mathcal{P}, g)$.*

The following lemma and its proof follow from a similar claim in [7].

**Lemma 3.** *For any $\epsilon > 0$, there are infinitely many instances $(G, \mathcal{P}, g)$ having infinitely many input sizes, such that $FirstFit(G, \mathcal{P}, g) > (3 - \epsilon) \cdot OPT(G, \mathcal{P}, g)$.*

Combining Theorem 1 and Lemma 3, we finally get the following theorem.

**Theorem 2.** *The approximation ratio of $FirstFit$ is between 3 and 4 in ring and path networks.*

## 3   Tree Networks

In this section we present an optimal algorithm $GreedyMatch$ for the case where the graph $G$ is a tree and $g = \infty$. Combining this algorithm and algorithm $FirstFit$ described in the previous section we obtain a 4-approximation algorithm for tree networks and any value of $g$.

### 3.1  $G, \mathcal{P}, \infty$ Instances

We first consider the special case of $g = \infty$, that will be useful in order to provide an approximation algorithm for general $g$. When $g = \infty$, any solution is a valid coloring. It remains to satisfy the no splitting condition. Therefore the problem becomes to partition $\mathcal{P}$ into no-split instances $NSI_1, NSI_2, \ldots$ such that $\sum_\lambda span(NSI_\lambda)$ is minimized.

Note that the span (lower) bound holds in this special case, i.e. $OPT(G, \mathcal{P}, \infty) \geq span(\mathcal{P})$.

Since $g = \infty$, we can assume that there is no path $P \in \mathcal{P}$ completely included in another path $P' \in \mathcal{P}$, because in this case we could remove $P$ from the input. In any solution of the remaining instance $P$ can be added to the NSI containing $P'$ without increasing the cost.

We introduce some additional notation.

- Two NSIs $NSI$ and $NSI'$ are said to be *compatible* if their union is also an NSI. We denote this fact as $NSI \sim NSI'$. Otherwise they are said to be *incompatible* and denoted as $NSI \nsim NSI'$.
- The overlap of two NSIs $NSI$ and $NSI'$ is $OV(NSI, NSI') \overset{def}{=} SPAN(NSI) \cap SPAN(NSI')$ and $ov(NSI, NSI') \overset{def}{=} |OV(NSI, NSI')|$.
- Two NSIs $NSI$ and $NSI'$ are overlapping if $ov(NSI, NSI') > 0$.
- An NSI is said to be *connected* if the union of its paths (as sets of edges) induces a connected graph.
- We say that $NSI \sqsubseteq NSI'$ if $\cup_{P \in NSI} P \subseteq \cup_{P \in NSI'} P$

Consider algorithm $GreedyMatch$; the following lemmata are needed for proving Theorem 3, in which it is shown that such an algorithm is optimal.

**Lemma 4.** *Every two NSIs in an optimal solution of $(G, \mathcal{P}, \infty)$ are either non-overlapping or incompatible.*

*Proof.* Assume, by contradiction that there are two NSIs that are both compatible and overlapping. Then they can be joined to form one NSI, and decrease the cost of the solution by the size of their overlap.                    □

**Lemma 5.** *At any given point of the execution of Algorithm GreedyMatch the sets $NSI_i$ are connected NSIs.*

*Proof.* The sets $NSI_i$ are trivially connected at the beginning of the algorithm. Moreover, since a new $NSI$ is constructed by unifying two compatible and overlapping NSIs, they are connected also at any point of the execution of Algorithm $GreedyMatch$.                    □

**Lemma 6.** *At any given point of the execution of Algorithm GreedyMatch, after step 1, consider the partition $\{NSI_1, NSI_2, \ldots\}$. There is an optimal solution $\{NSI_1^*, NSI_2^*, \ldots\}$ such that every $NSI_i$ is a subset of some $NSI_i^*$, or in other words the partition given by the algorithm is a refinement of the partition given by some optimal solution.*

---

**Algorithm 2.** $GreedyMatch(G, \mathcal{P}, \infty)$, $G$ being a tree

---

1: $\forall P_i \in \mathcal{P}, NSI_i \leftarrow \{P_i\}$                          ▷ Every path constitutes a connected NSI.
2: **while** there exist $NSI_i, NSI_j$ such that $NSI_i \sqsubseteq NSI_j$ **do** ▷ Eliminate inclusions
3:       $NSI_j \leftarrow NSI_j \cup NSI_i$
4:       $NSI_i \leftarrow \emptyset$
5: **end while**
6: **while** there exist two compatible $NSI_i, NSI_j$ such that $ov(NSI_i, NSI_j) > 0$ **do**
7:       Find two compatible NSIs $NSI_i, NSI_j$ maximizing $ov(NSI_i, NSI_j)$
8:       $NSI_j \leftarrow NSI_j \cup NSI_i$
9:       $NSI_i \leftarrow \emptyset$
10:      Eliminate inclusions (as in Steps 3-6) ▷ We will prove in Lemma 7 that this is unnecessary
11: **end while**

---

*Proof.* Without loss of generality we can assume that all the NSIs in an optimal solution of $(G, \mathcal{P}, \infty)$ are connected, because if we have a disconnected NSI $NSI$ we can replace $NSI$ with a connected NSI for each connected component of $NSI$. The claim is obviously true immediately after step 1 of the algorithm. Assume by contradiction that the claim is false and consider the first time during the execution of the algorithm that it becomes false. This can happen only after execution of step 8. $NSI_i$ and $NSI_j$ are overlapping and compatible, because they are chosen by the algorithm in step 7. They are also connected by Lemma 5. As the condition was true prior to the execution of step 8, there is some optimal solution $S^* = \{NSI_1^*, NSI_2^*, \ldots\}$ such that $NSI_i \subseteq NSI_i^*$ and $NSI_j \subseteq NSI_j^*$. Therefore $NSI_i^* \supset NSI_i$ and $NSI_j^* \supset NSI_j$ are overlapping. Also, by Lemma 4, $NSI_i^* \nsim NSI_j^*$.

As there are no inclusions, the $OV(NSI_i, NSI_j)$ is a proper subset of both $SPAN(NSI_i)$ and $SPAN(NSI_j)$ (see Figure 4 in [9]). Let $a, b, c, d \in V$ be four distinct nodes of the tree such that $SPAN(NSI_i)$ (resp. $SPAN(NSI_j)$) is the path between $b$ and $d$ (resp. $a$ and $c$). Then $OV(NSI_i, NSI_j)$ is the path between $b$ and $c$. The partition $\{NSI_1, \ldots\}$ is a refinement of the partition $\{NSI_1^*, \ldots\}$. Let $NSI_i^* = NSI_i \uplus NSI_{i_1} \uplus NSI_{i_2} \uplus \ldots$. We observe that for none of these sets $SPAN(NSI_{i_k})$ can intersect with both $a - b$ and $c - d$, because this would imply that $OV(NSI_i, NSI_j) \subsetneq OV(NSI_i, NSI_{i_k})$, a contradiction to the way $NSI_i$ and $NSI_j$ are chosen by the algorithm. Given this observation we partition the set $NSI_i^*$ into three sets $NSI_i, NSI_{ii}$ and $NSI_{ij}$ such that the sets $NSI_{i_k}$ spanning at least one edge of $c - d$ (resp. $a - b$) are in $NSI_{ii}$ (resp. $NSI_{ij}$), the rest are divided arbitrarily. We do the same for $NSI_j^*$.

$NSI_i, NSI_{ii}, NSI_{ij}$ are pairwise compatible, because they make part of $NSI_i^*$, and so are $NSI_j, NSI_{jj}$, and $NSI_{ji}$. Moreover $NSI_{ij} \sim NSI_{ji}$ and $NSI_{ii} \sim NSI_{jj}$, because the underlying graph is a tree and thus they can overlap only in the path $b - c$ in which there can not exist nodes with induced degree 3 or more.

We conclude the proof by case analysis. For each case we show how an optimal solution $S'^*$ can be built from $S^*$ such that $NSI_i$ and $NSI_j$ are contained in the same set of $S'^*$, a contradiction to the assumption that the condition became false.

Assume $NSI_{ji} \sim NSI_i$:

- $NSI_{ji} \sim NSI_{ii}$: In this case we can move $NSI_{ii}$ and $NSI_i$ into $NSI_j^{*}$ without increasing the cost of the solution.
- $NSI_{ji} \nsim NSI_{ii}$: In this case the node with induced degree more than 2 is necessarily beyond (in Figure 4 in [9], at the right of) the node $d$, proving that $SPAN(NSI_{ji})$ contains the path $c - d$. Therefore we can move $NSI_i$ into $NSI_j^{*}$ without increasing the cost of the solution.

After handling the case $NSI_{ij} \sim NSI_j$ symmetrically, it remains to solve the case $NSI_{ji} \nsim NSI_i$ and $NSI_{ij} \nsim NSI_j$. If $NSI_{ji}$ and $NSI_{ij}$ are overlapping than we can repartition these six sets into two sets $NSI_{ij} \cup NSI_{ji}$ and $NSI_i \cup NSI_j \cup NSI_{ii} \cup NSI_{jj}$ without increasing the cost. Otherwise we build three sets $NSI_{ij}$, $NSI_{ji}$ and $NSI_i \cup NSI_j \cup NSI_{ii} \cup NSI_{jj}$ without increasing the cost. □

We are now able to prove that Algorithm *GreedyMatch* is optimal.

**Theorem 3.** *When Algorithm GreedyMatch ends, the solution* $\{NSI_1, NSI_2, ...\}$ *is optimal.*

The following lemma shows that step 10 is redundant, and therefore can be removed from algorithm *GreedyMatch*.

**Lemma 7.** *When the Algorithm GreedyMatch reaches step 10, there are no inclusions.*

### 3.2   An Approximation Algorithm Scheme for Any Graph $G$ and Any Value of $g$

We propose the algorithm scheme $Combined(\mathcal{A}, (G, \mathcal{P}, g))$ for general graphs and any value of $g$, depending on Algorithm $\mathcal{A}$ working for the specific case in which $g = \infty$.

---

**Algorithm 3.** $Combined(\mathcal{A}, (G, \mathcal{P}, g))$

---

1: Partition $\mathcal{P}$ into NSIs $NSI_1, NSI_2, ...$ using algorithm $\mathcal{A}$ computed on the corresponding $(G, \mathcal{P}, \infty)$ instance.

2: For each $i$, let $G(NSI_i)$ be the graph induced by the paths of $NSI_i$. Split $NSI_i$ into sets $\mathcal{P}_{i,1}, \mathcal{P}_{i,2}, ...$ by solving the instance FirstFit$(G(NSI_i), NSI_i, g)$.

3: Assign each one of the sets $\mathcal{P}_{i,j}$ a distinct color $\lambda_{i,j}$.

---

**Lemma 8.** *Given any* $g \geq 1$, *if Algorithm $\mathcal{A}$ is a $\rho$-approximation algorithm for instance* $(G, \mathcal{P}, \infty)$, *then Algorithm Combined$(\mathcal{A}, (G, \mathcal{P}, g))$ is a $(\rho + 3)$-approximation algorithm for instance* $(G, \mathcal{P}, g)$.

*Proof.* In order to prove the correctness of the algorithm, it is sufficient to notice that every $NSI_i$ is a no-split instance, thus satisfies the no splitting condition. Therefore any subset $\mathcal{P}_{i,j}$ of it also satisfies the no splitting condition. Moreover, by the correctness of $FirstFit$ the output is a valid coloring.

By Lemma 2, for any instance $(SPAN(NSI_i), NSI_i, g)$ and any color $\lambda_{i,j}$ we have $span(\mathcal{P}_{i,j+1}) \leq \frac{3}{g} len(\mathcal{P}_{i,j})$.

Therefore

$$\sum_{i \geq 1, j \geq 2} span(\mathcal{P}_{i,j}) \leq \frac{3}{g} \sum_{i,j \geq 1} len(\mathcal{P}_{i,j})$$

$$= \frac{3}{g} \sum_{i \geq 1} len(NSI_i) = \frac{3}{g} \sum len(\mathcal{P}) \leq 3 \cdot OPT(G, \mathcal{P}, g).$$

On the other hand

$$\sum_{i \geq 1} span(\mathcal{P}_{i,1}) \leq \sum_{i \geq 1} span(NSI_i) \leq \rho \cdot OPT(G, \mathcal{P}, \infty) \leq \rho \cdot OPT(G, \mathcal{P}, g).$$

Combining we get

$$Combined(\mathcal{A}, (G, \mathcal{P}, g)) = \sum_{i,j \geq 1} span(p_{i,j}) \leq (\rho + 3) \cdot OPT(G, \mathcal{P}, g).$$

$\square$

### 3.3   The Approximation Algorithm for Tree Networks

By combining Theorem 3 with Lemma 8, we get the following theorem.

**Theorem 4.** *Given any $g \geq 1$, Algorithm $Combined(GreedyMatch, (G, \mathcal{P}, g))$ is a 4-approximation algorithm for instance $(G, \mathcal{P}, g)$, where $G$ is a tree network.*

The following lemma and its proof exploit arguments similar to the ones used in lemma 3.

**Lemma 9.** *For any $\epsilon > 0$, there are infinitely many instances $(G, \mathcal{P}, g)$ having infinitely many input sizes, such that $Combined(GreedyMatch, (G, \mathcal{P}, g)) > (3 - \epsilon) \cdot OPT(G, \mathcal{P}, g)$, where $G$ is a tree network.*

By combining Theorem 4 with Lemma 9 we get the following theorem.

**Theorem 5.** *The approximation ratio of $Combined(GreedyMatch)$ is between 3 and 4 in tree networks.*

## 4   Beyond Tree Networks: A Matching Technique

In this section we present a new technique to approximate $(G, \mathcal{P}, \infty)$ instances in any topology. In particular, we show a general technique able to reduce an instance of the general network to instances of ring and path networks. Using such a technique, and exploiting a reduction of the problem to an instance of the Maximum Weighted Matching on an auxiliary graph, we present an approximation algorithm for general topology.

## 4.1   The Endpoint Intersection Graph

In order to describe the matching technique, we need to define the *edge-weighted endpoint intersection graph* $EIG(G, \mathcal{P}) = (V', E')$ of $G$ and $\mathcal{P}$. $V'$ contains $2\,|\mathcal{P}|$ nodes $v_{1,1}, v_{1,2}, v_{2,1}, v_{2,2}, ..., v_{i,1}, v_{i,2}, ...$, one for each endpoint of a path $P_i \in \mathcal{P}$. There is an edge between two nodes $v_{i,k}, v_{j,k'}$ ($k, k' \in \{1, 2\}$) if $P_i \cup P_j$ is either a path or a ring and $P_i \cap P_j$ contains a path in $G$ with endpoints $v_{i,k}$ and $v_{j,k'}$.

The weight function $f : E' \to \mathbb{N}$ is defined as follows: $f(v_{i,k}, v_{j,k'})$ is the length of the path between $v_{i,k}$ and $v_{j,k'}$ in the intersection, minus one. As usual, the weight of a set of edges is defined as the sum of the weights of the edges belonging to it.

**Lemma 10.** *For every solution $w$ of $(G, \mathcal{P}, \infty)$ in any graph, there is a matching $M(w)$ of $EIG(G, \mathcal{P})$.*

*Proof.* Consider an NSI of a solution $w$. By definition, all the nodes of its paths, in particular their endpoints are in some subgraph of $G$ with maximum degree 2. This subgraph is the union of some paths and cycles of $G$.

Let us first consider a path $Q$ of this graph. We choose some arbitrary direction of $Q$, and number the paths of the NSI as $P_1, P_2, ..., P_l$ according to the order, in the chosen direction of their starting nodes. Let w.l.o.g. these nodes be $v_{1,1}, v_{2,1}, ....$ As the paths are inclusion-free, the order of the ending nodes in this direction is the same, namely $v_{1,2}, v_{2,2}, ...$ (upper part of Figure 2). For every two consecutive paths $P_i, P_{i+1}$, their intersection is the segment of the path between $v_{i,2}$ and $v_{i+1,1}$. Therefore $(v_{i,2}, v_{i+1,1})$ is an edge of $EIG(G, \mathcal{P})$, and the edges $(v_{1,2}, v_{2,1}), (v_{2,2}, v_{3,1}), ..., (v_{l-1,2}, v_{l,1})$ constitute a matching of $EIG(G, \mathcal{P})$ with $l - 1$ edges.

Now we consider a cycle $C$ of this graph. This case is similar to the previous case (consult the lower part of the figure), except that in this case $(v_{l,2}, v_{1,1})$ is also an edge of $EIG(G, \mathcal{P})$, and $(v_{1,2}, v_{2,1}), (v_{2,2}, v_{3,1}), ..., (v_{l-1,2}, v_{l,1}), (v_{l,2}, v_{1,1})$ constitute a matching of $EIG(G, \mathcal{P})$ with $l$ edges. $\qquad\square$

**Lemma 11.** $REG^w = len(\mathcal{P}) - f(M(w))$.

*Proof.* As in the proof of the previous lemma, consider an NSI that induces a path of $G$. Let without loss of generality this NSI be $\{P_1, P_2, ..., P_l\}$ in the chosen direction of the path, and assume that the endpoints of each path are indexed 1 and 2 in this direction. Then

$$span(NSI) = int(P_1) + (int(P_2) - f((v_{1,2}, v_{2,1}))) + \cdots$$
$$+ (int(P_l) - f((v_{l-1,2}, v_{l,1}))) = len(NSI) - f(M(w) \cap NSI).$$

The same result holds, similarly, for an NSI that induces a cycle. Summing up over all the NSIs, we get

$$REG^w = \sum_\lambda span(NSI_\lambda) = \sum_\lambda len(NSI_\lambda) - \sum_\lambda f(M(w) \cap NSI_\lambda)$$
$$= len(\mathcal{P}) - f(M(w)).$$

$\qquad\square$

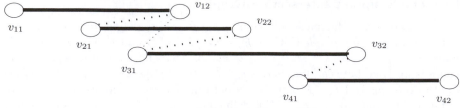

A path $Q$ induced by an NSI

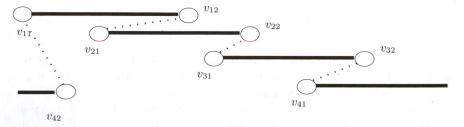

A cycle $C$ induced by an NSI

**Fig. 2.** Correspondence between feasible solutions of $(G, \mathcal{P}, \infty)$ and the matchings of $EIG(G, \mathcal{P})$

**Lemma 12.** *If $G$ is a ring or tree, then to every matching $M$ of $EIG(G, \mathcal{P})$ corresponds a solution $w(M)$ of the $(G, \mathcal{P}, \infty)$ instance.*

*Proof.* Consider a matching $M$ of $EIG(G, \mathcal{P})$. Consider also the *auxiliary* edges (not belonging to the edges of $EIG(G, \mathcal{P})$) $M' = \{(v_{i,1}, v_{i,2})\}$ of $EIG(G, \mathcal{P})$. Every node has degree at most 1 with respect to the edges in $M$, and degree exactly 1 with respect to the edges in $M'$, thus degree at most 2 with respect to the edges in $M \cup M'$. Therefore $M \cup M'$ can be partitioned into (alternating) paths and cycles. Note that a path ends with an auxiliary edge in $M'$ (because a node with degree 1 has its only incident edge in $M'$), thus has odd length; the cycles have even length.

We first consider a path of $M \cup M'$. It is of the form $v_{i_1,k_1} - v_{i_1,k_1'} - v_{i_2,k_2} - v_{i_2,k_2'} - \ldots - v_{i_l,k_l}, v_{i_l,k_l'}$, where $k_i \neq k_i'$ for all $i = 1, \ldots, l$. This corresponds to the sequence of paths $P_{i_1}, \ldots, P_{i_l}$ of $\mathcal{P}$. Since $\mathcal{P}$ is inclusion-free, these paths constitute an NSI of $\mathcal{P}$. The case of the cycle is similar.

By coloring the paths of each such component with a different color, we get the desired coloring $w(M)$.                                                                      □

It is worth noticing that as an immediate consequence of lemmata 10, 11 and 12 the following lemma, providing another optimal algorithm for the case in which $G$ is a tree and $g = \infty$, holds.

The following lemma is an immediate consequence of lemmata 10, 11 and 12.

**Lemma 13.** *If $G$ is a tree, then algorithm $MaxMatch$ runs in polynomial time and provides an optimal solution for any instance $(G, \mathcal{P}, \infty)$.*

---

**Algorithm 4.** $MaxMatch(G, \mathcal{P}, \infty)$

---

1: Construct the weighted endpoint intersection graph $EIG(G, \mathcal{P})$ of $G$ and $\mathcal{P}$ with the weight function $f$.
2: Calculate the maximum weighted matching $MM$ of $EIG(G, \mathcal{P})$ with weights $f$.
3: Return $w(MM)$.

---

### 4.2   Algorithm for General Networks

Unfortunately, as shown in the following theorem, the problem for $(G, \mathcal{P}, \infty)$ instances, with $G$ being a general network, is $NP$-hard. Therefore, approximation algorithm for solving it has to be provided.

**Theorem 6.** *The problem for $(G, \mathcal{P}, \infty)$ instances, with $G$ being a general network, is NP-hard.*

*Proof.* In order to prove the $NP$-hardness, we provide a polynomial reduction from the $TRIPART$ problem, known to be $NP$-complete (see [2]).

An instance of the $TRIPART$ problem is a simple graph $G' = (V'_{G'}, E'_{G'})$. The question is whether or not there is a partition of $E'_{G'}$ into triangles. Let $V'_{G'} = \{v'^1_{G'}, v'^2_{G'}, \ldots, v'^{n'}_{G'}\}$ and $E'_{G'} = \{e'^1_{G'}, e'^2_{G'}, \ldots, e'^{3q}_{G'}\}$ (note that if $|E'_{G'}|$ is not a multiple of 3, a partition does not exist and the answer is obviously NO).

From the above instance $G' = (V'_{G'}, E'_{G'})$ of $TRIPART$ we build the following instance $(G, \mathcal{P}, \infty)$ of the Regenerators Grooming Problem. $G = (V_1 \cup V_2, E_1 \cup E_2 \cup E_3 \cup E_4 \cup E_5)$, where $V_1 = \{a_i, b_i, c_i | i = 1, \ldots, n'\}$, $V_2 = \{d_{j,k}, e_{j,k}, f_{j,k} | j = 1, \ldots, 3q \wedge k = 1, \ldots, 3q+1\}$ and the edge sets are defined as follows. $E_1 = \{(a_i, b_i), (b_i, c_i) | i = 1, \ldots, n'\}$, $E_2 = \{(d_{j,k}, e_{j,k}), (e_{j,k}, f_{j,k}) | j = 1, \ldots, 3q \wedge k = 1, \ldots, 3q + 1\}$ and $E_3 = \{(f_{j,k}, d_{j,k+1}) | j, k = 1, \ldots, 3q\}$; moreover, for each edge $e'^j \in E'_{G'}$, connecting nodes $v'^i_{G'}$ and $v'^{i'}_{G'}$ ($i < i'$) we add to $E_4$ edges $(a_i, d_{j,1})$, $(c_i, d_{j,1})$, $(a_{i'}, f_{j,3q+1})$ and $(c_{i'}, f_{j,3q+1})$. Finally, $E_5 = \{(d_{j,k}, e_{j',k}), (f_{j',k}, d_{j,k+1}) | j, k = 1, \ldots, 3q \wedge 1 \leq j' < j\}$.

Now we are ready to define the paths of the instance. For each edge $e'^j \in E'_{G'}$, connecting nodes $v'^i_{G'}$ and $v'^{i'}_{G'}$ ($i < i'$), we add the following couple of paths (see Figure 5 in [9]): the *top* path $[a_i, b_i, c_i, d_{j,1}, e_{g(j,1),1}, f_{g(j,1),1}, d_{j,2}, e_{g(j,2),2}, f_{g(j,2),2}, \ldots, d_{j,3q}, e_{g(j,3q),3q}, f_{g(j,3q),3q}, d_{j,3q+1}, e_{j,3q+1}, f_{j,3q+1}, a_{i'}, b_{i'}, c_{i'}]$ and the *bottom* path $[c_i, b_i, a_i, d_{j,1}, e_{g(j,1),1}, f_{g(j,1),1}, d_{j,2}, e_{g(j,2),2}, f_{g(j,2),2}, \ldots, d_{j,3q}, e_{g(j,3q),3q}, f_{g(j,3q),3q}, d_{j,3q+1}, e_{j,3q+1}, f_{j,3q+1}, c_{i'}, b_{i'}, a_{i'}]$, where $g(j,k)$ is $j$ if edges $e'^j$ and $e'^k$ are not consecutive in $G'$, and is the minimum between $j$ and $k$ otherwise.

Notice that the top path and the bottom path relative to each edge $e'^j \in E'_{G'}$ cannot be put in a same NSI since otherwise nodes $d_{j,1}$ and $f_{j,3q+1}$ would have degree 3 (*Property 1*).

Moreover, any two (bottom or top) paths relative to non-consecutive edges $e'^j$ and $e'^k$ ($j < k$) of $G'$ (overlapping on edge $(e_{j,k}, f_{j,k})$), cannot be put in a same NSI since otherwise nodes $e_{j,k}$ and $f_{j,k}$ would have degree 3 (*Property 2*).

Finally, it can be easily verified that the only nodes in which it is possible to save regenerators are the $b$ nodes of $V_1$ (*Property 3*).

In order to prove the claim, it is sufficient to prove *(i)* that if the answer to the $TRIPART$ problem is YES, then there exists a solution of the constructed $(G, \mathcal{P}, \infty)$ instance in which it is possible to save $6q$ regenerators and, conversely, *(ii)* that if it is possible to save $6q$ regenerators in the constructed $(G, \mathcal{P}, \infty)$ instance, then the answer to the $TRIPART$ problem is YES.

In order to prove *(i)*, it is sufficient to notice that a triangle in $G'$ with vertices $v_{G'}^{\prime i}$, $v_{G'}^{\prime i'}$ and $v_{G'}^{\prime i''}$ ($i < i' < i''$) induces 6 paths in $\mathcal{P}$ that can be rearranged in 2 NSIs as follows (see Figure 6 in [9]): the top paths corresponding to edges $(v_{G'}^{\prime i}, v_{G'}^{\prime i'})$ and $(v_{G'}^{\prime i'}, v_{G'}^{\prime i''})$ and the bottom path corresponding to edge $(v_{G'}^{\prime i}, v_{G'}^{\prime i''})$ belong to an NSI, while the the bottom paths corresponding to edges $(v_{G'}^{\prime i}, v_{G'}^{\prime i'})$ and $(v_{G'}^{\prime i'}, v_{G'}^{\prime i''})$ and the top path corresponding to edge $(v_{G'}^{\prime i}, v_{G'}^{\prime i''})$ belong to the other NSI. Therefore, in such paths 6 regenerators (one per path, at nodes $b_i$, $b_{i'}$, $b_{i''}$) are saved. Since when the $TRIPART$ problem is YES $E'$ can be partitioned in $q$ triangles, $6q$ regenerators are saved in total.

It remains to prove *(ii)*. First of all, Property 3 ensures that regenerators can be only saved at $b$ nodes. By Property 2, only paths corresponding to edges of $G'$ sharing a node can be put in a same NSI, and moreover, by Property 1, the two paths corresponding to a same edge cannot be put in a same NSI. Therefore, regenerators can be saved only by putting in a same NSI the 2 paths corresponding to consecutive edges of $G'$ or the 3 paths corresponding to a triangle in $G'$. In the first case, a regenerator is saved ($\frac{1}{2}$ regenerator per path), whereas in the second case 3 regenerators are saved (1 regenerator per path). Since in $\mathcal{P}$ there are $6q$ paths, if it is possible to save $6q$ regenerators, then all the savings have to be due to $2q$ NSIs each containing 3 paths and in which 1 regenerator per path is saved; therefore, since at most 2 different NSIs correspond to a same triangle of $G'$, $q$ triangles have to be in $G'$ and the claim follows.   □

The following lemma provides an approximation algorithm for the $(G, \mathcal{P}, \infty)$ problem in general networks.

**Lemma 14.** *For every matching $M$ of $EIG(G, \mathcal{P})$ we can find in polynomial time a matching $\overline{M} \subseteq M$ such that $f(\overline{M}) \geq f(M)/2$ and there is a solution $w(\overline{M})$ of the $(G, \mathcal{P}, \infty)$ instance.*

*Proof.* We start as in the proof of Lemma 12. Consider a path or cycle of $M \cup M'$. Let $\{e_1, e_2, ...\}$ be the edges of $M$ in this path. We obtain a matching $M_o \subseteq M$ (resp. $M_e$) by removing the edges with odd (resp. even) indices in this paths. This breaks the paths into sub-paths of length three, in other words into paths containing exactly one edge of $M_o$ (resp. $M_e$), which in turn corresponds to a sequence of two paths $P_{i_1}, P_{i_2}$ of $\mathcal{P}$. These paths constitute an NSI of $\mathcal{P}$. Clearly $\max\{f(M_o), f(M_e)\} \geq f(M)/2$, thus either $M_o$ or $M_e$ is the claimed matching $\overline{M}$.   □

The following lemma, relates the approximation ratio of a solution with respect to the maximum matching problem to the one of the corresponding solution for our problem.

---

**Algorithm 5.** $MatchAndCut(G, \mathcal{P}, \infty)$

---
1: Construct the weighted endpoint intersection graph $EIG(G, \mathcal{P})$ of $G$ and $\mathcal{P}$ with the weight function $f$.
2: Calculate the maximum weighted matching $MM$ of $EIG(G, \mathcal{P})$ with weights $f$.
3: Calculate the matching $\overline{MM}$ of $EIG(G, \mathcal{P})$ as described in proof of Lemma 14.
4: Return $w(\overline{MM})$.

---

**Lemma 15.** *If a matching $M$ is a $\rho$-approximation to the maximum matching of $EIG(G, \mathcal{P})$ for some $\rho \geq 1$ and $w(M)$ exists, then $w(M)$ is a $(1/\rho + (1 - 1/\rho) g)$-approximation for the $(G, \mathcal{P}, g)$ instance.*

**Lemma 16.** *Algorithm $MatchAndCut$ runs in polynomial time and constitutes a $\left( \frac{1 + load(\mathcal{P})}{2} \right)$-approximation for any $(G, \mathcal{P}, \infty)$ instance.*

*Proof.* Clearly $g = \infty$ is equivalent to $g = load(\mathcal{P})$. By Lemma 14 $\overline{MM}$ is a 2-approximation to the maximum matching of $EIG(G, \mathcal{P})$. Substituting $\rho = 2$ and $g = load(\mathcal{P})$ in Lemma 15 we get $\rho' = (1 + load(\mathcal{P}))/2$ as the approximation ratio of $MatchAndCut$. □

Combining Lemma 16 with Lemma 8, we finally obtain the following theorem.

**Theorem 7.** *$Combined(MatchAndCut, (G, \mathcal{P}, g))$ is a $\left( \frac{7 + load(\mathcal{P})}{2} \right)$-approximation algorithm for any $(G, \mathcal{P}, g)$ instance.*

# 5    Conclusion and Future Work

In this paper we have studied an optimization problem in Optical Networks, that minimizes the use of regenerators when traffic grooming is exploited. We have considered the case in which a regenerator has to be placed at every internal node of every lightpath, and at most $g$ lightpaths can use the same regenerator. Starting from the 4-approximation algorithm of [7] that solves a closely related problem for a path topology, we have shown that it has the same approximation ratio for the ring topology. We have presented a greedy 4-approximation algorithm for tree networks that uses the mentioned algorithm as a subroutine. We have also introduced a new technique using matchings that can be used to obtain approximation algorithms for other topologies, and have used this technique for general topology to get an $\left( \frac{7 + load(\mathcal{P})}{2} \right)$-approximation.

A natural open problem is to discover the exact approximability of the problem. The problem is NP-complete already for $g = 2$ and networks with path topology. In this paper we have shown that the problem is in APX in tree networks. Determining whether the problem is in PTAS for these topologies and for particular cases is an open problem.

It would be also interesting to extend our result by considering $d > 1$, i.e. the case that regenerators do not have to be present at every node, or more involved cost functions taking into account other switching parameters (e.g., the ADMs - Add-Drop-Multiplexers - used at the endpoints of the lightpath). Finally, studying the on-line version of the problem is an intriguing future research direction.

# References

1. Chen, S., Ljubic, I., Raghavan, S.: The regenerator location problem. Networks 55(3), 205–220 (2010)
2. Dor, D., Tarsi, M.: Graph decomposition is np-complete: A complete proof of holyer's conjecture. SIAM Journal on Computing 26(4), 1166–1187 (1997)
3. Edmonds, J.: Paths, trees, and flowers. Canad. J. Math. 17, 449–467 (1965)
4. Fedrizzi, R., Galimberti, G.M., Gerstel, O., Martinelli, G., Salvadori, E., Saradhi, C.V., Tanzi, A., Zanardi, A.: A Framework for Regenerator Site Selection Based on Multiple Paths. In: Prooceedings of IEEE/OSA Conference on Optical Fiber Communications (OFC) (to appear, 2010)
5. Fedrizzi, R., Galimberti, G.M., Gerstel, O., Martinelli, G., Salvadori, E., Saradhi, C.V., Tanzi, A., Zanardi, A.: Traffic Independent Heuristics for Regenerator Site Selection for Providing Any-to-Any Optical Connectivity. In: Proceedings of IEEE/OSA Conference on Optical Fiber Communications (OFC) (to appear, 2010)
6. Flammini, M., Marchetti-Spaccamela, A., Monaco, G., Moscardelli, L., Zaks, S.: On the complexity of the regenerator placement problem in optical networks. IEEE/ACM Transactions on Networking (to appear, 2010)
7. Flammini, M., Monaco, G., Moscardelli, L., Shachnai, H., Shalom, M., Tamir, T., Zaks, S.: Minimizing total busy time in parallel scheduling with application to optical networks. Theoretical Computer Science 411(40-42), 3553–3562 (2010)
8. Flammini, M., Monaco, G., Moscardelli, L., Shalom, M., Zaks, S.: Approximating the traffic grooming problem with respect to adms and oadms. In: Luque, E., Margalef, T., Benítez, D. (eds.) Euro-Par 2008. LNCS, vol. 5168, pp. 920–929. Springer, Heidelberg (2008)
9. Flammini, M., Monaco, G., Moscardelli, L., Shalom, M., Zaks, S.: Optimizing regenerator cost in traffic grooming. Technical report, Faculty of Computer Science, Technion (September 2010), http://www.cs.technion.ac.il/users/wwwb/cgi-bin/tr-info.cgi/2010/CS/CS-2010-16
10. Kim, S.W., Seo, S.W.: Regenerator placement algorithms for connection establishment in all-optical networks. IEE Proceedings Communications 148(1), 25–30 (2001)
11. Mertzios, G.B., Sau, I., Shalom, M., Zaks, S.: Placing regenerator in optical networks: New model, hardness results and algorithms. In: Gavoille, C. (ed.) ICALP 2010, Part II. LNCS, vol. 6199, pp. 333–344. Springer, Heidelberg (2010)
12. Pachnicke, S., Paschenda, T., Krummrich, P.M.: Physical Impairment Based Regenerator Placement and Routing in Translucent Optical Networks. In: Optical Fiber Communication Conference and Exposition and The National Fiber Optic Engineers Conference (Optical Society of America, paper OWA2) (2008)
13. Sriram, K., Griffith, D., Su, R., Golmie, N.: Static vs. dynamic regenerator assignment in optical switches: models and cost trade-offs. In: Workshop on High Performance Switching and Routing (HPSR), pp. 151–155 (2004)
14. Winkler, P., Zhang, L.: Wavelength assignment and generalized interval graph coloring. In: SODA, pp. 830–831 (2003)
15. Yang, X., Ramamurthy, B.: Dynamic routing in translucent WDM optical networks. In: Proceedings of the IEEE International Conference on Communications (ICC), pp. 955–971 (2002)
16. Yang, X., Ramamurthy, B.: Sparse Regeneration in Translucent Wavelength-Routed Optical Networks: Architecture, Network Design and Wavelength Routing. Photonic Network Communications 10(1), 39–53 (2005)

# On Minimizing Average End-to-End Delay in P2P Live Streaming Systems

Fei Huang[1], Maleq Khan[2], and Binoy Ravindran[1]

[1] Dept. of Electrical & Computer Engineering
[2] Network Dynamics and Simulation Science Laboratory
Virginia Tech, Blacksburg, VA 24061, USA
{huangf,binoy}@vt.edu, maleq@vbi.vt.edu

**Abstract.** In this paper, we devise a streaming scheme, called *iStream*, to achieve the minimum average end-to-end P2P streaming delay by optimally allocating the bandwidth resource among peers. We first develop a generic analytical framework to model the minimum average delay P2P streaming problem, called the MADPS problem. We then present iStream to solve the MADPS problem. The core part of iStream is a fast approximation algorithm, called *iStream-APX*, based on primal-dual schema. We prove that the performance of iStream-APX is bounded by a ratio of $1 + \omega$, where $\omega$ is an adjustable input parameter. Furthermore, we show that the flexibility of $\omega$ provides a trade-off between the approximation factor and the running time of iStream.

## 1 Introduction

In the recent decade, P2P live media streaming applications have exhibited growing popularity, such as IPTV, VOIP, and video conferencing. By enabling efficient cooperation among end-users, P2P live streaming can distribute thousands of channels to millions of viewers simultaneously [1]. In these classes of applications, the delivery of real-time video content imposes rigorous constraints on the end-to-end delay. Obtaining assurances on meeting such delay constraints is a challenging problem, especially in highly dynamic and heterogeneous P2P network environments. The long playback latency has negatively affected the extensive commercial deployment of P2P systems. For example, IPTV deployment from commercial service providers is far below the industry expectation [2]. Motivated by these, in this paper, we focus on minimizing average end-to-end streaming delay in P2P networks.

Recently, layered coding has emerged as a viable solution for delivering real-time streaming content [3]. This technique not only provides an adaptive support for different downloading capacities on peers, but also allows IPTV service providers to deliver live content at diverse video definitions from the same coding process. For example, viewers may pay general fees for a standard service, or extra fees for 1080HD video or even 3D video. Unlike traditional IPTV service where viewers only download the multimedia content, under the P2P paradigm, substantial bandwidth may exist in viewers who pays only for a standard service,

C. Lu, T. Masuzawa, and M. Mosbah (Eds.): OPODIS 2010, LNCS 6490, pp. 459–474, 2010.
© Springer-Verlag Berlin Heidelberg 2010

while HD viewers may instead suffer bad streaming service due to the bandwidth deficit among them. To maximize the bandwidth utilization, we should enable peer cooperation among viewers of different service qualities. Toward that, the HD content can be forwarded through peers with standard service, but only the HD viewers receive the authorization key for viewing HD content. This raises a fundamental question: how to optimally distribute the video content and conduct sub-stream scheduling among peers with diverse service qualities, while achieving the minimum average end-to-end P2P streaming (or MADPS) delay. We call this problem, the MADPS problem.

Minimizing streaming delays for P2P live systems is not a trivial problem. This is due to the heterogeneous bandwidth requirements and network dynamics of P2P systems. Previous theoretical works on designing P2P live streaming usually assume a homogeneous service quality [4, 5]. Thus, obtaining optimal solutions to this problem for large-scale networks is expensive in terms of algorithmic computational costs [6]. Approximate or heuristic solutions with scalable costs are therefore highly desirable. In this paper, we focus on approximate algorithms because we target time-critical P2P applications (e.g., video conferencing, or cloud computing), for which assured bounds on end-to-end delays are more desirable than heuristic (or empirically-established) gains in end-to-end delays. In addition, the analytical foundation that is necessary for developing approximate algorithms can contribute to a greater understanding of the problem and can provide deeper insights on designing efficient algorithms, be they approximate or heuristic. We take the first such steps toward this. The paper is theory-oriented.

For a feasible solution, we start with the assumption of a static network—i.e., no churn. In this way, we can devise a framework which is analytically achievable. The method will be most suitable for the scenario where a service provider deploys a set-top box at viewers' homes. In that case, even when a viewer turns off the TV, the set-top box can still contribute its bandwidth to other viewers. For this scenario, we first develop an analytical model that formulates the MADPS problem as an optimization problem. Then we propose an algorithm called *iStream* to solve MADPS problem. Inspired by the primal-dual schema, we develop an approximation algorithm as the core of iStream, called *iStream-APX* for optimally utilizing the bandwidth among peers subscribing to different video qualities, while achieving the minimum average streaming delay. We show that iStream-APX's performance in terms of delay is bounded by a factor of $1 + \omega$, where $\omega$ is an input parameter. iStream's running time is also bounded. We show that there exists a trade-off between iStream-APX's approximation factor $\omega$ and its running time. The approximation factor is adjustable in the range of $(1, n]$, where $n$ is the number of peers in the network. This trade-off allows users to flexibly tune the performance bound according to running time requirements.

Thus, the paper's contribution is an approximation algorithm for the MADPS problem with bounded performance and running time (which can be traded-off, one for gains in the other), and its adaptive distributed version to operate in high-churn networks. iStream is the first approximation-based solution for the

MADPS problem, and we are not aware of any other past efforts on approximating the MADPS problem.

The rest of the paper is organized as follows. Section 2 overviews past and related works. In Section 3, we describe our network model and formulate the MADPS problem. Section 4 presents our proposed approximation algorithm and derives its performance. Section 5 concludes the paper.

## 2  Related Work

Theoretical works on the minimum delay P2P streaming problem are limited, though recently a growing number of studies have focused on P2P live streaming [4–9]. Due to the lack of formal theoretical bounds, intuitions and heuristics have driven the design of P2P schemes so far [4, 7]. For example, Ren *et al.* [4] propose a heuristic to reduce the delay on mesh topology, where peers select their parents based on the metric of link capacity divided by communication delay. In this algorithm, peers located at the edge of mesh may only download the data without uploading, which may lead to low bandwidth utilization in P2P networks. Thus, when the total uploading capacity is close to the downloading capacity in the P2P community, some peers may not be able to receive a live streaming.

In our previous work [5], we developed an approximation algorithm to minimize the maximum P2P streaming delay by clustering and filtering methods with an approximation bound of $O(\sqrt{\log n})$. The minimum delay P2P streaming problem (or MDPS) presented in [5] focuses on minimizing the maximum end-to-end streaming delay. The MDPS problem is significantly different from the problem of minimizing the average end-to-end delay problem which we focus here. For example, the simulation results in [5] show that minimizing the maximum delay does not necessarily minimize the average end-to-end delay. Furthermore, the work in [5] assumes a network model with a symmetric graph and satisfying the triangle inequality. In contrast, in this paper, we remove those assumptions in modeling the minimum average delay P2P streaming problem.

The MADPS problem that we focus has some similarity with the minimum-cost multi-commodity flow problem (or MCMF) [10, 11]. iStream is inspired by the primal-dual schema from Garg and Konemann [10]. However, previous approximation solutions to the MCMF problem generally assume flow conservation on nodes—i.e., incoming commodities and outgoing commodities are exactly equal in amount. This is not true in P2P streaming, where peers can reproduce whatever commodities they receive—i.e., flow conservation does not hold. In addition, the MCMF problem considers only the capacities on edges, whereas in P2P streaming, the capacities actually exist on nodes instead of edges. This distinction (for the MADPS problem) further requires optimal flow scheduling among edges departing from the same node. All these differences make the MADPS problem more complex than the MCMF problem. Our work tackles these complexities and achieves a solution with near-optimal performance bound.

# 3   Problem Formulation

In this section, we formally state the minimum average end-to-end delay P2P streaming (MADPS) problem and present the problem in linear programming (LP) framework.

## 3.1   Preliminaries and Modeling

We model an overlay network as a directed graph $G = (V, E)$, where $V$ is the set of vertices representing peer nodes, and $E$ is the set of overlay edges representing directed overlay links. Let $n$ represent the number of peers in the network, i.e. $n = |V|$. Each overlay link $(i, j) \in E$ is associated with a communication delay $l_{ij}$. In the rest of this paper, we define the length of edge $(i, j)$ as $l_{ij}$, $\forall (i, j) \in E$. For every peer $i \in V$, we define an upload capacity of $C_i$ units/second and a download capacity of $I_i$ units/second. For ease of presentation, we define *unit* as the minimum flow size in P2P streaming, which may vary in different applications [12, 13].

We consider a peer-to-peer streaming session to originate from a single source node $S$ to a set of receivers $R$, where $V = \{S\} \cup R$. Peers may receive the streaming data from the source node directly or indirectly from multiple P2P paths. In practical applications, receivers may pay for services of different streaming qualities, e.g., 720i/p and 1080i/p, which leads to different streaming rates correspondingly. Suppose peer $j$ selects a service that has a constant streaming rate of $d_j$ units/second. We denote $f_{ij}$ as the rate at which peer $i$ streams to peer $j$. If peer $j$ receives the aggregated non-identical streams at $d_j$ units/second from its parents, we call peer $j$ as *fully served* [4]. Mathematically, the fully served requirement of peer $j$ can be expressed as $\sum_{i:i \in L_j} f_{ij} = d_j$, where $L_j$ is the set of parents of peer $j$. We assume that a fully served peer can smoothly play back the streaming content at its original rate of $d_j$ units/second [4].

We call the stream from the source to one receiver $j$ as the *P2P unicast flow* to $j$. Each P2P unicast flow $U_j$ may consist of streams from multiple P2P paths, called *fractional flows* [6]. Each fractional flow $p \in U_j$ has the arrival latency $l(p)$ from the source to receiver, i.e., *end-to-end delay*, where $l(p) = \sum_{(i,j) \in p} l_{ij}$. We define the average end-to-end delay of the unicast flow $U_j$ as the weighted average of end-to-end latencies of all its fractional flows, where the weight is the portion of fractional flow rate to the total streaming rate. Denote $f(p)$ as the streaming rate of fractional flow $p$. For viewer $j$, the weighted average of end-to-end latencies can be expressed by

$$\frac{1}{d_j} \sum_{p \in U_j} l(p) f(p).$$

To stream multimedia content to multiple receivers, we can envision multiple unicast flows from the source to receivers. Thus, the *average end-to-end delay in*

*P2P streaming* is defined as the weighted average latency of all fractional flows to all receivers, which can be described by

$$\frac{1}{\sum_{j \in R} d_j} \sum_{p \in P} l(p) f(p), \tag{1}$$

where $P = \bigcup_{j \in R} U_j$. Since the term $\sum_{j \in R} d_j$ has no effect on the optimal solution, i.e., the solution that minimizes (1) also minimizes $\sum_{p \in P} l(p) f(p)$, we will focus on minimizing $\sum_{p \in P} l(p) f(p)$. It is easy see that removal of the term $\sum_{j \in R} d_j$ also preserves the approximation factor. For ease of presentation, we simply refer to $\sum_{p \in P} l(p) f(p)$ as *the cumulative delay* in the later sections.

To help understand the concept of average end-to-end delay, we use the term: "envision" in the above paragraph. In reality, there exists only one stream through each edge $(i, j)$ instead of multiple fractional flows and peer $j$ can reproduce any part of the stream content it receives and send it to other peers. Therefore, the actual data rate on an edge $(i, j)$ is $\max_{t \in R} \sum_{p \in P_{ij}^t} f(p)$, where $P_{ij}^t$ is the set of fractional flows through edge $(i, j)$ to receiver $t$.

Next we provide a formal description of the problem.

## 3.2  MADPS Problem

**Definition 1.** *Minimum Average End-to-End Delay P2P Streaming Problem (MADPS problem): Given the capacity and data rate constraints that are mentioned in this section, the MADPS problem is to devise a streaming scheme which minimizes the maximum average end-to-end streaming delay with all receivers fully served.*

There is no known efficient algorithm with a practically-feasible running time to solve this problem optimally. Therefore, we are motivated to develop a near-optimal approximation algorithm with significantly smaller running time.

To ensure a solution exists to the MADPS problem, it is reasonable to assume the total bandwidth resources in P2P networks is sufficient to support the full services on all the viewers. Hence, we deduct the bandwidth requirement in Corollary 1.

**Corollary 1.** *If the instance of MADPS problem has a solution, then the sum of the upload capacities, including source and receivers, must be no less than the sum of fully served streaming rates at all receivers, i.e.,*

$$\sum_{i \in V} C_i \geq \sum_{j \in R} d_j. \tag{2}$$

In addition, we presume that the download capacity $I_i \geq d_i, \forall i \in V$ for a smooth playback at the receiver.

## 4   Approximation Algorithm

In this section we devise an approximation algorithm to find the near-optimal solution with provable bounds on the worst-case performance and running time.

## 4.1   Overview of Techniques

There are two fundamental techniques used in this work, including *primal-dual schema* and *binary search* based on the result of primal-dual schema.

First, we describe *primal-dual schema* [10, 14]. Given a linear programming problem, also referred to as a *primal problem*, we can convert it to a *dual problem*. Due to space limitation, we do not present the detailed mechanics of this conversion here, which can be found at [14]. Primal and dual problems are in a "mirror" relation. If one problem is a maximization problem, the other problem is a minimization problem, and vise versa. Suppose we have a primal problem: $\max \mathbf{c}^{\mathrm{T}}\mathbf{x}$, and the corresponding dual problem:$\min \mathbf{b}^{\mathrm{T}}\mathbf{y}$. According to the weak duality theorem, if $\mathbf{X}$ and $\mathbf{Y}$ are feasible solutions for the primal and dual problems respectively, it follows that $\mathbf{c}^{\mathrm{T}}\mathbf{X} \leq \mathbf{b}^{\mathrm{T}}\mathbf{Y}$. Moreover, the primal and dual problems share the same optimum, denoted by OPT. Given an approximation factor $\rho$, $\rho$ bounds $\frac{\mathrm{OPT}}{\mathbf{c}^{\mathrm{T}}\mathbf{x}}$. Since any feasible solution to the dual also provides an upper bound on OPT, the approximation factor can be established by comparing the primal and dual solutions. In light of this, the primal-dual schema starts with a feasible solution for dual problem and relax the conditions for primal problem. Then, iStream iteratively improves the feasibility of primal conditions and the optimality of the dual solution. iStream winds up with feasible solutions for both primal and dual problems. So, the gap between them makes the approximation factor.

In detail, iStream employs the primal-dual schema to solve the delay-bounded maximum streaming rate problem (DBMSR problem) defined as follows.

**Definition 2.** *Delay-bounded Maximum Streaming Rate problem (DBMSR problem): Given a bound $L$ on the average delay, i.e., $\sum_{p \in P} l(p)f(p) \leq L$, the DBMSR problem is to devise a streaming scheme which maximizes $\lambda$, where $\sum_{p \in P^t} f(p) \geq \lambda d_t, \forall t \in R$.*

In the next step, we can do a binary search on $L$ to find the smallest $\lambda$ that satisfies $\lambda \geq 1$. Towards that purpose, a reasonable initial value of $L$ should be set in the range of $[\sum_{j \in R} d_j \cdot \min_{p \in P} l(p), \sum_{j \in R} d_j \cdot \max_{p \in P} l(p)]$. The result of this procedure leads to a near-optimal solution for MADPS problem.

In the rest of this section, we formulate the DBMSR problem by primal-dual schema. Then, we discuss the details of iStream and derive its performance bound.

## 4.2   Formulation about Primal and Dual

We refer to DBMSR problem as the primal problem here, or simply called primal. According to its definition, we formulate the primal as following.

**Primal**:

$$\max \lambda \qquad (3)$$

subject to

$$\sum_{p \in P_{ij}^t} f(p) \leq \sum_{p \in P_{ij}^j} f(p), \qquad \forall (i,j) \in E, \forall t \in R \qquad (4)$$

$$\sum_{j:(i,j)\in E} \sum_{p\in P_{ij}^j} f(p) \le C_i, \qquad\qquad \forall i \in V \qquad (5)$$

$$\sum_{p\in P^t} f(p) \ge \lambda d_t, \qquad\qquad \forall t \in R \qquad (6)$$

$$\sum_{p\in P} l(p)f(p) \le L, \qquad\qquad (7)$$

$$f(p) \ge 0, \lambda \ge 0. \qquad\qquad (8)$$

Equation (4) presents the fact that the amount of fractional flow through edge $(i,j)$ to any viewer will always be bounded by the total fractional flow sent to node $j$, i.e. $\sum_{p\in P_{ij}^j} f(p) = \max_{t\in R} \sum_{p\in P_{ij}^t} f(p)$. Because we attempt to utilize the bandwidth from peers scribing to the standard video quality, it is possible to see the amount of fractional flow to $j$ from all incoming edges of $j$ exceeds viewer $j$'s demand, i.e., $\sum_{i:(i,j)\in E} \sum_{p\in P_{ij}^j} f(p) \ge d_j$. Equation (5) ensures no conflicts in terms of the uploading capacities. In terms of the downloading capacities, which can be written as $\sum_{j:(j,i)\in E} \sum_{p\in P_{ji}^i} f(p) \le I_i, \forall i \in V$, we assume $I_i \ge \max_{j\in R} d_j$, which is practical with the wide deployment of high-speed internet. Since the actual flow sent to or relayed by node $i$ cannot be larger than the maximum service demand, expressed by $\max_{j\in R} d_j$, it is reasonable to remove the constraints on the downloading capacities in the LP expression without affecting the optimal solutions. Equation (6) means the objective of DBMSR problem is to maximize the minimum demand on nodes. Equation (7) puts a bound $L$ on the cumulative delay. As stated in Section 4.1, we can conduct a binary search on $L$ until $\lambda$ is very close to 1 to achieve a solution to the MADPS problem.

Next, we convert the primal to its dual problem, or simply called dual.

**Dual:**

$$\min \sum_{i\in V} C_i w_i + \varphi L \qquad\qquad (9)$$

subject to

$$\sum_t d_t z_t \ge 1, \qquad\qquad \forall t \in R \qquad (10)$$

$$\sum_{(i,j)\in p, i\ne i'} s_{ij}^t + w_{i'} + \varphi l(p) \ge z_t, \qquad (i',t)\in p, \forall t \in R,$$

$$\forall p \in P^t \qquad (11)$$

$$s_{ij}^t \ge 0, w_i \ge 0, z_t \ge 0, \varphi \ge 0, \qquad\qquad (12)$$

where $i'$ is the peer one hop away from the viewer $t$ on routed path.

Generally, there is no direct physical meaning to the dual problem because it comes from a mechanical conversion of the primal problem. To help the analysis on iStream, we hereby assign a logical explanation to the dual after investigating its formulation. We envision each edge $(i,j)$ has multiple copies

$(i,j)^1, (i,j)^2, \cdots, (i,j)^{|R|}$, where any copy $(i,j)^t$ exclusively represents to the usage of edge $(i,j)$ for flows to viewer $t$. Each edge $(i,j)^t$ is associated with a length metric $s_{ij}^t$, and each node $i$ is associated with a length metric $w_i$. Thus, we view $\sum_{(i,j)\in p, i\neq i'} s_{ij}^t + w_{i'} + \varphi l(p)$ as the length function associated with flow path $p$, where $\varphi$ is the weight associated with the delay metric $l(p)$. According to Equation (11), $z_t$ can be comprehended as the shortest length to node $t$ based on the length function.

## 4.3   Approximation Algorithm

iStream-APX is the core part of iStream, which is built with approximation algorithm. iStream-APX proceeds in phases. Each phase is completed by $|R|$ iterations with each iteration satisfy the demand of one viewer. Due to the constraints from LP conditions, each iteration may be completed by multiple steps. Inside each step, we route such amount of fractional flows that can ensure the constraints are not violated. At the end of all phases, iStream-APX will re-scale all the flows to ensure a feasible solution to the primal. We express the $k^{\text{th}}$ step in the $t^{\text{th}}$ iteration of $m^{\text{th}}$ phase by $(m,t,k)$. The initial status is marked by $(0,0,0)$, or simply $(0)$.

We start the algorithm with the following initial settings on length metrics.

$$w_i(0) = \delta/C_i, \qquad\qquad\qquad \forall i \in V \qquad\qquad (13)$$

$$s_{ij}^t(0) = w_i, \qquad\qquad\qquad \forall (i,j) \in E, \forall t \in R \qquad (14)$$

$$\varphi(0) = \delta/L, \qquad\qquad\qquad\qquad\qquad\qquad (15)$$

where $\delta$ is an input parameter. The proper assignment of it will be discussed in Section 4.4.

Throughout the execution of algorithm iStream-APX, it dynamically updates the length metrics, which are used to built the flowing path. Let $w_i(m,t,k)$, $s_{ij}^t(m,t,k), \varphi(m,t,k)$ be the length metrics at the end of step $(m,t,k)$. At step $(m,t,k)$, iStream-APX first computes the shortest path $p^*$ from $S$ to viewer $t$ in terms of the length function $\sum_{(i,j)\in p, i\neq i'} s_{ij}^t(m,t,k-1) + w_{i'}(m,t,k-1) + \varphi(m,t,k-1)l(p)$, where $(i',t) \in p, p \in P^t$. Then, it finds the minimum capacity $C_{\min}$ on nodes along the shortest path, which can be expressed by $C_{\min} = \min_{i\in p^*}\{C_i\}$. Since the previous steps may already route some flows to the viewer, let $\gamma_t$ be the residual amount of demands unsatisfied on node $t$, and $x(p) = \min\{\gamma_t, C_{\min}\}$. Next, we route $x(p)/\eta$ amount of flow to $t$, where $\eta = l(p)x(p)/L$ if $l(p)x(p) > L$; otherwise, $\eta = 1$. So the length bound $L$ and the capacities on the path are not violated in each step. At the end of this step, we update the length metrics as well as the residual demands according to Equations (16)-(19).

$$w_i(m,t,k) = w_i(m,t,k-1) \cdot [1 + \epsilon \cdot f(m,t,k)/C_i],$$
$$\forall i \in p^* \setminus \{t\} \qquad (16)$$

$$s_{ij}^t(m,t,k) = w_i(m,t,k), \forall i \in p^* \setminus \{t\}, \forall (i,j) \in E, \forall t \in R \qquad (17)$$

$$\varphi(m,t,k) = \varphi(m,t,k-1) \cdot \prod_{j \in p^* \cap R} [1 + \epsilon \cdot L_j(m,t,k)/L], \qquad (18)$$

$$\gamma_i(m,t,k) = \gamma_i(m,t,k-1) - f(m,t,k), \forall i \in p^* \setminus \{t\} \qquad (19)$$

where $f(m,t,k)$ is the amount of flow routed in current procedure $(m,t,k)$ and $L_j(m,t,k)$ means the cumulative delay of the routed flow through node $j$ which is on the path $p^*$ at step $(m,t,k)$. Mathematically, it can be expressed by $L_j(m,t,k) = l(p_j^*)x(p_j^*)$, where $p_j^*$ is the segmental path from $S$ to $j$ on path $p*$. We can observe in each step for every capacity-saturated node $i$ on the routing path, all the length metrics regarding $i$ increase by a factor of $1 + \epsilon$. Since the assignments of $s_{ij}^t$ are identical in Equation (17), we simply use $s_i$ to represent all $s_{ij}^t$.

We repeat the steps until the demand of viewer $t$ is fully satisfied. Then we call the end of iteration $t$, and start the iteration for next viewers which has positive residual demand in the current phase. After the last step of a phase, all viewers have no residual demands, i.e., $\gamma_t = 0, \forall t \in R$. Then, we start a new round of phase $m + 1$ after resetting the residual demands equal to viewer's actual demands, i.e., $\gamma_t = d_t, \forall t \in R$. The whole procedure completes as soon as $W(m,t,k) \geq 1$. Obviously, the cumulative flows routed in all phases may strongly violate the capacity and average delay constraints. Define $F(p)$ as the cumulative flows routed in all phases through path $p$. To obtain a feasible solution to the primal problem, we need to scale down each $F(p)$ by a factor of $\log_{1+\epsilon} 1/\delta$. We will justify the correctness of this scaling down factor in Section 4.4.

We continue a binary search on $L$ by repeating iStream-APX until $\lambda$ tends to 1, denoted as $\lambda \rightarrow 1$. The result of the binary search will provide a near-optimal solution to MADPS problem. The detailed procedures about the approximation algorithm are presented in [15].

## 4.4  Algorithm Analysis

In this section, we formally analyze the algorithm and prove the approximation factor. To facilitate the analysis, we make some definitions. Let $W = \sum_{i \in V} C_i w_i + \varphi L$ be the metric minimized by the dual. Let $\zeta_t$ be the shortest length from $S$ to $t$, i.e.,

$$\zeta_t = \min_{p \in P^t} \sum_{(i,j) \in p, i \neq i'} s_{ij}^t + w_{i'} + \varphi l(p). \qquad (20)$$

Here $\zeta_t$ actually represents and interprets the meaning of $z_t$. Besides, we define

$$\alpha = \sum_t \left( d_t \zeta_t \right). \qquad (21)$$

**Lemma 1.** *Denote the optimal solution to the dual by OPT(W). When OPT(W) is obtained, $\alpha$ is 1.*

*Proof.* We prove this lemma by contradiction. As we know, $\alpha$ represents $\sum_t d_t z_t$ in the dual. Let $W = W'$ when $\alpha = 1$. For the sake of contradiction, we assume

$W' > \text{OPT}(W)$, where $\text{OPT}(W)$ is achieved when $\alpha = \alpha^* > 1$. Then, we scale down $\alpha^*$ to 1. Towards that, we can divide all the $s_{ij}^t$ and $\varphi$ by a factor of $\sum_t d_t z_t$. As a result, $w_i$ will proportionally scale down the same factor. Consequently, it leads to an update on $W$ with a new value $W'$, where $W' = \text{OPT}(W)/\sum_t d_t z_t$. According to the assumption, $W'$ should be larger than $\text{OPT}(W)$. However, because $\sum_t d_t z_t > 1$, we have $W' = \text{OPT}(W)/\sum_t d_t z_t < \text{OPT}(W)$, which contradicts the assumption. Thus, the lemma follows.

Define $\beta$ as the minimum value of $W/\alpha$, i.e., $\beta = \min W/\alpha$. We conclude the following theorem.

**Theorem 1.** *The optimal solution to the dual, denoted as $\text{OPT}(W)$, is equivalently to the optimal solution $\beta$ under the same constraints in the dual.*

*Proof.* From the definition of $\beta$, we know that $\beta = \min W/\alpha$. Suppose $\beta$ is achieved when $\alpha = \alpha^* > 1$. We can always proportionally scale down all the $s_{ij}^t$ and $\varphi$ by multiplying a factor of $1/\alpha^*$. As a result, $\alpha = 1$. Since $W$ will scale down with the same factor, $W/\alpha$ will keep the optimal value $\beta$. That is to say we can always find the optimal solution $\beta$ with $\alpha = 1$.

According to Lemma 1, it follows that $\alpha = 1$ when $\text{OPT}(W)$ is achieved. Therefore, we can conclude the problem of finding $\text{OPT}(W)$ for the dual is equivalently to solving the optimization problem for $W/\alpha$. This completes the proof.

In iStream-APX, we update the length metrics $s_i, w_i, \varphi$ on the routing path. In terms of that, we can conclude the following.

**Lemma 2.** *$w_i$ increases at least by a factor of $1 + \epsilon$ for every $C_i$ units of flow through node $i, \forall i \in V$.*

*Proof.* Due to space limitation, we do not prove this lemma here. The detailed proof can be found at [15].

**Corollary 2.** *$s_i$ increases at least by a factor of $1 + \epsilon$ for every $C_i$ units of flow through node $i, \forall i \in V$.*

**Corollary 3.**

$$\log_{1+\epsilon} \frac{\varphi(m)}{\varphi(0)} \geq \sum_{p \ in P} l(p) f(p)/L,$$

*where $f(p)$ represents the cumulative amount of flows through path $p$ at the end of phase $m$.*

Given the assumption that the total bandwidth resources in P2P networks is sufficient to support the full services on all the viewers, we can do a binary search on $L$ so as to find the smallest $\lambda$ that satisfies $\lambda \geq 1$. According to the weak-duality theorem, it follows that $\beta \geq \lambda \geq 1$.

**Lemma 3.** *Given $\beta \geq 1$, we have*

$$\beta \leq \frac{\epsilon(M-1)}{(1-\epsilon)\ln\frac{1-\epsilon}{(|V|+1)\delta}}.$$

*Proof.* We start the proof by analyzing the change on $W$ on each step. At the end of this analysis, we will carry out the cumulative increment on $W$ when algorithms stops.

Let $p(m,t,k)$ be the shortest path found at procedure $(m,t,k)$, and $f(m,t,k)$ be the quantity of flow routed through path $p(m,t,k)$. Because in our algorithm we assign $s_i = w_i$ for any procedure $(m,t,k)$, we can simplify the length function as

$$\sum_{(i,j)\in p, i\neq i'} s_i^t + w_{i'} + \varphi l(p) = \sum_{(i,j)\in p}(w_i + \varphi l_{ij}), \tag{22}$$

where $(i',t) \in p$. Consequently, we can carry out the following.

Since the objective is to find the cumulative increment, we can think of the change on length metrics $w_i$ and $\varphi$ regarding node $i$ at procedure $(m,t,k)$, where $i \neq t$, will hold until procedure $(m,i,0)$ without loss on the final cumulative increment on $W$.

$$W(m,t,k) - W(m,t,k-1)$$

$$= C_{i'} \cdot \Big(w_{i'}(m,t,k) - w_{i'}(m,t,k-1)\Big) +$$

$$+ \Big(\varphi(m,t,k) - \varphi(m,t,k-1)\Big) \cdot L$$

$$\leq \sum_{i\in p(m,t,k)\setminus\{t\}} \Big(C_i \cdot w_i(m,t,k-1)\epsilon f(m,t,k)/C_i\Big) +$$

$$+ \Big(\varphi(m,t,k-1)\epsilon L(m,t,k)/L\Big) \cdot L$$

$$= \epsilon \cdot \Big[ \sum_{i\in p(m,t,k)\setminus\{t\}} \Big(w_i(m,t,k-1)f(m,t,k)\Big) +$$

$$+ \varphi(m,t,k-1)L(m,t,k)\Big].$$

Let $K_{mt}$ be the number of steps in a given iteration $t$ of phase $m$, $\zeta_t(m,t,k)$ be the shortest path at the end of procedure $(m,t,k)$, and $l(m,t,k)$ be the cumulative latency on path $p(m,t,k)$. We have

$$W(m,t+1,0) - W(m,t,0)$$

$$\leq \epsilon \cdot \sum_{k=1}^{K_{mt}} \Big[ \sum_{i\in p(m,t,k)\setminus\{t\}} \Big(w_i(m,t,k-1)f(m,t,k)\Big) +$$

$$+ \varphi(m,t,k-1)L(m,t,k)\Big]$$

$$= \epsilon \cdot \sum_{k=1}^{K_{mt}} \left[ f(m,t,k) \cdot \sum_{i \in p(m,t,k) \setminus \{t\}} \Big( w_i(m,t,k-1) \Big) + \right.$$
$$\left. + \varphi(m,t,k-1) l(m,t,k) \right]$$

$$= \epsilon \cdot \sum_{k=1}^{K_{mt}} \left[ f(m,t,k) \cdot \sum_{(i,j) \in p(m,t,k)} \Big( w_i(m,t,k-1) + \right.$$
$$\left. + \varphi(m,t,k-1) l_{ij} \Big) \right]$$

$$= \epsilon \cdot \sum_{k=1}^{K_{mt}} f(m,t,k) \cdot \zeta_t(m,t,k-1)$$

$$\leq \epsilon \cdot d_t \zeta_t(m,t,k).$$

For brevity on notations, we define $W(m)$ as the value of $W$ at the end of phase $m$, and make a similar definition for $\alpha(m)$. Then, it follows that

$$W(m) - W(m-1)$$
$$= W(m, |R|, K_{m|R|}) - W(m,0,0)$$
$$\leq \epsilon \cdot \sum_{t=1}^{|R|} \Big( d_t \zeta_t(m,t,K_{m|R|}) \Big)$$
$$\leq \epsilon \alpha(m). \tag{23}$$

Combining the property of $W(m)/\alpha(m) \geq \beta$ with Equation (23), we can carry out

$$W(m) \leq \frac{W(m-1)}{1 - \epsilon/\beta}.$$

In light of the initial settings, $w_i(0) = \delta/C_i$ and $\varphi(0) = \delta/L$. Thus, we obtain $W(0) = (|V|+1)\delta$.

Given $m \geq 1$ and $\beta \geq 1$, it follows that

$$W(m) \leq \frac{(|V|+1)\delta}{(1 - \epsilon/\beta)^m}$$
$$= \frac{(|V|+1)\delta}{1 - \epsilon/\beta} (1 + \frac{\epsilon}{\beta - \epsilon})^{m-1}$$
$$\leq \frac{(|V|+1)\delta}{1 - \epsilon/\beta} e^{\frac{\epsilon(m-1)}{\beta-\epsilon}}$$
$$\leq \frac{(|V|+1)\delta}{1 - \epsilon} e^{\frac{\epsilon(m-1)}{(1-\epsilon)\beta}}.$$

Let the last phase in the algorithm be numbered by $M$. It follows that $1 \leq W(M) \leq \frac{(|V|+1)\delta}{1-\epsilon} e^{\frac{\epsilon(M-1)}{(1-\epsilon)\beta}}$. Hence, we carry out

$$\beta \leq \frac{\epsilon(M-1)}{(1-\epsilon) \ln \frac{1-\epsilon}{(|V|+1)\delta}}.$$

Thus, the lemma follows.

**Lemma 4.** *iStream-APX generates a feasible streaming solution that makes* $\lambda \geq \frac{M-1}{\log_{1+\epsilon} 1/\delta}$.

*Proof.* At the end of the $(M-1)^{\text{th}}$ phase, $W(M-1) \leq 1$ for all node $i$. Thus, we deduct $s_i(M-1) = w_i(M-1) \leq 1/C_i$.

From Lemma 2 and Corollary 2, we know $w_i$ and $s_i$ increase at least by a factor of $1+\epsilon$ for every $C_i$ units of flow through node $i$. Denoting the total flow through node $i$ as $F_i$, we can carry out

$$F_i \leq C_i \log_{1+\epsilon} \frac{w_i(M-1)}{w_i(0)}$$

$$\leq C_i \log_{1+\epsilon} \frac{1/C_i}{\delta/C_i}$$

$$= C_i \log_{1+\epsilon} \frac{1}{\delta}.$$

Therefore, dividing all the flows through node $i$ by a scaling factor of $\log_{1+\epsilon} \frac{1}{\delta}$, we obtain feasible flows through $i$ without violating its uploading capacity $C_i$.

Applying the scaling factor, we can get feasible flows received by $t$ of a total value $(M-1)d_t / \log_{1+\epsilon} \frac{1}{\delta}$ units. Accordingly, a feasible $\lambda$ will follow

$$\lambda \geq \frac{(M-1)d_t / \log_{1+\epsilon} \frac{1}{\delta}}{d_t}$$

$$= \frac{(M-1)}{\log_{1+\epsilon} \frac{1}{\delta}}.$$

**Theorem 2.** *The result of iStream-APX follows the property of* $\sum_{p \in P} l(p) f(p) \leq L$.

*Proof.* According to Corollary 3, in our procedure every time we route every flow with a cumulative delay of $L$, we increase $\varphi$ by at least a factor of $1+\epsilon$.

Because $W(M-1) < 1$, we deduct that $\varphi(M-1) < 1/L$. Thus, in the first $M-1$ phases, the cumulative delay is at most $L \cdot \log_{1+\epsilon} \frac{\varphi(M-1)}{\varphi(0)} = L \cdot \log_{1+\epsilon} \frac{1}{\delta}$, i.e., $\sum_{p \in P} l(p) f(p) \leq L \cdot \log_{1+\epsilon} \frac{1}{\delta}$.

In the final procedure of the algorithm, we scale down all the flows proportionally by a scaling factor. Thus, applying the scaling factor of $\log_{1+\epsilon} \frac{1}{\delta}$, we have

$$\sum_{p \in P} l(p) f(p) \leq \frac{L \log_{1+\epsilon} \frac{1}{\delta}}{\log_{1+\epsilon} \frac{1}{\delta}}$$

$$= L.$$

The theorem follows.

**Theorem 3.** *The approximation factor, denoted as $\rho$, is $1 + \omega$.*

*Proof.* From Lemma 4, we have a feasible solution $\lambda = \frac{M-1}{\log_{1+\epsilon}\frac{1}{\delta}}$. It follows that

$$\frac{\beta}{\lambda} = \frac{\beta \log_{1+\epsilon}\frac{1}{\delta}}{(M-1)}$$

$$= \frac{\epsilon \ln \frac{1}{\delta}}{(1-\epsilon)\ln(1+\epsilon)\ln \frac{1-\epsilon}{(|V|+1)\delta}}.$$

Let $\delta = \left(\frac{1-\epsilon}{|V|+1}\right)^{1/\epsilon}$. We have

$$\frac{\beta}{\lambda} \leq \frac{\epsilon \ln \frac{1}{\delta}}{(1-\epsilon)\ln(1+\epsilon)\ln \frac{1-\epsilon}{(|V|+1)\delta}}$$

$$= \frac{\epsilon}{(1-\epsilon)^2 \ln(1+\epsilon)}$$

$$\leq \frac{\epsilon}{(1-\epsilon)^2(\epsilon - \epsilon^2/2)}$$

$$\leq (1-\epsilon)^{-3}.$$

According to the strong duality theorem, if the dual has the optimal solution $\beta$, the primal also has an optimal value, denoted as $\text{OPT}(\lambda)$, such that $\text{OPT}(\lambda) = \beta$. Therefore, the approximation factor $\rho$ can be obtained by

$$\rho = \max \frac{\text{OPT}(\lambda)}{\lambda}$$

$$= \max \frac{\beta}{\lambda}.$$

Now, we make an assignment of $\omega = (1-\epsilon)^{-3} - 1$. We have $\rho = 1 + \omega$. Thus, the proof is complete.

## 4.5   Running Time

In this section, we analyze the bound on running time. We define maximum binary search bound on $L$ as $\Gamma = \sum_{j \in R} d_j \cdot \max_{p \in P} l(p)$.

**Theorem 4.** *Suppose the shortest path algorithm employed will consume a running time of $\Psi$. The running time of iStream is $O(\epsilon^{-2}\Psi|V|\log|V|\log\Gamma)$.*

*Proof.* According to weak duality theorem, we have $\frac{\beta}{\lambda} \geq 1$, which deduces

$$\frac{\beta}{M-1}\log_{1+\epsilon}\frac{1}{\delta} > 1.$$

So the number of phases $M < 1 + \beta \log_{1+\epsilon} \frac{1}{\delta}$. Because $\delta = \left(\frac{1-\epsilon}{|V|+1}\right)^{1/\epsilon}$, it follows that

$$M = \lceil \frac{\beta}{\epsilon} \log_{1+\epsilon} \frac{|V|+1}{1-\epsilon} \rceil$$

If iStream-APX does not stop within $2\lceil \frac{1}{\epsilon} \log_{1+\epsilon} \frac{|V|+1}{1-\epsilon} \rceil$ phases, we must have $\beta \geq 2$. We know $\mathrm{OPT}(\lambda) = \beta$ and we are pursuing $\mathrm{OPT}(\lambda) = 1$. In the case of $\beta \geq 2$, we break the current call for iStream-APX, and continue the binary search on $L$. So each call for iStream-APX will have $2\lceil \frac{1}{\epsilon} \log_{1+\epsilon} \frac{|V|+1}{1-\epsilon} \rceil = O(\epsilon^{-2} \log |V|)$ phases.

In order to compute the total running time, we need to calculate the number of steps in each call for iStream-APX. It is easy to see at every step except the the last step in an iteration, we increase either $w_i$ of some node or $\varphi$ by a factor at least $1 + \epsilon$. So the number of steps exceeds the number of iterations by at most

$$|V| \log_{1+\epsilon} \frac{w_i(M-1)}{w_i(0)} = |V| \log_{1+\epsilon} \frac{1}{\delta} = O(\epsilon^{-2}|V| \log |V|). \tag{24}$$

Also, the maximum number of iterations in all phases is $|R| \cdot O(\epsilon^{-2} \log |V|) = O(\epsilon^{-2}|R| \log |V|)$. Combining this with Equation (24), we have the total number of steps in each call for iStream-APX is $O(\epsilon^{-2}(|V|+|R|)\log |V|)=O(\epsilon^{-2}|V| \log |V|)$.

Considering the number of calls for iStream-APX in binary search is bounded by $\log \Gamma$. Consequently, we can carry out the running time of iStream is bounded by $O(\epsilon^{-2}\Psi|V| \log |V| \log \Gamma)$. The theorem follows.

## 5   Conclusion

We present the design of iStream and derive a near-optimal approximation bound for its core component iStream-APX. To achieve a tractable theoretical analysis, we assume no network dynamics in the first stage of algorithm design. Although the assumption is strong in practical P2P applications, the value of this paper lies in the theoretical framework and analysis, which sheds light on the practical design. To reduce the complexity of the problem, we focus only on minimizing the communication delay. For packet scheduling, there exists a vast array of solutions. The mesh built from our algorithm can adopt any of these scheduling algorithms to yield low-delay streaming.

## References

1. Tomozei, D.-C., Massoulie, L.: Flow control for cost-efficient peer-to-peer streaming. In: 2010 Proceedings IEEE INFOCOM, 14-19, pp. 1–9 (2010)
2. Sentinelli, A., Marfia, G., Gerla, M., Kleinrock, L., Tewari, S.: Will IPTV ride the peer-to-peer stream? IEEE Communications Magazine 45(6), 86–92 (2007)

3. Liu, Z., Shen, Y., Panwar, S.S., Ross, K.W., Wang, Y.: Using layered video to provide incentives in p2p live streaming. In: Proceedings of the 2007 Workshop on Peer-to-Peer Streaming and IP-TV, P2P-TV 2007 (2007)

4. Ren, D., Li, Y.-T., Chan, S.-H.: On reducing mesh delay for peer-to-peer live streaming. In: INFOCOM 2008 (2008)

5. Huang, F., Ravindran, B., Kumar, V.A.: An approximation algorithm for minimum-delay peer-to-peer streaming. In: Peer-to-Peer Computing 2009 (2009)

6. Wu, C., Li, B.: rstream: Resilient and optimal peer-to-peer streaming with rateless codes. IEEE Transactions on Parallel and Distributed Systems 19(1), 77–92 (2008)

7. Bianchi, G., Blefari Melazzi, N., Bracciale, L., Lo Piccolo, F., Salsano, S.: Streamline: An optimal distribution algorithm for peer-to-peer real-time streaming. IEEE Transactions on Parallel and Distributed Systems (99), 1 (2010)

8. Chen, Z., Xue, K., Hong, P.: A study on reducing chunk scheduling delay for mesh-based P2P live streaming. In: GCC 2008 (2008)

9. Liu, Y.: On the minimum delay peer-to-peer video streaming: how realtime can it be? In: ACM Multimedia 2007 (2007)

10. Garg, N., Könemann, J.: Faster and simpler algorithms for multicommodity flow and other fractional packing problems. SIAM J. Comput. 37(2), 630–652 (2007)

11. Karakostas, G.: Faster approximation schemes for fractional multicommodity flow problems. ACM Trans. Algorithms 4(1), 1–17 (2008)

12. Hei, X., Liang, C., Liang, J., Liu, Y., Ross, K.: A measurement study of a large-scale P2P IPTV system. IEEE Transactions on Multimedia 9(8) (December 2007)

13. Hefeeda, M., Saleh, O.: Traffic modeling and proportional partial caching for peer-to-peer systems. IEEE/ACM Transactions on Networking 16(6), 1447–1460 (2008)

14. Vazirani, V.V.: Approximation Algorithm. Springer, New York (2007)

15. Fei Huang, B.R., Khan, M.: Technical report: On minimizing average end-to-end delay in p2p live streaming systems, Tech. Rep. (2010),
    http://staff.vbi.vt.edu/maleq/papers/APX_P2P_average_delay.pdf

# Monotonic Stabilization*

Yukiko Yamauchi[1] and Sébastien Tixeuil[2]

[1] Nara Institute of Science and Technology, Japan
[2] Université Pierre et Marie Curie - Paris 6, France
y-yamauchi@is.naist.jp, Sebastien.Tixeuil@lip6.fr

**Abstract.** Self-stabilization guarantees *convergence* to a legitimate configuration in every execution starting from any initial configuration. However, during convergence, most self-stabilizing protocols make unnecessary output changes that do not directly contribute to the progress of convergence.

We define and study *monotonic stabilization*, where *every* output change is a step toward convergence. That is, any output change at a process $p$ gives the final output of $p$ in the legitimate configuration to be reached. It turns out that monotonic stabilization requires additional information exchange between processes, and we present task dependent tradeoff results with respect to the locality of exchanged information.

**Keywords:** Distributed system, fault-tolerance, self-stabilization, fault-containment, locality.

## 1 Introduction

One of the most versatile techniques to ensure forward recovery of distributed systems and networks is that of *self-stabilization* [1–3]. A distributed algorithm is self-stabilizing if after faults and attacks hit the system and place it in some arbitrary global state, the system recovers from this catastrophic situation without external (*e.g.* human) intervention in finite time. As self-stabilization makes no hypothesis about the nature or the extent of the faults (self-stabilization only deals with the effect of the faults), it can also be used to deal with other transient changes while the network is being operated (topology change, message loss, spontaneous resets, etc.).

There exist self-stabilizing protocols for many usual tasks in distributed computing, such as leader election, center finding, vertex coloring, maximal matching, or tree construction. Nevertheless, in many cases, processes may change their output several times during the stabilization (*a.k.a.* convergence) phase, which induces unnecessary state changes and information exchange. For example, in the well known *min+1* protocol for self-stabilizing BFS tree construction [4], the corruption of a single process that is close to the tree root may lead to unnecessary changes in all its subtrees, inducing large waste of system resources, lack of fault containment, etc.

* This work is supported in part by JSPS Grant-in-Aid for Young Scientists (Start-up) (21800031), ANR projects SHAMAN, ALADDIN, and R-DISCOVER.

C. Lu, T. Masuzawa, and M. Mosbah (Eds.): OPODIS 2010, LNCS 6490, pp. 475–490, 2010.

The classical performance metrics for distributed protocols are the *time complexity* and the *message complexity*. However, those metrics relate to the implementation of a particular protocol, not to the problem that the protocol is supposed to solve. The amount of information a process needs to take a correct local state defines the *locality* of the problem (*i.e.*, the specification), and is independent of the implementations (*i.e.*, the protocols) of the solutions to the problem. Of course, the intrinsic cost of solving a problem in a distributed system relates to the locality of the problem. Locality of problems has been discussed in many papers since Linial first proposed the locality of graph problems in distributed systems [5].

*Related work.* In the context of self-stabilization, Awerbuch, Patt-Shamir, and Varghese proposed the notion of *locally checkable and locally correctable* property of problems and proposed to transform a locally checkable and locally correctable problem to equivalent self-stabilizing protocols [6]. Awerbuch, Patt-Shamir, Varghese, and Dolev extended the transformer that is applicable for *locally checkable* problems by using global reset [7]. Based on these studies, Beauquier and Delaët proposed a classification of problems using locality for checking and locality for correction [8]. Beauquier, Delaët, Dolev and Tixeuil proposed *transient fault detector* that detects inconsistencies of the system configuration [9]. They defined locality with the history and the view that are necessary at each process so that at least one process finds inconsistency in the current configuration. However, the cost for recovery against transient faults are not addressed, since a process that finds inconsistency is not always the process that should change its state to optimally regain consistency of the system.

Another related approach is *fault-containment* of self-stabilizing protocols [10]. A fault-containing protocol contains the effect of the fault in a limited number of processes and/or limited time after corruption. A configuration that is obtained after a transient fault corrupting $f$ processes in a legitimate configuration is called $f$-*faulty configuration*. Then, the number of processes that change their states during recovery and/or the recovery time depends on just $f$, not the number of the entire processes in the system. The design of fault-containment is also based on local checking at each process. Because the number of faulty processes is limited, inconsistency is detected in an $f$-faulty configuration. Then, the faulty processes (and possibly, their neighbors) change their states to regain consistency in a local correction manner. However, most of the existing fault-containing self-stabilizing protocols are designed for 1-faulty configuration [11–14].

To the best of our knowledge, the optimality of recovery in self-stabilization was never discussed before. Optimal recovery can be defined in two ways: *progress optimality* and *goal optimality*. When *every* move during recovery makes the system progress towards a legitimate configuration, the recovery is progress optimal. When the legitimate configuration that is eventually reached is actually (one of) the nearest legitimate configuration(s) from the initial configuration, the recovery is goal optimal. Progress optimality guarantees no oscillation during the recovery phase since unnecessary and/or repeated moves are not allowed.

The addition of goal optimality to progress optimality brings optimal transient fault-containment by promising recovery without perturbation. Obviously, implementing optimal recovery (be it progress optimal and/or goal optimal) in self-stabilizing distributed systems entails information exchange among distant processes. However, no appropriate metrics exists for measuring this trade-off.

*Our contribution.* In this paper, we define and study *monotonic stabilization*, where *every* output change is a step toward convergence. That is, any output change by a process $p$ is the final output of $p$ in the legitimate configuration to be reached. Monotonic stabilization guarantees progress optimality, and we also define the optimal variant of monotonic stabilization to consider the case where goal optimality is also required (*e.g.* for fault containment purposes). To achieve (optimal) monotonic stabilization, each process should collect information about distant processes and previous state changes at other processes. We formalize those notions as *history locality* and *view locality* and examine these two localities in several problems. The obtained results and classification are useful when designing and evaluating fault-tolerant distributed protocols with respect to stabilization and fault-containment.

## 2   Preliminaries

*System model.* A system is a collection of processes connected by communication links which is represented by an undirected graph $G = (V, E)$ where the vertex set $V$ is a set of processes and the edge set $E$ is a set of bidirectional communication links. Each process has a unique identity. Process $p$ is a neighbor of process $q$ if there is a communication link $(p, q) \in E$. A set of direct neighbors of $p$ is denoted by $N_p$. Let $N_p^1$ be $N_p$ and for each $i \geq 2$, $N_p^i = N_p^{i-1} \cup \bigcup_{q \in N_p^{i-1}} N_q \setminus \{p\}$. The set of processes in $N_p^i$ is called $i$-neighbors of $p$. The distance between $p$ and $q$ ($p \neq q$), denoted by $dist(p, q)$, is $i$ if and only if $q \notin N_p^{i-1}$ and $q \in N_p^i$. The $i$-neighbor of $p$ is the set of processes whose distances from $p$ is smaller or equals to $i$ excluding $p$. The *eccentricity* of process $p$ is $\max_{q \in V} dist(p, q)$. The *diameter*, denoted by $d$, is the maximum value of eccentricity of a process in the system. The *radius*, denoted by $r$, is the minimum value of eccentricity of a process in the system. A *center* is the process(es) with the eccentricity which equals to the radius.

Each process maintains local variables and each local variable has its own domain. A *state* of a process is defined by the values of its local variables. Local variables consist of *output variables* and *inner variables*. For simplicity and without loss of generality, we assume each process $p$ maintains a single output variable denoted by $v_p$. A *configuration* of a system is a tuple of local states of all processes. We denote the value of $v_p$ in configuration $C$ by $v_{p|C}$ and the set of values of $v_p$ in configurations $C, C', \cdots$ by $dom(v_{p|\{C,C',\cdots\}})$. The distance between two configurations $C$ and $C'$ is defined by the number of processes whose output values are different, *i.e.*, $D(C, C') = |\{p \in V : v_{p|C} \neq v_{p|C'}\}|$.

We use the same definition of view and history as in [9].

## Definition 1 (View)

The $view_p^k$ at process $p$ contains (i) the subgraph induced by $N_p^k \cup \{p\}$ on $G$, and (ii) the values of output variables at each $q$ in $N_p^k$.

## Definition 2 (History)

The $history_p^k[0..\ell]$ at process $p$ is a sequence of $\ell + 1$ consecutive $k$-views at $p$. Given an execution $E = C_0, C_1, \cdots$, in $C_i$, $history_p^k[0]$ is the current $view_p^k$ (in $C_i$) and $history_p^k[j]$ is $view_p^k$ in $C_{i-j}$. When $i < j$, $history_p^k[j]$ is undefined and denoted by $\perp$.

We call $view_p^k$ $k$-view at $p$ and $history_p^k[0..\ell]$ $(k, \ell)$-history at $p$.

A protocol is a function $f$ that returns a value for $v_p$ and other inner variables when given $history_p^k[0..\ell]$ for some $k$ and $\ell^1$. We assume that $f$ is uniform in the sense that $f$ at each process is identical. We also assume that $f$ is deterministic in the sense that when given an identical history, $f$ returns the identical output. We consider the following execution model: In a computation step, the daemon selects process(es) and the selected process(es) changes its state according to the result of $f$. The *distributed daemon* selects non-empty subset of processes at a time. Whenever process $p$ is selected by the scheduler, $history_p^k[0..\ell]$ is given to $p$ and $p$ changes the value of $v_p$ and its internal variables according to $f(history_p^k[0..\ell])$. Even when $f$ does not change the value of the output variable, we consider that the output variable takes a new value and the move changed the values of some internal variables. We say a process takes a *move* when it updates its state with $f(history_p^k[0..\ell])$. An *execution* is an infinite sequence of configurations $E = C_0, C_1, \cdots$ such that $C_{i+1}$ is obtained by applying one computation step to $C_i$.

A *problem* (*task*) is defined by a set of legitimate configurations. The set of legitimate configuration for $G$ are denoted by $\mathcal{C}_L(G)$. When $G$ is clear, we omit $G$.

## Definition 3 (Self-stabilization)

A protocol $\mathcal{A}$ is self-stabilizing if and only if for any execution $E = C_0, C_1, \cdots$ of $\mathcal{A}$, the following two properties are satisfied: (i) convergence: there exists $i \geq 0$ such that $C_i \in \mathcal{C}_L$, and (ii) closure: for any $j \geq i$, $C_j \in \mathcal{C}_L$.

## Definition 4 (Monotonic Convergence)

A self-stabilizing protocol $\mathcal{A}$ is monotonically converging if and only if for any execution $E = C_0, C_1, \cdots, C_\ell, \cdots$, where $\ell = \min\{j | C_j \in \mathcal{C}_L\}$, when process $p$ is selected by the daemon in $C_0, \cdots, C_{\ell-1}$, it changes $v_p$ to $v_{p|C_\ell}$.

Monotonicity guarantees that each transition $(C_i, C_{i+1})$ satisfies $D(C_{i+1}, C_\ell) \leq D(C_i, C_\ell)$, in other words, each transition makes the system to approach to a legitimate configuration. Additionally, during a convergence, each process updates its output at most once. Monotonically converging self-stabilizing protocols

---

[1] The history $history_p^k[0..\ell]$ is not the part of the local state of $p$ but the information given by the oracle. The goal of this paper is not to show the implementation for providing $history_p^k[0..\ell]$ to each process $p$, but to show the size of $k$ and $\ell$.

is called *monotonically self-stabilizing protocol* for short. We say an execution is monotonic if it satisfies the condition in Definition 4. A move is said to be *monotonic* if it decreases or keeps the distance to the target legitimate configuration.

Given a monotonic execution $E = C_0, C_1, \cdots, C_i$, a configuration $C$ is *monotonically reachable* from $C_i$ if and only if $C$ appears in a monotonically converging execution whose prefix is $E$. We denote the set of monotonically reachable legitimate configurations for $E$ by $\mathcal{C}_{MRL}(E)$. The set of $\mathcal{C}_{MRL}(E)$ for $E$ depends on not only $C_i$ but the entire prefix of $C_0, C_1, \cdots, C_i$ because the states of the processes selected by the daemon in this prefix are the states of these processes in each $C \in \mathcal{C}_{MRL}(E)$. Given a configuration $C_0$, a legitimate configuration $C_{Opt} \in \mathcal{C}_L$ is *optimal* for $C_0$ if $C_{Opt}$ satisfies $D(C_0, C_{Opt}) = \min_{C' \in \mathcal{C}_L} \{D(C_0, C')\}$. Note that $C_{Opt} \in \mathcal{C}_{MRL}(C_0)$ and $C_{Opt}$ is not necessarily uniquely defined by $C_0$.

**Definition 5 (Optimal monotonic convergence)**
*A monotonically converging self-stabilizing protocol is* optimal *if and only if for any execution $E = C_0, C_1, \cdots$, it monotonically converges to an optimal legitimate configuration for $C_0$.*

The goal of this paper is to show the size of history that is necessary at each process to provide (optimal) monotonic convergence.

We now define our complexity metrics for monotonic stabilization variants.

**Definition 6 (Locality)**
*Problem $\mathcal{P}$ is $(k, \ell)$-local for optimal monotonic (monotonic, respectively) stabilization if and only if we have $f_{\mathcal{P}}$ with $(k, \ell)$-history at each process that satisfies optimal monotonically (monotonically, respectively) self-stabilization property, while we do not have any $f'_{\mathcal{P}}$ with views at distance at most $k - 1$ or at most $\ell$ consecutive views at each process that satisfies optimal monotonically (monotonically, respectively) self-stabilization property.*

*Problem statements.* In this paper, we determine the locality of the following seven problems. The definitions consist of the values of output variables at processes in a legitimate configuration.

1. *Leader election*: Each process $p$ maintains a local output Boolean variable $l_p$ that takes *true* if $p$ is elected and *false* otherwise. There is exactly one process $p$ which is elected ($l_p = true$).
2. *Center finding*: Each process $p$ maintains a local output Boolean variable $c_p$ that takes *true* if and only if $p$ is the center of $G$. If there are multiple center processes, all center processes outputs *true*.
3. *Majority consensus*: Each process $p$ maintains a Boolean local output variable $v_p$. All output variables are equal and the value is the majority value in the initial configuration. If there is no majority in the initial configuration, all output variables are equal to a value that is held at some process in the initial configuration.
4. *Vertex coloring*: Each process $p$ maintains a local output variable $col_p$ that takes an integer as its color. No two neighboring processes have the same color.

5. *Maximal matching*: Each process $p$ maintains a local output variable $m_p$ which is a pointer to one of its neighbors chosen as its matching process. A *matching* of a graph is a subset of edges in which no two edges are adjacent. A matching $M$ is is said to be *maximal* if and only if no proper subset of $M$ is also a matching. Hence, if and only if all neighbors form matchings with other processes, $m_p$ takes $\perp$.

6. *Rooted tree construction*: Each process $p$ maintains a local output variable $par_p$ which is a pointer to one of its neighbors chosen as its parent on the tree. A single process $r$ in the system which is called the *root* has $par_r = \perp$, while the values of pointers at other processes define a tree rooted at $r$.

7. *Shortest path tree construction*: Each process $p$ maintains a local output variable $par_p$ that satisfies the definition of the rooted tree construction problem. Additionally, the distance between each process $p$ and the root process $r$ on the rooted tree is same as the one in the original topology $G$.

## 3   Optimal Monotonic Convergence

In this section, we show the locality of problems defined in Section 2 for optimal monotonic convergence. We first show Theorem 1 that gives a basic approach to determine the locality of these problems.

**Lemma 1.** *Given a monotonic execution $C_0, C_1, \cdots$, for any $i, j$ ($0 < i < j$)*

$$\mathcal{C}_{MRL}(C_{i-1}, \cdots, C_j) \subseteq \mathcal{C}_{MRL}(C_i, \cdots, C_j).$$

*Proof.* We show the remark by contradiction. Assume $\mathcal{C}_{MRL}(C_i, \cdots, C_j) \subset \mathcal{C}_{MRL}(C_{i-1}, \cdots, C_j)$. If the transition from $C_{i-1}$ to $C_i$ does not change an output of any process, $\mathcal{C}_{MRL}(C_i, \cdots, C_j) = \mathcal{C}_{MRL}(C_{i-1}, \cdots, C_j)$. Hence, the transition changes some process's output. Let this process be $p$.

During the execution $C_i, \cdots, C_j$, some processes fix its output value or no process changes its output value. Let the number of these processes be $n'(\geq 0)$. Consequently, we have the following statements:

1. With the monotonic execution $C_i, \cdots, C_j$, the output of $n'$ processes are fixed. Hence, in each legitimate configuration $C \in \mathcal{C}_{MRL}(C_i, \cdots, C_j)$, these $n'$ processes takes the value in $C$.

2. With the monotonic execution $C_{i-1}, C_i, \cdots, C_j$, the output of $n'+1$ processes are fixed. Hence, in each legitimate configuration $C \in \mathcal{C}_{MRL}(C_{i-1}, \cdots, C_j)$, these $n' + 1$ processes takes the value in $C$.

Clearly, there exists no configuration $C'$ such that

$$C' \in \mathcal{C}_{MRL}(C_{i-1}, \cdots, C_j) \setminus C \in \mathcal{C}_{MRL}(C_i, \cdots, C_j).$$

This is the contradiction and we have the lemma.                    $\square$

Let $\mathcal{C}_{Opt}(C_0)$ be the set of optimal legitimate configurations for an initial configuration $C_0$.

For distributed problems, the synchronization among processes is the key to achieve monotonicity. From Definition 4, each process updates its state at most once during the convergence. Hence, each transition defines the set of monotonically reachable legitimate configurations. When the number of such legitimate configurations is decreased by a process move, synchronization among some processes is necessary.

The following theorem defines the worst case of the history locality and view locality of problems for optimal monotonic convergence.

**Theorem 1.** *There exists no optimal monotonic self-stabilizing protocol by giving each process views at distance $k$ or $\ell$ consecutive views if at least one of the following conditions hold:*

**Condition 1 (view).** *There exists a monotonic execution $E = C_0, C_1, \cdots$ on $G$ and a monotonic execution $E' = C'_0, C'_1, \cdots$ on $G'$ such that there exists $C_i$ in $E$ and $C'_j$ in $E'$ that satisfies $history^k_{p|C_i}[0..i] = history^k_{q|C'_j}[0..i]$ and $v_p|C_{opt} \neq v_p|C'_{opt}$ for some $C_{opt} \in \mathcal{C}_{Opt}(C_0) \cap \mathcal{C}_{MRL}(C_0, \cdots, C_i)$ and $C'_{opt} \in \mathcal{C}_{Opt}(C'_0) \cap \mathcal{C}_{MRL}(C'_0, \cdots, C'_j)$.*

**Condition 2 (history).** *There exists an optimal monotonic execution $E_i = C_0, C_1, \cdots, C_i$ and a process $p$ such that*

$$dom(v_{p|\mathcal{C}_{Opt}(C_0) \cap \mathcal{C}_{MRL}(C_{i-\ell}, \cdots, C_i)}) \neq dom(v_{p|\mathcal{C}_{Opt}(C_0) \cap \mathcal{C}_{MRL}(C_{i-\ell-1}, \cdots C_i)}).$$

*Proof.* Starting from an initial configuration $C_0$, in an optimal monotonic convergence, each process $p$ takes $v_{p|C_{opt}}$ for a legitimate configuration $C_{opt} \in \mathcal{C}_{Opt}(C_0)$ when it is selected by the scheduler.

First, we show the proof for Condition 1. Consider the case where Condition 1 holds at process $p$ in $C_i$. In $C_i$, $f$ cannot determine with $(k, \ell)$-history whether it is called at $p$ in $C_i$ or at $q$ in $C'_j$ because $history^k_{p|C_i}[0..\ell]$ is identical to $history^k_{q|C'_j}[0..\ell]$. Hence, though the values that $p$ takes in possible goal configurations ($\mathcal{C}_{Opt}(C_0) \cap \mathcal{C}_{MRL}(C_0, \cdots, C_i)$) are different from those that $q$ takes ($\mathcal{C}_{Opt}(C'_0) \cap \mathcal{C}_{MRL}(C'_0, \cdots, C'_j)$), $f$ returns the same value at $p$ and $q$. This value violates the monotonicity of one of the two executions.

Consequently, when Condition 1 holds, there is no possibility to construct a monotonic self-stabilizing protocol with views at distance $k$.

Next, we show the proof for Condition 2. Condition 2 addresses the following case: the set of monotonically reachable legitimate configuration with $\ell$ consecutive views is different from that with $\ell - 1$ consecutive views and the sets of values that $v_p$ takes are different ($dom(v_{p|\mathcal{C}_{Opt}(C_0) \cap \mathcal{C}_{MRL}(C_{i-\ell}, \cdots, C_i)}) \neq dom(v_{p|\mathcal{C}_{Opt}(C_0) \cap \mathcal{C}_{MRL}(C_{i-\ell-1}, \cdots C_i)})$).

From Lemma 1, when Condition 2 holds, $\mathcal{C}_{MRL}(C_{i-\ell-1}, \cdots, C_i)$ is a subset of $\mathcal{C}_{MRL}(C_{i-\ell}, \cdots, C_i)$. When Condition 2 holds, there exists a configuration $C' \in \mathcal{C}_{Opt}(C_0) \cap \mathcal{C}_{MRL}(C_{i-\ell}, \cdots, C_i) \setminus \mathcal{C}_{Opt}(C_0) \cap \mathcal{C}_{MRL}(C_{i-\ell-1}, \cdots, C_i)$. With $\ell$ consecutive views, the set of goal legitimate configuration at $p$ is $\mathcal{C}_{Opt}(C_0) \cap \mathcal{C}_{MRL}(C_{i-\ell}, \cdots, C_i)$ and this results in $f$ may returns $v_{p|C'}$. Then the execution is no more monotonic.

Consequently, when Condition 2 holds, there is no possibility to construct a monotonic self-stabilizing protocol with $\ell$ consecutive views.     □

*Remark 1.* Condition 1 of Theorem 1 is degenerated as follows: There exists an initial configuration $C_0$ on $G$ such that there exists an initial configuration $C_0'$ on $G'$ such that there exists a process $p \in G$ and a process $q \in G$ where $view^k_{p|C_0} = view^k_{q|C_0'}$, and $v_{p|C_{Opt}} \neq v_{q|C_{Opt}'}$ hold. ($C_{Opt}$ ($C_{Opt}'$, respectively) is the optimal legitimate configuration for $C_0$ ($C_0'$).)

Note that Remark 1 is the case where $i = j = 0$ in Condition 1 $(i)$. For some graph problems, we can easily find a specific topology that defines the optimal legitimate configuration for a specific initial configuration. This means that a process cannot take an optimal move without global information and that the optimal monotonic solution is impossible on an arbitrary graph for these problems.

The following remark shows the trivial upper bound of locality for a complete graph.

*Remark 2.* On a complete graph, any problem is $(1, \infty)$ local for optimal monotonic convergence.

On a complete graph, each process can check the states of all other processes with direct links. Note that Theorem 1 is the necessary condition and Remark 2 is the sufficient condition for a problem of locality $(d, \infty)$.

In the following, we determine the history locality and view locality of problems based on Theorem 1.

We first show the locality of the leader election problem which is one of the global problems.

**Lemma 2.** *The leader election problem is $(d, \infty)$-local for optimal monotonic convergence.*

*Proof.* We call a process is *leader* if its output is *true*. In an optimal monotonic convergence, if there is a leader process, no process is allowed to change its output from *false* to *true* because it increases the distance to the optimal legitimate configuration defined by the initial configuration. On the topology shown in Figure 1, with $(d-1)$-view, the leftmost process cannot check the existence of the leader (the rightmost process) and may change its output from *false* to *true*. Hence, distance $d$ view is necessary for the leader election problem.

Next, we show the history locality of the leader election problem. Consider an monotonic execution $C_0, C_1, \cdots$ and a configuration $C_i$ where there are multiple leader processes. Let $p$ be a leader process in $C_i$ and the daemon selects $p$ in $C_i$. Then, $dom(l_{p|C_i})$ is $\{true, false\}$. However, if another leader processes $q$ was selected by the daemon and kept its output *true* in the execution $C_0, C_1, \cdots, C_{i-1}$, then, $p$ cannot take *true* because it causes $q$ to change its state again. (Note that this $q$'s move is allowed under optimal monotonic convergence assuming that no other leader process keeps its output before this move.) Hence, $dom(l_{p|C_0, C_1, \cdots, C_i}) \subset dom(l_{p|C_i})$.

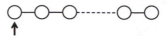

**Fig. 1.** Optimal Monotonic Convergence: Leader election

Consider a configuration where there is no leader process. In this configuration, $(d, \infty)$-history is necessary for each process. Because if all the other processes decided to not to change its output before, then a selected process has to take *true*. The monotonicity does not force a selected process to change its state even when the state change reduces the distance to a legitimate configuration.

Consequently, $\infty$ consecutive view is necessary for the leader election problem.

$\square$

**Lemma 3.** *The center finding problem is $(r+1,0)$-local for optimal monotonic convergence.*

*Proof.* For the center finding problem the legitimate configuration is defined by the topology. A center process finds that there are two processes such that the distance between them is $2r$ but there are no two processes such that the distance between them is greater than $2r$. On the other hand, other processes should check if they have a process at distance $(r+1)$. With $r$-view, these processes cannot check the existence of such processes.

The set $dom(c_p)$ at process $p$ is fixed by the topology and not by the execution. Hence, the current view is sufficient for the center finding problem.     $\square$

**Lemma 4.** *The majority consensus problem is $(d,1)$-local for optimal monotonic stabilization.*

*Proof.* The optimal legitimate configuration is determined by the majority of output values in the initial configuration. In the initial configuration $C_0$ shown in Figure 2(a), the leftmost process cannot find the difference between a configuration $C_0'$ shown in Figure 2(c) with distance $2k$ view. The optimal legitimate configurations for $C_0'$ are the one where all output variables are 0 or the one where all output values are 1. Hence, $f$ may return 0 and 1. While the correct output value for the leftmost process in $C_0$ is 0, $f$ may return 0 and 1 in $C_0$. With distance $d$ view, $f$ differentiate these two configurations. Hence, distance $d$ view is necessary.

Next, we show the history locality. For the majority consensus problem, the first transition determines the goal legitimate configuration in a monotonic convergence. For example, if the set of processes enabled in the initial configuration change their outputs to 1 (or keep their outputs 1), then all the following transitions do not allow a process to change (or keep) its output to 0. Consequently, by checking the previous transition, each process recognizes the allowed state changes. Hence, the 1 consecutive view is necessary.

Note that if there is a majority of output values in an initial configuration, the history locality is $(d,0)$. This is because the optimality does not allow any

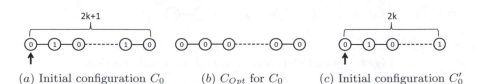

(a) Initial configuration $C_0$    (b) $C_{Opt}$ for $C_0$    (c) Initial configuration $C_0'$

**Fig. 2.** Optimal Monotonic Convergence: Majority consensus

process to change its state from a majority value to a minority value. The majority of output values does not change during the convergence and each process determines it by $(d, 0)$-history. □

Next, we show the locality of the rooted tree construction problem and the shortest path tree construction problem that shows a significant difference in history locality.

**Lemma 5.** *Spanning tree construction is $(d, \infty)$-local and shortest path tree construction is $(d, 0)$-local for optimal monotonic stabilization.*

*Proof.* We first show the view locality of rooted tree construction. Consider the two initial configurations $C_0$ (Figure 3 (a)) and $C_0'$ (Figure 3 (c)). The optimal legitimate configuration for $C_0$ is shown in Figure 3 (b) and the optimal legitimate configuration for $C_0'$ is the configuration where all process point to the other neighbor than in Figure 3 (b). However, in $C_0$, process $p$ cannot determine the optimal legitimate configuration without $d$ view. Consequently, the view locality for the rooted tree construction problem is $d$.

Next, we show the history locality of the rooted tree construction problem. Consider a configuration $C_i$ shown in Figure 3 (d) that appears in an execution $C_0, C_1, \cdots$. During the execution $C_0, C_1, \cdots, C_{i-1}$, it is possible that some processes except $p$ and the root has decided not to change its state in configuration $C_j$ where $t$ is selected by the daemon. Let this process $t$. In this case, $p$ cannot point to the upper side because it then makes $t$ to change its pointer again during the convergence. Consequently, $dom(par_{p|C_j, C_{j+1}, \cdots, C_i}) \neq dom(par_{p|C_i})$. Hence, $(d, \infty)$-history is necessary for process $p$.

We show the locality of shortest path tree construction. Each process $p$ should find a neighbor that is on a shortest path from $p$ to the root. With $d - 1$-view, $p$ cannot find the shortest path to the root. Hence, distance $d$ view is necessary.

In a legitimate configuration, it is guaranteed that each process points to a neighbor on a shortest path tree to the root. In other words, for process $p$, each neighbor process $q$ points to a neighbor on a shortest path tree from $q$ to the root. During a monotonic convergence, each process changes its pointer to point such a neighbor. Hence, the $dom(par_p)$ is not changed by any state change. Consequently, the history locality is 0. □

For problems that are considered to have some local property, global information is necessary to achieve optimal monotonic convergence.

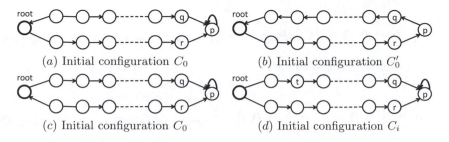

(a) Initial configuration $C_0$  (b) Initial configuration $C_0'$

(c) Initial configuration $C_0$  (d) Initial configuration $C_i$

**Fig. 3.** Optimal Monotonic Convergence: Rooted tree construction

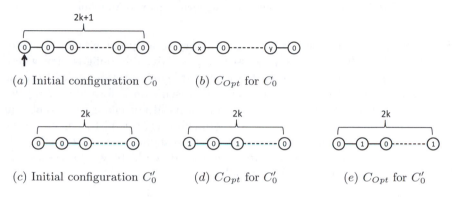

(a) Initial configuration $C_0$  (b) $C_{Opt}$ for $C_0$

(c) Initial configuration $C_0'$  (d) $C_{Opt}$ for $C_0'$  (e) $C_{Opt}$ for $C_0'$

**Fig. 4.** Optimal Monotonic Convergence: Vertex coloring

**Lemma 6.** *The vertex coloring problem and the maximal matching problem are $(d, \infty)$-local for optimal monotonic convergence.*

*Proof.* We prove the view locality by showing the topology where each process cannot make a optimal monotonically converging move without $d$-view.

Figure 4 shows an example for the vertex coloring problem. The processes that change their output values depends on the length of the line. With distance $d$ view, the leftmost process cannot find the difference between the configuration $C_0$ (Figure 4(a)) and $C_0'$ (Figure 4(c)) while it can take 0 and 1 in optimal legitimate configurations for $C_0'$ (Figure 4(d) and (e)). In vertex coloring, these processes take an arbitrary color that is different from 0.

Figure 5 shows an example for the maximal matching problem. In this case also, the leftmost process cannot determine its output value with distance $d$ view. In an initial configuration $C_0$ (Figure 5 (a)), the leftmost process cannot find the difference between $C_0$ and $C_0'$ (Figure 5 (c)) with distance $3k$ view.

Next, we show the history locality of the two problems. In the vertex coloring problem, the history of all previous moves is necessary when given a configuration where two processes takes the same color. If one of them has decided to keep its color before, the other process should change its color during a monotonic convergence.

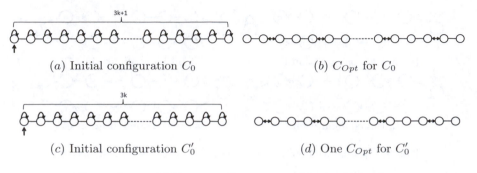

**Fig. 5.** Optimal Monotonic Convergence: Maximal Matching

In the maximal matching problem, the history of all previous moves is necessary. Consider a line of three processes $p$, $q$, $r$. Given a configuration where $IS_p = IS_r = q$ and $IS_q = \perp$, to reach a legitimate configuration, $q$ has to point one of its neighbors. However, as in the same discussion before, $dom(IS_q)$ depends on the former state change at $p$ and $r$. If $p$ did not change its pointer in the previous step, then $q$ has to point to $r$. (In this case, this history is necessary for $r$ to determine its output value since during a monotonic convergence, $r$ is not allowed to point to $p$ after this step.) Hence, $(d, \infty)$-history is necessary and sufficient for the maximal matching problem.                          □

## 4    Monotonic Convergence

In this section, we show the locality of problems for monotonic convergence. We first show Theorem 2 that refers to the problems whose locality for monotonic convergence is identical to that for optimal monotonic convergence.

**Theorem 2.** *For a problem $\mathcal{P}$ that is $(k, \ell)$-local for monotonic convergence, if the legitimate configuration is uniquely defined by $G$ and the initial configuration, then $\mathcal{P}$ is $(k, \ell)$-local for optimally monotonic convergence.*

*Proof.* For the problems that satisfy the condition of Theorem 2, finding a legitimate configuration is finding the optimal legitimate configuration.                          □

From Theorem 2, the center finding problem is $(r + 1, 0)$ local for monotonic self-stabilization. The rooted tree construction problem and the shortest path tree construction problem also fall in this category, because for some topology, these trees are uniquely defined by the topology.

The remaining is the problems that have multiple legitimate configurations for some initial configurations, *i.e.*, the majority consensus problem, the leader election problem, the vertex coloring, and the maximal matching problem.

The majority consensus problem has only two legitimate configurations for any topology. During a monotonic convergence, if a process changes its output variable from 1 to 0, it means all the succeeding state changes should not change the values of output variables of other processes from 0 to 1. In this paper, we do not restrict the strategy of $f$: it is not the majority computation, nor one-way resetting. Additionally, because the goal is not an optimal legitimate configuration, a process cannot determine a new state by the majority of output values. Hence, only with the current view, a process cannot find the process that has taken a move.

Next, we show Theorem 3 that is the counterpart for Theorem 1.

**Theorem 3.** *There exists no monotonic self-stabilizing protocol by giving each process views at distance $k$ or $\ell$ consecutive views if one of the following conditions hold:*

**Condition 1 (view).** *There exists a monotonic execution $E_i = C_0, C_1, \cdots, C_i$ such that there exists a monotonic move at process $p$ in $C_i$ that changes the configuration to $C_{i+1}$ that satisfies at process $q \in V \setminus (N_p^k \cup \{p\})$*

$$dom(v_q|_{\mathcal{C}_{MRL}(E_{i+1})}) \subset dom(v_q|_{\mathcal{C}_{MRL}(E_i)})$$

**Condition 2 (history).** *There exists a monotonic execution $E_i = C_0, C_1, \cdots, C_i$ and a process $p$ such that*

$$dom(v_p|_{\mathcal{C}_{MRL}(C_{i-\ell}, \cdots, C_i)}) \neq dom(v_p|_{\mathcal{C}_{MRL}(C_{i-\ell-1}, \cdots C_i)}).$$

*Proof.* To monotonically converge to a legitimate configuration, each transition make an enabled process to choose a new state that is consistent with former transitions.

We first show the proof for Condition 1. Let $p$ and $q$ be two processes at distance $k$. Condition 1 addresses about the following case: A state change at process $p$ excludes some entry in $dom(v_q|_{\mathcal{C}_{MRL}(E_i)})$ at process $q$. Let one of these excluded value be $v'$. Suppose that immediately after $p$'s state change $q$ is enabled. The $k$-view at $q$ is same as $C_i$ because there is no transition after $C_{i+1}$. With $k$-view, $q$ cannot find the difference between $C_i$ and $C_{i+1}$, and this can results in $p$ taking $v'$ as its output value. Then the execution is no more monotonic.

Consequently, when Condition 1 holds, there is no possibility to construct a monotonic self-stabilizing protocol with views at distance $k$.

We omit the proof for Condition 2 because it is the same as the proof for Condition 2 of Theorem 1. □

Both the Condition 1 of Theorem 1 and the Condition 1 of Theorem 3 refer to the domain of possible output at processes. While Condition 1 of Theorem 1 is based on the difference between optimal legitimate configurations, Condition 1 of Theorem 3 is based on the difference between any legitimate configurations.

From Theorem 3, we derive the locality of the problems.

**Lemma 7.** *The majority consensus problem is $(d, 1)$-local for monotonic stabilization.*

*Proof.* Consider a line of $n$ processes. Let the leftmost process be $p$ and the rightmost process be $q$. In an initial configuration $C_0$, $dom(v_{r|\mathcal{C}_{MRL}}) = \{0, 1\}$ for any process $r$ in $V$. If the central daemon first selects $p$ and $v_p$ takes $f(view^k_{p|C_0})$, $dom(v_{q|\mathcal{C}_{MRL}})$ is restricted to $\{f(view^k_{p|C_0})\}$ for each process $q$ in $V \setminus \{p\}$.

For the majority consensus problem, the first transition determines the goal legitimate configuration. During the convergence, $\mathcal{C}_{MRL}$ is unique, *i.e.*, a configuration where each process $r$ takes $\{f(view^k_{p|C_0})\}$. Hence, for any $i$ and any process $p$, we have

$$dom(v_{p|\mathcal{C}_{MRL}(C_{i-1}, C_i)}) = dom(v_{p|\mathcal{C}_{MRL}(C_{i-2}, C_{i-1}, C_i)})$$
$$dom(v_{p|\mathcal{C}_{MRL}(C_i)}) \neq dom(v_{p|\mathcal{C}_{MRL}(C_{i-1}, C_i)}).$$

As a result, the majority consensus problem is $(d, 1)$-local for monotonic convergence. □

**Lemma 8.** *The leader election problem is $(d, \infty)$-local for monotonic stabilization.*

*Proof.* Consider a line of $n$ processes where $l_p = false$ holds at each process $p$ in $V$. Let the leftmost process be $p$ and the rightmost process be $q$. In an initial configuration, $dom(v_{r|CMRL}) = \{false, true\}$ for any process $r$ in $V$. If the central daemon first selects $p$ and $l_p$ takes $true$, $dom(v_{q|CMRL})$ is restricted to $\{false\}$ for each process $q$ in $V \setminus \{p\}$. Hence, the view locality of the leader election problem is $d$.

Consider another configuration $C_0$ of a line of $n$ processes where $l_p = l_q = true$ holds at the two endpoint processes $p$ and $q$ and all other processes output $false$. Let $p$ kept $l_p = true$ when it is first selected by the scheduler. Hence, the goal legitimate configuration is a configuration where only $p$ outputs $true$. However, after this, with constant number of consecutive views, $q$ cannot find that $q$ should change $l_q$ to $false$ because the scheduler does not select $q$ for an arbitrary long time. Consequently, the leader election problem is $(d, \infty)$ local for monotonic stabilization. □

We omit the detailed proofs for Lemma 9 and Lemma 10 due to page restriction. The view locality of these two problems are derived from the fact that on a line topology, a state change of a process shrinks the domain of the output value at other processes at distance smaller than or equals to 1 in the vertex coloring problem and 3 in the maximal matching problem, respectively. The history localities are derived from the fact that each process cannot recognize whether other processes have changed their outputs without the entire history.

**Lemma 9.** *The vertex coloring problem is $(1, \infty)$-local for monotonic stabilization.*

**Lemma 10.** *The maximal matching problem is $(3, \infty)$-local for monotonic stabilization.*

We note that Hedetniemi's self-stabilizing vertex coloring protocol in [15] is monotonically self-stabilizing under centralized daemon and $(1, 0)$ local. In that protocol, if a process finds a conflict of colors with its neighboring processes, then the process takes a new color that is not used by its neighboring processes. This protocol proposes a simple strategy to reduce the history locality for the vertex coloring problem.

## 5    Concluding Remarks

We proposed new metrics for optimality of convergence of self-stabilizing protocols. We showed the locality of problems for optimal monotonic convergence and monotonic convergence. Table 1 presents our results together with the locality of fault detectors from [9].

**Table 1.** Localities of problems

| Problem | Optimal monotonicity | Monotonicity | Fault detector [9] |
|---|---|---|---|
| Leader election | $(d, \infty)$ | $(d, \infty)$ | $(r, 0)$ |
| Center finding | $(r + 1, 0)$ | $(r + 1, 0)$ | $(r, 0)$ |
| Majority consensus | $(d, 1)$ | $(d, 1)$ | - |
| Rooted tree | $(d, \infty)$ | $(d, \infty)$ | $(\lceil n/4 \rceil, 0)$ |
| Shortest path tree | $(d, 0)$ | $(d, 0)$ | - |
| Vertex coloring | $(d, \infty)$ | $(1, \infty)$ | $(1, 0)$ |
| Maximal matching | $(d, \infty)$ | $(3, \infty)$ | $(1, 0)$ |

It turns out that the localities of the majority of problems are $(d, \infty)$. However, the view localities of the vertex coloring problem and the maximal matching problem are reduced to a constant for (simple) monotonicity compared to optimal monotonicity. This result shows that the requirement of optimality often translates a (supposedly) local problem into a global task. We observe that the history localities of the vertex coloring problem and the maximal matching problem are $\infty$.

Another interesting result is shown by the locality of the rooted tree construction problem and the shortest path tree construction problem. Though a shortest path tree is an instance of a rooted tree, the history locality becomes constant. So, a more constrained problem may lead to a less costly optimally converging solution. This has some intuitive explanation, since more constrained solutions tend to reduce the number of possible legitimate configurations, it is then easier to coordinate processes to reach a single one of them.

Finally, the locality results of this paper show a significant difference between the locality results of transient fault detectors [9]. Our results confirm that the resources necessary for optimal convergence are much more significant than for simply detecting some inconsistency.

An interesting open question raised by our work is the definition of a suitable notion of monotonic convergence for reactive tasks (such as token passing or propagation of information with feedback).

# References

1. Dijkstra, E.W.: Self-stabilizing systems in spite of distributed control. ACM Commun. 17(11), 643–644 (1974)
2. Dolev, S.: Self-stabilization. MIT Press, Cambridge (March 2000)
3. Tixeuil, S.: Self-stabilizing Algorithms. Chapman & Hall/CRC Applied Algorithms and Data Structures. In: Algorithms and Theory of Computation Handbook, 2nd edn., pp. 26.1–26.45. CRC Press, Taylor & Francis Group (November 2009)
4. Delaët, S., Ducourthial, B., Tixeuil, S.: Self-stabilization with r-operators revisited. Journal of Aerospace Computing, Information, and Communication (2006)
5. Linial, N.: Locality in distributed graph algorithms. SIAM Journal of Computing 21(1), 193–201 (1992)
6. Awerbuch, B., Patt-Shamir, B., Varghese, G.: Self-stabilization by local checking and correction. In: Proceedings of 32nd Annual Symposium on Foundations of Computer Science, pp. 268–277 (October 1991)
7. Awerbuch, B., Patt-Shamir, B., Vargheese, G., Dolev, S.: Self-stabilization by local checking and global reset. In: Tel, G., Vitányi, P.M.B. (eds.) WDAG 1994. LNCS, vol. 857, pp. 326–339. Springer, Heidelberg (1994)
8. Beauquier, J., Delaet, S.: Classes of self-stabilizing protocols. In: Proceedings of the 4th FTDCS, pp. 361–365 (September 1993)
9. Beauquier, J., Delaet, S., Dolev, S., Tixeuil, S.: Transient fault detectors. Distributed Computing 20, 39–51 (2007)
10. Ghosh, S., Gupta, A., Herman, T., Pemmaraju, S.V.: Fault-containing self-stabilizing distributed protocols. Distributed Computing 20(1), 53–73 (2007)
11. Ghosh, S., Gupta, A.: An exercise in fault-containment: self-stabilizing leader election. Information Processing Letters 59(5), 281–288 (1996)
12. Ghosh, S., Gupta, A., Pemmaraju, S.: Fault-containing network protocols. In: Proceedings of the 12th ACM Symposium on Applied Computing, pp. 431–437 (February 1997)
13. Ghosh, S., He, X.: Fault-containing self-stabilization using priority scheduling. Information Processing Letters 73, 145–151 (2000)
14. Lin, J., Huang, T.C.: An efficient fault-containing self-stabilizing algorithm for finding a maximal independent set. IEEE Transactions on Parallel and Distributed Systems 14, 742–754 (2003)
15. Hedetniemi, S.T., Jacobs, D.P., Srimani, P.K.: Linear time self-stabilizing colorings. Information Processing Letters 87, 251–255 (2003)

# Upper and Lower Bounds of Space Complexity of Self-Stabilizing Leader Election in Mediated Population Protocol

Ryu Mizoguchi, Hirotaka Ono, Shuji Kijima, and Masafumi Yamashita

Department of Informatics, Kyushu University
744, Motoka, Nishi-ku, Fukuoka, 819-0395, Japan
{ryu,ono,kijima,mak}@tcslab.csce.kyushu-u.ac.jp

**Abstract.** This paper investigates the space complexity of a *self stabilizing leader election* in a mediated population protocol (*SS-LE MPP*). Cai, Izumi and Wada (2009) showed that SS-LE in a *population protocol* (SS-LE *PP*) for $n$ agents requires at least $n$ agent-states, and gave a SS-LE PP with $n$ agent-states for $n$ agents. MPP is a model of distributed computation, introduced by Chatzigiannakis, Michail and Spirakis (2009) as an extension of PP allowing an extra memory on every agents pair. While they showed that MPP is stronger than PP in general, it was not known if a MPP can really reduce the space complexity of SS-LE with respect to agent-states. We in this paper give a SS-LE MPP with $(2/3)n$ agent-states and a single bit memory on every agents pair for $n$ agents. We also show that there is no SS-LE MPP with any constant agent-states and any constant size memory on each agents-pair for general $n$ agents.

**Keywords:** Mobile agents, anonymous, population protocols, self-stabilization, leader election.

## 1 Introduction

*Population Protcol* (*PP*), proposed by Angluin et al. [1], is a model of distributed computation consisting of agents and communication links among them, and *Mediated Population Protocol* (*MPP*) proposed by Chatzigiannakis et al. [7], is an extended model of PP allowing memories on communication links. PP and MPP are models of sensor networks consisting of *passively* mobile agents with limited computational resources, motivated by practical networks such as networks of smart sensors attached to cars or animals, synthesis of chemical materials, complex biosystems, and so on (cf. [1,7]).

In MPP, every agent is identically programmed as a finite state machine, and every communication link is equipped with a (finite) buffer. The agents sequentially interact with each other updating their states; a pair of agents chosen by a scheduler updates their own *agent-states* and *edge-states* between them in an interaction. The order of interactions of agent-pairs is unpredictable, and is

C. Lu, T. Masuzawa, and M. Mosbah (Eds.): OPODIS 2010, LNCS 6490, pp. 491–503, 2010.

scheduled by an adversarial scheduler satisfying a *fairness condition*; the scheduler must accept any possible interaction within a finite time if a configuration in which the interaction can arise should appear infinitely many times.

Angluin et al. [3] discussed the *leader election* in a population protocol, which is a fundamental problem in distributed computing, and introduced the problem of *self stabilizing leader election* in a *population protocol* (*SS-LE PP*, for short). In a SS-LE PP, any initial configuration of agent-states eventually have to reach at a configuration whose successive configurations contain exactly one leader. Thus a SS-LE PP should be equipped with seemingly conflicting functions; the protocol has to decrease the number of leaders if a configuration contains two or more leaders, while the protocol has to appoint an agent to be a leader if a configuration does not contain a leader. This causes some difficulties on SS-LE, as it is usual with self-stabilizing distributed problems.

Angluin et al. [2] discussed that no SS-LE PP with any constant number of agent-states exists for general $n$ agents on some types of interaction graphs. Fischer and Jiang [9] discussed SS-LE PP assuming a (global) oracle for leader detector on complete communication graphs. Canepa and Potop-Butucaru [6] discussed SS-LE PP on any communication graph in the same assumption with [9]. Cai, Izumi and Wada [5] discussed SS-LE on complete interaction graph without any (global) oracles, and showed that SS-LE for $n$ agents requires at least $n$ agent-states, and presented a SS-LE PP with $n$ agent-states for $n$ agents.

This paper is concerned with the space complexity of *self stabilizing leader election* in a *mediated population protocol* (*SS-LE MPP*, for short) for $n$ agents, where we assume that an interaction graph is complete as did Cai, Izumi and Wada [5]. We present a SS-LE MPP with roughly $(2/3)n$ agent-states and two edge-states for $n$ agents. We also show that there is no SS-LE MPP with any constant agent-states and any constant edge-states for general $n$ agents. As far as the authors know, this is the first result on SS-LE MPP. One may say it obvious that the number of agent-stats decreases in MPP comparing with PP due to extra memories on edges. In fact, it is clear that $n$ is also sufficient for the number of agent-states in SS-LE MPP for $n$ agents. However, extra memories on edges in MPP, which are expected to resolve the issue of conflicting functions in the self-stabilizing setting instead of a certain number of agent-states, may cause another issue of increasing possible (bad) initial configurations in the self-stabilizing setting.

This paper is organized as follows; in Section 2, we describe the detail of our model. To explain our basic idea for reducing agent-states, we in Section 3 give a SS-LE MPP with $n-1$ agent-states and 2 edge-states for $n$ agents. In Section 4, we present a SS-LE MPP with $(2/3)n$, $\lfloor (2/3)n \rfloor + 1$ in precise, agent-states and two edge-states for $n$ agents. In Section 5 we give lower bounds of agent-states of a SS-LE MPP.

## 2    Model Description — SS-LE MPP

A *mediated population protocol* is defined by 3-tuple $(Q, S, \delta)$, where $Q$ denotes a finite set of agent-states, $S$ denotes a finite set of edge states, and $\delta: Q \times Q \times S \rightarrow$

$Q \times Q \times S$ denotes a transition function. Let $A$ denote the set of anonymous agents and let $n = |A|$, and let $\mathcal{C} \stackrel{\text{def}}{=} Q^A \times S^{\binom{A}{2}}$ denote *all configurations*. A transition from a configuration $C \in \mathcal{C}$ to the next configuration $C' \in \mathcal{C}$ is defined as follows. An arbitrary pair of agents $a_i, a_j \in A(a_i \neq a_j)$ is chosen by a scheduler, thus an interaction graph is complete in our model. States of the agents $a_i$ and $a_j$, and a state of an edge $\{a_i, a_j\}$ are updated according to a transition function $\delta$. Let $r : (p, q, s) \mapsto (p', q', s')$ denote a specified transition rule of $\delta$, and let $C \xrightarrow{r; a_i, a_j} C'$ denote a transition from $C \in \mathcal{C}$ to $C' \in \mathcal{C}$ in which agents $a_i$ and $a_j$ interact and their states $p, q$ and edge-state $s$ between them are updated to $p', q', s'$ according to the rule $r$ of $\delta$. We simply write $C \xrightarrow{r} C'$ without confusing. An *execution* of a protocol is represented by an infinite sequence of configurations and transitions $C_0, r_0, C_1, r_1, \ldots$, where $C_0$ is an initial configuration and $C_i \xrightarrow{r_i} C_{i+1} (i \geq 0)$.

We assume that a scheduler in a MPP is *adversarial* but (globally) *fair*, as usual (cf. [5]). Thus we have to think that an adversarial scheduler schedules the order of interactions in a worst case scenario for us, but it is forced to satisfy that if a configuration $C \in \mathcal{C}$ appears infinitely often in an execution, a configuration $C' \in \mathcal{C}$ must also appear infinitely often in an execution, where $C'$ is a configuration obtained by an arbitrary transition $r \in \delta$ which arises in $C$. We say that $C$ *eventually transits* to $C'$, denoted by $C \xrightarrow{*} C'$, if $C'$ must appear after $C$ by the adversarial but globally fair scheduler in MPP. In addition, we describe a sequence of transitions as the *trace T*.

*Leader election* in a MPP is to assign a special state, representing a "leader", in $S$ to exactly one agent. We say a configuration $C \in \mathcal{C}$ is *legal* if $C$ contains exactly one agent with the leader state, and so does any configuration $C'$ satisfying $C \xrightarrow{*} C'$. Let $\mathcal{L}$ denote the set of all legal configurations. We say a protocol for the leader election (for a distributed problem, in general), is *self-stabilizing* if $C \xrightarrow{*} C', C' \in \mathcal{L}$ hold for any $C \in \mathcal{C}$. We simply say *SS-LE MPP* as a mediated population protocol for the leader election which is self stabilizing.

Our goal is to give upper and lower bounds of the sizes of the agent-states $Q$ and edge-states $S$ for SS-LE MPP concerning the number of agents $n$. Main results of the paper are to give a SS-LE MPP with $|Q| = \lfloor (2/3)n \rfloor + 1$ and $|S| = 2$ for $n$ agents in Section 4, and to show that there is no SS-LE MPP with constant sizes of $Q$ and $S$ for general $n$ agents in Section 5. To describe our basic idea for reducing the number of agent-states, we in Section 3 give a SS-LE MPP with $|Q| = n - 1$ and $|S| = 2$ for $n$ agents.

## 3   Simple SS-LE MPP with $n - 1$ Agent-States

In this section, we show the following.

**Theorem 1.** *There exists a SS-LE MPP with $n - 1$ agent-states and 2 edge-states for $n(\geq 4)$ agents.*

We give a constructive proof. In particular, we show that Protocol $P_1$, defined as follows, is a SS-LE MPP.

## Protocol $P_1$

$Q = \{q_0, q_1, \ldots, q_{n-2}\}$, where $q_0$ denotes the leader state.

$S = \{s_0, s_1\}$,

$\delta = \{$

$r_1: (q_0, q_0, s) \mapsto (q_0, q_{n-2}, s_0)$   for   $s \in S$,

$r_2: (q_1, q_1, s) \mapsto (q_1, q_2, s_0)$     for   $s \in S$,

$r_3: (q_2, q_2, s_0) \mapsto (q_2, q_2, s_1)$,

$r_4: (q_2, q_2, s_1) \mapsto (q_2, q_1, s_0)$,

$r_5: (q_2, q_1, s_1) \mapsto (q_2, q_0, s_0)$   $((q_1, q_2, s_1) \mapsto (q_0, q_2, s_0),$ symmetrically$)$,

$r_6: (q_i, q_i, s) \mapsto (q_i, q_{i-1}, s_0)$   for   $i \geq 3, s \in S$,

$r_7: (q_j, q_k, s) \mapsto (q_j, q_k, s_0)$     for   $j \neq k, s \in S$, except for the case of $r_5$

$\}$.

**Remark.** *Except for Transition $r_5$, the state of an agent can change only when the agent interacts with another agent in the same state.*

Let $\gamma_k(C)$ for $k \in \{0, 1, \ldots, n-2\}$ denote the number of agents with state $q_k$ in a configuration $C \in \mathcal{C}$. We define a set of configurations $\mathcal{L} \subset \mathcal{C}$ by

$$\mathcal{L} \stackrel{\text{def}}{=} \left\{ C \in \mathcal{C} \,\middle|\, \begin{array}{l} \gamma_k(C) > 0 \text{ for } k \in \{0, 1, \ldots, n-2\}, \; \gamma_1(C) + \gamma_2(C) = 3, \\ \text{both ends of an edge with state } s_1 \text{ are agents with state } q_2. \end{array} \right\}.$$

Note that the number of edges with state $s_1$ in $C \in \mathcal{L}$ is at most one since $\gamma_2(C)$ is at most two from the definition of $\mathcal{L}$.

In the following, we claim that $\mathcal{L}$ is the set of legal configurations for Protocol $P_1$. Let $H$ denote a subconfiguration of $C \in \mathcal{L}$ consisting of three agents with states $q_1$ or $q_2$ and three edges among them. Then $H$ can be one of three types of subconfigurations $H_1, H_2, H_3$ of six possible types $H_1, H_2, H_3, H_4, H_5, H_6$ in Fig.1 which satisfy that the number of edges with state $s_1$ is at most one, $\gamma_1(C) > 0$ and $\gamma_2(C) > 0$. First, we show that $\mathcal{L}$ is "closed" under the transition function $\delta$.

**Lemma 2.** *If configurations $C$ and $C'$ satisfy $C \in \mathcal{L}$ and $C \stackrel{*}{\to} C'$, then $C' \in \mathcal{L}$ and $C' \stackrel{*}{\to} C$ hold.*

*Proof.* Transition $r_1$ cannot arise in a configuration $C \in \mathcal{L}$ by the condition $\gamma_0(C) = 1$ in $\mathcal{L}$. By the condition $\gamma_1(C) + \gamma_2(C) = 3$ and $\gamma_k(C) > 0$ for $k \in \{0, 1, \ldots, n-2\}$, $C$ satisfies $\gamma_k(C) = 1$ for $k \in \{3, 4, \ldots, n-2\}$. Thus, Transition $r_6$ cannot arise in $C \in \mathcal{L}$.

Now we show that Transition $r_5$ cannot arise in $C$. Since the edge-state $s_1$ appears only within the subconfiguration $H$ of $C \in \mathcal{L}$, it is enough to show that Transition $r_5$ cannot arise in subconfigurations $H_1, H_2, H_3$ of $C$.

Case 1. $H = H_1$: Consider an agent with state $q_1$ as $a$ and agents with state $q_2$ as $b, c$, and consider every state of every edge among them as state $s_0$. Then Transitions $r_3$ or $r_7$ can arise in $H_1$, that is $H_1 \stackrel{r_3}{\longrightarrow} H_2$ or $H_1 \stackrel{r_7}{\longrightarrow} H_1$.

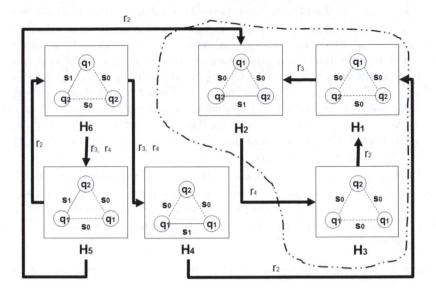

**Fig. 1.** $\{H_1, H_2, H_3\}$ is closed

**Case 2. $H = H_2$:** Consider an agent with state $q_1$ as $a$ and agents with state $q_2$ as $b, c$, and consider a state of the only one edge between the two agents $b, c$ as state $s_1$. Then Transitions $r_4$ or $r_7$ can arise in $H_2$, that is $H_2 \xrightarrow{r_4} H_3$ or $H_2 \xrightarrow{r_7} H_2$.

**Case 3. $H = H_3$:** Consider agents with state $q_1$ as $a, b$ and an agent with state $q_2$ as $c$ and consider every state of every edge among them as state $s_0$. Then Transitions $r_2$ or $r_7$ can arise in $H_3$, that is $H_3 \xrightarrow{r_2} H_1$ or $H_3 \xrightarrow{r_7} H_3$.

Therefore, if $C \in \mathcal{L}$ and $C'$ satisfies $C \xrightarrow{*} C'$, then $C' \in \mathcal{L}$ and $C' \xrightarrow{*} C$ hold. □

Next, we show that any configuration $C \in \mathcal{C}$ eventually transits to a configuration $C' \in \mathcal{L}$, in Lemma 5. To show Lemma 5, we show Lemmas 3 and 4.

**Lemma 3.** *If a configuration $C \in \mathcal{C}$ satisfies $\gamma_k(C) > 0$ for $k \in \{0, 2, 3, \ldots, n - 2\}$, and $C \xrightarrow{*} C'$, then the configuration $C'$ also satisfies $\gamma_k(C') > 0$.*

*Proof.* After an agent with state $q_k$ for $k \in \{0, 2, 3, \ldots, n - 2\}$ interacts with any other agent, $\gamma_k(C)$ decreases at most one in any transition. In fact, $\gamma_k(C)$ decreases only when the agent interacts with another agent in the same state $q_k$. This implies that $\gamma_k(C)$ never decreases from one to zero by any transition. □

**Lemma 4.** *If configurations $C, C' \in \mathcal{C}$ satisfy $\gamma_0(C) = 0$, $C \xrightarrow{*} C'$ and $\gamma_0(C') = 0$, then the followings hold;*

1. $\sum_{i=1}^{k} \gamma_i(C') \geq \sum_{i=1}^{k} \gamma_i(C)$ *for any $k \in \{2, 3, \ldots, n - 2\}$.*
2. *If $\gamma_i(C) > 0$, then $\gamma_i(C') > 0$.*

*Proof.* 1. If $\gamma_0(C) = 0$ and $\gamma_0(C') = 0$ hold, Transition $r_5$ cannot have arisen on $C \xrightarrow{*} C'$. Note that Transition $r_7$ does not change any agent-state. Since $n - 2$ states are assigned to $n$ agents, there exists a pair of agents and they are in a common state $q_i$. When $i \geq 3$, Transitions $r_6$ or $r_7$ can arise in $C$ and exactly one of the agents changes its state from $q_i$ to $q_{i-1}$. Thus $\gamma_{i-1}(C') + \gamma_i(C') = \gamma_{i-1}(C) + \gamma_i(C)$ and $\gamma_{i-1}(C') = \gamma_{i-1}(C) + 1$. When $i = 1, 2$, Transitions $r_2, r_3$ or $r_4$ can arise in $C$ except for Transitions $r_5$ and $r_7$ and their transitions does not change $q \in \{q_1, q_2\}$ to $q' \notin \{q_1, q_2\}$. Therefore, for any $r \in \{r_2, r_3, r_4\}$ a configuration $C'$ of $C \xrightarrow{r} C'$ satisfies that $\gamma_1(C) + \gamma_2(C) = \gamma_1(C') + \gamma_2(C')$. That indicates $\sum_{i=1}^{k} \gamma_i(C') \geq \sum_{i=1}^{k} \gamma_i(C)$.

2. By Lemma 3, if $\gamma_0(C) = 0$ and $\gamma_0(C') = 0$, Transition $r_5$ cannot have arisen on $C \xrightarrow{*} C'$. In arbitrary transitions except for $r_5$, $\gamma_1(C)$ decreases at most one in a transition. $\gamma_1(C)$ decreases only when a pair of agents with same state $q_1$ interact. This implies that $\gamma_1(C)$ never decreases from one to zero by any transition. □

**Lemma 5.** *For any configuration $C \in \mathcal{C}$, there exists a configuration $C' \in \mathcal{L}$ and $C \xrightarrow{*} C'$.*

*Proof.* Case 1. $\gamma_0(C) = 0$
We show that for any configuration $C \in \mathcal{C}$, there exists a configuration $C' \in \mathcal{C}$ satisfying that $C \xrightarrow{*} C'$ and $\gamma_0(C') > 0$.
Case 1.1. $\gamma_1(C) + \gamma_2(C) \leq 3$
Since $n - 4$ states $q_3, q_4, \ldots, q_{n-2}$ are assigned to at least $n - 3$ agents, there exists a pair of agents and their states are common $q_i$. When $i \geq 3$, Transitions $r_6$ or $r_7$ can arise in $C$ and exactly one of the agents changes its state from $q_i$ to $q_{i-1}$. By Lemma 4, $\gamma_i(C)(> 0)$ does not become zero by any transition and $\sum_{i=1}^{k} \gamma_i(C)$ does not decrease by any transition, thus there exists a configuration $C' \in \mathcal{C}$ satisfying that $C \xrightarrow{*} C'$ and $\gamma_1(C') + \gamma_2(C') > 3$.
Case 1.2. $\gamma_1(C) + \gamma_2(C) > 3$

Suppose Transition $r_5$ cannot have arisen on $C \xrightarrow{*} C'$, then Transitions $r_2, r_3, r_4$, or $r_7$ can arise in $C$ except for Transition $r_5$. It implies that $C$ eventually transits to a configuration $C' \in \mathcal{C}$ satisfying $\gamma_2(C') \geq 3$, thus configurations satisfying $\gamma_2(C') \geq 3$ infinitely often appear. Consider three agents $a, b, c$ with state $q_2$. A trace $(r_3; a, b)$, $(r_3; b, c)$, $(r_4; a, b)$, $(r_5; c, b)$ can infinitely often arise in $C'$. Therefore, it contradicts the assumption of the global fairness, the configuration eventually transits to Case 2.
Case 2. $\gamma_0(C) > 0$
If $\gamma_0(C) > 1$, Transition $r_1$ can infinitely often have arisen by fairness condition. Hence $C$ eventually transits to a configuration $C' \in \mathcal{C}$ satisfying $\gamma_0(C') = 1$. If $\sum_{i=0}^{k} \gamma_i(C) \geq k+3$ for $k \geq 3$, in a similar way as Case 1, $C$ eventually transits to a configuration $C'' \in \mathcal{C}$ and $\gamma_0(C'') \geq 2$, and $\sum_{i=0}^{k} \gamma_i(C'')$ decreases again after Transition $r_1$ arises. Since $n - 1$ states are assigned to $n$ agents and $\gamma_j(C)(> 0)$ except for $j = 1$ does not become zero by any transitions, $C$ eventually transits to $C' \in \mathcal{C}$ satisfying $\gamma_2(C') = 2$ and $\gamma_j(C') = 1$ except for $j = 2$. By Lemma 3

such a configuration $C'$ infinitely often appears, therefore Transition $r_7$ can have arisen until no edge with state $s_1$ remain $C'$. It is clear that such a configuration is included in $\mathcal{L}$.     □

We obtain Theorem 1 by Lemmas 2 and 5.

## 4   SS-LE MPP with $(2/3)n$ Agent-States for $n$ Agents

In this section, we prove the following theorem.

**Theorem 6.** *There exists a SS-LE MPP with*

$$m = \left\lfloor \frac{2n}{3} \right\rfloor + 1 \equiv \begin{cases} \frac{2n+1}{3} & (if \ n \equiv 1 \pmod 3) \\ \frac{2n+2}{3} & (if \ n \equiv 2 \pmod 3) \\ \frac{2n+3}{3} & (if \ n \equiv 0 \pmod 3) \end{cases}$$

*agent-states and 2 edge-states for $n$ agents.*

We give a constructive proof. In particular, we show that Protocol $P_2$, defined as follows, is a SS-LE MPP. For simplicity of arguments, we first consider the case $n \bmod 3 = 1$.

**Protocol $P_2$ for $n \equiv 1 \pmod 3$**

$Q = \{q_0, q_1, \ldots, q_{m-1}\}$, where $q_0$ denotes the leader state.
$S = \{s_0, s_1\}$.
$\delta = \{$

   $r_1 \colon (q_0, q_0, s) \mapsto (q_0, q_{m-1}, s_0)$   for   $s \in S$,

   $r_2 \colon (q_{2i-1}, q_{2i-1}, s) \mapsto (q_{2i-1}, q_{2i}, s_0)$   for $s \in S$ and $i \in \{1, 2, \ldots, \frac{m-1}{2}\}$,

   $r_3 \colon (q_{2i}, q_{2i}, s_0) \mapsto (q_{2i}, q_{2i}, s_1)$   for $i \in \{1, 2, \ldots, \frac{m-1}{2}\}$,

   $r_4 \colon (q_{2i}, q_{2i}, s_1) \mapsto (q_{2i}, q_{2i-1}, s_0)$   for $i \in \{1, 2, \ldots, \frac{m-1}{2}\}$,

   $r_5 \colon (q_{2i}, q_{2i-1}, s_1) \mapsto (q_{2i}, q_{2i-2}, s_0)$,

   $\Big( (q_{2i-1}, q_{2i}, s_1) \mapsto (q_{2i-2}, q_{2i}, s_0), \text{ symmetrically} \Big)$   for $i \in \{1, 2, \ldots, \frac{m-1}{2}\}$,

   $r_6 \colon (q_j, q_k, s) \mapsto (q_j, q_k, s_0)$   for $j \neq k$ and $s \in S$, except for the case of $r_5$.

   $\}$.

**Remark.** *Except for Transition $r_5$, the state of an agent can change only when the agent interacts with another agent in the same state.*

We define a set of configurations $\mathcal{L} \subset \mathcal{C}$ by

$$\mathcal{L} \stackrel{\text{def}}{=} \left\{ C \in \mathcal{C} \,\middle|\, \begin{array}{l} \gamma_k(C) > 0 \text{ for } k \in \{0, 1, \ldots, m-1\}, \\ \gamma_{2i-1}(C) + \gamma_{2i}(C) = 3 \text{ for } i \in \{1, 2, \ldots, \frac{m-1}{2}\}, \\ \text{both ends of an edge with state } s_1 \text{ are agents with state } q_{2i}. \end{array} \right\}.$$

Let $H(i)$ for $i \in \{1, 2, \ldots, \frac{m-1}{2}\}$ denote a subconfiguration of $C \in \mathcal{L}$ consisting of three agents with states $q_{2i-1}$ or $q_{2i}$ and three edges among them. A subconfiguration $H(i)$ for $i \in \{1, 2, \ldots, \frac{m-1}{2}\}$ corresponds to $H$ in Section 3, and can be one of three types $H_1', H_2', H_3'$ corresponding to $H_1, H_2, H_3$ in which states $q_1, q_2$ is replaced by $q_{2i-1}, q_{2i}$ respectively. Then, the edge-state $s_1$ can appear only within $H(i)$ for $i \in \{1, 2, \dots, \frac{m-1}{2}\}$, and the number of edges with state $s_1$ in $H(i)$ is at most one.

**Lemma 7.** *If configurations $C$ and $C'$ satisfy $C \in \mathcal{L}$ and $C \xrightarrow{*} C'$, then $C' \in \mathcal{L}$ and $C' \xrightarrow{*} C$ hold.*

*Proof.* Transition $r_1$ cannot arise in $C \in \mathcal{L}$ by the condition $\gamma_0(C) = 1$ in $\mathcal{L}$. In a similar way as the proof of Lemma 2, $H(i)$ is closed as $H_1' \xrightarrow{r_3} H_2' \xrightarrow{r_4} H_3' \xrightarrow{r_2} H_1'$ under transitions except for $r_5$. From the condition "the both ends of an edge with state $s_1$ are $q_{2i}$", Transition $r_5$ cannot arise in $C \in \mathcal{L}$. Thus we obtain the claim. □

Next, we show that any configuration $C \in \mathcal{C}$ eventually transits to a configuration $C' \in \mathcal{L}$, in Lemma 10. To show Lemma 10, we show Lemmas 8 and 9.

**Lemma 8.** *If a configuration $C$ satisfies $\gamma_{2i-1}(C) + \gamma_{2i}(C) \geq 4$, then Transition $r_5$ can eventually arise.*

*Proof.* Let $C'$ be an arbitrary configuration which $C$ eventually transits to, and suppose Transition $r_5$ cannot have arisen on $C \xrightarrow{*} C'$. Transitions $r_2, r_3, r_4$ or $r_6$ can arise except for Transitions $r_1, r_5$. Note that Transitions $r_3$ and $r_6$ do not change any agent-states. Exactly one of two agents with a common state changes its state from $q_{2i-1}$ to $q_{2i}$ by Transition $r_2$, and also changes its state from $q_{2i}$ to $q_{2i-1}$ by Transition $r_4$. This implies that $C$ eventually transits to a configuration $C' \in \mathcal{C}$ satisfying $\gamma_{2i}(C') \geq 3$. Thus some configurations satisfying $\gamma_{2i}(C') \geq 3$ infinitely often appear. Consider three agents with state $q_{2i}$ as $a, b, c$. A trace $(r_3; a, b)$, $(r_3; b, c)$, $(r_4; a, b)$, $(r_5; c, b)$ can infinitely often arise in $C'$ see also Fig. 2. $D_k(i)$ for $k = 1, 2, \dots, 5$ denote a subconfiguration of $C \in \mathcal{C}$ consisting of four agents whose states are $q_{2i}, q_{2i-1}$ or $q_{2i-2}$. Since it contradicts the assumption of the global fairness, we obtain the claim. □

**Lemma 9.** *If a configuration $C \in \mathcal{C}$ satisfies $\gamma_0(C) = 0$ and a configuration $C' \in \mathcal{C}$ which $C$ eventually transits to satisfies $\gamma_0(C') = 0$, then $\sum_{i=1}^{2k} \gamma_i(C') \geq \sum_{i=1}^{2k} \gamma_i(C)$ for all $k \in \{1, 2, \dots, \frac{m-1}{2}\}$.*

*Proof.* If $\gamma_0(C) = 0$ and $\gamma_0(C') = 0$ hold, Transitions $r_1, r_5$ cannot have arisen on $C \xrightarrow{*} C'$. Note that Transition $r_6$ dose not change any agent-state. Since $m-1$ states are assigned to $\frac{3}{2}(m-1) + 1$ agents, there exists a subconfiguration $H(j)$ of the configuration $C$ satisfying $\gamma_{2j-1}(C) + \gamma_{2j}(C) > 3$ for $j \in \{1, 2, \dots, \frac{m-1}{2}\}$. By Lemma 8, an agent in $C$ eventually changes its state from $q_{2j-1}$ to $q_{2j-2}$, and let $C' \in \mathcal{C}$ denote the configuration, then $\sum_{i=1}^{2j} \gamma_i(C') = \sum_{i=1}^{2j} \gamma_i(C)$ and $\gamma_{2j-2}(C') = \gamma_{2j-2}(C) + 1$ hold. Let $C''$ be an arbitrary configuration which $C$ eventually transits to and suppose Transitions $r_1, r_5$ cannot have arisen on $C \xrightarrow{*} C'$. $\gamma_{2j-1}(C') + \gamma_{2j}(C') = \gamma_{2j-1}(C) + \gamma_{2j}(C)$ holds for $C \xrightarrow{*} C''$. Therefore, $\sum_{i=1}^{2k} \gamma_i(C') \geq \sum_{i=1}^{2k} \gamma_i(C)$ for all $k \in \{1, 2, \dots, \frac{m-1}{2}\}$. □

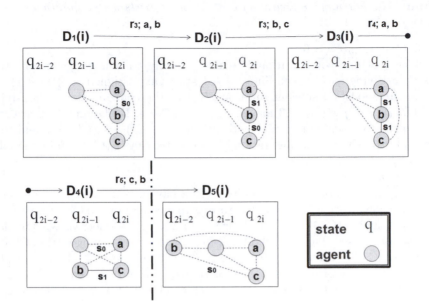

**Fig. 2.** A trace of a subconfiguration of $C$ which satisfies $\gamma_{2i-1}(C) + \gamma_{2i}(C) \geq 4$

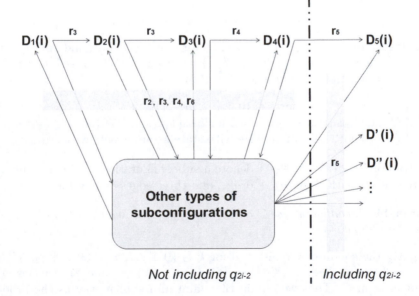

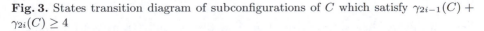

**Fig. 3.** States transition diagram of subconfigurations of $C$ which satisfy $\gamma_{2i-1}(C) + \gamma_{2i}(C) \geq 4$

**Lemma 10.** *For any configuration $C \in \mathcal{C}$, there exists a configuration $C' \in \mathcal{L}$ and $C \xrightarrow{*} C'$.*

*Proof.* Case 1. $\gamma_0(C) = 0$

We show that any configuration $C \in \mathcal{C}$ eventually transits to a configuration $C' \in \mathcal{C}$ satisfying $\gamma_0(C') > 0$. Since $m - 1$ states are assigned to $\frac{3}{2}(m-1) + 1$ agents, there exists a subconfiguration $H(j)$ of the configuration $C$ satisfying $\gamma_{2j-1}(H(j)) + \gamma_{2j}(H(j)) > 3$ for $j \in \{1, 2, \ldots, \frac{m-1}{2}\}$. By Lemma 9, $C$ eventually transits to a configuration $C' \in \mathcal{C}$ satisfying $\gamma_1(C') + \gamma_2(C') > 3$, and then eventually transits to a configuration $C'' \in \mathcal{C}$ satisfying $\gamma_0(C'') > 0$ in a similar way as the proof of Lemma 5.

Case 2. $\gamma_0(C) > 0$

If $\gamma_0(C) > 0$, Transition $r_1$ can infinitely often have arisen under fairness condition. The configuration $C$ eventually transits to a configuration $C' \in \mathcal{C}$ satisfying $\gamma_0(C') \geq 2$, and $\gamma_0(C) + \sum_{i=1}^{2k} \gamma_i(C)$ for $k \in \{1, 2, \ldots, \frac{n-1}{2}\}$ decreases again after Transition $r_1$ arises. Thus, $C$ infinitely often transits to a configuration $C'' \in \mathcal{C}$ satisfying $\gamma_0(C'') = 1$ and $\gamma_{2i-1}(C'') + \gamma_{2i}(C'') = 3$ for every $i \in \{1, 2, \ldots, \frac{m-1}{2}\}$. Since $C''$ infinitely often appears, Transition $r_6$ can have arisen until no edge with state $s_1$ remain $C''$. It is clear that such a configuration is included in $\mathcal{L}$. $\square$

By Lemmas 7 and 10, we obtain a SS-LE MPP with $m$ agent-states and 2 edge-states for $n$ agents in case of $n \equiv 1 \pmod 3$.

Next we give Protocol $P_2$ for $n \equiv 2 \pmod 3$ agents. We define Protocol $P_2$ for $n \equiv 2 \pmod 3$ by adding

$$r_7 : (q_{m-1}, q_{m-1}, s) \mapsto (q_{m-1}, q_{m-2}, s_0)$$

to the transition function $\delta$ of Protocol $P_2$ for $n \equiv 1 \pmod 3$, and appropriately modify the domain of $i$ as $\{1, 2, \ldots, \frac{m-2}{2}\}$ in Protocol $P_2$ for $n \equiv 1 \pmod 3$.

We define a set of configurations $\mathcal{L} \subset \mathcal{C}$ by

$$\mathcal{L} \stackrel{\text{def}}{=} \left\{ C \in \mathcal{C} \left| \begin{array}{l} \gamma_k(C) > 0 \text{ for } k \in \{0, 1, \ldots, m - 1\}, \\ \gamma_{2i-1}(C) + \gamma_{2i}(C) = 3 \text{ for } i \in \{1, 2, \ldots, \frac{m-2}{2}\}, \\ \text{both ends of an edge with state } s_1 \text{ are agents with state } q_{2i}. \end{array} \right. \right\}.$$

Let $H(i)$ for $i \in \{1, 2, \ldots, \frac{m-2}{2}\}$ denote a subconfiguration of $C \in \mathcal{L}$ consisting of three agents with states $q_{2i-1}$ or $q_{2i}$, and three edges among them.

**Lemma 11.** *If configurations $C$ and $C'$ satisfy $C \in \mathcal{L}$ and $C \xrightarrow{*} C'$, then $C' \in \mathcal{L}$ and $C' \xrightarrow{*} C$ hold.*

*Proof.* By the conditions $\gamma_k(C) > 0$ for $k \in \{0, 1, \ldots, m - 1\}$ and $\gamma_{2i-1}(C) + \gamma_{2i}(C) = 3$ for $i \in \{1, 2, \ldots, \frac{m-2}{2}\}$, $\gamma_0(C) = \gamma_{m-1}(C) = 1$ and Transitions $r_1, r_7$ cannot arise in $C$. Thus we obtain the claim in a similar way as the proof of Lemma 7. $\square$

Next, we prove that any configuration $C \in \mathcal{C}$ eventually transits to a configuration $C' \in \mathcal{L}$ (Lemma 12). If $\gamma_0(C) = 0$, since $m - 1$ states are assigned to

$\frac{3}{2}(m-2)+2$ agents, there exists a subconfiguration $H(i)$ of the configuration $C$ satisfying $\gamma_{2i-1}(H(i))+\gamma_{2i}(H(i)) > 3$ for $i \in \{1,2,\ldots,\frac{m-2}{2}\}$, or there exists a pair of agents with the common state $q_{m-1}$ in $C$. Transition $r_7$ arises in a configuration $C \in \mathcal{C}$ and exactly one of the agents changes its state from $q_{m-1}$ to $q_{m-2}$. Using (a type of) Lemmas 8 and 9, we can show the following.

**Lemma 12.** *For any configuration $C \in \mathcal{C}$, there exists a configuration $C' \in \mathcal{L}$ and $C \xrightarrow{*} C'$.*

*Proof.* Case 1. $\gamma_0(C) = 0$
We show that any configuration $C \in \mathcal{C}$ eventually transits to a configuration $C' \in \mathcal{C}$ satisfying $\gamma_0(C') > 0$. Since $m-1$ states are assigned to $\frac{3}{2}(m-2)+2$ agents, there exists a subconfiguration $H(i)$ of the configuration $C$ satisfying $\gamma_{2i-1}(H(i))+\gamma_{2i}(H(i)) > 3$ for $i \in \{1,2,\ldots,\frac{m-2}{2}\}$ or there exists a pair of agents and their states are common $q_{m-1}$. In the former case, $C$ eventually transits to a configuration $C' \in \mathcal{C}$ satisfying $\gamma_1(C')+\gamma_2(C') > 3$ by Lemma 9. In the latter case, Transition $r_7$ can arise in $C$ and exactly one of two agents with common $q_{m-1}$ changes its state to $q_{m-2}$. Therefore, $C$ eventually transits to a configuration $C' \in \mathcal{C}$ satisfying $\gamma_1(C')+\gamma_2(C') > 3$, and then eventually transits to a configuration $C'' \in \mathcal{C}$ satisfying $\gamma_0(C'') > 0$.
Case 2. $\gamma_0(C) > 0$
Shown in a similar way as Case 2 of the proof of Lemma 10.    □

Finally we present our SS-LE MPP for $n$ agents in case of $n \equiv 0 \pmod 3$. We define Protocol $P_2$ for $n \equiv 0 \pmod 3$ by adding

$$r_8 : (q_{m-2}, q_{m-2}, s) \mapsto (q_{m-2}, q_{m-3}, s_0)$$

to the transition function $\delta$ of Protocol $P_2$ for $n \equiv 2 \pmod 3$ and appropriately replace the domain of $i$ as $\{1,2,\ldots,\frac{m-3}{2}\}$ in Protocol $P_2$ for $n \equiv 2 \pmod 3$. Let $H(i)$ for $i \in \{1,2,\ldots,\frac{m-3}{2}\}$ denote a subconfiguration of $C \in \mathcal{L}$ consisting of three agents with states $q_{2i-1}$ or $q_{2i}$ and three edges among them.

It is not difficult to see $\mathcal{L}$, appropriately modifying the domain of $i$, is the set of legal configurations of the case, in a similar way as the case of $n \equiv 2 \pmod 3$. Now we obtain Theorem 6.

## 5    Lower Bounds for SS-LE MPP

In this section, we give two lower bounds of the number of agent-states.

**Theorem 13.** *For general $n$ agents, any SS-LE MPP with a constant number of agent-states and a constant number of edge-states does not exist.*

*Proof.* Suppose $P(n)$ is a SS-LE MPP with a constant agent-states and a constant edge-states for $n$ agents. We show that $P(n)$ cannot be a SS-LE MPP for $n'$ agents where $n' \neq n$. Without loss of generality we may assume that $n > n'$. Let $C$ be a legal configuration of $P(n)$ and let $D$ be a subconfiguration of $C$ where

$D$ consists of $n'$ non-leader agents. Then $P(n)$ cannot create leader states by any interactions within $D$, since $D$ is a subconfiguration of a legal configuration $C$ of $P(n)$. This means that $P(n)$ is not a SS-LE MPP for $n'$ agents, because $D$ cannot reach at a legal configuration for $n'$ agents.

When possible agent-states and edge-states are finite, the number of components included in the transition function is also finite. In fact, for $c$ agent-states and $d$ edge-states, the number of possible protocols is (at most) $(c^2 d)^{c^2 d}$. Since the number of agents $n$ can be fairly larger than them, we obtain the claim.   □

**Theorem 14.** *If the number of agent-states is 2, then no SS-LE MPP exists for $n(> 2)$ agents, even when the number of edge-states is infinitely large.*

*Proof.* With an assumption that a legal configuration exists and we derive a contradiction. When $n > 2$, the number of an agent with a leader state $q_0$ included in a legal configuration $\mathcal{L}$ is one and the number of agents with non-leader state $q_1$ is at least two. Therefore it is clear that the state of agents with state $q_1$ in a configuration $C \in \mathcal{L}$ does not change. Let $C' \in \mathcal{C}$ be a configuration which is constructed by the 3 tuple of two agents and an edge among them which are included in $C$, then $C'$ cannot transit to a configuration which includes a agent-state $q_0$. This contradicts any initial configuration eventually transits to a configuration including a leader state.   □

## 6   Conclusion

We gave a SS-LE MPP with roughly $(2/3)n$ agent-states and two edge-states for $n$ agents, while we showed that there is no SS-LE MPP with any constant agent-states and any constant edge-states for general $n$ agents. We conjecture that our upper bound is almost tight. A future work is to analyze SS-LE MPP with a constant edge-states. Analyses on other interaction graphs may be another future work.

## References

1. Angluin, D., Aspnes, J., Diamadi, Z., Fischer, M.J., Peralta, R.: Computation in networks of passively mobile finite-state sensors. Distributed Computing 18, 235–253 (2006)
2. Angluin, D., Aspnes, J., Fischer, M.J., Jiang, H.: Self-stabilizing population protocols. In: Anderson, J.H., Prencipe, G., Wattenhofer, R. (eds.) OPODIS 2005. LNCS, vol. 3974, pp. 103–117. Springer, Heidelberg (2006)
3. Angluin, D., Aspnes, J., Eisenstat, D.: Stably computable predicates are semilinear. In: Prasanna, V.K., Iyengar, S.S., Spirakis, P.G., Welsh, M. (eds.) DCOSS 2005. LNCS, vol. 3560, pp. 63–74. Springer, Heidelberg (2005)
4. Aspnes, J., Ruppert, E.: An introduction to population protocols. Bulletin of the EATCS 93, 98–117 (2007)
5. Cai, S., Izumi, T., Wada, K.: Space complexity of self-stabilizing leader election in passively-mobile anonymous agents. In: Kutten, S., Žerovnik, J. (eds.) SIROCCO 2009. LNCS, vol. 5869, pp. 113–125. Springer, Heidelberg (2010)

6. Canepa, D., Potop-Butucaru, M.G.: Stabilizing leader election in population protocols (2007) (unpublished), http://hal.archives-ouvertes.fr/docs/00/16/66/52/PDF/RR-6269.pdf
7. Chatzigiannakis, I., Michail, O., Spirakis, P.G.: Mediated population protocols. In: Albers, S., Marchetti-Spaccamela, A., Matias, Y., Nikoletseas, S., Thomas, W. (eds.) ICALP 2009. LNCS, vol. 5556, pp. 363–374. Springer, Heidelberg (2009)
8. Chatzigiannakis, I., Michail, O., Spirakis, P.G.: Recent advances in population protocols. In: Královič, R., Niwiński, D. (eds.) MFCS 2009. LNCS, vol. 5734, pp. 56–76. Springer, Heidelberg (2009)
9. Fischer, M.J., Jiang, H.: Self-stabilizing leader election in networks of finite-state anonymous agent. In: Shvartsman, M.M.A.A. (ed.) OPODIS 2006. LNCS, vol. 4305, pp. 395–409. Springer, Heidelberg (2006)

# Improving Space Complexity of Self-stabilizing Counting on Mobile Sensor Networks

Keigo Kinpara[1], Tomoko Izumi[2], Taisuke Izumi[1], and Koichi Wada[1]

[1] Graduate School of Engineering, Nagoya Institute of Technology, Nagoya, 466-8555, Japan
keigo@phaser.elcom.nitech.ac.jp, t-izumi@nitech.ac.jp, wada@nitech.ac.jp
[2] College of Information Science and Engineering, Ritsumeikan University,
Kusatsu, 525-8577 Japan
izumi-t@fc.ritsumei.ac.jp

**Abstract.** We consider a problem on a passively-mobile sensor network with a base station; the base station counts the number of sensors in the network. In [6], these passively-mobile sensor networks are modeled by extending the model of population protocols and self-stabilizing protocols to count the number of existing sensors, where self-stabilizing counting means from any initial states of sensors and some initialization of the base station (unless the base station is initialized, this problem can not be solved in general), the base station eventually counts the exact number of sensors in the system. In this setting, Beauquier et al.[6] show several protocols to solve the self-stabilizing counting (See Table 1). In this paper, we focus on space complexity of the self-stabilizing counting protocols (that is, the number of states sensors can possess, denoted by $\alpha(P)$, where $P$ is an upper bound of the number of states) and improve it by showing self-stabilizing counting protocols using $\alpha(P) = 2P$ and $\alpha(P) = 3P/2$, respectively. Since previous best known protocol needs $\alpha(P) = 4P$ and a lower bound of $\alpha(P)$ is $P$, we can shrink the gap lying that feasibility.

## 1   Introduction

*Background* A *passively-mobile* system is a collection of agents that move in a certain region but have no control over how they move. Since the communication range of each agent is quite small compared to the size of the region, two agents can communicate only when they are sufficiently close to each other. Passive mobility appears in many real systems. A representative example is a network of smart sensors attached to cars or animals. Recently, as a model for such passively-mobile systems, *population protocols* are introduced [4]. A population protocol consists of a number of agents, to which some program (protocol) is deployed. Following the deployed protocol, each agent changes its state by *pairwise interactions* to other agents (that is, two agents come closer to each other in the region and update their states by exchanging information).

In this paper, we are interested in a variant of population protocols in which, given a base station and a number of mobile sensors, we consider the *self-stabilizing counting* problem on that model. The counting problem requires that the base station counts and outputs the exact number of sensors in the system. The self-stabilizing counting must achieve the counting from any configuration where only the base station can be initialized. This problem is motivated by the following scenario : A group of birds(petrels in

C. Lu, T. Masuzawa, and M. Mosbah (Eds.): OPODIS 2010, LNCS 6490, pp. 504–515, 2010.

[6]) carries on their body a small sensor and a base station wants to count the remaining birds alive. When a bird is close enough to the base station, its sensor interacts with the base station, which can control the memory of bird's sensor and make some computation. Also the sensors of birds can interact with each other when two birds approach close enough. Since each sensor is exposed in wild nature, its state can be corrupted by its surroundings. This setting of the problem is first introduced by Beauquier et al.[6], and several protocols, each of which works on slightly different models, are presented.

*Our Contribution.* In this paper we consider an improvement of previous protocols. In particular, we focus on space complexity of the self-stabilizing counting protocols on the symmetric petrels-to-base-station-and-to-petrels (STBTP) model. The STBTP model is a variation of population protocols with base stations, where sensors have no mechanism to break symmetry. That is, it allows only interactions between two sensors such that the two sensors necessarily have the same poststate if their states before interaction are the same. Because of weak capability of sensors, any protocol is required to be designed with small resource of sensor-side algorithm. Thus, our primary interest is how we can reduce the memory space of sensor nodes. In the prior work for this problem [6], a protocol is developed by using $4P$ states for each sensor, where $P$ is the known upper bound for the number of sensors. It also shows that there is no protocol using $P - 1$ states or less. The feasibility of self-stabilizing counting between $P$ and $4P - 1$ still remains open[1].

Our contribution is to shrink the gap lying on that feasibility. We present two protocols for self-stabilizing counting on STBTP models. In the first protocol, each sensor uses $2P$ states, and its convergence time is $O(\log n)$, The second protocol uses $3P/2$ states and the convergence time is finite. Table 1 shows the comparison between prior work and our results. Interestingly, we can find the trade-off between space and time complexity.

**Table 1.** Previous result and our result

| Number of states of sensors | Number of states of the base station | convergence time | reference |
| --- | --- | --- | --- |
| $\alpha(P) < P$ | impossible | | [6] |
| $\alpha(P) = 3P/2$ | 3 states array of size $P$ | finite number of rounds | this paper |
| $\alpha(P) = 2P$ | 3 states array of size $P$ | $O(\log n)$ rounds | this paper |
| $\alpha(P) = 4P$ | 3 states array of size $2P$ | 3 rounds | [6] |

*Related Work.* The population protocol model is originally introduced by Angluin et al.[1][2]. Those papers focus on the predicates computable of that model, and show that semilinear predicates are the necessary and sufficient class of computable predicates. Triggered by those seminal papers, a number of studies about population protocols are now emerging.

The leader election is one of the popular problems considered in several papers [3, 8–10] since the original study [1] uses the leader election algorithm as a building block of the predicate computation. All of them include the interest to self-stabilization. The

---

[1] This lower bound holds even if the symmetricity condition is removed from the STBTP model [6].

paper by Angluin et al. [3] is the first paper considering the self-stabilizing aspects of population protocols. The following paper by Fischer and Jiang [9] considers the design of self-stabilizing leader election protocol on a population protocol model with restricted communications. Cai et al. [8] investigates the space complexity of self-stabilizing leader election. A relaxed variant of self-stabilizing leader election is studied by Sudo et al. [10]. Under a certain kind of synchrony assumption, a generic scheme to make a population protocol self-stabilizing is introduced by Beauquier et al. [7].

The computation with the base station is first considered by Beauquier et al. [6], which consider the self-stabilizing counting. The aggregation problem is also considered in the same model [5].

*Roadmap.* The paper is organized as follows: Section 2 gives necessary definition and terminology including the system model and the problem definition. In Section 3, we provide a self-stabilizing counting protocol using $2P$ states and its correctness. The second protocol using $3P/2$ states and its correctness are provided in Section 4. We present the conclusion in Section 5.

## 2  Preliminaries

### 2.1  System Model

A mobile sensor network consists of a base station and $n$ undistinguishable mobile sensors. A mobile sensor is modeled by an anonymous finite state machine (i.e., it has no identifier and executes the same algorithm), and the base station is a special finite state machine which executes an algorithm different from mobile sensors. A sensor (or the base station) changes its own state by pairwise interaction with other sensors or the base station. The base station always outputs the computation results according to its internal state. Formally a protocol of passively mobile sensor networks is specified by a tuple $(Q_s, Q_B, A, B, B_0, \omega, \delta)$. Each entry is defined as follows:

- $A = \{A_1, \ldots, A_n\}$: A finite set of sensors.
- $B$: The base station.
- $Q_A$: A finite set of states of sensors.
- $Q_B$: A finite set of states of the base station.
- $B_0$: An initial state of the base station (the system initiates the state of the base station by $B_0 \in Q_B$)
- $\omega$: The output maps from $Q_B$ to $Y$ ($Y$ is a set of output values).
- $\delta : Q_A^2 \to Q_A^2$ or $Q_A \times Q_B \to Q_A \times Q_B$: A state transition function.

In this paper, we adopt the STBTP model [6]. Informally, that model does not allow any symmetry breaking mechanism between two sensors. It is formally defined as the following restriction to transition functions:

**Definition 1.** *A protocol can run on STBTP model if its transition function $\delta$ satisfies that for any $x \in Q_A$, $\delta(x, x) = (y, z)$ implies $y = z$.*

We assume that the system has the knowledge about the upper bound of the number of all sensors, say $P$. In addition, we allow the value of $|Q_A|$ and $|Q_B|$ to depend on the upper bound $P$. For a given protocol, let $\alpha(P)$ be the number of states of sensors for upper bound $P$.

## 2.2   Execution

An interaction between two sensors (or a sensor and the base station) which have states $x_i$ and $x_j$ respectively is represented by an *interaction event* (or simply interaction) $e = \{x_i, x_j\}$. Configuration $C$ of the system consists of the states of the base station and sensors. That is, $C = (b, s_1, s_2, \ldots, s_n)$, where b is the state of the base station and $s_i$ is the state of the sensor $A_i$.

If a configuration $C$ becomes $C'$ by an interaction $e$ it is denoted by $C \xrightarrow{e} C'$. A configuration $C'$ is *reachable* from $C$ if there is a sequence of the configurations $C = C_k, C_{k+1}, \ldots, C_\ell = C'$ such that $C_i \xrightarrow{e} C_{i+1}(k \leq i < \ell)$. We describe $\#(p) = x$ at a configuration $C$ if the number of sensors with state $p$ is x at $C$.

An interaction $e = \{x_i, x_j\}$ is *applicable* at configuration $C$ if $\#(x_i) > 0$ and $\#(x_j) > 0$ at $C$. An execution of a protocol is an infinite sequence $(C_0, e_0), (C_1, e_1), \ldots$ such that $e_j$ is applicable to $C_j$ and $C_j \xrightarrow{e_j} C_{j+1}$ holds.

## 2.3   Fairness

Fairness is an assumption restricting the possible behavior of systems. Formally, it is defined as a constraint for executions. Throughout this paper, we introduce the *global fairness*, which is defined as follows:

**Definition 2 (Global fairness).** *An execution $E = (C_0, e_0), (C_1, e_1), \cdots$ is globally fair: for every pair of $C$ and $C'$ such that there exists an interaction $e$ satisfying $C \xrightarrow{e} C'$, if $C = C_i$ for infinitely many i, then $C_i = C$ and $C_{i+1} = C'$ for infinitely many i.*

Intuitively, global fairness guarantees any possible execution from $C$ actually occurs if $C$ appears infinitely often in the execution. Thus, it prevents the occurrence of livelock caused by some looped execution.

## 2.4   Complexity Measure

In this paper, we measure the time complexity by *asynchronous rounds* (or simply *rounds*). A round is defined as the shortest fragment of an execution in which each sensor interacts with all of other sensors and the base station at least once.

## 2.5   Self-stabilizing Counting Problem

The counting problem requires that the base station eventually counts and outputs the exact number of sensors in the system. The self-stabilizing property guarantees that the configuration eventually converges to and stabilizes at legitimate configurations from any initial configuration within finite rounds. In this paper, we consider a weaker form self-stabilization property, which allows the initialization of the base station. This is because if we allow any initial configuration including the base station and sensors, the self-stabilizing counting problem cannot be solved [6]. The set of legitimate configurations is defined by the output value of the base station. A configuration $C$ is legitimate if the output value of the base station is equal to $n$.

# 3   A self-stabilizing Counting Protocol Using 2P States

## 3.1   Protocol

In this section, we present a self-stabilizing counting protocol with $2P$ sensor states (i.e., $\alpha(P) = 2P$). Throughout this section, we call it $2P$-protocol.

Algorithm 1 shows the pseudo-code of our $2P$-protocol. The basic principle of the $2P$-protocol is that the base station gives a unique name $i \in [1..P]$ to each sensor. For each name $i$, two different states $S_i$ and $D_i$ are prepared. Thus, totally the number of possible sensor states is $2P$. Informally, the state $S_i$ implies that the sensor with name $i$ is expected to be unique. It should be noted that we just"expect" the uniqueness of name $i$, and actually two or more sensors with that name can exist during executions. The state $D_i$ means that at least two agents have the same name $i$[2]. For each name $i$, the base station prepares one slot $M[i]$, which has three possible states Free($F$), Taken($T$), and Wait($W$). The intuitive meaning of each state can be described as follows:

- Free : The base station does not find the sensor with the corresponding name.
- Taken : The base station takes one sensor with the corresponding name.
- Wait : The base station is willing to take the sensor at next interaction.

Our protocol is designed so that in any legitimate configuration $M[i] = T$ implies that exactly one sensor has name $i$ and its state is $S_i$. Thus, the base station always outputs the number of slots with state $T$ as the counted number of sensors.

We explain the behavior of our protocol and the idea behind it. First, all slots of the base station are initialized by $F$. When a sensor $A_x$ with name $i$ (whose state is either $S_i$ or $D_i$) interacts with the base station, the base station counts it by setting $M[i] = T$ and changing the state of $A_x$ to $S_i$. If no other sensor with name $i$ exists, this slot must be closed. Thus, when a sensor with state $S_i$ interacts with the base station of $M[i] = T$, no update of states for name $i$ occurs. On the other hand, if two or more sensors with names $i$ exist, we have to give different names to them except for one. For that mechanism, the state $D_i$ is introduced: If two sensors having state $S_i$ interact with each other, their states are updated to $D_i$. When a sensor $A_x$ with state $D_i$ interacts with the base station of $M[i] = T$, it is given a different name $j$, which is selected from all indices such that the corresponding slot has state $F$. After giving a different name $j$, the base station changes the state of slot $M[i]$ to $W$, and newly counts the name $j$ by setting $M[j] = T$. Since the state $W$ is set to $M[i]$, the sensor $A_y$ with name $i$ (other than $A_x$) is counted again at its next interaction with the base station. Consequently, by the above process, one agent $A_x$ takes a new name different from $i$. Repeating this process eventually gives each agent a unique name.

## 3.2   Correctness

We prove the correctness of $2P$-protocol by showing that the output of the base station converges to the exact number of the sensors in the network. In the proof, we do not describe interaction which do not change configuration, for example interaction between a sensor with state $S_i$ and a sensor with state $D_i$.

---

[2] The only exceptional case when exactly one sensor has the name $i$ with state $D_i$ initially.

---

**Algorithm 1.** $2P$-protocol

---

**variable of sensor :**
$$state \in \{S_0, \ldots, S_{P-1}, D_0, \ldots, D_{P-1}\}$$

**variables of the base station :**
$M$ : array of size $P$ (the initial value of each entry $F$)

**When a sensor interacts with the base station :**

> **if** $state = S_i$ **then**
>> $M[i] \leftarrow T$
>
> **else if** $state = D_i$ **then**
>> **if** $M[i] = T$ **then**
>>> Let h be the index such that $M[h] = F$
>>> $M[i] \leftarrow W$
>>> $state \leftarrow S_h$
>>> $M[h] \leftarrow T$
>>
>> **else if** $M[i] \in \{W, F\}$ **then**
>>> $state \leftarrow S_i$
>>> $M[i] \leftarrow T$
>>
>> **end if**
>
> **end if**

**When two sensors interact :**

> **if** both states are $S_i$ **then**
>> $state \Leftarrow D_i$
>
> **end if**

---

**Lemma 1.** *If $M[i] \neq F$ and $\#(S_i) + \#(D_i) > 0$ at $C_h$, the following two properties hold:*

1. *$\#(S_i) + \#(D_i)$ does not increase in any execution from $C_h$.*
2. *There exists an execution from $C_h$ where $\#(S_i) + \#(D_i)$ decreases by one if $\#(S_i) + \#(D_i) > 1$.*

*Proof.* (The proof of 1.) Since $M[i] \neq F$ holds at $C_h$, this property clearly holds because the number of $\#(S_i) + \#(D_i)$ can increase only if $M[i] = F$.

(The proof of 2.) We first show that if $M[i] = W$ at $C_h$, we can obtain the execution from $C_h$ where the state of the base station eventually satisfies $M[i] = T$. Since there exists at least one sensor $A_x$ with state $S_i$ or $D_i$, the base station can interact with $A_x$, which changes the state of $M[i]$ to $T$. This interaction is the execution we find.

Since we also have $\#(S_i) + \#(D_i) > 1$, there are two sensors $A_y$ and $A_z$ whose states are $S_i$ or $D_i$. If both of them have state $S_i$, we can create the state $D_i$ by making them interact. Consequently, we can have an execution that takes $C_h$ to the configuration $C'$ with $M[i] = T$ and $\#(D_i) > 0$. If the sensor with state $D_i$ interacts with the base station at $C'$, $\#(D_i)$ decreases by one. The lemma is proved. □

**Lemma 2.** *A configuration satisfying $M[i] = T$, $\#(S_i) = 0$ and $\#(D_i) = 1$ for all $i$ is not reachable from any initial configuration.*

*Proof.* Suppose by contradiction that there is a configuration $C_k$ where $\#(S_i) = 0$, $\#(D_i) = 1$ and $M[i] = T$ in some possible execution. Without loss of generality, we assume $k$ is the smallest index satisfying those conditions. That is, $C_{k-1}$ does not satisfy at least one of them. Let $e$ be the interaction changing $C_{k-1}$ to $C_k$. If $e$ changes the state of $M[i]$ to $T$, it also increases the number of sensors with state $S_i$ by one , which contradicts the fact that $\#(S_i) = 0$ at $C_k$. If $e$ does not change the state of $M[i]$. Then, $M[i] = T$ also holds at $C_{k-1}$ because $M[i] = W$ at $C_{k-1}$ implies $\#(S_i) > 0$ at $C_k$. If $e$ is the interaction with the base station, the only possible situation is a sensor with state $S_i$ interacts at $e$, which contradicts $\#(S_i) = 0$ at $C_k$. If $e$ is the interaction between two sensors, $\#(S_i) + \#(D_i) \geq 2$ must hold at $C_k$. This is also a contradiction.    □

**Lemma 3.** *For any configuration reachable from some initial configuration, if $M[i] \in \{W, T\}$, $\#(S_i) + \#(D_i) \geq 1$.*

*Proof.* Suppose by contradiction that $\#(S_i) = \#(D_i) = 0$ at a configuration $C_k$. We can assume $C_k$ is the first configuration satisfying that condition. That is, either $M[i] \in \{T, W\}$ or $\#(S_i) + \#(D_i) \geq 1$ does not hold at $C_{k-1}$. If $M[i] = F$ at $C_{k-1}$, $\#(S_i) + \#(D_i) > 1$ at $C_{k-1}$ necessarily holds. A contradiction. If $\#(S_i) + \#(D_i) \geq 1$ holds at $C_{k-1}$, one sensor $A_x$ with state either $S_i$ or $D_i$ must disappear by interaction $e_{k-1}$. This implies that $A_x$ has state $D_i$ and interacts with the base station of $M[i] = T$. However, it follows that $\#(S_i) = 0$, $\#(D_i) = 1$, and $M[i] = T$ hold at $C_{k-1}$. It contradicts Lemma 2.    □

**Lemma 4.** *There is an index $j$ such that $M[j] = F$ when a sensor of state $D_i$ interacts with the base station satisfying $M[i] = T$.*

*Proof.* We can trivially deduce this lemma from Lemma 3 and the pigeonhole principle.    □

**Theorem 1.** *Any initial configuration is reachable to a legitimate configuration and any legitimate configuration can reach only legitimate configurations.*

*Proof.* If $M[i] = F$ and $\#(S_i) + \#(D_i) > 0$ at $C_h$, the system eventually reaches configuration $C_k$ such that $M[i] = T$ because the base station interacts a sensor with state $S_i$ or $D_i$. By Lemma 1 and 3, if $M[i] \neq F$ holds, the system eventually stabilizes on a configuration $C$ where $\#(S_i) + \#(D_i) = 1$. Lemma 2 implies that $C$ satisfies $\#(S_i) = 1$ and $M[i] = T$. So the system converges to a legitimate configuration.    □

### 3.3    Convergence Time

We can show that the number of sensors with the same name decreases by half in two rounds and thus it converges to a legitimate configuration in $O(\log n)$ rounds, where $n$ is the number of sensors.

**Lemma 5.** *If the number of sensors is $\#(S_i) + \#(D_i) = k (k > 1)$ at the configuration $C_h$, then the number of sensors satisfies $\#(S_i) + \#(D_i) \leq k - \lfloor \frac{k}{2} \rfloor$ after two rounds.*

*Proof.* Consider two consecutive asynchronous rounds $r$ and $r + 1$. Let $E = e'_1, e'_2, e'_3, \cdots e'_m$ be the sequence of interactions between a sensor with state $D_i$ and the base station in the order of their occurrences. Since $M[i] = W$ at interaction $e'_l$ implies $M[i] = T$ at interaction $e'_{l+1}$. Thus, at least $\lfloor m/2 \rfloor$ sensors obtain a name different from $i$. The remaining part of the proof is to show $m \geq k - 1$. Let $T_1$ be the set of sensors such that it has state $D_i$ at the interaction with the base station at round $r$, and $T_2$ be the set of sensors with name $i$ not belonging to $T_1$. If $|T_2| = 1$, we obtain $m \geq |T_1| = k - 1$. Thus we assume $|T_2| > 1$. Since any sensor in $T_2$ interacts with some other sensor in $T_2$ during round $r$, its state is $D_i$ at the beginning of round $r + 1$. Exceptionally, if $|T_2|$ is odd, one sensor keeps state $S_i$. Consequently, at least $|T_2| - 1$ sensors has state $D_i$ at the beginning of round $r + 1$. This implies that any sensor in $T_2$ has state $D_i$ at the interaction with the base station during round $r + 1$. That is, they appear in $E$, and thus $m \geq |T_1| + |T_2| - 1 \geq k - 1$.                                          □

The above lemma directly implies the following theorem.

**Theorem 2.** *The convergence time of* 2*P-protocol is* $O(\log n)$.

## 4    A self-stabilizing Counting Protocol Using $3P/2$ States

### 4.1    Protocol

In this section, we improve the self-stabilizing counting 2$P$-protocol to one with $3P/2$ sensor states (i.e., $\alpha(P) = 3P/2$). Throughout this section, we call it $3P/2$-protocol and without loss of generality, we assume that $P$ is an even number.

In the 2$P$-protocol, we used 2$P$ sensor states and we prepared two states $S_i$ and $D_i$ for each sensor name $i$. Note that $S_i$ and $D_i$ show the uniqueness and the duplication of a sensor name $i$, respectively. If we reduce the number of sensor states to less than 2$P$, such $D_i$'s can not be prepared for each name $i$. Since we design $3P/2$-protocol, for each sensor name $i$ we prepare $D_i$ for $S_{2i}$ and $S_{2i+1}$.

Algorithm 2 shows the pseudo-code of the $3P/2$-protocol. Its basic idea is similar to that of the 2$P$-protocol but it is more complicated because we do not have a sufficient number of sensor states. Like the 2$P$-protocol, for each name $i$ the base station prepares one slot $M[i]$, which has three possible states Free($F$), Taken($T$), and Lock($L$). The intuitive meanings of $F$ and $T$ are the same as those in the 2$P$-protocol but that of $L$ is that the base station waits a sensor with state $S_{2i}$, $S_{2i+1}$ or $D_i$ and locks transition $S_{2i}$ or $S_{2i+1}$ from other state of a sensor.

In order to explain the behavior of the $3P/2$-protocol, we introduce a notion called *block*. A *block* $i$ is defined as the numbers of sensors with $S_{2i}$, $S_{2i+1}$, and $D_i$ and the values of slots $M[2i]$ and $M[2i + 1]$ in the base station and is denoted by $< \#(S_{2i}), \#(S_{2i+1}), \#(D_i), M[2i], M[2i + 1] >$ and $\#(S_{2i}) + \#(S_{2i+1}) + \#(D_i)$ is denoted by $\sigma(i)$. The configuration of our protocol can be denoted by $(block_0, \ldots, block_{P/2-1})$.

*special states of blocks:* If $block_i$ is $< 1, 1, 0, T, T >$, $block_i$ is said to be *completely-closed*. If $block_i$ is either $< 1, 0, 0, T, F >$ or $< 0, 1, 0, F, T >$, $block_i$ is said to be *semi-closed*. If $block_i$ is $< 0, 0, 0, F, F >$, it is said to be *empty*. If $block_i$ is completely-closed, semi-closed or empty, it is said to be *closed*.

If $block_i$ is either $< 1, 0, 1, T, T >$ or $< 0, 1, 1, T, T >$, it is celled *error-block*.

If all blocks are closed, the base station can count the exact number of sensors in the network. Thus, legitimate configurations of the $3P/2$-protocol are defined to be configurations such that all blocks are closed.

We explain the behavior of the $3P/2$-protocol. Like the 2$P$-protocol, all slots of the base station are initialized to $F$. The behavior of the $3P/2$-protocol depends on the number of sensors in $block_i$ (that is $\sigma(i)$). When the base station interacts with a sensor with state $S_{2i}$ (or $S_{2i+1}$), the corresponding slot is set to $T$. The role of the state $D_i$ is similar to the case for the $2P$-protocol but in the $3P/2$-protocol, for the both cases that two sensors with state $S_{2i}$ (and $S_{2i+1}$) interact with each other their states are updated to the same $D_i$. When the base station interacts with a sensor with state $D_i$, it decides the slots and the state of the sensor based on the slot values of $M[2i]$ and $M[2i + 1]$.

If $M[2i] = M[2i + 1] = T$, the base station notices that $\sigma(i) \geq 3^3$ and sets both $M[2i]$ and $M[2i + 1]$ to $Ls$ meaning that the base station wait a sensor in $block_i$ and locks transition $S_{2i}$ and $S_{2i+1}$ from other state of a sensor and chooses a next state of the sensor if there is a free index $j$ such that $M[j] = F$. In this case, $\sigma(i)$ decreases by one. If there is no free slot, the base station sets $M[2i] = F$ and the state of the interacting sensor is set to $S_{2i}$. In this case, since it will be proved that if $M[j] = T$ then there is only one sensor with $S_j$, $block_i$ will become completely-closed.

If $M[2i] = M[2i + 1] = L$, the base station notices that there are at least two sensors with $S_{2i}$ and $S_{2i+1}$ and the state of one of them is changed to $S_j$ such that $M[j] = F$ and it sets $M[2i] = T$ and $M[2i + 1] = F$. By repeating this case and the above case, $\sigma(i)$ will eventually become at most two and thus, $block_i$ will become closed.

Otherwise, the base station notices that $\sigma(i) \leq 2$ for $block_i$. It can be easily verified that $block_i$ will eventually become closed.

## 4.2    Correctness

We prove the correctness of $3P/2$-protocol i.e. the output of the base station converges to the correct number of sensors. Since we can easily verify the closure of the $3P/2$-protocol, that proof is omitted.

**Lemma 6.** *If block i is error-block and there is no interaction e such that $\sigma(i)$ increases, it eventually becomes semi-closed.*

*Proof.* We prove the case that the block $i$ is error-block $< 1, 0, 1, T, T >$ in a configuration $C_h$. Another case that $block_i =< 0, 1, 1, T, T >$ can be treated similarly. Since the base station is $M[2i] = M[2i + 1] = T$, $\sigma(i)$ does not increase and if the base station interacts with a sensor with state $S_i$, the configuration do not change. By the fairness assumption, the base station eventually interacts with a sensor with state $D_i$, and the block $i$ reaches $< 1, 0, 0, L, L >$. Since $M[2i] = M[2i + 1] = L$, the number of sensors in block $i$ does not increase. And the base station eventually interacts with the sensor with state $S_i$. The block $i$ becomes semi-closed.                                          □

**Lemma 7.** *If a configuration $C_h$ has $block_i$ such that $\sigma(i) \geq 2$, $C_h$ eventually reaches a configuration $C_k$ $(k > h)$ where the base station has $M[2i] = T$ and $M[2i + 1] = T$.*

*Proof.* We assume that there is more than one sensor in the block $i$ at the configuration $C_h$. If two sensors have different states $S_{2i}$ and $S_{2i+1}$, the base station eventually interacts

---

[3] $M[2i] = M[2i + 1] = T$ means the base station has assigned $S_{2i}$ and $S_{2i+1}$ to states of its interacting sensors. Therefore, since the base station interacts with a sensor with state $D_i$, which initially exists or is newly created via interaction of two sensors with state $S_{2i}$ or $S_{2i+1}$, it notices that there is at least a sensor with state $D_i$ in addition to those with $S_{2i}$ and $S_{2i+1}$.

---

**Algorithm 2.** $3P/2$-Protocol

---

**variable of sensor :**

    $state \in \{S_0, \ldots, S_{P-1}, D_0, \ldots, D_{P/2-1}\}$

**variables of the base station :**

    $M$ : array of size $P$ (initial value is $F$)

**Transition when the base station interacts with sensor:**

    **if** $state = S_{2i}$ **then**

        $M[2i] \Leftarrow T$

        **if** $M[2i + 1] = L$ **then**

            $M[2i + 1] \Leftarrow F$

        **end if**

    **else if** $state = S_{2i+1}$ **then**

        $M[2i + 1] \Leftarrow T$

        **if** $M[2i] = L$ **then**

            $M[2i] \Leftarrow F$

        **end if**

    **else if** $state = D_i$ **then**

        **if** $M[2i] = T \land M[2i + 1] = T$ **then**

            // The base station knows $\sigma(i) \geq 3$

            **if** there is index $h$ such that $M[h] = F$ **then**

                $M[2i] \Leftarrow L$

                $M[2i + 1] \Leftarrow L$

                $state \Leftarrow S_h$

            **else**

                $M[2i] \Leftarrow T$

                $M[2i + 1] \Leftarrow F$

                $state \Leftarrow S_{2i}$

            **end if**

        **else if** $M[2i] \neq T$ **then**

            $M[2i] \Leftarrow T$

            $state \Leftarrow S_{2i}$

        **else**

            $M[2i + 1] \Leftarrow T$

            $state \Leftarrow S_{2i+1}$

        **end if**

    **end if**

**Transition when two sensors interact each other :**

    **if** both sensors' state is $S_i$ **then**

        $state \Leftarrow D_{\lfloor \frac{i}{2} \rfloor}$

    **end if**

---

with these sensors, and hence block $i$ reaches a configuration such that $M[2i] = T$ and $M[2i+1] = T$. If there are sensors with the same state $S_{2i}$ or $S_{2i+1}$, these sensors interact with each other and change their states to $D_i$. Since the base station will interact with two $D_i$'s, block $i$ reaches a configuration where $M[2i] = T \land M[2i + 1] = T$.     □

**Lemma 8.** *Assume that a configuration has* $block_i$ *such that* $\sigma(i) \geq 3$. *Then there must be an interaction* $e$ *where a sensor with state* $D_i$ *interacts with the base station with* $M[2i] = T$ *and* $M[2i + 1] = T$.

*Proof.* Since the block $i$ has more than one sensors, the system reaches configuration $C_k$ in which the base station's state is $M[2i] = T$ and $M[2i + 1] = T$ by Lemma 7. If there is no sensor with state $D_i$ in $C_k$, then there are more than two sensors with state $S_{2i}$ or $S_{2i+1}$. These sensors with the same state interact with each other by fairness assumption. So the system is reachable to the block $i$ having a sensor with state $D_i$ and thus the interaction $e$ occurs.    □

**Lemma 9.** *If* $\sigma(i) = 2$ *and no an interaction* $e$ *where a sensor state is changed into neither* $S_{2i}$ *nor* $S_{2i+1}$ *occur.* $block_i$ *will eventually become closed.*

*Proof.* If $\sigma(i)$ decreases, $block_i$ will become error-block and thus semi-closed by Lemma 6. If $\sigma(i)$ does not change, we consider three cases. First, when two sensors having different states interact with the base station, there are sensors with states $S_{2i}$ and $S_{2i+1}$ and the base station with state $M[2i] = T$ and $M[2i] = T$. Second, when there are sensors with a same state, these sensors interact with each other and change their states to $D_i$. Then the base station interacts with these sensors, and block $i$ becomes completely-closed. Third, if the state of the base station is $M[2i] = T$ and $M[2i + 1] = T$, but a sensor with state $D_i$ interacts with the base station and the state of the interacting sensor is changed to $S_{2i}$. If the state of the other sensor is $S_{2i+1}$ or $D_i$, the sensor eventually interacts with the base station and the block $i$ become completely-closed. Otherwise the state of the other sensor is $S_{2i}$, two sensors have same state $S_{2i}$. these sensors interact with each other and two sensors with state $D_i$ appear. Thin case can be reduced to the second case.    □

**Lemma 10.** *Let* $C_k$ *be a configuration with* $k$ *non-closed blocks. If the number of non-closed blocks does not decrease, a configuration reachable from* $C_k$ *satisfies that the number of sensors in non-closed blocks are at most* $2k$.

*Proof.* Assume that the number of non-closed blocks will not decrease from $C_k$. If the number of sensors in non-closed blocks more than $2k$, there is a block where the number of sensors is more than two and there must be a semi-closed block among the closed blocks. By Lemma 8, the base station interacts with a sensor with state $D_i$. The interacting sensor will eventually reach a semi-closed block.    □

**Theorem 3.** *Any configuration* $C_i$ *eventually reaches a legitimate configuration* $C_\ell$.

*Proof.* We prove that the number of non-closed blocks decreases from $C_i$. We assume the number of non-closed blocks is $k$ at $C_i$ and the number of non-closed blocks will not decrease from $C_i$. When the number of non-closed blocks does not decrease, non-closed blocks such that $\sigma(i) = 1$ must increase the number of sensors because if the number of sensors does not increase, these blocks become semi-closed. And non-closed blocks such that $\sigma(i) = 2$ must not decrease the number of sensors by Lemma 6 and must increase the number of sensors because if the number of sensors in block do not increase, these sensors become completely-closed by Lemma 9. Since $C_i$ will eventually reach $C'$ such that $\forall i \sigma(i) \geq 2$. Then $C'$ will eventually reach $C''$ which the number of sensors in non-closed blocks are at most $2k$ and $\forall i \sigma(i) \geq 2$. If all non-closed blocks are $\sigma(i) = 2$ in $C''$, closed block will eventually increase by Lemma 9. If any non-closed block is $\sigma(i) \geq 3$ in $C''$, it is a contradiction. $\qquad\square$

## 5   Concluding Remarks

We have presented space-efficient self-stabilizing counting protocols on a passively-mobile sensor network with the base station. It is an interesting open problem to obtain more space-efficient protocols on the STBTP model or better lower bounds on STBTP model. On the ATBTP model allowing symmetry breaking mechanism between two sensors, that is, a transition function $\delta$ can be $\delta(x, x) = (y, z)(y \neq z)$, a protocol with $\alpha(P) = P$ can be constructed and it can attain the lower bound [6]. However, it is not known on the STBTP model we must use some property about the symmetricity of transition functions in order to get better lower bounds.

## Acknowledgement

This work is supported in part by KAKENHI No. 21680001 and 22700010.

## References

1. Angluin, D., Aspnes, J., Diamadi, Z., Fischer, M.J., Peralta, R.: Computation in networks of passively mobile finite-state sensors. Distributed Computing, 235–253 (2006)
2. Angluin, D., Aspnes, J., Eisenstat, D., Ruppert, E.: The computational power of population protocols. Distributed Computing 20, 279–304 (2007)
3. Angluin, D., Aspnes, J., Fischer, M.J., Jiang, H.: Self-stabilizing population protocols. In: Anderson, J.H., Prencipe, G., Wattenhofer, R. (eds.) OPODIS 2005. LNCS, vol. 3974, pp. 103–117. Springer, Heidelberg (2006)
4. Aspnes, J., Ruppert, E.: An introduction to population protocols. Bulletin of the European Association for Theoretical Computer Science 93, 98–117 (2007)
5. Beauquier, J., Burman, J., Clement, J., Kutten, S.: On utilizing speed in networks of mobile agents. In: Proceeding of the 29th ACM SIGACT-SIGOPS Symposium on Principles of Distributed Computing, pp. 305–314 (2010)
6. Beauquier, J., Clement, J., Messika, S., Rosaz, L., Rozoy, B.: Self-stabilizing counting in mobile sensor networks with a base station. In: Pelc, A. (ed.) DISC 2007. LNCS, vol. 4731, pp. 63–76. Springer, Heidelberg (2007)
7. Beauquier, J., Burman, J., Kutten, S.: Making population protocols self-stabilizing. In: Guerraoui, R., Petit, F. (eds.) SSS 2009. LNCS, vol. 5873, pp. 90–104. Springer, Heidelberg (2009)
8. Cai, S., Izumi, T., Wada, K.: Space complexity of self-stabilizing leader election in passively-mobile anonymous agents. In: Kutten, S., Žerovnik, J. (eds.) SIROCCO 2009. LNCS, vol. 5869, pp. 113–125. Springer, Heidelberg (2010)
9. Fischer, M., Jiang, H.: Self-stabilizing leader election in networks of finite-state anonymous agents. In: Shvartsman, M.M.A.A. (ed.) OPODIS 2006. LNCS, vol. 4305, pp. 395–409. Springer, Heidelberg (2006)
10. Sudo, Y., Nakamura, J., Yamauchi, Y., Ooshita, F., Kakugawa, H., Masuzawa, T.: Loosely-stabilizing leader election in population protocol model. In: Kutten, S., Žerovnik, J. (eds.) SIROCCO 2009. LNCS, vol. 5869, pp. 295–308. Springer, Heidelberg (2009)

# Author Index

Printing: Mercedes-Druck, Berlin
Binding: Stein+Lehmann, Berlin